WATER JUSTICE

Water justice is becoming an ever-more pressing issue in times of increasing water-based inequalities and discrimination. Megacities, mining, forestry, industry, and agribusiness claim an increasingly large share of available surface and groundwater reserves. Water grabbing and pollution generate poverty and endanger ecosystems' sustainability. Beyond large, visible injustices, the book also unfolds the many "hidden" water world injustices, subtly masked as "rational," "equitable," and "democratic." It features critical conceptual approaches, including analysis of environmental, social, cultural, and legal issues surrounding the distribution and management of water. Illustrated with case studies of historic and contemporary water injustices and contestations around the world, the book lays new ground for challenging current water governance forms and unequal power structures. It also provides inspiration for building alternative water realities. With contributions from renowned scholars, this is an indispensable book for students, researchers, and policy makers interested in water governance, environmental policy and law, political geography, and cultural anthropology.

RUTGERD BOELENS is Professor of Water Governance and Social Justice in the Department of Environmental Sciences at Wageningen University and Professor of Political Ecology of Water in Latin America with CEDLA (Centre for Latin American Research and Documentation) and the University of Amsterdam. He is also Visiting Professor at the Catholic University of Peru and the Central University of Ecuador and coordinates the international Justicia Hídrica/Water Justice alliance. His work focuses on water rights, water grabbing, hydrosocial territories, legal pluralism, cultural politics, governmentality, and social mobilization.

TOM PERREAULT is Professor of Geography at Syracuse University in New York. His research examines the intersections of political ecology, resource governance, agrarian transformation, and indigenous and *campesino* social mobilization in the central Andes and western Amazon. His research has been funded by Fulbright, the Inter-American Foundation, the United Nations, and the US National Science Foundation.

JEROEN VOS is Assistant Professor of Water Governance in the Department of Water Resources Management at Wageningen University where he teaches on agricultural water governance and political ecology of water. As a water policy advisor he has also worked in Peru and Bolivia with different international development organizations. His current research interests are the dynamics and discourses of water use and governance by agribusinesses in Latin America.

"This is a major book on the political ecology of water conflicts by the top experts in the field. It defines a new field of study, 'water justice.' It's a great addition to the study of local and global movements against environmental injustice with a focus on water grabbing and unequal access to water for irrigation, mining, urban sanitation, and hydroelectricity."
– *Joan Martinez-Alier, Emeritus Professor of Economics and Economic History, ICTA, Autonomous University of Barcelona*

"Boelens, Perreault and Vos have assembled a genuinely impressive set of authors to tackle the nature, meaning, and drivers of water injustices across the world, and to explore the possibilities of water justice. While the picture is far from rosy, the book provides rich theoretical and empirical perspectives through which to understand the inequities surrounding the control and use of water and to imagine alternative futures. This text will be a point of reference for many years to come."
– *Anthony Bebbington, Australian Laureate Fellow, University of Melbourne, and Milton P. and Alice C. Higgins Professor of Environment and Society, Clark University*

"This timely and engaging volume by some of the world's foremost scholars on water constitutes a loud sound of alarm. Not only that, it shows why liberal and neoliberal water rationalities … won't work. Proposed instead is a sophisticated approach to the question of water as nature, and of its relation to justice, from which emerges a powerful framework for alternative hydrosocialities. By reminding us that what is at stake … is people's very right to exist, *Water Justice* enables us to imagine and construct other paths for fair and wise water policies."
– *Arturo Escobar, Professor of Anthropology, University of North Carolina at Chapel Hill*

"It would be difficult to overstate the global significance of water injustice, which continues to be a major obstacle preventing millions of human beings from enjoying a dignified life. *Water Justice* addresses key aspects of this complex problem, bringing together a unique international team of scholars. This is not only a timely collection, but also one that provides access to rich theoretical arguments and empirical examples, allowing an in-depth treatment of the topic. The book is a welcome contribution for academics, students, and practitioners, and will attract a wider readership among those concerned with the future of civilized human life."
– *José Esteban Castro, Emeritus Professor, Newcastle University*

"'Water justice!' is the rallying cry of this book. It explores in an illuminating and comprehensive way the multiple dimensions of water injustice and the diverse struggles to change them."
– *Cristóbal Kay, Emeritus Professor, International Institute of Social Studies, Erasmus University Rotterdam, and Professorial Research Associate, Department of Development Studies, SOAS University of London*

WATER JUSTICE

Edited by

RUTGERD BOELENS
Wageningen University and University of Amsterdam, The Netherlands

TOM PERREAULT
Syracuse University, New York

and

JEROEN VOS
Wageningen University, The Netherlands

CAMBRIDGE
UNIVERSITY PRESS

University Printing House, Cambridge CB2 8BS, United Kingdom

One Liberty Plaza, 20th Floor, New York, NY 10006, USA

477 Williamstown Road, Port Melbourne, VIC 3207, Australia

314–321, 3rd Floor, Plot 3, Splendor Forum, Jasola District Centre, New Delhi – 110025, India

79 Anson Road, #06-04/06, Singapore 079906

Cambridge University Press is part of the University of Cambridge.

It furthers the University's mission by disseminating knowledge in the pursuit of education, learning, and research at the highest international levels of excellence.

www.cambridge.org
Information on this title: www.cambridge.org/9781107179080
DOI: 10.1017/9781316831847

© Cambridge University Press 2018

This publication is in copyright. Subject to statutory exception and to the provisions of relevant collective licensing agreements, no reproduction of any part may take place without the written permission of Cambridge University Press.

First published 2018

Printed in the United Kingdom by Clays, St Ives plc

A catalogue record for this publication is available from the British Library.

Library of Congress Cataloging-in-Publication Data
Names: Boelens, Rutgerd, editor. | Perreaultm, Tom, editor. | Vos, Jeroen, editor.
Title: Water justice / edited by Rutgerd Boelens, Wageningen University and University of Amsterdam, The Netherlands, Tom Perreault, Syracuse University, New York, and Jeroen Vos, Wageningen University, The Netherlands.
Description: Cambridge, United Kingdom: New York, NY, USA: Cambridge University Press, [2018] | Includes bibliographical references and index.
Identifiers: LCCN 2017046247 | ISBN 9781107179080 (hardback)
Subjects: LCSH: Water supply – Social aspects. | Water rights – Social aspects. | Water security.
Classification: LCC HD1691.W3245 2018 | DDC 333.91–dc23
LC record available at https://lccn.loc.gov/2017046247

ISBN 978-1-107-17908-0 Hardback

Cambridge University Press has no responsibility for the persistence or accuracy of URLs for external or third-party internet websites referred to in this publication and does not guarantee that any content on such websites is, or will remain, accurate or appropriate.

Contents

List of Contributors		*page* viii
1	Introduction: The Multiple Challenges and Layers of Water Justice Struggles RUTGERD BOELENS, JEROEN VOS, AND TOM PERREAULT	1
Part I	Re-Politicizing Water Allocation	33
	Introduction TOM PERREAULT, RUTGERD BOELENS, AND JEROEN VOS	34
2	Water Governance as a Question of Justice: Politics, Rights, and Representation DIK ROTH, MARGREET ZWARTEVEEN, K. J. JOY, AND SEEMA KULKARNI	43
3	Water Grabbing: Practices of Contestation and Appropriation of Water Resources in the Context of Expanding Global Capital GERT JAN VELDWISCH, JENNIFER FRANCO, AND LYLA MEHTA	59
4	De-Politicized Policy Analysis: How the Prevailing Frameworks of Analysis Slight Equity in Water Governance ANDREA K. GERLAK AND HELEN INGRAM	71
5	Urban Water and Sanitation Injustice: An Analytic Framework BEN CROW	89
Part II	Hydrosocial De-Patterning and Re-Composition	107
	Introduction RUTGERD BOELENS, TOM PERREAULT, AND JEROEN VOS	108
6	"… And Not a Single Injustice Remains": Hydro-Territorial Colonization and Techno-Political Transformations in Spain ERIK SWYNGEDOUW AND RUTGERD BOELENS	115

7 Making Space for the Cauca River in Colombia: Inequalities and
  Environmental Citizenship 134
  RENATA MORENO-QUINTERO AND THERESA SELFA

8 Reconfiguration of Hydrosocial Territories and Struggles for Water Justice 151
  LENA HOMMES, RUTGERD BOELENS, BIBIANA DUARTE-ABADÍA, JUAN PABLO
  HIDALGO-BASTIDAS, AND JAIME HOOGESTEGER

9 Large-Scale Dam Development and Counter Movements: Water Justice
  Struggles around Guatemala's Chixoy Dam 169
  BARBARA ROSE JOHNSTON

Part III Exclusion and Struggles for Co-Decision 187

Introduction 188
JEROEN VOS, TOM PERREAULT, AND RUTGERD BOELENS

10 Indigenous Peoples and Water Governance in Canada: Regulatory
   Injustice and Prospects for Reform 193
   KAREN BAKKER, ROSIE SIMMS, NADIA JOE, AND LEILA HARRIS

11 Sanitation Justice? The Multiple Dimensions of Urban Sanitation Inequalities 210
   MARIA RUSCA, CECILIA ALDA-VIDAL, AND MICHELLE KOOY

12 Uniting Diversity to Build Europe's Right2Water Movement 226
   JERRY VAN DEN BERGE, RUTGERD BOELENS, AND JEROEN VOS

13 Everyday Water Injustice and the Politics of Accommodation 246
   FRANCES CLEAVER

14 Sharing Our Water: Inclusive Development and Glocal Water Justice
   in the Anthropocene 259
   JOYEETA GUPTA

Part IV Governmentality, Discourses and Struggles over Imaginaries
and Water Knowledge 275

Introduction 276
TOM PERREAULT, RUTGERD BOELENS, AND JEROEN VOS

15 Neoliberal Water Governmentalities, Virtual Water Trade, and Contestations 283
   JEROEN VOS AND RUTGERD BOELENS

16 Critical Ecosystem Infrastructure? Governing the Forests–Water Nexus
   in the Kenyan Highlands 302
   CONNOR JOSEPH CAVANAGH

17 The Meaning of Mining, the Memory of Water: Collective Experience
    as Environmental Justice   316
    TOM PERREAULT

18 New Spaces for Water Justice? Groundwater Extraction and Changing
    Gendered Subjectivities in Morocco's Saïss Region   330
    LISA BOSSENBROEK AND MARGREET ZWARTEVEEN

19 Conclusions: Struggles for Justice in a Changing Water World   346
    TOM PERREAULT, RUTGERD BOELENS, AND JEROEN VOS

Index   361

Contributors

Cecilia Alda-Vidal is a PhD researcher at the University of Manchester, School of Environment, Education, and Development, and Sustainable Consumption Institute. In her PhD project she examines the production of unequal sanitation and hygiene experiences in the peri-urban areas of Lilongwe, Malawi. Prior to commencing her PhD, she worked and conducted research in different water and sanitation projects in Latin America and Africa.

Karen Bakker is a professor and Canada Research Chair at the University of British Columbia. She is the Co-Director of the University's program on Water Governance. Her research covers a range of issues related to water justice, including water privatization, water access, water as a human right, and Indigenous water governance. Her books include *Privatizing Water* (2010), and the co-edited *Water Security: Principles, Practices and Perspectives* (2013). She acts as an advisor to a wide range of international organizations, and is currently a Board Member of the International Institute for Sustainable Development.

Jerry van den Berge is an independent researcher and consultant on water and sanitation for sustainable development. Previously he was policy officer in the European Federation of Public Services Unions, based in Brussels, dealing with water, energy and waste services, EU and corporate policies, European works councils, and climate change and sustainable development issues. He coordinated the first successful European Citizens' Initiative, "Water is a human right!" He has been involved in the international trade union movement since 2004 and has a drive for social justice as the pillar for sustainable development. His work involves linking social and environmental justice, forming alliances between trade unions and non-governmental organizations.

Rutgerd Boelens is Professor of Water Governance and Social Justice in the Department of Environmental Sciences at Wageningen University, and Professor of Political Ecology of Water in Latin America with CEDLA (Centre for Latin American Research and Documentation), the Department of Geography, Planning and Institutional Development, and the Faculties of Social Sciences and Humanities at the University of Amsterdam. He is also Visiting Professor at the Catholic University of Peru and the Central University of Ecuador, and coordinates the international Justicia Hídrica/Water Justice alliance. He is the main scientific editor of the book series *Water and Society*, bringing together various Latin American publishing houses. His work includes many articles and books focusing on water rights, water grabbing, hydrosocial territories, legal pluralism, cultural politics, governmentality, and social mobilization, in Latin America and Spain. His most recent book is *Water, Power and Identity: The Cultural Politics of Water in the Andes* (2015).

Lisa Bossenbroek obtained her PhD in rural sociology at the University of Wageningen, the Netherlands. As part of her research she studied the role of young people in agrarian dynamics and the interactions of processes of agrarian change and gender relations. Currently, she works as a post-doc at the Faculty of Governance, Economics and Social Sciences (EGE–RABAT), Morocco. She is conducting comparative research that focuses on the interaction of agrarian change and gender relations and subjectivities in Morocco and India.

Connor Joseph Cavanagh is a research fellow in the Department of International Environment and Development Studies (Noragric), at the Norwegian University of Life Sciences. Recent articles have appeared in *Environment and Planning D: Society and Space, Antipode, Journal of Peasant Studies, Forest Policy and Economics,* and *Geoforum*.

Frances Cleaver is Professor in Environment and Development in the Geography Department at the University of Sheffield, UK. Her work is centered on understanding the ways in which institutions shape access to natural resources (especially water and land). Her interests link theoretical and methodological advances with practical policy application. Main research themes are: institutions, collective action and participatory natural resource management; water governance, poverty and wellbeing; and the everyday politics of natural resource access and gendered livelihoods. She has published several books including *Development Through Bricolage: Rethinking Institutions and Natural Resource Management* (2012) and many scientific articles on these subjects.

Ben Crow is Professor of Sociology at the University of California, Santa Cruz. His research areas include international development, sociology of water, famine and hunger, political economy, environment and development. He trained and worked as an engineer in London and Africa, and was an activist and volunteer in South Asia, before becoming a social scientist. His PhD is from Edinburgh University, and he has taught at the Open University (UK), Stanford and UC Berkeley. He has written several books, including *Sharing the Ganges: The Politics and Technology of River Development*; and *The Atlas of Global Inequalities* (with Suresh Lodha). Recent work explores water access by low-income households in urban settlements. He is currently helping to organize a four-city, collaborative research initiative on "capabilities and domestic work, dignity and deprivation in entangled urbanization."

Bibiana Duarte-Abadía is a research fellow at the University of Amsterdam and associated researcher with Wageningen University, the Netherlands, and the Alexander von Humboldt Research Institute, Bogota. Her research interests include: water governance, political ecology, water conflicts analysis, landscape ecology, biophysical studies in river basins and watersheds, geographical information systems and participatory action research. Her latest book (in Spanish) (with Cristina Yacoub and Rutgerd Boelens) is *Water and Political Ecology: Extractivism in Agribusiness, Mining and Hydropower Development in Latin America* (2015).

Jennifer Franco is a researcher in the Agrarian and Environmental Justice program at the Transnational Institute, the Netherlands, and an Adjunct Professor at the China Agricultural University in Beijing. Her regional focus is Southeast Asia, particularly Myanmar and the Philippines. Her research interests include: social movements, agrarian change, rural democratization, and the political dynamics of regulating access and control of land, water, fisheries, and forests.

Andrea K. Gerlak is an associate professor in the School of Geography and Development and Associate Research Professor at the Udall Center for Studies in Public Policy at the University of Arizona. Her research examines cooperation and conflict in water governance, including questions of institutional change and adaptation to climate change in river basins, and human rights and equity issues in water governance.

Joyeeta Gupta is Professor of Environment and Development in the Global South at the University of Amsterdam and IHE Institute for Water Education. Her research fields include changing geopolitics and multi-level governance patterns in relation to environment and development with a focus on climate change and water. She is editor-in-chief of *International Environmental Agreements: Politics, Law and Economics*. Her publications include many books and scientific articles. Her book *History of Global Climate Governance* was published by Cambridge University Press in 2014 and won the Atmospheric Science Librarians International (ASLI) Choice Award for 2014 in its history category. She is presently co-chair of UNEP's Global Environmental Outlook.

Leila Harris is an associate professor at IRES Institute on Resources Environment and Sustainability and the Institute for Gender, Race, Sexuality and Social Justice; faculty associate at UBC Department of Geography; faculty associate at Peter Wall Institute for Advanced Studies, as well as Co-Director of the Program on Water Governance; and member of EDGES (Environment and Development: Gender, Equity, and Sustainability) Research Collaborative, University of British Columbia. Trained as a political and sociocultural geographer (PhD Minnesota), her work examines social, cultural and political-economic dimensions of environmental issues. Her latest book (with J. Goldin and C. Sneddon) is *Contemporary Water Governance in the Global South: Scarcity, Marketization and Participation* (2013).

Juan Pablo Hidalgo-Bastidas is a research fellow at the Center for Latin American Research and Documentation (CEDLA), and the Department of Geography, Planning and International Development, University of Amsterdam, and associated researcher with the Department of Water Resources Management, Wageningen University. His current research is on hydrosocial territories, power, the politics of implementing mega-hydraulic projects, and environmental transformations in Ecuador.

Lena Hommes is a research fellow with the Department of Water Resources Management, Environmental Sciences, at Wageningen University. Her current research focuses on hydrosocial territories, rural–urban water connections, water justice, and socio-environmental transformations in Turkey and Peru.

Jaime Hoogesteger is an assistant professor with the Department of Water Resources Management, Environmental Sciences, at Wageningen University. His research focuses on themes of (ground)water policies and governance, democracy and equity which he relates to processes of agrarian change. His research experience is in Ecuador, Peru, Mexico, Iran and Mozambique. As part of his research he has made two documentaries, published book chapters, a monograph, two co-edited books and peer-reviewed journal articles. At present he is chair of the Netherlands Association of Latin American and Caribbean Studies (NALACS) and member of the Mexican Network of Researchers (SNI).

Helen Ingram is Professor Emerita at the University of California at Irvine, and Research Fellow at the Southwest Center at the University of Arizona. Her publications span the fields of public policy theory and water resources governance. Fairness, equity and democracy are major themes in her work, beginning with her collaboration on the book *Water and Equity in the Southwest* with F. Lee Brown in 1984. Her most recent publication, co-authored with Margaret Wilder, is "Knowing Equity When We See It: Water Equity in Contemporary Global Contexts" in Ken Conca and Erika Weinthal (eds.), *The Oxford Handbook of Water Politics and Policy* (2016).

Nadia Joe (Gugula) is Nlaka'pamux on her mother's side and southern Tutchone/Tlingit – belonging to the Crow Clan of the Champagne and Aishihik First Nations – on her father's side. She was privileged to receive an education both on the land and in the classroom. She currently works as an environmental scientist specializing in water management and wastewater treatment and supports Indigenous communities across Canada to bridge cultural divides over water governance.

Barbara Rose Johnston is Michigan State University Adjunct Professor of Anthropology and the Senior Research Fellow at the Center for Political Ecology (Santa Cruz, California). Her research and publications examine the intersect between environmental quality, health, human rights and social justice, work that has led to numerous scientific and public policy advisory appointments in international, national, and community-based forums. Major publications on water justice include *Water, Culture, Power: Local Struggles in Global Context* (with John Donahue, 1997) and *Water, Cultural Diversity and Global Environmental Change* (editor-in-chief, 2012). Her work on reparation and the right to remedy for Chixoy dam-affected communities serves as an evidentiary source that supports the historic reparation agreement for indigenous communities in Guatemala.

K. J. Joy is a senior fellow at the Society for Promoting Participative Ecosystem Management (SOPPECOM), Pune, India. He has been an activist-researcher for more than 30 years, working on people's rights to natural resources at grassroots and policy levels.

His research interests also include drought-proofing, participatory irrigation and river basin management, water conflicts and people's movements. He coordinates "Forum for Policy Dialogue on Water Conflicts in India," a national level network in India that engages with various types of water conflicts. He has published extensively on water-environment-development issues including a co-edited book, *Water Conflicts in India: A Million Revolts in the Making*.

Michelle Kooy is an associate professor of the Politics of Urban Waters at the IHE-Delft Institute for Water Education, and a faculty associate at the Department of Human Geography, Planning, and International Development at the University of Amsterdam, Netherlands. She was trained as a human geographer (PhD University of British Columbia) and she applies her academic interest in the politics of urban planning, urban water infrastructure systems and social justice through her work with international development agencies and non-governmental organizations. She is an associate editor of Wiley Interdisciplinary Reviews on Water, and a member of the Advisory Committee to the UNESCO International Hydrological Program in Cities.

Seema Kulkarni is a senior fellow at the Society for Promoting Participative Ecosystem Management (SOPPECOM), Pune, India. She holds an MA degree in Social Work from Tata Institute of Social Sciences, Mumbai. She has over 20 years of experience of working in drought-prone rural areas of Maharashtra. Her key area of interest is gender and natural resource management. She is associated with the women's movement in Maharashtra, working on peasant women's claims and rights over land and water resources. She has contributed many articles for various journals and conferences about gender and rural livelihoods.

Lyla Mehta is a professorial fellow at the Institute of Development Studies, UK, and a visiting professor at Noragric, Norwegian University of Life Sciences. A sociologist working in development studies, she uses the case of water and sanitation to focus on rights and access to resources, resource-grabbing, the politics of scarcity, gender, power and policy processes. Her work also concerns gender, displacement and resistance and climate change and uncertainty from 'below'. She has extensive research and field experience in India and southern Africa. She has published about 90 scientific publications, including the most recent edited book, *Flows and Practices: The Politics of Integrated Water Resources Management in Southern and Eastern Africa*.

Renata Moreno-Quintero is a lecturer-researcher in the Faculty of Humanities at the Universidad Autónoma de Occidente in Cali, Colombia. She holds a PhD degree in Environmental Policy from the State University of New York, College of Environmental Science and Forestry (SUNY-ESF), USA, where she was a Fulbright fellow from 2011 to 2016.

Tom Perreault is Professor and Director of Graduate Studies in the Department of Geography at Syracuse University in Syracuse, New York, USA. He received his PhD from the University of Colorado and has two decades of research experience in Ecuador and Bolivia. His research focuses on the intersections of political ecology, resource governance, indigenous *campesino* social mobilization, and agrarian change. He is editor of *Minería, Agua y Justicia Social en los Andes: Experiencias Comparativas de Perú y Bolivia* (2014), and lead editor of *The Handbook of Political Ecology* (2015).

Dik Roth is an associate professor at the Department of Sociology of Development and Change, Wageningen University, the Netherlands. His research areas are the anthropology of law and policy, legal plurality and complexity, property rights and justice, natural resources, resource conflicts, management and governance, with a focus on land and water. His regional focus is on South Asia/Southeast Asia and the Netherlands. He is editor-in-chief of the *Journal of Legal Pluralism and Unofficial Law*.

Maria Rusca is a Marie Skłodowska-Curie Research Fellow at King's College London. Her work examines urban socio-ecological dynamics of (waste)water governance in various geographical contexts and scales. Her research themes include multiple aspects of water justice, including informality in the urban waterscape, gendered water supply and everyday hygiene practices, the production of uneven drinking water contamination and understanding water governance challenges in the everyday. She believes that understanding socio-ecological dynamics requires interdisciplinary approaches and promotes research that equally accounts for the ecological and socio-political dimensions of water governance.

Theresa Selfa is Associate Professor of Environmental Studies at the State University of New York, College of Environmental Science and Forestry (SUNY ESF), USA. Her research focuses on environmental governance and politics of agrifood, energy and water systems, and on household livelihood impacts of land use change, in Latin America and the USA.

Rosie Simms holds an MA from the Institute for Resources, Environment and Sustainability at UBC, and a BA&Sc from McGill University. Throughout her studies, Rosie's work focused on intersections between water governance and Indigenous rights and governance. She currently works as a water law and policy researcher at the University of Victoria.

Erik Swyngedouw is Professor of Geography at the University of Manchester, UK. He previously held a professorship at Oxford University. He was also a visiting professor at the University of Ghent, Belgium and at Sciences Po, Paris. His research focuses on political ecology and political economy, with a particular interest in theorizing the society-nature articulation from a broadly historical-geographical materialist analysis. He has worked extensively on issues of water politics and the political ecology of water and urbanization. Recent work focuses on new political movements. He has authored many books, among which: *Social Power and the Urbanization of Water* (2004), *The Post-Political and its Discontent* (edited with J. Wilson, 2014) and *Liquid Power: Contested Hydro-Modernities in Twentieth-Century Spain* (2015).

Gert Jan Veldwisch is an assistant professor with the Department of Water Resources Management, Wageningen University, the Netherlands. His areas of research include participatory irrigation and development, water grabbing, small-scale irrigation, agrarian change, and issues around water justice. Currently he works mostly in Southern and Eastern Africa, but he has also worked in Latin America and Central Asia. He has published several articles on water grabbing.

Jeroen Vos is an assistant professor in the department of Water Resources Management at Wageningen University, the Netherlands. As a water policy adviser he worked for a decade in Peru and Bolivia with different international development organizations. He was editor of two Spanish language books on water management in Latin America. His current research interests are the dynamics and discourses of water use by agribusinesses in Latin America. He has published on the effects of virtual water trade and private water stewardship certification.

Margreet Zwarteveen is Professor of Water Governance with the Integrated Water Systems and Governance Department at IHE Delft Institute for Water Education, and with the Governance and Inclusive Development group, University of Amsterdam. Approaching water governance from perspectives of equity and justice, she defines water governance as the practices of coordination and decision-making between different actors around contested water distributions. She is concerned both with looking at actual water distribution practices and with analyzing the different ways in which water distributions can be regulated (through technologies, markets and institutions), justified (decision-making procedures) and understood (expertise and knowledge). Her work is based on an interdisciplinary approach that sees water distributions as interactions between technology, nature and society.

1

Introduction: The Multiple Challenges and Layers of Water Justice Struggles

Rutgerd Boelens, Jeroen Vos, and Tom Perreault

1.1 Introduction

Water is a resource that triggers profound conflicts and close collaboration, a source of deep injustices, and fierce struggles for life. In many regions of the world, rising demand and declining availability of adequate-quality water foster severe competition and ferocious clashes among different water uses and users. People also suffer from flooding; contamination caused by industry and mining; privatization of public water utilities; corruption; and displacement by large dam projects. Climate change intensifies most human-made water problems. In struggles for water security, the poor tend to lose (e.g. Crow *et al.*, 2014; Escobar, 2006; Harvey, 1996; Perreault *et al.*, 2011).

Through exemplary cases, the chapters in this book show how new competitors – including megacities, mining, forestry, and agribusiness companies – demand and usurp a mounting share of available surface and groundwater resources (e.g., Donahue and Johnston, 1998; GRAIN, 2012). Water deprivation and water insecurity affect marginalized urban households, and rural smallholder families and communities. In many regions, this poses profound threats to environmental sustainability and local and national food security (e.g., Escobar, 2008; Mehta *et al.*, 2012; Mena *et al.*, 2016).

Such proliferating problems of material and social "water injustices" provide the backdrop for this book. Distribution of access water rights and water-related decision-making is extremely skewed. Smallholder communities' water-based livelihoods and rights in many countries of the global South are constantly threatened by bureaucratic administrations, market-driven policies, and top-down project intervention practices.

Despite the fact that water injustices have existed throughout human history, water justice problems and related policy interventions have changed rapidly over recent decades (Zwarteveen and Boelens, 2014). For instance, rather than focusing on simply enlarging water flows through new hydraulic engineering projects, new perspectives focus on water saving and conservation (Vos and Marshall, 2017; Zwarteveen, 2015). New scientific fields and water professionals have entered the water policy-making and intervention worlds to accompany (increasingly high-tech) hydraulic engineering

(Buscher and Fletcher, 2015; Goldman, 2007, 2011). Also, climate change threats and water-related disasters have changed science and policy debates and water funding projects related to issues such as "mitigation and adaptation," flood control and drought prevention (Heynen et al., 2007; Lynch, 2012; Martínez-Alier, 2002). Further, global neoliberalism has assured that water development and governance are no longer seen as the exclusive realm of the state, with water knowledge and authority concentrated in powerful public agencies (Hommes et al., 2016; Loftus, 2009; Zwarteveen, 2015). Water governance scales have changed: the nation-state has lost territorial sovereignty in water control. Civil-society organizations and, particularly, multinational companies and global policy institutes have entered the water governance scene (Molle et al., 2009; Perreault, 2015; Swyngedouw, 2004). In practice, this has shifted accountability relations, from publicly-elected governments or local water user groups to non-democratic multilateral financial institutions (Zwarteveen, 2015; see also Bakker, 2010; Swyngedouw, 2004).

An important starting point of the book is the authors' shared recognition that understanding and challenging water injustices requires conceptual tools to recognize the power and politics of water use, management and governance. Beyond their expression in laws, explicit rules and formal hierarchies, the book calls attention to how power and politics also significantly work through more invisible norms and rules that present themselves as naturally or technically ordered. These rules are part of established water development intervention procedures and practices, and are embedded in water expert communities' cultural codes of behavior (Zwarteveen and Boelens, 2014). Therefore, in addition to dealing with the urgent issues such as water grabbing and dam building, the book's attention goes beyond such overt water injustices and open conflicts, showing how unfairness and injustices are intrinsic to standard ways of knowing and governing.

Understanding how water injustices are embedded and situated, and possible ways to remedy them, is a central aim of this book. This entails an acknowledgment of diversity and plurality – in views, knowledge, rights systems, ideas and norms about fairness etc. – without embracing a stance of cultural relativism or denying the broader similarities across specific instances of injustice (Roth et al., 2005).

This introductory chapter provides some starting points for the water justice explorations that the book will elaborate on. As we argue in the next sections, the evolving field of water's political ecology builds on transdisciplinarity (Perreault et al., 2015). As such, it treats nature, technology and society as mutually constitutive (Haraway, 1991; Latour, 1993; Swyngedouw, 2009), forming hydrosocial networks that establish how water and decision-making power over water control are (to be) distributed. By deconstructing technical discourses of efficiency, economists' stories of productivity and naturalized ideas of scarcity, it searches for new insights to challenge unequal power structures as manifested in and through water. The sections examine the multiple layers of water injustices, ranging from the brutal, visible practices of water grabbing and pollution to the subtle powers and politics of misrecognition and exclusion, and covert equalization and subjugation techniques.

1.2 Examining Water Justice

The combination of intensified resource extraction, land and water degradation, increasing competition over water access and control, and growing reliance on market forces and forms of water expertocracy, have profound implications for debates over water rights and justice. On the one hand, it is increasingly clear that water scarcity and insecurity are not so much related to the absolute availability of fresh and clean water, but rather are expressions of how water, and water services, are unequally distributed among societal groups. Unequal water distribution and exposure to contaminated water, flooding and failed water projects often reveal elite capture of the state and related biased policies and corrupt practices. In other words, the so-called "water crisis" is less a consequence of generalized scarcity than a manifestation of uneven power geometries (UNDP, 2006). On the other hand, the mainstream water policy community tends to avoid scrutinizing the root causes of water problems. Instead, in accordance with its own positivist, universalist epistemologies and its belief in expert knowledge systems, formal legal structures and market forces, it blames the victims: local water user groups, communities and their "chaotic, inefficient plural rights systems" (Boelens and Zwarteveen, 2005).

Recently implemented global water reforms tend to ascribe water inequities and unsustainability to incomplete implementation of the universalistic, market-based expert model (Achterhuis *et al.*, 2010). Therefore, paradoxically, the remedy that is often prescribed is to follow the rationality and forces that largely have caused the problem in the first place: to increase free-market rules in local communities, and give more leeway to outside and private-interest groups (Bauer, 1997; Heynen *et al.*, 2007; Perreault *et al.*, 2015).

Such policy practices form part of a larger phenomenon in the water world: most international policy models and national water laws are not adapted to local populations' contexts, assuming that it is these local populations, rather than official plans, laws and theories, that need to adapt. These models aim to create their own, utopian water world. Consciously or subconsciously, such policies hold that local water territories are basically unruled – or at least unruly: disorganized humans, irrational values, unproductive ecologies, inefficient resource use, and continual water conflicts. Existing water norms and practices are misrecognized by overlooking water values, identities, rights systems, and users on the ground. Mainstream water policy-makers then construct imaginary water users, with identities that conveniently fit the models, with needs and rationales matching the interests and knowledge of those in power, shored up in their science, technology and policy towers. This way, policy models justify dramatic interventions, even when well-intended (Boelens, 2015a).

It is for these reasons that we base our understandings of "water justice" on a notion that sees environmental governance not as the "governance of nature" but "as 'governance *through* nature' – that is, as the reflection and projection of economic and political power via decisions about the design, manipulation and control of socio-natural processes" (Bridge and Perreault, 2009: 492). More specifically, we situate "water justice" conceptually and politically in the field of the "political ecology of water," which may be defined as: "the politics and power relationships that shape human knowledge of and intervention in the water world, leading to forms of governing nature and people, at once and at different scales, to produce particular hydro-social order" (Boelens 2015a: 9). This political ecology of water thus focuses on unequal distribution of benefits and burdens, access to and control over water, winners and losers, and disputed water rights, knowledge, and culture. It is also about practical and theoretical efforts to build alternative water realities. Therefore, our questions address fundamental issues regarding how water scarcity is being constructed by dominant agents, and how power relations influence water knowledge and development to produce particular claims to truth. Our questions also intrinsically engage research and transdisciplinary social action, focusing for instance on how knowledge production can contribute to strategies that contest water dispossession and accumulation; and how the knowledge systems of scholars, activists and water users can be mutually enriching and complementary.

Approaching such questions requires an understanding of "justice" as based on a complex set of notions and dynamic principles that are grounded in particular social realities. It means that we must deviate from prevailing liberal political-philosophical theories that have tried to present justice as a universal, transcendent concept (Lauderdale, 1998; Roth *et al.*, 2005). We therefore differ with positivist traditions, such as the utilitarian philosophy of eighteenth-century political economist Jeremy Bentham, who defined justice as that particular societal order that would bring the greatest happiness to the greatest number of citizens. To this end, the rights and happiness of some may be sacrificed – generally, this means society's most vulnerable social groups. Bentham sought to establish a system "that aims to construct happiness societally by means of reason and law" (1988 (1781): 1–2), whereby happiness could be exactly calculated. Echoing the current water expertocracy, this calculated design of happiness and overall wellbeing would be the task of moral and justice experts; common people would lack reason. Utilitarian justice as defined by Mill (1874, 1999) – advocating legal rationalization and the use of economic theory in political decision-making to, ultimately, devise a politics oriented by human happiness – also means excluding "irrational deviants" from (Western positivist) justice. Most legal justice constructs deploy variations of these liberal-universalist ideas and theoretical ideals of justice.

We also differ with "social contract" notions of distributional justice based on Rawls (1971), which stress "procedural fairness" and "ethics-based autonomous decision-making." Rawlsian justice takes place behind abstract, illusory "veils of ignorance" (which supposedly allow people to make just decisions without knowing the impact these decisions will have on themselves), but ignores actually existing class, gender, education and ethnic inequality structures. And in the same vein, we challenge liberal-individualist or socialist-collectivist theories that concentrate only on distributive justice but overlook sources of everyday injustices based on discrimination, misrecognition, and exclusion from decision-making. Young (1990), Fraser (2000), Schlosberg (2004) and Escobar (2008) have shown how such (universalistic) distributive models and procedures fail to "examine the social, cultural, symbolic and institutional conditions underlying poor distributions in the first place" (Schlosberg, 2004: 518). Next, we are profoundly distant from libertarian entitlement (e.g. Nozick, 1974) and neoliberal appropriation theories (e.g. Hayek, 1944, 1960; Friedman, 1962, 1980) that stress the relationship between individual freedom and private property maximization. Hayek and Friedman see no conceptual or empirical problems in building "justice" precisely on expanding economic-distributive inequalities and further dis-protection of the vulnerable: equality is defined as all individuals' freedom to become rational market actors (Swyngedouw 2005; Ahlers and Zwarteveen 2009).

For these reasons, differing with these universalistic (mis)understandings of justice, we deploy a relational perspective (see also Boelens, 2015a; Perreault, 2014; Roth *et al.*, 2005, 2014; Zwarteveen and Boelens, 2014): to understand the embeddedness of particular ideals of justice, and the way these get constituted through social practices, requires a grounded, comparative and historical approach (Lauderdale, 1998). Such critical, grounded justice perspectives must understand how diverse people see and define justice within a specific context, history and time (Joy *et al.*, 2014; Perreault, 2014; Zwarteveen and Boelens, 2014). They also examine the effects that particular definitions of justice have on how a society distributes wealth and authority (Roth *et al.*, 2005). Justice proposals based solely on abstract, universalistic criteria, have been unable

> to respond to indigenous and peasants throughout the world who are still experiencing the full presence of injustice in the form of poverty, landlessness, dispossession, political and religious oppression, and genocide. Philosophical formulas become hollow without systematic explorations of the sources of injustice, including those within indigenous and peasant societies.
>
> (Lauderdale, 1998: 5–6)

Consequently, we argue for the need to analyze, in all their diversity, how living people experience injustice, facing political oppression, cultural discrimination and economic marginalization. We relate these injustice experiences to, on the one hand, locally prevailing perceptions of equity and, on the other, hegemonic discourses, constructs and procedures of formal justice. Moreover, we also call for an analysis of the actors who develop or impose these views, and why certain perspectives on justice or equity are promoted while others are ignored, plus the effects of these views and conceptualizations for specific groups.

As Fraser (2000) has argued, injustice combines issues of distribution with those of (cultural) recognition, in often complex and sometimes paradoxical ways (also see Schlosberg, 2004; Young, 1990). Cultural, ethnic and gender discrimination often constitute the (implicit or explicit) foundation to privilege allocation of water rights to some over others. For example, in many African countries, a common feature of irrigation modernization projects is that they have cut off women from any possibility to control land or water. In Mali, after 50 years of investment in irrigation, only 12 of the 2,500 farmers under the Office du Niger were women. In Burkina Faso, all land titles granted by the Volta Valley Authority went to male household heads. In Senegal, women own less than 4 percent of the newly irrigated areas. In Mauritania, nearly 20 percent of the households in the river area are headed by women, and yet women comprise only 5 percent of participants in new schemes (Dankelman and Davidson, 2013; Zwarteveen, 2006).

Exclusion from decision-making often has direct effects on unequal allocation of and access to water. In turn, decision-making authority is determined by economic power relations and cultural and behavioral norms that interlink with how particular forms of water knowledge are legitimized and privileged. Indeed, questions of participation, recognition and distribution are intimately linked to water control. Further, in addition to Fraser's three domains of justice struggle ("recognition" and "participation" and "distribution"), a fourth domain of water justice may be expressed as "socio-ecological justice." This refers to the ways in which water-allocation decisions and struggles are embedded in sensitive, dynamically shaped socio-natural environments, seeking to sustain livelihood security for contemporary and future generations (Boelens, 2015a; Zwarteveen and Boelens, 2014; Escobar, 2008). Before returning to this relational, engaged understanding of water justice, and what we see as important ingredients of an approach to identifying, understanding, challenging and defying water injustices (in Section 6), we first consider some examples of water injustice in practice.

1.3 The Cruel Face of Water Injustice: Some Expressions in Practice

1.3.1 Agribusiness Water Grabbing

The world is experiencing a boom in transnational agricultural produce trade. Exports of fresh vegetables, fruits, and flowers have doubled in the last decade (Vos and Hinojosa, 2016). Governmental policies support large agribusiness companies that buy up land in countries in the South on a massive, unprecedented scale (Rulli and D'Odorico, 2013; Woodhouse, 2012). Land grabs of this sort lead to competition for water with local communities, degrade local ecosystems, jeopardize local food security, and profoundly alter existing modes of production and income distribution (Van der Ploeg, 2008; Zoomers and Kaag, 2014). The land purchased is worth little if not accompanied with access to water. In most cases, therefore, land grabs are in fact water grabs, a process that dispossesses and displaces existing water users (Mehta *et al.*, 2012; Woodhouse, 2012).

Based on the data set of *The Land Matrix* (2012), Rulli and D'Odorico (2013) calculated that total land deals reported by foreign companies amounted to some 43 million ha. Water used in agricultural products can be seen as "virtual water" embedded in those products. Export of crops from the 43 million ha would represent some 497 billion cubic meters of virtual water exported to rich consumers. This would increase current virtual water exports by one-third, as calculated by Hoekstra and Mekonnen

(2012: Table 1). Some 22 percent of these reported land deals was under production in 2013.

Clearly, this "hydro-colonialism" goes beyond classic North-South opposition: companies from Asia have in recent years bought more than 8 million ha in the Nile basin, to grow export crops that need water far beyond the entire water availability of the basin. According to GRAIN (2012: 8) this is "hydrological suicide: four countries alone already have irrigation infrastructure established for 5.4 million ha of land and have leased out a further 8.6 million. Irrigating just these lands would require much more water than is available to all ten countries in the Nile basin." As GRAIN reported, the Ethiopian Government aims to evict 1.5 million people from their territories to make irrigated land available (GRAIN, 2012: 18).

National policies often allocate water to where "its marginal returns are highest" and link this to promotion of commercial (export) crops, which replace staple crops. This may endanger food security. Gaybor (2011) provides an illustration of this for Ecuador. Nationally, according to official registration, the large-scale export sector represents 1 percent of the farms, but has concessions for 67 percent of the total available irrigation water. Peasant and indigenous producers in community irrigation systems represent 86 percent of the water users, but own only 22 percent of irrigated land and have access to only 13 percent of the total allocated irrigation water. In some provinces, water allocation inequality is outright appalling. In Imbabura Province in the north, for example, a small number of large landholdings (>100 ha) account for 91 percent of the total allocated volume of water (Gaybor, 2011: 200). Actual water distribution is even worse than official figures show, as more than half of the water that is used by large-scale agribusiness companies is not registered and is illegally tapped.

In Peru, we can witness similar practices. In the dry Ica Valley, with fertile soils and strategically located near Lima, rainfall is close to zero. Groundwater, therefore, is a vital resource for thousands of small farmers. For the past decade, however, the aquifer has been dramatically over-pumped, with its water table dropping by nearly one meter per year (Progressio, 2010). New agro-export companies have purchased most valley land to produce high water-consumption export crops such as grapes and asparagus. Small and medium farmers, who are unable to compete with these large owners' powerful water pumping technologies, have seen their wells run dry. Agro-exporters, who constitute 0.1 percent of the users, consume 36 percent of the water. Small farmers, who account for 71 percent of all users, have access to only 9 percent of the water (Cárdenas, 2012). As in India, Mexico, Chile and other regions around the world, only those who can afford to purchase powerful pumps and ever-deeper wells are able to access groundwater. The resulting inequality is a major source of conflict (Joy *et al.*, 2014; Roth *et al.*, 2005, 2015).

This also places the dominant neoliberal logic regarding the benefits of virtual water export in a different light. The discourse on virtual water efficiency assumes that, through global trade liberalization, virtual water flows from water-rich to water-poor areas (Vos and Hinojosa, 2016). In many cases, however, this is simply incorrect. Water-poor countries such as India and China, Kazakhstan, Australia, and Tanzania are net exporters of virtual water. Water-rich countries such as the Netherlands, UK and Switzerland are net importers of virtual water. The NAFTA agreement between Mexico and USA led to virtual water flow from dry areas in Mexico to the USA. As mentioned above, asparagus and grapes exported by large-scale agribusiness, from the desert coast of Peru, deprive local communities of water and income. Flower production for the USA and Europe in vulnerable areas of Kenya and the Andean mountains of Colombia and Ecuador profoundly affects the quantity and quality of local community water sources, as well as overall livelihood conditions (Mena *et al.*, 2016; Vos and Boelens, 2014).

1.3.2 Extractive Industries

Encroachment on water territories by extractive industries provides another illustrative example of brutal water injustice. In many parts of the world, mining companies make use of water in headwater catchments, thereby diverting and polluting the downstream flows on which peasant and indigenous communities (and sometimes entire cities) depend. In the highlands of the Andean countries, for instance, mining companies buy up water rights and gain de facto control over water resources, which sometimes leads to conflict (e.g. Budds, 2010; Preciado-Jeronimo *et al.*, 2015; Sosa and Zwarteveen, 2012). In the lowlands, hydrocarbon industries are increasingly dominating water control. As Bebbington *et al.* (2010) show, in Ecuador's Amazon region, approximately half of the total area is allotted in concessions to oil companies (see also Lu *et al.*, 2017). In neighboring Peru, it is even worse: nearly three-quarters of the Amazon region is allocated or subject to leasing to hydrocarbon transnationals (Bebbington *et al.*, 2010: 309–11).

A telling illustration is a Canadian gold mine intervention in San Luis Potosí, Mexico. Cerro de San Pedro is an ecological reserve and cultural heritage site. Water is fundamental for local livelihoods and the large city of San Luis Potosí. New water extraction is entirely forbidden in this desert region. In 2007, however, international laws and Mexican politics led the mining company to circumvent all local rules, annexing the previously untouchable communal land and water rights, with disastrous effects for the ecological reserve (Peña and Herrera, 2008; Stoltenborg and Boelens, 2016). Cerro de San Pedro has been converted into a large, toxic cyanide dump. Land and waterscapes have been destroyed, and the river has stopped flowing. National politicians forced the local mayor to accept the mine. He had no alternative: the former mayor, his father, had opposed the mine and was murdered. Internationally, however, the Canadian company is recognized for its corporate social responsibility, suggesting deep concern for community development and the environment. It has been issued a Conflict-Free Gold Certificate. International agreements actively support the mine's plunder. The NAFTA water charter forced the local population to accept the mine. Complaints from local communities receive no consideration. They have no right to participate in decisions about their own future.

Meanwhile, Latin American governments increasingly invoke anti-terrorist laws and discourses – initially a response to civil wars in the 1980s but given new impetus by the global war on terror – to label and imprison protesting villagers as "environmental terrorists." A Peruvian environmental movement leader said: "We now have a state that no longer protects people's rights and instead protects investment" (The Guardian, 2014). In 2014, the Inter-American Human Rights Commission (IACHR, 2014) investigated 22 large-scale Canadian mining projects in nine Latin American countries, concluding that they all caused profound environmental impacts, contaminating rivers, displacing people, impoverishing communities, and dispossessing water rights. Protesters have been killed. As the report observes, development cooperation increasingly promotes mining; Canada has advised Latin American governments on how to circumscribe protective laws and curtail civil rights to facilitate mining. China, Australia, Europe, and the United States may follow suit (Stoltenborg and Boelens, 2016).

1.3.3 Hydropower and Mega-Hydraulic Development

Water injustice also originates from hydropower development, large irrigation schemes and other mega-hydraulic infrastructure, triggering broad societal struggle (e.g. Baviskar, 2007; Kaika, 2006; McCully, 2001). Mega-dams drastically change hydrological regimes, flood important cultural landscapes, and often alter local rural livelihoods irreversibly. In many places around the world, dispossessed or resettled people bear the burdens, while benefits accrue to distant cities, financial institutions and construction, hydropower and mining industries (see Duarte *et al.*, 2015; Hidalgo *et al.*, forthcoming; Hommes *et al.*, 2016).

Notwithstanding growing public criticism of this top-down, supply-driven hydraulic development, these projects have gained new impetus worldwide, since they are portrayed as key ingredients of the new "green economy" (Goldman, 2011; Hommes and Boelens, 2017; Huber and Joshi, 2016; Sneddon and Fox, 2008). Hydropower generation as "clean development" is currently a basic justification for dam projects. However, new mega-works often ignore the lessons of past decades, also disregarding these projects' contribution to climate change (Jasanoff, 2010; Moore *et al.*, 2010). The nexus among state bureaucracies and politicians, private companies, engineering schools and global capital lending steers policy, giving preference to large-scale hydraulic works over context-sensitive, less capital-intensive, interactively designed alternatives (e.g. Hommes *et al.*, 2016; Moore *et al.*, 2010; Sneddon and Fox, 2008).

Mega-hydraulic development tends to be neatly integrated with a market-based capitalist model of economic growth, triggering rights and resources accumulation by some players and the simultaneous dispossession of vulnerable groups.

A horrifying example is the Chixoy Dam in Guatemala, detailed by Barbara Johnston in Chapter 9. In order to construct the dam, the Achi Maya indigenous population living there was labeled a "backward people" without territorial rights or homestead properties, and the dam site was labeled "unruled, empty space." Project documentation ignored the Achi Maya's strong cultural-productive roots in their territory. The project blended participatory jargon with racist ideas to explain why these ignorant people resisted displacement. An Inter-American Development Bank report states: "In the native peoples' world view, traditional lifestyles and agricultural practices are expected to remain changeless for evermore, which explains why native *campesinos* ... have proven resistant to change and innovation" (IDB, 1991, annex II-2:1, cited by Lynch, 2006: 14). When Achi Maya communities peacefully resisted displacement from their homes, the World Bank, donor governments and international consultants actively ignored state-sponsored military violence (Johnston, 2005; Lynch, 2006). As a consequence, many years of intimidating, torturing, and raping the local population left 440 men, women, and children dead and displaced thousands of local families (Aguirre, 2014).

1.3.4 Rural-Urban Transfers and Intra-Urban Water Provision Inequality

In many places in the world, the expanding thirst of cities and industries is quenched at the expense of rural communities and smallholder families. Ironically, supply-oriented engineering projects that divert water from increasingly distant rural areas to urban areas are justified, among others, by references to the Human Right to Water, the Millennium Development Goal of ensuring safe drinking water access for all, and the national importance of megacities (Hommes and Boelens, 2017). While such references may be well-intended, water transfers are represented as neutral, scientifically justified options, while the societal power relations inscribed in such technologies generate very unequal outcomes for different groups (Bakker, 2010; Yacoub *et al.*, 2015). They often evoke the image of water-supplying watersheds and forests as being uninhabited or even virgin lands, in which water is freely accessible for "high-priority demands." Historically grounded customary water rights are often considered clandestine.

An exemplary case is Peru's capital, Lima. As Hommes and colleagues have shown, much of its drinking water is transferred from the Andean highland territories, and any opposition by smallholder communities is characterized as backwardness, ignorance or stubbornness, to be resolved by "awareness-raising." As the national drinking water agency writes in its public relations book, tellingly entitled *The Land of the Lagoons*,

> Living in a natural paradise, they [communities affected by water extraction] find themselves a bit distant from the reality which our country lives, and even more so from the necessities of other Peruvian regions. Accordingly, their initial attitude was one of indifference towards the great project that will benefit the regions of Lima and Callao with new water sources ... Nevertheless, SEDAPAL planned actions to benefit the community, such as this book, which ... has the value of striving to integrate the most remote communities and those least informed about the country's reality ... Explanations about the project's kind-heartedness ... conquered local leaders' resistance.
> (SEDAPAL, 1998: 17–18, cited in Hommes and Boelens, 2017: 77)

Drinking water extraction is sustained and legitimized by a discourse and policy that present the countryside as embodied by abundant resources and backwardness, and the city as a place of civilized society and progress that has natural water scarcity. At the same time, the discourse naturalizing water scarcity, legitimizes ever-larger rural–urban water transfers without touching upon the fundamental issues of obsolete water infrastructure and low efficiency inside the city, plus huge inequality in water access within the city's neighborhoods (Hommes and Boelens, 2017; Ioris, 2016). As elsewhere, Lima's water scarcity is referred to as a natural problem caused by its arid environment and by climate change, rather than as a problem of distribution or of uneven power relations (Bakker, 2007; Linton, 2010; Lynch, 2012). Around one million inhabitants in Lima lack access to public drinking water and sanitation systems, but in the wealthy neighborhoods pools are filled and parks intensively irrigated (Ioris, 2016). As can be witnessed in Lima, as in many of the world's megacities, "water transfers are promoted as charitable 'water for all' projects even though the water often does not reach those most in need" (Hommes and Boelens, 2017: 78).

1.3.5 Water Service Privatization

Neoliberal thinkers and policy-makers advocate treating water as an economic good. According to neoliberal logic, policy measures such as privatizing water and water service provision, granting concessions to operate distribution networks, and implementing full-cost recovery in water service pricing would lead to improved water service, increased investments in infrastructure, and more efficient operation and maintenance. However, several studies have shown that privatizing public utilities has often failed to benefit water users; rather, tariffs hiked, investments in infrastructure lagged behind, quality of service provision did not improve, and the environment was jeopardized. Companies also faced disappointing returns and now retreat from selected countries and intensify privatization in more profitable regions (Bakker, 2010, 2013; Van den Berge, *et al.*, Chapter 12 of this book). In recent years, protests have been organized in various parts of the world to stop privatization of drinking water utilities or demand cancelation of these contracts: e.g. in Dar es Salaam in Tanzania, Jakarta in Indonesia, and in different cities in South Africa, India, Brazil and Spain. Because of these social protests, the meager service provision results for the people, and lower-than-expected profits for companies, many drinking water companies have been "re-municipalized." By 2014, over 180 water utilities worldwide had been returned to public management (Lobina *et al.*, 2014).

Water services are often privatized by means of public-private partnerships (PPPs). However, in many cases the public partners in a PPP assume a relatively higher share of the burdens and risks, while the private partners take a higher part of the benefits. The "commons" partners are not even considered in such alliances, which exclude local water-management collectives from decision-making about their own systems or territories. A PPP example is the recently built irrigation system in Olmos, on Peru's desert coast, promoted internationally as a high-tech, modern project. Locally, it met with resistance from communities that envisioned completely different hydro-territorial development. Building the dam, tunnel and irrigation canal cost an estimated US$800 million, of which the Peruvian state put in US$450 million and the Brazilian company, Odebrecht, that constructed the infrastructure put in US$350 million (Eguren, 2014). The 43,000 ha of land that will get irrigation water was sold to ten agribusiness companies at very low prices. Amnesty International reported gross violations of human rights when local farmers and goat farmers were evicted from the land claimed by the project (Amnesty International, 2013). Two major companies acquired large tracts of land: the Peruvian agribusiness company Grupo Gloria (15,600 ha) and Odebrecht itself (18,000 ha). The average cost per hectare was only US$4,723 (implying US$3,370 state subsidy per hectare, and far below market value). Eguren (2014) calculated that, after 50 years of operating the Olmos irrigation system, Odebrecht would have made a net profit of US$464 million by selling land, water and energy, and the Peruvian state would be left with a loss of US$328 million (at current market prices). This loss could be seen as an investment in water infrastructure that would create jobs for poor people. However, the total value of income for field laborers generated over these 50 years would hardly amount to this "investment." After Odebrecht CEO Marcelo Odebrecht was sentenced to 19 years of prison for acts of corruption in Brazil, the Odebrecht company sold their share in the Olmos system to Suez in December 2016. The press release by Suez CEO Jean-Louis Chaussade on this deal stated:

We are proud to bring our expertise and our solutions to a project that is vital to the development of the Olmos region and its inhabitants. In a world of scarce resources, the agricultural sector needs sustainable, efficient solutions in order to nourish expanding populations. *It is therefore crucial that we work to distribute water more equally.*

(SUEZ, 2016, emphasis added)

Similar modernist promotion and elite capture of the state (resulting in vast subsidies for agribusiness) happen in many parts of the world (e.g. Vos and Marshall, 2017).

1.4 The Subtleties of Water Injustice

Although water injustices sometimes become manifest through large or even violent conflicts, they more often occur in less visible ways, where resistance or disputes may (seem to) be absent altogether. For example, the fierce global policy effort to make water rights transferable by formalizing and standardizing rights systems typically results in silent water take-overs, rather than open disputes. Use of technological or policy innovations, such as deep tube wells, or financializing the water sector, can also induce silent water take-overs.

Throughout the world, we can witness how social norms and scientific standards in water governance naturalize and normalize injustices and inequities (Boelens and Vos, 2012), with water policies often sanctioning rather than questioning concentration of water rights in the hands of a few private powerful actors (e.g. Loftus, 2009; Swyngedouw, 2005; Venot and Clement, 2013). Neoliberal discourses have become so dominant in framing the terms of water debate that they have come to be accepted as normal or inevitable, making it difficult to recognize them for what they are: deeply ideological ideas (Achterhuis *et al.*,

2010). Eduardo Galeano underscored the subtlety of this process in *News about the Nobodies:*

> Up till recently, poverty was the fruit of injustice. But times have changed greatly: now, poverty is the just punishment that inefficiency deserves, or simply a way of expressing the natural order of things. The world has never been so unfair in dividing up resources, but the system that governs the world – now discreetly called "the market economy" – takes a daily dip in the bath of impunity.
>
> (Galeano, 1995: 1)

Water (in)justices involve both quantities and qualities of water, the modes of accessing and distributing water, and the meanings, discourses, truths and knowledge that shape water control (Zwarteveen and Boelens, 2014). Therefore, water conflicts include questions about decision making, authority and legitimacy, which extend into questions of culture, territory and identity.

1.4.1 Equalization, Commensuration and Inclusion

Modernist water policies emphasize unity and uniformity in water governance, whereby the state is increasingly instrumentalized to protect and enforce market rules and forces. At the same time, the state's monopoly on water rule-making, rule-enforcement and dispute-solving overrides all other tribunals or rights frameworks. A fundamental principle is blanket enforcement throughout national territory, based on the proclaimed equality of all citizens. Though the referent model of "being equal" is, in practice, often based on the class, gender and cultural standards and interests of a powerful minority, the image of a neutral legal-justice framework is strong.

The diversity of context-based, "intangible" water rights systems in most countries poses a tremendous problem for water bureaucrats, planners, and international companies. The diverse authorities, territorial autonomies and community rules make state domination or free-market operation very difficult (Achterhuis *et al.*, 2010; Boelens, 2015b). To bring about a uniform property framework, the construction and functioning of law in social action tends to be conveniently ignored. Participation and consensus-seeking policy-making presume the commensurability of values and equal power of social groups to voice their ideas and preferences. Formal water laws and institutions are presented as objective, rational systems for designing societal life, rather than as deeply cultural and political products, developed and enacted by societal groups, classes, and governmental agents who ply their strategies to foster their interests (Benda-Beckmann *et al.*, 1998; Roth *et al.*, 2015; cf. Sousa-Santos, 1995).

Mainstream water policies and discourses tend to pay much attention to the issues of "participation," "integration," and "recognition of local rights and cultures," appealing to common-sense notions of justice and equality. The hidden principle, however, is the active destruction of "inconvenient otherness" through subtle strategies of "managed multiculturalism" (Baud, 2010; Boelens, 2015b; Hale, 2002), while "convenient expressions" of local water-rights pluralism are, as much as possible, included in the modern private property market economy. Compared to the earlier top-down state-centric and neoliberal policy interventions, we see here how current ideas about redistribution, private property rights and market-based governance represent a shift. Rather than being based on explicit top-down hierarchies, visible rulers, exclusion, and sometimes brutal violence, modern equality ideologies aim to subtly seduce, include, and make equal. Indeed, in modern water policies everybody is potentially equal and *should be* equal.

Evidence from around the world regarding water allocation and administration makes clear, however, that this ideology of "equality of all" is not used to abolish the enormously unequal distribution of water property or stop water grabbing. Rather, making water users equal means: oppressing their *deviation* from the formal rules, norms and rights. Modern water policies impose "equalization." Following universalistic good governance discourse, governments differentiate "responsible water citizens," who are state- and market-compatible, from "irrational water spoilers," who devise their own rights systems. Nowadays, all too often, "making water use and rights rational" has become a missionary process of supplanting relationships of community, local property, knowledge and ethics, often in combination with large-scale water transfer and grabbing practices.

1.4.2 Knowing Water, Naturalizing Water Solutions, and Expertocracy

Water policy plans and intervention models commonly rely on professional-discipline knowledge and the expertise of international water research centers, and are implemented by established water bureaucracies (Linton, 2010; Molle *et al.*, 2009; Whatmore, 2009). Water problems are increasingly framed in global expert terms, promoting standardized expert solutions, assuming that these have generalizable answers and global applicability (see e.g. GWP, 2000; UNDP-CLEP, 2008; World Bank, 1999, 2012). At the same time, emerging proposals for dealing with water management issues increasingly look to private actors. Assuming that water has globally commensurable meanings and values, and treating water as a scarce and "therefore" economic good, is closely coupled with this tendency to extend expert roles and involve the private sector, even in water allocation and management functions (e.g. Duarte and Boelens, 2016; Mollinga, 2001). This shows how water knowledge production and implementation is deeply political.

When examining water (in)justice practices it is therefore important to consider that knowledge about water, including scientific knowledge, does not spring from natural reality but instead helps to construct these realities. Water knowledge and truth claims are internal to the socio-natural networks that constitute reality (e.g. Foucault, 1980; Whatmore, 2009). The choice and classification of concepts and their interrelationships do not represent the nature of water control, but the human intentions to tame and order water affairs. As Haraway (1991) argued, they sprout from situated knowledge. Water knowledge, power and truth all depend on and reproduce each other. As Foucault (1975) argued, power cannot be exercised without knowledge, and knowledge necessarily engenders power. Power, therefore, produces water reality and knowledge claims.

Naturalizing one version of "water reality" helps justify and depoliticize unequal water orders – as sedimented hegemonic practices (Mouffe, 2005, 2007). Dominant water-governance discourses, for instance, aim to unequivocally present *the* water problems and solutions. They tend to invalidate other types of knowledge, making it difficult or impossible to see other, "inconvenient" (non-dominant) water realities. Global discourses and transnational relationships influence the articulation of water problems and promote authorized water knowledge and governance models, applying concepts that often obscure the contextual and political nature of water management. Universalizing policy concepts such as "good governance," "rational and efficient water use," "decentralization," "transparency and accountability," or "best practices," often conceals and reproduces inequalities and misrecognition (Boelens and Vos, 2012). These presumably value-free, depoliticized concepts, cornerstones of leading water-policy models, erase context, situatedness and power.

1.4.3 Some Important Expressions in Water Use and Governance Practice

1.4.3.1 Formalizing Local Water Rights amidst Legal Plurality and Divergent Water Securities

Water rights express the legitimacy of claims to water and to water management decision-making. Rights need endorsement by an authority that has legitimacy in the eyes of users and non-users and that is able to enforce these rights. State officials commonly equate "legal" and "legitimate" water rights, but local user groups usually differentiate between the two and challenge this conflation: in many water-control settings around the world, water-user collectives consider that they have several authorities, both state and non-state, simultaneously – each representing different socio-legal systems and often taking divergent positions on the legitimacy of water-use claims (Perreault, 2008; Rasmussen, 2015). These different water-rights regimes coexist, complement or even contradict each other. In this way, users actively produce inter-legality and pluralism. Everyday water control is a product of this pluralism (cf. Cleaver and de Koning, 2015; Roth *et al.*, 2005; Sousa-Santos, 1995).

Despite this empirical, context-based heterogeneity of what constitutes a "water right," water rights and property relations in modern global expert centers (and government institutes and intervening agencies that follow their advice) tend to consider water rights as merely standard black boxes that juxtapose the frameworks of positivist technical and economists' water science (e.g. GWP, 2000; Ringler *et al.*, 2000; World Bank, 2012). Habitually, water law and rights are seen both as instruments to "engineer" water society and as the standards according to which existing water reality is judged (Roth *et al.*, 2005). Indeed, this follows from a long tradition in which water rights have been treated under the paradigm of state-defined, centralized water control. Today, this state-centric water rights model is fused with a market-focused neoliberal paradigm.

One enduring supposition of modernist water policy programs is that standardized rule-making will benefit all and produce efficient rights, mutually beneficial exchange, and rational organization (Boelens, 2009). In direct relation to this, there is a widespread assumption in policymaking that "formalizing local water rights" is the key to increasing water security for local user groups – as also attested by international financing institutions' worldwide support for numerous large water-rights formalization programs (e.g. Soussan, 2004; World Bank, 2012). Hernando de Soto, the influential World Bank policy scholar, for example, explains that the lack of such universal norms in "closed" countries in the South is the main reason they cannot fully enter the world system of capitalist exchange. Thus, the civilizing mission of the academic community would be "to help governments in developing countries build formal property systems that embrace all their people" (De Soto, 2000: 180).

Not just mainstream policies but equally many critically engaged policy scholars and benevolent "pro-poor advocates" assume that formally recognizing customary water rights will directly enhance water security for marginalized communities. Nevertheless, many in-depth studies have shown how the widespread (techno-economic and rationalized-legalistic) recognition of local water access and water control rights contradicts existing use and allocation practices, authority, and management modes. This might weaken rather than strengthen water security, with a negative impact for food and livelihood security (Boelens and Seemann, 2014; Lankford *et al.*, 2013; Seemann, 2016; Zeitoun *et al.*, 2016).

As one country example out of many, Peru has received US$200 million from the Inter-American Development Bank to foster water rights security by formalization while battling the country's "limited water culture" and "irrational water use" (Ministerio de Economia y Finanzas, 2007: 3, 24). In the modernist minds of the Bank's formalizers and national elites, these two are seen as two facets of a single objective. That is no coincidence, since local understandings of water-rights autonomy and water security tend to be a primary obstacle for formal rule-makers and intervening agents. Their multi-faceted, dynamic character makes them intangible and unrecognizable in positivist, bureaucratic, neoliberal frameworks.

1.4.3.2 Payment for Environmental Services

In many of the world's regions, national governments and international policy, development and funding agencies have worked to re-scale water governance structures: upwards to transnational governance scales and simultaneously downwards to local governments operating in public-private partnerships. Cities, often situated in downstream areas, seek regular, reliable supplies of sufficiently clean water, which governments, drinking water utilities and industries increasingly want to secure through Payment for Environmental Services (PES) schemes, which have boomed in Costa Rica, Colombia, Ecuador, Mexico, South Africa, China, and the Philippines, among others. The idea behind PES schemes is that downstream users pay upstream land managers to implement land and water conservation measures, such as erosion control, afforestation programs, reduced use of pesticides, and nature reserves around water sources. These measures should increase base flows, reduce peak flows and increase water quality. PES schemes are portrayed as "win-win" deals: city dwellers and industries pay for a necessary service and upstream farmers receive extra income (Büscher and Fletcher, 2015; Duarte-Abadía and Boelens, 2015; Rodríguez-de-Francisco et al., 2013). PES schemes are presented as alternatives to state-imposed land-use planning and conservation in the catchment areas, applying voluntary free-market principles of supply and demand for ecosystem services. This principle reduces water security to a monetary value relationship (Castro, 2007; Robertson, 2007).

In practice, many of these schemes do not function as predicted. City dwellers, water utilities and industries are unwilling to pay for conservation measures upstream. This is partly because increased water security is attained only in the long run, and effects of conservation on water flows are hard to measure (Schröter et al., 2014). Many PES schemes receive large subsidies and conservation measures rely more on imposed conservation regulation than on free-market initiatives (Schomers and Matzdorf, 2013). On balance, PES favors the largest landowners but tends to have negative effects on most upstream communities, particularly the poorest families, who lose their livelihoods (Rodríguez-de-Francisco et al., 2013). Moreover, these PES schemes are usually imposed non-democratically, favoring the companies that install them. Policy discourses highlight win-win neoliberal "PES-speech" in the foreground, commodifying production/reproduction relations, and sidelining alternative ways to organize conservation. In many cases, PES deeply transforms vernacular community reciprocity bonds (cf. Li, 2011; Neumann, 2004; Rodríguez-de-Francisco and Boelens, 2015; Sullivan, 2009).

1.4.3.3 Water Rationality and Efficiency

Many of today's water deprivations are justified or presented on grounds of privileging efficient uses and users over inefficient ones (Achterhuis et al., 2010; Bakker, 2010; Ioris, 2016; World Bank, 2012). However, concepts such as irrigation efficiency, water productivity, or crop water requirements are not socially neutral (Roa-García, 2014; Zwarteveen, 2006). These dominant analytical/policy tools are developed in particular scientific/policy settings. They have political, material, and discursive force.

Policy documents often relate the need for water efficiency to the necessity to produce more food for the growing population, easily leading to promotion of "efficient" technologies such as drip irrigation. However, irrigation water that percolates beyond a crop's root zone is often not "lost": it is used downstream, or pumped again from the groundwater. Consequently, installing drip technology concentrates water for the early implementer, but does not necessarily generate more crops per drop (Seckler, 1996; Venot and Clement, 2013).

Similar problems arise when solely "economic water efficiency" criteria are applied, to "increase water productivity" by introducing water pricing and marketing, and "maximize water allocation efficiency" from a neoclassical economist's perspective. Policies based on such notions generally entail full-cost water pricing to encourage water saving and reallocate water to the economically "most efficient" user. Also here, different stakeholders' normative frameworks are likely to hold different notions of values, risks, costs and benefits.

Reallocating water (rights) to gain "productive efficiency" implies that some groups win and others lose access to water (Boelens and Vos, 2012; Budds, 2009; Moore, 1989).

Aside from technical and economic reasoning, efficiency/inefficiency labels imply moral judgment. Blaming inefficient farmers is a powerful discursive practice with political consequences. For example, Diemer and Slabbers (1992, 7) found that many project planners classified African farmer-managed irrigation systems as "unscientific and wasteful." According to Gelles (2010), project planners in the Majes project in Peru found that local farmers lacked water culture and were morally backward. In general, in many places around the world, irrigation modernization and economic development is promoted as a civilization project based on moral superiority/inferiority relationships.

1.4.3.4 Corporate Social Responsibility and Sustainability Certification Schemes

As we have argued above, transnational agro-export companies have depleted and contaminated water sources the world round. They have accumulated land- and water-use rights at local users' expense, and appropriated water without formal use rights. Partly in response to critics, and partly to secure stable supplies, food industry and retail companies from the Global North demand environmental and social certification of producers that export fresh products to Europe and the US. The standards increasingly include criteria that inhibit and prescribe certain water management practices. The certifications form part of a wider politics of corporate social responsibility, that alters local-global relations of production and water use. Multinational export companies proudly display the multiple certifications on signs at the entrances to their production units and on their websites. However, the standards are problematic because they do not take into account local diversity in social and biophysical conditions: they are expensive to obtain and thus exclude smallholders from the export market, they seek to standardize smallholder practices, they are non-democratic, and in many instances they fail to prevent depletion or contamination of water sources (Vos and Boelens, 2015; Vos and Hinojosa, 2016).

Private environmental and social standards are defined by a variety of organizations, which are dominated by major retail companies. These dominant standards reshape knowledge frameworks and truth claims about water realities (Goldman, 2011). Producers' compliance with production standards is monitored by third-party private audit companies that usually inspect production facilities once a year. Competition between the various standards and also among the audit companies contributes to superficial inspections and permissive enforcement (Vos and Boelens, 2016).

Retailers and the food industry have the power to set norms and reshape local and global food production (Roth and Warner, 2008), so ideas and norms regarding "good" agriculture change, increasingly externalizing water communities' knowledge, production and governance rationalities (Boelens and Zwarteveen, 2005; Van der Ploeg, 2008). This way, water certification regimes become gauges to detect and "correct" deviations from the universal norms (Moore, 1989; Venot and Clement, 2013).

1.5 Water Governmentalities

As the previous sections have shown, producing water knowledge, rules, policies and technology concentrates increasingly on aligning people, their mind-sets, identities and resources with the interests of dominant water-sector groups. Modernist water development projects deploy forms of governmentality through water. They re-pattern water space and territory, which reshapes rules and authority; redirects labor and production; induces new norms and values; and rearranges people in new, externally driven techno-political constellations. Many designs underlying these water-development projects, far beyond just installing a new hydraulic technology, introduce new management hierarchies, commoditized (or privatized) water services, new legal frameworks, often resulting in a new socio-nature hostile to the autonomy or even survival of existing water cultures and user collectives. New hydraulic power grids, commonly linked to nation-state authority, markets and companies, de-pattern and re-pattern local water control systems. So, natural resource governance efforts are based on truth regimes that aim to (re)produce socio-natural order and acceptance via the particular positioning of and control over natural resources, infrastructure, investments, knowledge, and ultimately, whole population groups (e.g. Harris, 2012; Scott, 1990; Swyngedouw, 2009).

As Foucault (1991) argued when examining these "government-mentalities" (i.e. the rationality and strategies of dominant groups to conduct subjects' conduct), rulers increasingly deploy governance tactics to economically manage and direct society instead of legal-bureaucratic regimes based on sovereign power (cf. Dean, 1999). Thus, aside from the direct rule of law, two forms of governmentality are prominent in water governance: disciplinary and neoliberal governmentality. Disciplinary governmentality works through normalizing power (Foucault, 1975). Deviant thinking and acting is oppressed, where possible through self-correction based on internalized norms. Disciplinary power "produces" a model water user: efficient, responsible and modern.

Neoliberal governmentality works by directing people's thinking and acting according to "rational" economic principles. People are approached as rational actors who strategically pursue their personal interests, based on calculated costs and benefits (Boelens *et al.*, 2015; Fletcher, 2010). Neoliberal principles such as private water rights, decentralized decision-making and volumetric water pricing are based on the assumption that maximum welfare will be reached if all citizens behave as profit-maximizers seeking the right incentives. In water governance, the assumption is that neoliberal incentives will automatically yield maximum investments and efficient, productive water use. In neoliberal logic, the state's role is to install market rationalities in all spheres of society (Foucault, 2008; cf. Harvey, 2003; Hayek, 1960). Indeed, neoliberal water governance, far from laissez-faire, builds on aggressive state vigilance and intrusion. Or as Bourdieu (1998: 86) stated, "what is portrayed as an economic system governed by iron laws of a social nature is actually a political system that can be set up only with official political powers' active or passive complicity."

Currently practiced combined modes of disciplinary and neoliberal governmentalities present political choices (e.g. distributive and representational questions) as technical-managerial options. Denying any connections between power and knowledge, and assuming new-institutionalist rationality (viewing humans as rational individuals pursuing only self-interested goals), have pervaded mainstream water-policy discourses: wide-ranging redistributions of water and authority seem natural, inevitable and scientifically rational (Espeland, 1998; McCarthy and Prudham, 2004).

1.6 Water Conflicts and Water Justice Struggles: Entwining Different Layers, Scales and Actors

1.6.1 Conflicts over Resources, Rights, Authority and Discourses

Water control conflicts are everywhere. Disputes and struggles may occur over how water is to be used, distributed, managed, treated or talked about (e.g. Donahue and Johnston, 1998; Dimitrov, 2002). What follows from the previous section is that they cannot always easily be witnessed. Water conflicts may be open and visible, but often also happen in subtler, less directly visible ways. Moreover, as Frances Cleaver explains in Chapter 13, marginalized user groups appear to accept the large-scale environmental injustices inflicted upon them. In those cases, they avoid opposition and instead accommodate unequal water-based relationships, trying to give them meaning in local historical, cultural and political constellations. Such accommodation of water injustices may be based on mechanisms of control over grassroots groups (e.g. resulting from oppression by political, economic and military powers, or from disciplining through symbolic violence and discursive powers), or on grassroots groups' strategies of how to deal with the asymmetrical interdependencies and power relationships they experience vis-à-vis dominant private and state actors.

In overt and covert water conflicts and struggles for water justice, there is more at stake than just water distribution. We distinguish four interrelated echelons ("Echelons of Rights Analysis," Boelens, 2015b; Boelens and Zwarteveen 2005; Zwarteveen *et al.*, 2005). At a basic level, there is the dispute concerning *access to and use of water-related resources*: which users and use sectors have access to water, hydraulic infrastructure, and the material and financial means to use and manage water resources. At the next level, there is contestation over the *contents of rules and rights*: formulation and substance of water rights, management rules and laws that determine water distribution and allocation. And at a third echelon, we see the struggle over the *authority and legitimacy* to make and enforce those water rights and rules: who has decision-making power about questions of water use, allocation and governance. And fourth, there is the conflict among *discourses*: the power-knowledge regimes that articulate water problems and solutions, and that defend or impose particular water policies and water hierarchies. As we have argued above, water policy and scientific discourses make fixed linkages and standard logical relations among concepts, actors, objects, defining their identity, position and hierarchies, and forcefully defining problems and their solutions.

These four echelons are intrinsically related; conflict and outcomes at one echelon define the contents and contestations at the next echelon. The struggle over discourses, the fourth echelon, is about inducing a coherent regime of representation that strategically links the previous echelons together and makes their contents and linkages appear natural, as the morally or scientifically best "order of things." For example, a particular discourse will also defend the decision-making arrangements and authorities it considers convenient, who in turn will formulate and enforce the rules; according to which the resources are to be distributed. Therefore, contestations range from opposing current distributive inequalities and undemocratic forms of representation to challenging the very politics of truth themselves, including the identities that are imposed upon marginalized water cultures and user groups by state and market-based governmentalities.

1.6.2 Water Justice Interlinking Multiple Dimensions, Knowledge, Scales and Actors

Attention to water rights' cultural embeddedness, plurality and complexity requires a shift of focus, away from exclusive attention to formal structures and regulations towards an interest in how and by whom water rights and governance forms are produced, reproduced and transformed in particular ecological and cultural settings. It examines how people experience law in the context of their own local society and use it as a crucial resource in their day-to-day aspirations and struggles (Benda-Beckmann *et al.*, 1998; Roth *et al.*, 2005, 2015). Therefore, local water societies often see water rights framed as instruments to arrange their systems and as weapons to defend themselves. Far from egalitarian microsocieties, they are an effort, a process and a capacity to merge collectivity with diversity and to exercise mutual dependence on nature and on each other (Boelens *et al.*, 2014).

In their struggles, these water cultures continually reinvent rules and identities and traditions. Water user collectives and federations know that their existence depends on defending their water rights and rule-making spaces and will continue to create "non-conformity" and "complexities," while at the same time trying to conquer representation and achieve changes in the policy institutes, intervention projects, and the state institutional network.

Most water-user communities integrate with national and international policies, markets and partnerships, embedding local in global and global in local. Conflicts over water governmentality involve community-state contradictions and conflicts among local smallholders and new water lords, as well as the transnational extractive industries and globalized policy-making that operates across spatial scales (cf. McCarthy, 2005; Perreault, 2015). These processes and relationships comprise patterns of multiple actors, scales, and trans-local networks arising in many places – "the continuous reorganization of spatial scales is an integral part of social strategies to combat and defend control over limited sources and/or a struggle for empowerment" (Swyngedouw and Heynen, 2003: 912–13). In many regions, grassroots organizations build multi-actor federations to contest the neoliberalization of water, the negative effects of dams, water pollution, separation of water rights and decision-making powers from local livelihoods, and policies and actions that attack rights pluralism, polycentrism and the integrity of their territories (e.g. Bebbington *et al.*, 2010; Hoogesteger and Verzijl, 2015; Romano, 2017). Such networks also show that state, scientific, and policy-making communities are not monolithic, but reflect the track records of their social conquests. Many state employees, professionals and scientists struggle "from within," forming alliances with water-user groups to capture cross-scale opportunities. Social movements also need to frame their demands in ways that align with values and ideas of national political parties and/or the general public (Benford and Snow, 2000).

Therefore, fundamentally, struggles over water are contests over resources and legitimacy, the right to exist as water-control communities, and the ability to define the nature of water problems and solutions. By connecting material with cultural-political struggles, they demand both the right to be equal *and* the right to be different. Increasingly, affected water user communities combine their struggle against highly unequal resource distribution with their demands for greater autonomy and sharing in water authority. The intimate connection among people, water, space, and identity fuses their struggles for material access and control of water-use systems (distributive justice) and ecological defense of neighborhoods and territories (socio-ecological integrity) with their battle over the right to culturally define and politically organize these socio-natural systems (cultural and representational justice) (cf. Fraser, 2000; Martínez-Alier, 2002; Schlosberg, 2004; Young, 1990). Therefore, to understand "water justice," as we did when starting this chapter, we move from universalist, descriptive theories that prescribe what water justice "should be," to focus on understanding how people on-the-ground experience and define water justice. In the formal water policy and governance world, liberal, socialist, or neoliberal models of "equality" have generally tended to reflect the dominant water society's elitist, capitalist or scientific-expert mirror – ignoring peasant, indigenous and women's interests and views. Beyond abstract, de-humanized models, but also beyond localized romanticism, we urge a systematic exploration of the sources of water injustice, local views on fairness, and the impacts of formal laws and justice policies on human beings and ecosystems. Indeed, understanding water justice calls for a contextual, grounded, relational approach (Joy *et al.*, 2014; Perreault, 2014; Roth *et al.*, 2005; Zwarteveen and Boelens, 2014).

As the following chapters demonstrate, appeals for greater water justice call for combining grassroots, academic, activist, and policy action: engagement across differences (Schlosberg, 2004). Accordingly, we may understand "water justice" as:

the interactive societal and academic endeavor to critically explore water knowledge production, allocation and governance and to combine struggles against water-based forms of material dispossession, cultural discrimination, political exclusion and ecological destruction, as rooted in particular contexts.

(Boelens, 2015a: 34)

Water justice research and action, therefore, engages diverse water actors, to see multiple water truths and world views and to co-create transdisciplinary knowledge about understanding, transforming and distributing nature. It explores connections among the diverse ways of struggling for water justice. Water justice research involves critical engagement with water movements, dispossessed water societies, and interactive design of alternative hydrosocial orders. These alternatives cannot be engineered by scientists or policy-makers; they result from interweaving cross-cultural water knowledge and cross-societal pressures from the bottom up.

1.7 The Book's Contents

The following 18 chapters aim to provide a detailed understanding of the questions, complexities and opportunities for research and action regarding the issue of "water justice." Four sections address a broad variety of themes, approaches, geographical regions, and research, policy and action strategies. Even though most authors take a political ecology perspective, the book does not advocate one overall perspective on water justice. Nor does it suggest the opposite, the relativist trap that gives equal value to all particular views on social justice. As we have argued above, the book's chapters and authors take seriously the idea of "engagement across notions of justice – something crucial to notions of justice as recognition and political process" (Schlosberg, 2004: 532). Water-justice theories, scholars and movements bring together a critical plurality of contexts, experiences, views, tools and strategies. What is common to all our authors is that they expressly engage and identify with those groups in society that have the least rights and power over water access and decision-making. They all aim to support their water security struggles.

The book is divided into four sections, which examine different water justice themes and their associated social and political struggles. Each section begins with an introductory essay to introduce and contextualize key themes in the section's chapters.

The chapters in Part I deal with the theme of "Repoliticizing water allocation": they provide insight into the multi-layered contents and everyday working rationality of on-the-ground water rights and governance systems, and unfair water distribution and water-grabbing. These chapters highlight water injustices in common rural or urban water management frameworks and cultural realities that are often omitted from scientific water studies, legal frameworks, and policy proposals. Other chapters tell about the overt and covert ways in which intervening agents and elites take over water resources.

The chapters in Part II examine dominant policies and intervention projects fostering "Hydrosocial de-patterning and re-composition," and struggles to build alternative socio-natural and techno-political configurations. State, market, and expert networks use water interventions to reshape existing water societies according to their imageries or ideologies, often favoring specific interests and promoting specific developmental pathways. These chapters explain how these changes or clashes may provoke more or less open water conflicts, and unravels how such conflicts evolve in contexts of highly differentiated power relationships.

Part III chapters scrutinize cases and theories regarding "Exclusion and struggles for co-decision." The authors identify exclusion mechanisms and possible responses to and solutions for water-injustice problems, inspired by the ways in which local user collectives, sometimes through multi-level alliances with others (water citizen groups, professionals, rights coalitions, tribunals, scholars and policy-makers), strategize to defend, reclaim and re-embed their water rights, knowledge systems and governance forms.

Finally, Part IV chapters focus on theories and empirical cases that delve more deeply into notions of "Governmentality, discourses and struggles over imaginaries and water knowledge." Clashes between discourses and imaginaries constitute an important dimension of water justice conflicts. These struggles to protect and secure water resources as well as water communities, identities, territories and cultures provide the creative, pragmatic ingredients of strategies towards a more water-just world.

In short, the book does not promise easy one-size-fits-all analyses or silver-bullet solutions, but instead explicitly engages with the complex linkages between ecosystems and societies that characterize questions of what is fair, equitable and sustainable in water. By identifying with those who stand to lose or remain marginal in contemporary water development and policy reform processes, the book provides ingredients for new ways of thinking about and acting on water that make visible the many entanglements among culture, power and knowledge.

References

Achterhuis, H., Boelens, R. and Zwarteveen, M. (2010). Water property relations and modern policy regimes: Neoliberal utopia and the disempowerment of collective action. In R. Boelens, D. Getches and A. Guevara (eds.), *Out of the Mainstream: Water Rights, Politics and Identity*. London and New York: Earthscan, pp. 27–55.

Aguirre, M. (2014). "Reparations for the Maya Achi Chixoy Dam affected" (16/01/2014) and "Healing begins for the Maya Achi people of Guatemala" (17/11/2014) International Rivers Network blog, www.internationalrivers.org/blogs/223.

Ahlers, R. and Zwarteveen, M. (2009). The water question in feminism: Water control and gender inequities in a neo-liberal era. *Gender, Place & Culture*, 16(4), 409–26.

Amnesty International Peru (2013). "Fuente de vida – Visita a la comunidad La Algodonera en Olmos" (video), www.youtube.com/watch?v=Zq_1jyASiec.

Bakker, K. (2007). The "commons" versus the "commodity": Alter-globalization, anti-privatization and the Human Right to water in the Global South. *Antipode*, 39(3), 430–55.

Bakker, K. (2010). *Privatizing Water: Governance Failure and the World's Urban Water Crisis*. Ithaca, NY: Cornell University Press.

Bakker, K. (2013). Neoliberal versus postneoliberal water: Geographies of privatization and resistance. *Annals of the Association of American Geographers*, 103(2), 253–60.

Baud, M. (2010). Identity politics and indigenous movements in Andean history. In R. Boelens, D. Getches and A. Guevara-Gil (eds.), *Out of the Mainstream: Water Rights, Politics and Identity*, London and Washington, DC: Earthscan, pp. 99–118.

Bauer, C. (1997). Bringing water markets down to earth: The political economy of water rights in Chile, 1976–95. *World Development,* 25(5), 639–56.
Baviskar, A. (2007). *Waterscapes: The Cultural Politics of a Natural Resource.* Delhi: Permanent Black.
Bebbington, A., Humphreys, D. and Bury, J. (2010). Federating and defending: Water, territory and extraction in the Andes. In R. Boelens, D. Getches and A. Guevara (eds.), *Out of the Mainstream: Water Rights, Politics and Identity.* London and Washington, DC: Earthscan, pp. 307–27.
Benda-Beckmann, F. von, von Benda-Beckmann, K. and Spiertz, J. (1998). Equity and legal pluralism: Taking customary law into account in natural resource policies. In R. Boelens and G. Dávila (eds.), *Searching for Equity.* Assen: Van Gorcum, pp. 57–69.
Benford, R. D. and Snow, D. A. (2000). Framing processes and social movements: An overview and assessment. *Annual Review of Sociology,* 26(1), 611–39.
Bentham, J. (1988 (1781)). *The Principles of Morals and Legislation.* Amherst, NY: Prometheus Books.
Boelens, R. (2015a). *Water Justice in Latin America: The Politics of Difference, Equality, and Indifference.* Amsterdam: CEDLA and University of Amsterdam.
Boelens, R. (2015b). *Water, Power and Identity: The Cultural Politics of water in the Andes.* London: Earthscan, Routledge.
Boelens, R. and Seemann, M. (2014). Forced engagements: Water security and local rights formalization in Yanque, Colca Valley, Peru. *Human Organization,* 73(1), 1–12.
Boelens, R. and Vos, J. (2012). The danger of naturalizing water policy concepts: Water productivity and efficiency discourses from field irrigation to virtual water trade. *Agricultural Water Management,* 108, 16–26.
Boelens, R. and Zwarteveen, M. (2005). Prices and politics in Andean water reforms. *Development and Change,* 36(4), 735–58.
Boelens, R., Hoogesteger, J. and Baud, M. (2015). Water reform governmentality in Ecuador: Neoliberalism, centralization and the restraining of polycentric authority and community rule-making. *Geoforum,* 64, 281–91.
Boelens, R., Hoogesteger, J. and Rodriguez de Francisco, J. C. (2014). Commoditizing water territories: The clash between Andean water rights cultures and Payment for Environmental Services policies. *Capitalism Nature Socialism,* 25(3), 84–102.
Bourdieu, P. (1998). *Acts of Resistance against the Tyranny of the Market.* New York: New Press.
Bridge, G. and Perreault, T. (2009). Environmental governance. In N. Castree, D. Demeritt, D. Liverman, and B. Rhoads (eds.), *Companion to Environmental Geography.* Oxford: Blackwell, pp. 475–97.
Budds, J. (2009). Contested H2O: Science, policy and politics in water resources management in Chile. *Geoforum,* 40(3), 418–30.
Budds, J. (2010). Water rights, mining and indigenous groups in Chile's Atacama. In R. Boelens, D. Getches and A. Guevara (eds.), *Out of the Mainstream: Water Rights, Politics and Identity.* London and Washington, DC: Earthscan, pp. 197–211.
Büscher, B. and Fletcher, R. (2015). Accumulation by conservation. *New Political Economy,* 20(2), 273–98.
Cárdenas, A. (2012). *La carrera hacia el fondo: Acumulación de agua subterránea por empresas agroexportadoras en el Valle de Ica, Perú.* Wageningen: Justicia Hídrica.
Castro, J. E. (2007). Poverty and citizenship: Sociological perspectives on water services and public–private participation. *Geoforum,* 38(5), 756–71.
Cleaver, F. and de Koning, J. (2015). Furthering critical institutionalism. *International Journal of the Commons,* 9(1), 1–18.

Crow, B., Lu, F., Ocampo-Raeder, C., Boelens, R., Dill, B. and Zwarteveen, M. (2014). Santa Cruz declaration on the global water crisis. *Water International*, 39(2), 246–61.

Dankelman, I. and Davidson, J. (2013). *Women and the Environment in the Third World: Alliance for the Future*. London: Routledge.

Dean, M. (1999). *Governmentality: Power and Rule in Modern Society*. London: Sage Publications.

De Soto, H. (2000). *The Mystery of Capital: Why Capitalism Triumphs in the West and Fails Everywhere Else*. New York: Basic Books.

Diemer, G. and Slabbers, J. (eds.) (1992). *Irrigators and Engineers*. Amsterdam: Thesis.

Dimitrov, R. (2002). Water, conflict and security: A conceptual minefield. *Society & Natural Resources*, 15, 677–91.

Donahue, J. M. and Johnston, B. R. (1998). *Water, Culture and Power, Local Struggles in a Global Context*. Washington, DC: Island Press.

Duarte-Abadía, B. and Boelens, R. (2016). Disputes over territorial boundaries and diverging valuation languages: The Santurban hydrosocial highlands territory in Colombia. *Water International*, 41(1), 15–36.

Duarte-Abadía, B., Boelens, R. and Roa-Avendaño, T. (2015). Hydropower, encroachment and the repatterning of hydrosocial territory: The case of Hidrosogamoso in Colombia. *Human Organization*, 74(3), 243–54.

Eguren, L. (2014). "Estimación de los subsidios en los principales proyectos de irrigación en la costa peruana." Lima: CEPES, www.cepes.org.pe/sites/default/files/Eguren-Lorenzo_Subsidios-proyectos-irrigacion-costa_2014.pdf.

Escobar, A. (2006). Difference and conflict in the struggle over natural resources: A political ecology framework. *Development*, 49(3), 6–13.

Escobar, A. (2008). *Territories of Difference: Place, Movements, Life, Redes*. Durham, NC: Duke University Press.

Espeland, W.N. (1998). *The Struggle for Water: Politics, Rationality, and Identity in the American Southwest*. Chicago: University of Chicago Press.

Fletcher, R. (2010). Neoliberal environmentality: Towards a poststructuralist political ecology of the conservation debate. *Conservation & Society*, 8, 171–81.

Forsyth, T. (2003). *Critical Political Ecology: The Politics of Environmental Science*. London: Routledge.

Foucault, M. (1975). *Discipline and Punish: The Birth of the Prison*. New York: Vintage Books.

Foucault, M. (1980). *Power/Knowledge: Selected Interviews and Other Writings 1972–1978*, ed. C. Gordon. New York: Pantheon Books.

Foucault, M. (1991 (1978)). Governmentality. In G. Burchell, C. Gordon and P. Miller (eds.), *The Foucault Effect: Studies in Governmentality*. Chicago: University of Chicago Press, pp. 87–104.

Foucault, M. (2008). *The Birth of Biopolitics*. New York: Palgrave Macmillan.

Fraser, N. (2000). Rethinking recognition. *New Left Review*. May/June, 107–20.

Friedman, M. (1962). *Capitalism and Freedom*. Chicago: University of Chicago Press.

Friedman, M. (1980). *Free to Choose*. Harcourt and New York: Harvest Book.

Galeano, E. (1995). Noticias de los nadies. *Revista Brecha*, Dec. 1995, www.brecha.com.uy/numeros/n522/contra.html.

Gaybor, A. (2011). Acumulación en el campo y despojo del agua en el Ecuador. In R. Boelens, L. Cremers and M. Zwarteveen (eds.), *Justicia Hídrica: Acumulación, Conflicto y Acción Social*. Lima: IEP, pp. 195–208.

Gelles, P. H. (2010). Cultural identity and indigenous water rights in the Andean Highlands. In R. Boelens, D. Getches and A. Guevara-Gil (eds.), *Out of the Mainstream: Water Rights, Politics and Identity*. London and Washington, DC: Earthscan, pp.119–44.

GRAIN. (2012). "Squeezing Africa dry: Behind every land-grab is a water-grab," www.grain.org/e/4516.
GWP. (2000). *Integrated Water Resources Management*. TAC Background Papers No. 4. Stockholm: GWP.
Goldman, M. (2007). How "Water for all!" policy became hegemonic: The power of the World Bank and its transnational policy networks. *Geoforum*, 38(5), 786–800.
Goldman, M. (2011). The birth of a discipline: Producing authoritative green knowledge, World Bank-style. *Ethnography*, 2(2), 191–217.
Hale, C. (2002). Does multiculturalism menace? Governance, cultural rights, and the politics of identity in Guatemala. *Journal of Latin America Studies*, 34(3), 485–524.
Haraway, D. (1991). *Simians, Cyborgs and Women: The Reinvention of Nature*. New York: Routledge.
Harris, L. (2012). State as socionatural effect: Variable and emergent geographies of the state in Southeastern Turkey. *Comparative Studies of South Asia, Africa and the Middle East*, 32(1), 25–39.
Harvey, D. (1996). *Justice, Nature and the Geography of Difference*. Cambridge and Oxford: Blackwell Publishers.
Harvey, D. (2003). *The New Imperialism*. Oxford: Oxford University Press.
Hayek, F. A. (1944). *The Road to Serfdom*. London: George Routledge.
Hayek, F. A. (1960). *The Constitution of Liberty*. Chicago: University of Chicago Press.
Hendriks, J. (1998). Water as private property: Notes on the case of Chile. In R. Boelens and G. Davila (eds.), *Searching for Equity: Conceptions of Justice and Equity in Peasant Irrigation*. Assen, The Netherlands: Van Gorcum, pp. 297–310.
Heynen, N., McCarthy, J., Prudham, S. and Robbins, P. (2007). *Neoliberal Environments: False Promises and Unnatural Consequences*. New York: Routledge.
Hidalgo, J. P., Boelens, R. and Isch, E. (forthcoming). The Daule-Peripa Multipurpose Hydraulic Scheme: Technocratic reconfiguration of a hydro-social territory and dispossession in coastal Ecuador. *Latin American Research Review*.
Hoekstra, A. Y. and Mekonnen, M. M. (2012). The water footprint of humanity. *Proceedings of the National Academy of Sciences*, 109(9), 3232–37.
Hoogesteger, J. and Verzijl, A. (2015). Grassroots scalar politics: Insights from peasant water struggles in the Ecuadorian and Peruvian Andes. *Geoforum*, 62, 13–23.
Hoogesteger, J., Boelens, R. and Baud, M. (2016). Territorial pluralism: Water users' multi-scalar struggles against state ordering in Ecuador's highlands. *Water International*, 41(1), 91–106.
Hommes, L. and Boelens, R. (2017). Urbanizing rural waters: Rural-urban water transfers and the reconfiguration of hydrosocial territories in Lima. *Political Geography*, 57, 71–80.
Hommes, L., Boelens, R. and Maat, H. (2016). Contested hydrosocial territories and disputed water governance: Struggles and competing claims over the Ilisu Dam development in southeastern Turkey. *Geoforum*, 71, 9–20.
Huber, A. and Joshi, D. (2015). Hydropower, anti-politics, and the opening of new political spaces in the Eastern Himalayas. *World Development*, 76, 13–25.
Inter-American Human Rights Commission (IACHR). (2014). *The Impact of Canadian Mining in Latin America and Canada's Responsibility*. Working Group on Mining and Human Rights in Latin America. Organization of American States (OAS), Washington, DC.
Ioris, A. (2016). Water scarcity and the exclusionary city: The struggle for water justice in Lima, Peru. *Water International*, 41(1), 125–39.
Jasanoff, S. (2010). A new climate for society. *Theory, Culture & Society*, 27(2–3), 233–53.
Johnston, B. R. (2005). *Chixoy Dam Legacy Issues Study*, Vol. 1–5. International Rivers Network. www.irn.org/programs/chixoy/index.php?id=ChixoyLegacy.2005/03.findings.html.

Joy, K. J., Kulkarni, S., Roth, D. and Zwarteveen, M. (2014). Re-politicizing water governance: Exploring water reallocations in terms of justice. *Local Environment*, 19(9), 954–73.

Kaika, M. (2006). Dams as symbols of modernization: The urbanization of nature between geographical imagination and materiality. *Annals of the Association of American Geographers*, 96(2), 276–301.

Lankford, B., Bakker, K., Zeitoun, M. and Conway, D. (eds.) (2013). *Water Security: Principles, Perspectives, and Practices*. London: Earthscan.

Latour, B. (1993). *We Have Never Been Modern*. Cambridge, MA: Harvard University Press.

Lauderdale, P. (1998). Justice and equity: A critical perspective. In R. Boelens and G. Dávila (eds.), *Searching for Equity: Conceptions of Justice and Equity in Peasant Irrigation*. Assen: Van Gorcum, pp. 5–10.

Li, T. (2011). Rendering society technical: Government through community and the ethnographic turn at the World Bank in Indonesia. In Mosse, D. (ed.), *Adventures in Aidland: The Anthropology of Professionals in International Development*. Oxford: Berghahn, pp. 57–80.

Linton, J. (2010). *What is Water? The History of a Modern Abstraction*. Vancouver: University of British Columbia Press.

Lobina, E., Kishimoto, S. and Petitjean, O. (2014). "Here to stay: Water remunicipalisation as a global trend," Public Services International Research Unit (PSIRU), Transnational Institute (TNI) and Multinational Observatory, www.tni.org/files/download/hereto-stay-en.pdf.

Loftus, A. (2009). Rethinking political ecologies of water. *Third World Quarterly*, 30(5), 953–68.

Lu, F., Valdivia, G. and Silva, N. L. (2017). *Oil, Revolution, and Indigenous Citizenship in Ecuadorian Amazonia*. London: Palgrave Macmillan.

Lynch, B. D. (2006). *The Chixoy Dam and the Achi Maya: Violence, Ignorance, and the Politics of Blame*. Mario Einaudi Centres, Ithaca, NY: Cornell University.

Lynch, B. D. (2012). Vulnerabilities, competition and rights in a context of climate change toward equitable water governance in Peru's Rio Santa Valley. *Global Environmental Change*, 22, 364–73.

Martínez-Alier, J. (2002). *The Environmentalism of the Poor*. Cheltenham, UK and Northampton, MA: Edward Elgar.

McCarthy, J. (2005). Scale, sovereignty and strategy in environmental governance. *Antipode*, 37(4), 731–53.

McCarthy, J. and Prudham, S. (2004). Neoliberal nature and the nature of neoliberalism. *Geoforum*, 35, 275–83.

McCully, P. (2001). *Silent Rivers: The Ecology and Politics of Large Dams*. London: Zed.

Meehan, K. (2013). Disciplining de facto development: Water theft and hydrosocial order in Tijuana. *Environment and Planning D*, 31, 319–36.

Mehta, L., Veldwisch, G. J. and Franco, J. (2012). Water-grabbing? Focus on the (re)appropriation of finite water sources. *Water Alternatives*, 5(2), 193–207.

Mena-Vásconez, P., Boelens, R. and Vos, J. (2016). Food or flowers? Contested transformations of community food security and water use priorities under new legal and market regimes in Ecuador's highlands. *Journal of Rural Studies*, 44, 227–38.

Mill, J. S. (1874). *A System of Logic*. New York: Harper & Brothers.

Mill, J. S. (1999). *On Liberty*. Peterborough, Canada: Broadview Press.

Ministerio de Economía y Finanzas [Government of Peru]. (2007). Documento conceptual de proyecto "Programa de Recursos Hídricos I," PE-L 1024. Ministerio de Economía y Finanzas. Lima, Perú, April 9, 2007.

Molle, F., Mollinga, P. and Wester, F. (2009). Hydraulic bureaucracies and the hydraulic mission: Flows of water, flows of power. *Water Alternatives*, 3(2), 328–49.

Mollinga, P. (2001). Water and politics: Levels, rational choice and South Indian canal irrigation. *Futures*, 33, 733–52.

Moore, M. (1989). The fruits and fallacies of neoliberalism: The case of irrigation policy. *World Development*, 17(11), 1733–50.

Moore, D., Dore, J. and Gyawali, D. (2010). The World Commission on Dams + 10: Revisiting the large dam controversy. *Water Alternatives* 3(2), 3–13.

Mouffe, C. (2005). *On the Political*. London: Routledge.

Mouffe, C. (2007). Artistic activism and agonistic spaces. *Art & Research*, 1(2), 1–5.

Neumann, R. P. (2004). Nature-state-territory: Toward a critical theorization of conservation enclosures. In R. Peet and M. Watts (eds.), *Liberation Ecologies: Environment, Development and Social Movements*. London: Routledge, pp. 179–99.

Nozick, R. (1974). *Anarchy, State, and Utopia*. New York: Basic Books.

Peña, F. and Herrera, E. (2008). El litigio de Minera San Xavier: Una cronología. In M. C. Costero-Garbarino (ed.), *Internacionalización Económica, Historia y Conflicto Ambiental en la Minería. El caso de Minera San Xavier*. San Luis Potosí: COLSAN.

Perreault, T. (2008). Custom and Contradiction: Rural water governance and the politics of usos y costumbres in Bolivia's irrigators' movement. *Annals of the Association of American Geographers*, 98(4), 834–54.

Perreault, T. (2014). What kind of governance for what kind of equity? Towards a theorization of justice in water governance. *Water International*, 39(2), 233–45.

Perreault, T. (2015). Beyond the watershed: Decision making at what scale? In E. Norman, C. Cook and K. Furlong (eds.), *Negotiating Water Governance: Why the Politics of Scale Matter*. London: Ashgate, pp. 117–24.

Perreault, T., Bridge, G. and McCarthy, J. (eds.) (2015). *The Handbook of Political Ecology*. London: Routledge.

Perreault, T., Wraight, S. and Perreault, M. (2011). *The Social Life of Water: Histories and Geographies of Environmental Injustice in the Onondaga Lake Watershed*. New York. Justicia Hídrica, www.justiciahidrica.org.

Ploeg, J. D. Van der (2008). *The New Peasantries: Struggles for Autonomy and Sustainability in an Era of Empire and Globalization*. London: Earthscan

Preciado-Jeronimo, R., Rap, E. and Vos, J. (2015). The politics of land use planning: Gold mining in Cajamarca, Peru. *Land Use Policy*, 49, 104–17.

Progressio. (2010). *Drop by Drop, Understanding the Impacts of the UK's Water Footprint through a Case Study of Peruvian asparagus*. London: Progressio, CEPES, WWI.

Rasmussen, M. B. (2015). *Andean Waterways: Resource Politics in Highland Peru*. Seattle: University of Washington Press.

Rawls, J. (1971). *A Theory of Justice*. Cambridge, MA: Harvard University Press.

Ringler, C., Rosegrant, M. and Paisner, M. S. (2000). *Irrigation and Water Resources in Latin America and the Caribbean*. EPTD Discussion Paper 64. Washington, DC: EPTD, IFPRI.

Robertson, M. (2007). Discovering price in all the wrong places: The work of commodity definition under neoliberal environmental policy. *Antipode*, 39(3), 500–26.

Roa-García, M. C. (2014). Equity, efficiency and sustainability in water allocation in the Andes: Trade-offs in a full world. *Water Alternatives*, 7(2), 298–319.

Rodriguez-de-Francisco, J. C. and Boelens, R. (2015). Payment for environmental services: Mobilising an epistemic community to construct dominant policy. *Environmental Politics*, 24(3), 481–500.

Rodriguez-de-Francisco, J. C., Budds, J. and Boelens, R. (2013). Payment for environmental services and unequal resource control in Pimampiro, Ecuador. *Society & Natural Resources*, 26, 1217–33.

Romano, S. (2017). Building capacities for sustainable water governance at the grassroots: "Organic empowerment" and its policy implications in Nicaragua. *Society & Natural Resources*, DOI: 10.1080/08941920.2016.1273413.

Roth, D. and Warner, J. (2008). Virtual water: Virtuous impact? The unsteady state of virtual water. *Agriculture and Human Values*, 25, 257–70.

Roth, D., Boelens, R. and Zwarteveen, M. (eds.) (2005). *Liquid Relations: Contested Water Rights and Legal Complexity*. New Brunswick, NJ and London: Rutgers University Press.

Roth, D., Boelens, R. and Zwarteveen, M. (2015). Property, legal pluralism, and water rights: The critical analysis of water governance and the politics of recognizing "local" rights. *Journal of Legal Pluralism and Unofficial Law*, 47(3), 456–75.

Rulli, M. C. and D'Odorico, P. (2013). The water footprint of landgrabbing. *Geophysical Research Letters*, 40, 6130–35.

Schlosberg, D. (2004). Reconceiving environmental justice: Global movements and political theories. *Environmental Politics*, 13(3), 517–40.

Schomers, S. and Matzdorf, B. (2013). Payments for ecosystem services: A review and comparison of developing and industrialized countries. *Ecosystem Services*, 6, 16–30.

Schröter, M., Zanden, E. H., Oudenhoven, A. P., Remme, R. P., Serna-Chavez, H. M., Groot, R. S. and Opdam, P. (2014). Ecosystem services as a contested concept: A synthesis of critique and counter-arguments. *Conservation Letters*, 7(6), 514–23.

Scott, J. (1990). *Domination and the Arts of Resistance: Hidden Transcripts*. New Haven, CT: Yale University Press.

Seckler, D. (1996). *The New Era of Water Resources Management*. Research Report 1. International Irrigation Management Institute (IIMI), Colombo, Sri Lanka.

Seemann, M. (2016). *Water Security, Justice and the Politics of Water Rights in Peru and Bolivia*. Houndmills, Basingstoke: Palgrave Macmillan.

Sneddon, C. and C. Fox (2008). Struggles over dams as struggles for justice: The World Commission on Dams and anti-dam campaigns in Thailand and Mozambique. *Society & Natural Resources*, 21(7), 625–40.

Sosa, M. and Zwarteveen, M. (2012). Exploring the politics of water-grabbing: The case of large mining operations in the Peruvian Andes. *Water Alternatives*, 5(2), 360–75.

Sousa-Santos, B. de (1995). *Toward a New Common Sense: Law, Science and Politics in the Paradigmatic Transition*. London and New York: Routledge.

Soussan, J. (2004). *Poverty and Water Security: Understanding How Water Affects the Poor*. Manila: Asian Development Bank.

Stoltenborg, D. and Boelens, R. (2016). Disputes over land and water rights in gold mining: The case of Cerro de San Pedro, Mexico. *Water International*, 41(3), 447–67.

SUEZ. (2016). Press release, Paris, December 14, 2016. "SUEZ announces agreement for two concessions involved in irrigation and agricultural production project in Peru's Olmos valley," http://newsroom.suez-environnement.fr/wp-content/uploads/2016/12/PR-suez-irrigation-agricultural-production-olmos-peru-20161214-EN.pdf.

Sullivan, S. (2009). Global enclosures: An ecosystem at your service. *The Land*, 2008/9, 21–23.
Swyngedouw, E. (2004). Globalisation or "glocalisation"? Networks, territories and rescaling. *Cambridge Review of International Affairs*, 17(1), 25–48.
Swyngedouw, E. (2005). Dispossessing H2O: The contested terrain of water privatization. *Capitalism, Nature, Socialism*, 16(1), 81–98.
Swyngedouw, E. (2009). The political economy and political ecology of the hydro-social cycle. *Journal of Contemporary Water Research & Education*, 142, 56–60.
Swyngedouw, E. and Heynen, H. (2003). Urban political ecology, justice and the politics of scale. *Antipode*, 35(5), 898–918.
The Guardian (D. Hill). (May 14, 2014). "Canadian mining doing serious environmental harm, the IACHR is told," www.theguardian.com/environment/andes-to-the-amazon/2014/may/14/canadian-mining-serious-environmental-harm-iachr.
The Land Matrix. (2012). The Land Matrix Database, available at landmatrix.org.
United Nations Development Program (UNDP). (2006). *Beyond Scarcity: Power, Poverty and the Global Water Crisis*. Human Development Report 2006. Houndmills, NY: Palgrave Macmillan.
UNDP-CLEP. (2008). *Making the Law Work for Everyone*. Report of the Commission on Legal Empowerment of the Poor Volume I. UNDP-CLEP, chairs M. Albright and H. De Soto. New York: United Nations Development Program.
Valladares, C. and Boelens, R. (2017). Extractivism and the rights of nature: Governmentality, "convenient communities" and epistemic pacts in Ecuador. *Environmental Politics*, 26, 6, 1015–34.
Venot, J. P. and Clement, F. (2013). Justice in development? An analysis of water interventions in the rural South. *Natural Resources Forum*, 37, 19–30.
Vos, J. and Boelens, R. (2014). Sustainability standards and the water question. *Development and Change*, 45(2), 205–30.
Vos, J. and Boelens, R. (2015). Water in global agrifood chains. In K. Albala (ed.), *The SAGE Encyclopedia of Food Issues*. Thousand Oaks: SAGE Publications, pp. 1459–63.
Vos, J. and Boelens, R. (2016). The politics and consequences of virtual water export. In P. Jackson, W. Spiess and F. Sultana (eds.), *Eating, Drinking: Surviving*. Cham, Switzerland: Springer, pp. 31–40.
Vos, J. and Hinojosa, L. (2016). Virtual water trade and the contestation of hydrosocial territories. *Water International*, 41(1), 37–53.
Vos, J. and Marshall, A. (2017). Conquering the desert: Drip irrigation in the Chavimochic system in Peru. In J. P. Venot, M. Kuper, and M. Zwarteveen (eds.), *Drip Irrigation for Agriculture: Untold Stories of Efficiency, Innovation and Development*. London and New York: Routledge, pp. 134–50.
Whatmore, S. (2009). Mapping knowledge controversies: Science, democracy and the redistribution of expertise. *Progress in Human Geography*, 33, 5, 1–12.
Woodhouse, P. (2012). New investment, old challenges: Land deals and the water constraint in African agriculture. *Journal of Peasant Studies*, 39(3–4), 777–94.
World Bank. (1999). *Tradable Water Rights: A Property Rights Approach to Resolving Water Shortages and Promoting Investment*. Washington, DC: World Bank.
World Bank. (2012). *Evaluating the Impacts of the Formalization of Water Right for Agriculture Use: Water Rights in Peru*. Washington, DC: World Bank.
Yacoub, C., Duarte, B. and Boelens, R. (eds.) (2015). *Agua y ecología política. El extractivismo en la agro-exportación, la minería y las hidroeléctricas en Latino América*. Quito: Abya-Yala.

Young, I. M. (1990). *Justice and the Politics of Difference*. Princeton, NJ: Princeton University Press.

Zeitoun, M., Lankford, B., Krueger, T., Forsyth, T., Carter, R., Hoekstra, A., Taylor, R., Varis, O., Cleaver, F., Boelens, R., Swatuk, L., Tickner, D., Scott, C., Mirumachi, N. and Matthews, N. (2016). Reductionist and integrative research approaches to complex water security policy challenges. *Global Environmental Change*, 39, 143–54.

Zoomers, A. and Kaag, M. (eds.) (2013). *Beyond the Hype: A Critical Analysis of the Global "Land Grab."* London: ZED.

Zwarteveen, M. Z. (2006). "Wedlock or deadlock? Feminists. Attempts to engage irrigation engineers," PhD thesis. Wageningen: Wageningen University.

Zwarteveen, M. Z. (2015). Regulating water, ordering society: Practices and politics of water governance. Inaugural lecture, University of Amsterdam.

Zwarteveen, M. and Boelens, R. (2014). Defining, researching and struggling for water justice: Some conceptual building blocks for research and action. *Water International*, 39(2), 143–58.

Zwarteveen, M., Roth, D. and Boelens, R. (2005). Water rights and legal pluralism: Beyond analysis and recognition. In D. Roth, R. Boelens and M. Zwarteveen (eds.), *Liquid Relations*. New Brunswick, NJ: Rutgers University Press, pp. 254–68.

Part I
Re-Politicizing Water Allocation

Mural painting in Cochabamba, Bolivia.
Photo by Tom Perreault.

Introduction: Re-Politicizing Water Allocation

Tom Perreault, Rutgerd Boelens, and Jeroen Vos

Flint

Who poisoned Flint? This question, easy to ask but harder to answer, draws attention to the complex of political and economic institutions, processes, social relations and hydraulic infrastructures involved in urban water governance. Flint, Michigan, a city of 100,000 people northwest of Detroit, is a city in decline. Like other cities and towns in the US "rust belt," Flint's landscape is marked by abandoned, decaying factories, brownfields (urban spaces too polluted for redevelopment) and crumbling infrastructure. Flint was the founding place of General Motors, which still operates a plant there. Industry downsizing sacked thousands of workers in the 1980s, driving the city's poverty rate up sharply. De-industrialization and population loss have left the city with a diminished tax base, which barely covers operating costs for basic city services, much less for environmental cleanup and infrastructure upgrades. Some 40 percent of the city's population lives below the federal poverty line, and city residents, some 52 percent of whom are African-American, had experienced more than their share of problems when, in 2014, they were confronted with another: poisoned water.

Prior to that year, Flint received its drinking water from Detroit's water system, which in turn drew from Lake Huron, part of North America's Great Lakes system. Flint had plans to connect to a new water pipeline, to be completed in 2016 or 2017, which would provide Lake Huron water directly, at lower cost. As a further cost-saving measure, in April 2014, Flint's governor-appointed Emergency Manager decided that the city should no longer draw water from Detroit's system, and should instead begin drawing water from the Flint River, which flows through the city. Like many waterways in and around rust belt cities, the Flint River is acutely contaminated by decades of industrial waste and lax regulation, as well as municipal waste seepage from aging sewer systems (cf. Perreault *et al.*, 2012). As early as August 2014, drinking water tests showed elevated levels of fecal coliform bacteria. Residents were instructed to boil water and the city began treating the water system with chlorine. Between June 2014 and November 2015, Flint also experienced an outbreak of legionnaires' disease, which eventually claimed the lives of ten people.

Bacterial contamination was not even the gravest danger. In what proved to be catastrophic neglect of standard protocol, the city's water managers failed to add anti-corrosion agent to the water system, a failure with dramatic consequences for the city's out-of-date water infrastructure (Davidson, 2016). In the absence of anti-corrosion agent, the large amounts of chlorine they added to the water system damaged its pipes, which in turn released heavy metals into the water system. By October 2015, corrosion had become so bad that the local General Motors plant announced plans to disconnect from the city water mains and build its own water system, because the water was damaging its machinery. For the city's population, and particularly its 9,000 children under age six (two-thirds of whom are African-American), the greatest threat came from lead contamination (Grevatt, 2016; Welburn and Seamster, 2016). Lead poisoning is most dangerous for the very young and the very old. It can cause miscarriages or delay the growth of newborns. In young children, lead can slow brain development and cause permanent cognitive impairment and developmental disabilities. In older adults, lead poisoning is associated with memory loss, circulatory problems and a host of other ailments (Davidson, 2016).

Local outrage was met with institutional indifference. Officials at Michigan's Department of Environmental Quality (DEQ) sought to downplay, denigrate and dismiss residents' concerns and even studies conducted by non-aligned scientists (Osnos, 2016; Welburn and Seamster, 2016). Efforts by state and city agencies to test lead levels were marked by incompetence or gross negligence. The *Detroit Free Press* reported that DEQ officials pressured city testers to find and test samples with low lead levels, which they managed to do only by concentrating 25 percent of the samples in a single neighborhood that had updated water lines, and by excluding one sample that was extremely high (though it was later found to be accurate; Egan, 2015). By August 2015, researchers at Virginia Tech University reported that over one-third of the water samples they analyzed had elevated lead levels – including samples that contained lead levels 16 times higher than permissible limits (Davidson, 2016; Welburn and Seamster, 2016).

In January 2016, Michigan Governor Rick Snyder, under pressure to resign, finally offered an apology to the people of Flint. A legal settlement in March 2017 has forced Michigan to spend US$87 million to upgrade water infrastructure throughout the city (Bosman, 2017). While surely a welcome investment, this money comes fully three years after the decision to disconnect from Detroit's water system, and too late for the thousands of children exposed to lead. An entire generation of children in Flint will grow up with the fear of lead-related disabilities. This is no minor concern. In addition to the direct and indirect health effects of lead poisoning, the long-term social implications of lead poisoning (and related cognitive disabilities) can be devastating.

Snyder, an accountant by training, was elected governor in 2010 on a platform of fiscal austerity and pragmatic governance. He pursued his conservative economic program in part through Michigan's Emergency Manager Law, an explicitly undemocratic measure that allows the governor to appoint city managers with broad decision-making powers and the authority to override local governments. Perhaps not surprisingly, Michigan's Emergency Manager Law has had especially pernicious effects on African-Americans. Approximately half the state's African-American population now lives in a municipality under Emergency Manager control (compared with only 2 percent of whites). Flint, a majority African-American city, has had six different Emergency Managers in the past 13 years. Emergency Managers are accountable to the governor but not to the residents of the city they manage, and they are primarily concerned with imposing fiscal discipline. Given their mandate for cost recovery and economic efficiency, emergency managers look to save money by increasing user fees, cutting costs and reducing operation and maintenance services, including for drinking water. Throughout Michigan, Emergency Managers have treated municipal water systems as a source of revenue, raising water fees, reducing service, privatizing systems, and cutting off service to those behind in their bills (Warikoo, 2017; Welburn and Seamster, 2016). For example, in Detroit, one of the poorest major cities in the US, residents pay roughly twice the national average for their water service (US$75/month for a family of four), and fees continue to rise. In reaction, some residents choose not to pay their water bills, which has led to a sharp rise in service cutoffs (Clark, 2014). Similarly, Flint residents were already paying the highest water service rates in the state when the lead crisis began in 2014. As in Detroit, some residents stopped paying their water bills altogether, either because they could no longer afford the fees or because they refused to pay for water that they could not drink. And in another echo of Detroit's management style, in 2016 Flint officials announced they would begin shutting off water service to some 1,800 residents who were behind on their water payments (Grevatt, 2016).

For cities like Flint and Detroit, the global financial crisis of 2008 compounded the long-term economic decline associated with de-industrialization. These cities and their residents have experienced capital flight and population loss (Detroit's population declined from nearly 2 million in 1950 to under 680,000 in 2015). The US's peculiar brand of neoliberalism has meant that cities are largely left to succeed or fail on their own, with only minimal financial support from states or the federal government. Environmental quality and basic services such as drinking water and sanitation are among the sectors most impacted by fiscal austerity. In a city such as Flint, the cold logic of neoliberal urban governance, which systematically disadvantages low-income communities of color, is yet another form of structural racism, layered on top of histories and geographies of uneven capitalist development, de-industrialization, urban abandonment, white flight and racial segregation (Pulido, 2000, 2015).

Hidden Politics of Water Allocation: Justice, Equity and Rights

"Thirsty for justice" – so read one sign at a protest rally in Flint. But how do we understand water justice in the context of Flint's ongoing crisis? How do we make sense of the obvious inequities at play? As a field of social action and academic inquiry, the "environmental justice" concept finds its roots in North America and struggles by African-American and Latino communities, disproportionately burdened by pollution, or by industrial or other undesirable land-uses, or by a lack of access to environmental amenities, such as green spaces, clean air or water (e.g. Bullard, 1990; Holifield, 2015; see also Pulido, 2000, 2015). Understandings of environmental justice have traditionally been rooted in three complementary concepts of justice: socio-economic justice ("distribution"), representational justice ("participation"), and cultural justice ("recognition") (see Fraser, 2000; Schlosberg, 2004). Distributive or socio-economic justice refers to the fair, equitable distribution of environmental "goods" and "bads" in society. Procedural and representational justice refers to equal access to and treatment by the formal institutions of law and state, or by local and customary law systems. Are claimants given due process? Are they treated fairly by "the system?" Surely these are necessary conditions, but far from sufficient for achieving social justice, and other forms of justice are therefore necessary. Justice as "recognition," or the "right to have rights," ("cultural justice") considers the need for the powerless, socially marginalized and non-dominant cultures to be legitimately recognized as holding valid political, social, and cultural standing, and for alternative values, norms, rights repertoires, forms of organizing, and rule systems, to be considered legitimate. Recognition, in this sense, is a precondition for other forms of justice (cf. Boelens, 2009).

In the context of Flint, we can see each of these forms of (in)justice at play. Lead poisoning is not distributed evenly in Flint, and differentially affects lower-income and African-American communities to a greater extent than wealthier, whiter neighborhoods. At broader spatial and social scales, environmental justice patterns map onto the geographies of uneven capitalist development, locally, regionally, nationally and globally. Residents of Flint have also been denied representational and procedural justice, as state officials (particularly Michigan's Governor and the Department of Environmental Quality) spent months ignoring or dismissing their claims and those of non-aligned researchers, all the while prolonging lead exposure for Flint's children and seniors. While a court ruling eventually forced the state of Michigan and the city of Flint to spend US$87 million to upgrade water infrastructure in the city, this decision came three years after the initial decision to draw water from the Flint river, and it will take several more years before Flint's water pipes are fully replaced. Finally, Flint residents struggled to have their claims recognized as legitimate. Michigan state officials initially treated complaints about water quality as little more than a nuisance, even when confronted at community meetings by residents carrying jugs of brown tap water. Officials at the DEQ callously noted that environmental laws do not require them to regulate the aesthetic qualities of water. Brownish water was to be expected, they noted, because Flint has aging infrastructure and a polluted river (Davidson, 2016). Never did they question why the river was allowed to become so contaminated in the first place, nor did they consider the fact that Flint is no older than many cities in Michigan that enjoy far better water systems. The obvious difference was left unmentioned but widely understood: Flint is a low-income, largely African-American city and therefore is considered undeserving of full recognition or state accountability.

Useful as the concepts of distributive, representational and cultural justice are, we would do well to broaden our understandings of justice to include Sen's (2001) notion of "capability." By this he means an individual's capacity to achieve certain basic needs in society, as mediated by institutional frameworks such as law, rights, and societal norms. A capabilities approach to justice emphasizes *positive freedoms* – the right, ability, or capacity to do or achieve something – as opposed to *negative freedoms*, or freedom from an external constraint. Sen argues that governments should be measured according to their citizens' actual capabilities – their freedoms to achieve desired ends – as opposed to the idealized formal rights they are legally accorded. In this view, justice maximizes everyone's human potential, by providing for basic material needs (water, food, shelter) and the social institutions necessary to attain them. As such, supplying water and other basic necessities is a means to an end – not the end itself (Bakker, 2010). This, for Sen, is the essence of development – the freedom to fulfill one's capabilities infers other rights, and cannot be viewed in isolation from the institutional arrangements or hydraulic infrastructures through which water is delivered (see also the chapters by Roth *et al.* and Crow, in this volume).

The institutional arrangements of water governance draw our attention to questions of water rights, and the relationship among rights, justice and equity. Mirosa and Harris (2012) distinguish between different uses of the term "rights" in relation to water. In their reading, the "right to water" refers to formal, legal recognition of an individual or group's right to water. For instance, the 2009 Bolivian constitution recognizes the right of all Bolivian citizens to basic water service (whether this is achieved in practice, however, is another issue altogether). This rendering is distinct from the "human right to water," which refers to all people's general right to water sufficient to satisfy basic needs and human dignity. This has been a popular rallying cry for water activists worldwide. Finally, Mirosa and Harris define "water rights" as the individualized right to a limited and clearly defined quantity of water, and directly connected to property. Insofar as water rights in this narrow sense constitute a form of property, and property entails the right to exclude others, it is worth noting that specific water rights for some may in effect deny water to others. This is especially the case for marginalized, vulnerable populations, as historically demonstrated with the so-called First in Time, First in Right principle, which characterizes water law in Western North America (Hicks, 2011; Wilkinson, 2011; see the chapter by Bakker *et al.*, this volume).

Beyond such limited conceptualizations of "right to water," "human right to water," and "water rights," which basically refer to legitimate access to a particular volume of water (in terms of quantity, quality, duration, etc.), we suggest an alternative understanding of "water rights" as implicating and entwining the domains of distribution, participation and recognition. Water rights are, necessarily, multi-layered entities (e.g. Benda-Beckmann and Benda-Beckmann, 2000; Roth *et al.*, 2015; Zwarteveen and Boelens, 2014). Water rights are historically, contextually and culturally specific demands to access and use (part of) a water flow and join (particular spheres of) water-control operation and decision-making, while including certain restrictions, obligations and penalties that are adjunct to this authorization. Water rights, in this sense, are defined by and embedded in particular ecological, political and cultural settings. This is very different from mainstream definitions of water rights as just state-centered (e.g. concessions) or that have a normative view (what "real" water rights "should be"). It is also different from the presumably "clear and enforceable" water rights that are presented as tools and conditions for installing water markets. The latter state- and market-based water rights require uniformity and universality. By contrast, we argue that locally, state- or market-defined water rights are never politically neutral: "Water rights both embody and shape social and power relations (because they organize inclusion and exclusion) and they are shaped by the way power is socially and culturally organized in water governance practice" (Boelens, 2015: 5, 6)

For similar reasons, caution is warranted. Unlike the notion we have outlined above, the liberal conception of rights, as atomistic and individualized, is subsumed within the logic of capital (Harvey, 2008). For instance, as Bakker (2007, 2010) has pointed out, accepting the human right to water does not preclude its privatization, commercialization or commodification. In its liberal, individualized form, the human right to water contradicts visions of collective rights to water and other resources. Indeed, international financial institutions have used the human right to water as a justification to privatize and marketize water service, on the assumption that the poor can access water most efficiently through market-based mechanisms. Whether or not water is considered a human right, then, has little bearing on the institutional arrangements through which water is allocated. It must be recognized, however, that the discourse and politics of the "human right to water" carry potent symbolic meaning (Bakker, 2010; Mirosa and Harris, 2012; see the chapter by Van den Berge *et al.*, in this volume).

In her critique of "human right to water" debates, Bakker (2007) suggests that viewing drinking water as a commons, as opposed to a commodity, shifts both the terms of debate and the possibilities for achieving equity in water governance. For those who consider water to be a collective good, water should be viewed in its social context (Bakker, 2010). In Sen's terms, cross-subsidizing drinking water and sanitation services for the poor would help secure their capabilities, an end whose attainment would surely be aided by tactically designating drinking water access as a human right (Bakker, 2010; Mirosa and Harris, 2012). In contrast to individualized water rights as a form of exclusionary private property, the human right to water recognizes water's fundamentally *universal* character. These concepts are profoundly at odds with one another. Moreover, we argue that, although having drinking water *access* is a universal need, the water *control* privileges that come with water rights (drinking water, irrigation, or other uses) are *not* universal but are, rather, context- and culture-specific. Such understandings highlight the political potential of rights claims. As Mitchell (2003: 25, emphasis in original) notes, "Rights establish an important *ideal* against which behavior by state, capital, and other powerful actors must be measured – and held accountable." In this sense, the ideal of rights, whatever its flaws, is a *political* ideal: "at once a means of organizing power, a means of contesting power, and a means of adjudicating power" (ibid.: 22). For the people of Flint, then, the concept of rights – their right to water and their rights as citizens – represents a potent ideal for holding the state accountable, for contesting its authority and for mobilizing resistance.

Structure of Part I

The chapters in Part I examine issues of water allocation and its (re-)politicization. They do so through a variety of empirical cases, drawing on a range of conceptual frameworks. Following this introductory essay, Chapter 2 ("Water Governance as a Question of Justice: Politics, Rights and Representation"), by Dik Roth, Margreet Zwarteveen, K.J. Joy, and Seema Kulkarni, proposes a reconceptualization of contemporary water issues as justice problems, which are centrally concerned with water distribution, claims recognition, and political participation. Drawing on empirical evidence from India, the authors call for actively re-politicizing water service and allocation: making explicit the distributional assumptions of water governance. These underlying assumptions are too often obscured by mainstream approaches to water governance, which are deeply committed to apolitical technical solutions and universalized notions of market-based water allocation mechanisms. The authors further argue that calls for water justice must be anchored in a sound understanding of water's specific characteristics as a resource, as well as the particular social relations and norms that shape its access and control.

This is followed by Chapter 3 ("Water Grabbing: Practices of Contestation and Appropriation of Water Resources in the Context of Expanding Global Capital"), by Gert Jan Veldwisch, Jennifer Franco, and Lyla Mehta, which examines the relationship between land grabbing and water justice. The authors demonstrate that water's fluid nature and hydrologic complexity often obscure the ways that water grabbing takes place. Water's fluid material properties interact with the "slippery" nature of grabbing, such as unequal power relations, unclear administrative jurisdictions, and fragmented negotiation. Moreover, as the authors show, water grabbing often takes place in social and political contexts marked by pluri-legalism and diverse forms of water knowledge. As the authors demonstrate, uneven power relations involved in global governance have increased local-level uncertainties, converting local complexities into mechanisms to exclude poor and marginalized peoples.

In Chapter 4 ("De-Politicized Policy Analysis: How the Prevailing Frameworks of Analysis Slight Equity in Water Governance"), authors Andrea K. Gerlak and Helen Ingram examine the ways that water governance is represented and analyzed in the academic field of public policy. Public policy was established as an academic field in the mid-twentieth century to foster better policy-making by democratic governments, and should be attentive to the uneven distribution of political power and access to water-related decision-making. However, as the authors demonstrate, most public-policy frameworks largely sidestep equity issues and water governance participation. While some well-established frameworks emphasize grassroots, decentralized decision-making, they do not explain how minorities and the poor, who have little voice at the local level, are to be represented. Moreover, as the chapter shows, although discursive approaches to policy analysis do recognize equity and differences in access among different populations, they largely fail to offer workable policy solutions.

Finally, Chapter 5 ("Urban Water and Sanitation Injustice: An Analytical Framework"), by Ben Crow, looks at questions of urban water (in)justices in the global South. Urban water systems rarely have appropriately traditional or plural institutional frameworks to guide water allocation, and poor households typically have unequal, unjust relations with municipal agencies. Water injustices are particularly widespread in informal settlements and other low-income residential areas, which are commonly excluded from basic urban services. Access to household water and sewerage is among a range of capabilities required for city life and livelihood, alongside access to sanitation, garbage collection, electricity and digital connections.

References

Bakker, K. (2007). The "commons" versus the "commodity": Alter-globalization, anti-privatization and the human right to water in the global South. *Antipode*, 39, 430–55.

Bakker, K. (2010). *Privatizing Water: Governance Failure and the World's Urban Water Crisis*. Ithaca, NY: Cornell University Press.

Benda-Beckmann, F. von and Benda-Beckmann, K. von (2000). Gender and the multiple contingencies of water rights in Nepal. In R. Pradhan, F. von Benda-Beckmann and K. von Benda-Beckmann (eds.), *Water, Land and Law*. Kathmandu: FREEDEAL.

Boelens, R. (2009). The politics of disciplining water rights. *Development and Change*, 40, 307–31.

Boelens, R. (2015). *Water, Power and Identity: The Cultural Politics of Water in the Andes*. London and Washington, DC: Routledge/Earthscan.

Bosman, J. (2017). "Michigan allots $87 million to replace Flint's tainted water pipes." *The New York Times*, March 27, www.nytimes.com/2017/03/27/us/flint-water-lead-pipes.html.

Bullard, R. D. (1990). *Dumping in Dixie: Race, Class, and Environmental Quality*. Boulder, CO: Westview Press.

Clark, A. (2014). "Going without water in Detroit." *The New York Times*, July 3, 2014, www.nytimes.com/2014/07/04/opinion/going-without-water-in-detroit.html.

Davidson, A. (2016). "The contempt that poisoned Flint's water." *The New Yorker*, 22 January, www.newyorker.com/news/amy-davidson/the-contempt-that-poisoned-flints-water.

Fraser, N. (2000). Rethinking Recognition. *New Left Review*. May/June, 107–20.

Grevatt, M. (2016). "Flint demands water justice." *International Action Center*, 19 January, http://iacenter.org/environment/flint012016/.

Harvey, D. (2008). The right to the city. *New Left Review*, 53, 23–40.

Hicks, G. (2011). Acequias of the south-western US in tension with state water law. In R. Boelens, D. Getches and A. Guevara-Gil (eds.), *Out of the Mainstream: Water Rights, Politics and Identity*. London: Routledge, pp. 223–34.

Holifield, R. (2015). Environmental justice and political ecology. In T. Perreault, G. Bridge and J. McCarthy (eds.), *The Routledge Handbook of Political Ecology*. London: Routledge, pp. 585–97.

Mirosa, O. and Harris, L. M. (2012). Human right to water: Contemporary challenges and contours of a global debate. *Antipode*, 44, 932–49.

Mitchell, D. (2003). *The Right to the City: Social Justice and the Fight for Public Space*. New York: Guilford.

Osnos, E. (2016). "The crisis in Flint goes deeper than the water." *The New Yorker*, 20 January, www.newyorker.com/news/news-desk/the-crisis-in-flint-goes-deeper-than-the-water.

Perreault, T., Wraight, S. and Perreault, M. (2012). Environmental justice in the Onondaga Lake waterscape, New York. *Water Alternatives*, 5, 485–506.

Pulido, L. (2000). Rethinking environmental racism: White privilege and urban development in southern California. *Annals of the Association of American Geographers*, 90(1), 12–40.

Pulido, L. (2015). Geographies of race and ethnicity I: White supremacy vs. white privilege in environmental racism research. *Progress in Human Geography*, 39(6), 809–17.

Roth, D., Boelens, R. and Zwarteveen, M. (2015). Property, legal pluralism, and water rights: The critical analysis of water governance and the politics of recognizing "local" rights. *Journal of Legal Pluralism and Unofficial Law*, 47(3), 456–75.

Schlosberg, D. (2004). Reconceiving environmental justice: Global movements and political theories. *Environmental Politics*, 13(3), 517–40.

Sen, A. (2001). *Development as Freedom*. New York: Alfred Knopf.

Warikoo, N. (2017). "200 clergy members protest hike in Detroit water and drainage fees." *Detroit Free Press*, February 28, www.freep.com/story/news/local/michigan/detroit/2017/02/28/200-clergy-members-protest-hike-detroit-water-and-drainage-fees/98528480/.

Welburn, J. and Seamster, L. (2016). "How a racist system has poisoned the water in Flint, Mich." *The Root*, 9 January, www.theroot.com/how-a-racist-system-has-poisoned-the-water-in-flint-mi-1790853824.

Wilkinson, C. (2011). Indian water rights in conflict with state water rights: The case of the Pyramid Lake Piute tribe in Nevada, US. In R. Boelens, D. Getches and A. Guevara-Gil (eds.), *Out of the Mainstream: Water Rights, Politics and Identity*. London: Routledge, pp. 213–22.

Zwarteveen, M. and Boelens, R. (2014). Defining, researching and struggling for water justice: Some conceptual building blocks for research and action. *Water International*, 39(2), 143–58.

2

Water Governance as a Question of Justice: Politics, Rights, and Representation

Dik Roth, Margreet Zwarteveen, K. J. Joy, and Seema Kulkarni

2.1 Introduction

Policy discourses – at the heart of water governance – are seldom explicit about the distributional assumptions and consequences underlying water policies, technologies, and institutions. They treat water problems as natural problems affecting all of us, and proposed solutions are "rendered technical" (Li, 2007) or leave allocation to anonymous markets. This makes it difficult, if not impossible, to recognize that water governance is significantly about justice. Therefore, this chapter shows how making water justice issues visible significantly hinges on defining water governance through water distribution and water rights (see also Zwarteveen, 2015). This starts by acknowledging and teasing out how the socio-environmental processes of change that water interventions (involving institutions, technologies and markets) entail alter existing water stocks, flows, quantity and quality, and create new access patterns and mechanisms, establish new rights and forms of in/exclusion, and thus new constellations of winners and losers (Swyngedouw and Heynen, 2003). Contestation and conflict are intrinsic to such changes, which is why "rational organization of dissent" (see Mollinga, 2008) is essential to water governance approaches that take justice seriously. Debate and disagreement may concern direct physical control over water resources; rules and laws governing water allocation, use and management; authority and power to define, decide upon, and enforce such rules; or the discourses and knowledge used to frame or make sense of society–water relations (Boelens and Zwarteveen, 2005; Zwarteveen *et al.*, 2005).

To understand water governance in terms of justice, we recognize that many current water governance reforms are part of broader capitalist transformation under globalization. Dominant water governance language and logic are so deeply infused with neoliberalism that it has become difficult to see and recognize them as part of an ideology or belief rather than a (natural or economic) given or a necessity (see Achterhuis *et al.*, 2010; Ahlers and Zwarteveen, 2009; Boelens and Zwarteveen, 2005). The following section shows how India's rapid economic growth is partly driven by equally rapid (although neither new nor recent) capitalization of nature, increasingly allocating water resources to supposedly more productive uses – industries and private companies – at the expense of supposedly less-productive users, including smallholder farmers or the urban poor. The state actively supports and facilitates this, reforming water law to standardize and privatize water rights (Cullet *et al.*, 2010a).

After a rough sketch of India's context, focusing on Odisha and Maharashtra, the rest of the chapter further discusses implications of these two foundational elements (the central role of distribution, and recognizing that water is part of overall capitalist transformation) to conceptually disentangle and understand water governance in terms of (in)justice. Anchored in a broad framework of water rights[1] and access, we suggest to re-politicize water debates as a necessary first step towards a more explicit discussion of justice (Section 3). We continue by proposing a conceptualization of water justice that combines a particular ontology of water with insights from broader environmental justice approaches. A central question here is how to reconcile (recognition of) the intrinsic localness of all definitions of justice, with the desire for more universal and overarching normative frameworks. We end the chapter with a short conclusion.

2.2 Water Reallocations in India as Forms of Resource Accumulation

India has always had water inequalities, inequities and injustices, with huge extremes between the water-rich and the water-poor across class, caste, gender and regional divisions, and between cities and rural areas. Indeed, class, caste, gender, tribe, and ethnicity are and have always been the basis for hierarchical social power relations that also co-determine access to water, mapped onto spatial and locational water availability, making it much easier for some to access and control water than for others (Joy and Paranjape, 2004). For irrigation, this has happened, for instance, through a specific regional concentration of irrigation facilities (often constructed with public funds), with vast areas left without access to irrigation. The country's 170 most marginalized districts practice rainfed agriculture (Government of India, 2008); irrigation crucially co-determines a region's prosperity (next to education and literacy, see Bandyopadhyay, 2011). However, long-standing water-based inequities are worsening, and new ones are rapidly emerging. Since the 1990s, development by increasing industrialization and urbanization, spurred by economic reforms (liberalization, privatization, and globalization) has achieved rapid economic growth rates in India (10–11 percent annually). There is a booming middle class, skyrocketing demand for food, consumer products and services. Agriculture, once the economy's backbone, now contributes only 25 percent to the national GDP, while 75 percent of the population depends on it.

Official claims that this growth decreases poverty overall and enhances wealth are increasingly countered by statistics breaking down averages to show that there are large, growing categories of the Indian population, especially in rural areas, left behind by economic growth (Aiyer, 2007). On the contrary, impressive growth in some regions and sectors parallels and sometimes results from stagnation in other regions and sectors. Such wider inequities result from, and shape, water-based inequities. Water (re)allocations from agriculture to industry and cities, for instance, leave some areas with less water, increasing water-related conflicts in the last decade (Joy *et al.*, 2008). Almost every dam and reservoir originally built for irrigation is now also used to meet urban-industrial demands, to farmers' dismay. The Hirakud dam in Odisha is a case in point: faced with a shrinking water supply, farmers are actively opposing water diversion from the dam to industry (Choudhury et al., 2012).

Cropping pattern changes – often from subsistence and food crops to more profitable and water-intensive export crops – or the conversion of public water to private water, with increased reliance on groundwater, are more difficult to trace, but still exclude from access to and benefit from water and water-related income. A striking example of such new water injustices is sugarcane cultivation co-existing with severe drought in Maharashtra. A recent study of the Waghdad and Kukdi irrigation schemes in the same state show that only 10 to 20 percent of the original canal infrastructure network area is now being irrigated, even though per-hectare water use exceeds formal entitlements: most irrigation now uses groundwater from wells recharged by canal water. Only farmers with land rights and who can afford to invest in wells and pumps can access this water, undermining the equity aims originally guiding public investment in surface irrigation systems. Chennai is shifting from public piped water to private groundwater, where leaky drinking water supply pipes recharge the aquifers that only people with expensive private wells can tap into (Srinivasan and Kulkarni, 2014).

Economic and technological changes mirror changes in (water and land) tenure and labor relations, all deeply shaped by existing gendered caste and class institutions. Women of low castes and poor households are worse-positioned to secure fair shares (of water or water-related benefits and incomes) or protect their livelihoods, as they typically have no land rights and therefore no water rights, and are also excluded from membership in water users' associations (SOPPECOM, 2008, 2010). Recent NSSO data reveal that 40 percent of all households in India are landless (NSSO 59th round), while the agricultural census shows that only 12 percent of all women are operational landholders (Agricultural Census, 2010–11). This shapes access relations, for instance in the water-scarce and labor-short Ahmednagar district of Maharashtra: upper-caste households provide water for domestic use to *dalit* households in exchange for farm labor (SOPPECOM, 2010), exacerbating existing dependencies and inequities. Another example: in Nasik, the grape export district of Maharashtra, the shift to commercial production has increased women's work on family farms, without reducing their domestic workload or giving women any control over the fruits of their labor. Recent droughts in Marathwada particularly affected poor girls' lives most, kept home from school to help collect water (SOPPECOM, 2010). In general, poor, low-caste women and girls have to walk ever-further to fetch water. This erodes their livelihood options, making them vulnerable to all kinds of exploitation, notably by loan providers and micro-credit institutions who charge ever-higher interest rates. Poor women (and some men) use such micro-loans not for productive investment but for survival, risking a downward spiral of indebtedness, misery and despair, pushing many to the verge of suicide (Kulkarni, 2016).

All these examples of water-based injustices underscore the importance of not accepting contemporary reallocations as an ultimately beneficial by-effect of, or a necessary condition for, an evolutionary or inevitable process of demographic and economic growth or progress.

Instead, the explicit questioning of the model of growth that justifies or provokes such reallocations needs to form part of the theorization of water governance in terms of justice. Theories that are critical of capitalist growth – such as regarding capitalization of nature (O'Connor, 1994) and accumulation by dispossession (Harvey, 2003) – are germane here. Though current water governance reforms and reallocations in India are blurred, messy and full of contradictions, the clear concentration of land and water rights by ever-fewer owners does ratify Harvey's theorization that contemporary resource reallocations consist of new "enclosures of the commons" releasing low-cost labor and water. Irrigated area on larger-sized farms, for instance, expands much faster than on small, marginal farms (Vaidyanathan, 2005), while water use concentration in ever-smaller areas (cf. Waghad and Kukdi) also reveals dispossession. Whether and how this is feeding new forms of capitalist accumulation remains to be further investigated, but this clearly treats water as a commodity subject to "rational calculation of production and exchange" (Baudrillard, 1981:188, cited in O'Connor, 1994).

The state's crucial role backs and facilitates this re-conceptualization of water as a commodity: by amending regulations, legal frameworks and institutions, it clearly favors, privileges and even promotes capitalist accumulation as economic growth, progress and development. The state also more actively appropriates rural land and water resources to benefit cities and industry, as when the high-powered Committee of Ministers of Maharashtra transferred 1,500 million m^3 of irrigation water (in 2003) to industrial and urban uses. This transfer denied water to about 300,000 ha of land, creating misery for thousands of people who depend on this land for their livelihoods (Prayas, 2010). The Maharashtra State Water Policy of 2003 also actively reversed water-use priorities: until about the year 2000, national and state-level policies gave priority to drinking and domestic uses, followed by agriculture and industry. The 2003 policy reversed this order, giving industry second and agriculture third priority. However, civil-society action forced the Government of Maharashtra to restore second priority to agriculture.

These water reallocations are difficult to identify as obvious injustice and dispossession, because they do not use brutal coercion or extra-economic force and methods. Instead, the modern Indian state justifies them as public good or the only desirable road to development and wellbeing for all, often inspired by neoliberal utopian thinking. As Cullet *et al.* (2010b) conclude, international financial institutions' prominent role in the water sector has strongly promoted a specific package of measures: cost recovery and efficiency, uniform water rights, reducing the state's role, and increasing supply costs (Cullet *et al.*, 2010b) as core elements of the water governance reform agenda. Hence, it is no coincidence that reforms in Maharashtra – a range of new policies and regulations for water transferability and marketability – directly followed a World Bank loan to the irrigation sector. These new policies make water rights partially tradable, either within a particular use or across different uses, and recognize only productive-use water rights, thereby denying domestic water uses and users any legal water security. Reforms also favor water entitlements and rights for landowning farmers – in proportion to the land they own – excluding landless people from irrigation systems.[2] The 2005

Maharashtra Management of Irrigation Systems by Farmers Act states that, if part of the service area under a Water Users Association is taken up for contract farming, that land's water entitlements also go to the contract farmers (mainly companies). These reforms all actively concentrate water entitlements and rights in those considered productive and economically efficient, dispossessing anyone who does not meet these criteria (see Cullet et al., 2010a).

Water reforms, and the development model informing them, do not go uncontested. In addition to conflicts over reallocations from dams, civil-society observers and activists are more generally critiquing the idea of allocating water (exclusively) by market forces, or based on productivity and efficiency criteria. They argue that water allocation should, instead, be based on needs. According to them, water is not (just) a productive resource and commodity, but (also) a public good. Treating the right to water as a human right alongside other human rights (e.g. food, housing) is an important part of their counter-proposal: it is state responsibility to provide a social minimum to all, irrespective of property rights. Activists and civil-society groups also advocate re-conceptualizing water use prioritization from the current proportional system to a sequential one: unless a higher priority is met, water does not move to a lower priority. This would guarantee that water needs for domestic (drinking, cooking, washing, hygiene, sanitation, and livestock), ecosystems, and basic livelihoods are met first, before water is allocated to other uses and needs (e.g. hydropower, industrial, commercial, recreational), which could then be prioritized proportionally.

There are many more creative critiques and engagements with alternatives: mass mobilizations for equitable distribution; innovative experiences in participatory irrigation management (PIM) addressing concerns of gender equity, sustainability and democratization; women and *dalit* movements fighting for rights over land; struggles against privatization; stakeholder organization and conflict resolution around water pollution; and innovative watershed development experiences. all elements of a more widespread plea to make justice more central in water governance.

2.3 Re-Politicizing Water

Popular approaches to studying and solving water problems are often narrowly system- or intervention-oriented and supply-focused (Lahiri Dutt and Wasson, 2008; Molden, 2007), prioritizing efficiency over equity. This can be partly explained by the engineering and natural-science tradition of much water resource expertise, which relies on positivism to justify its claims to truth and authority. This appeal to universal reason conceals connections between power and knowledge, and the fact that water interventions imply political choices about water allocation, and distribution of water-related incomes, costs and risks. Much normal water expertise thus makes far-reaching water redistributions and reallocations appear natural, inevitable or (scientifically and technologically) rational (cf. Donahue and Johnston, 1998), through combined mechanisms of technification (rendering technical), naturalization and universalization.

First, by making them technical (Li, 2007), complex and wicked problems become framed as clearly bounded, researchable areas of a knowable, controllable world. Molle (2008) illustrates, for instance, how the use of technical "nirvana concepts" such as integrated water resources management (IWRM) simplifies complex problems into manageable solutions. By focusing on how to produce consensus (e.g. multi-stakeholder approaches), these concepts also make it difficult to recognize conflicts of interests or existing social hierarchies, which make it easier for some than for others to get what they want. Naturalization is a second mechanism by which contentious distribution issues are screened from public deliberation and debate: flood or drought problems are presented as (essentially) *natural* problems, rather than as (also) problems of distribution or social power relations. Mehta (2003), for instance, has shown for the Kutch region in arid western India how treating water scarcity as a natural phenomenon served as a meta-narrative to legitimize a large politically contested dam (the Sardar Sarovar Narmada project). Calls to use river basins as the natural water-management unit can likewise be seen as a form of naturalization (Warner *et al.*, 2008; also see Bakker, 1999; Wester, 2008), as can references to climate change as a primarily natural process. A third depoliticization mechanism is universalization, which has a scale element (framing water problems as global or universal) and a definitional element (treating water as similar anywhere irrespective of context) (Linton and Budds, 2014). Commonly heard calls to govern water on a global scale (e.g. Pahl-Wostl *et al.*, 2008) erroneously and dangerously suggest that water problems affect "all of us" equally, and makes local social-ecological contexts and histories seem unimportant (Donahue and Johnston, 1998; Lahiri Dutt and Wasson, 2008).

An important first step towards effectively addressing justice and equity in water governance, therefore, is to challenge these processes of technification, naturalization and universalization, to retrieve its basically political character. This is a proposal we share with many political ecologists (see Loftus, 2009; Mollinga, 2003; Mosse, 2003) and more generally with political theorists (e.g. Mouffe, 1993, 2005). Re-politicizing water significantly hinges on critical awareness that water problems are essentially allocation and distribution problems (see Boelens *et al.*, 2010; Roth *et al.*, 2005; Zwarteveen, 2015), and actively recognizes how water is always both natural and social. Re-politicization teases out how specific combinations and entanglements of environmental and sociopolitical relationships produce (re-)distributions of water. This requires interdisciplinary approaches, which recognize that pressures on water resources and resulting scarcities are the outcome of specific water resource usage histories and practices. Human activity and biophysical factors interact to shape environments that are dynamic and continuously contested in the political economy of access to and control over resources (Harvey, 1996; Swyngedouw, 1999, 2003). Terms such as "waterscapes" (Baviskar, 2007) or the "hydrosocial cycle" (Linton and Budds, 2014) have been proposed to help this understanding, to replace definitions of water that reduce it either to its economic properties (as a scarce natural resource) or to its chemical or physical characteristics, with understanding that water is both natural and social (Baviskar, 2007; Swyngedouw, 1999; also see Castree and Braun, 2001).

2.4 Theorizing Water Justice

2.4.1 Environmental Justice

A particular domain of research and social action was coined as "environmental justice" in the 1980s in the USA (Holifield, 2001; Holifield et al., 2009). This early work mainly concerned distribution of environmental burdens such as toxic waste disposal, with race as the primary focus of analysis (Holifield et al., 2009; Walker, 2009b). It sought statistical substantiation of linkages among space (proximity), inequality and politics of siting – assumed to be intentionally discriminatory (Schroeder et al., 2008; Walker, 2009b). Environmental justice consequently attempted to equitably distribute such environmental "bads" without discrimination. This bias towards the distributive side of environmental justice neglected or poorly theorized other dimensions – primarily procedural issues such as recognition and participation (Schlosberg, 2004; see Islar, 2012).

The concept travelled to other regions and times, acquiring new meanings and connecting with new issues, gradually pluralizing environmental justice analysis between social practices and theory (Walker, 2009a; see Olson and Sayer, 2009).[3] Holifield (2001), for instance, argues for contextual understanding of environmental justice, allowing for time and place-specific definitions by different actors experiencing (in)justices in various ways. The gradually growing field of research and action improved understanding and conceptualizations of the complex "geography of burdens/benefits" (Schroeder et al., 2008: 551), with more attention not only to the "bads" but also to the "goods" involved (see Holifield, 2001). The larger questions of "who gains, who loses?" became central in environmental justice debates; also guiding recent work on water resources by political ecologists and political geographers.

For an operational definition of water justice, Schlosberg (2004: 518) takes inspiration from authors such as Fraser (2000, 2005, 2008). Schlosberg proposes a "trivalent conception of justice" (Schlosberg, 2004: 521)[4] which complements distribution with dimensions of recognition (e.g. of specific cultural identities, rights, and practices) and participation (in decision-making). According to Schlosberg (2004: 528), justice "requires not just an understanding of unjust distribution and a lack of recognition, but, significantly, the way the two are tied together in political and social processes."

Conceptualizations of water justice should take into account scalar and place dimensions. Water's biophysical characteristics co-determine how it is or can be harvested, stored, used and managed, fundamentally co-shaping access and control patterns. For instance, spatiality and scale are an issue between upstream and downstream communities in a watershed: fair or unfair behavior by upstream communities is fundamental in water negotiations and governance.[5] Water access conflicts and contestations need to be understood along the river basin's spatial continuum (e.g. Joy et al., 2008), where spaces and scales may be manipulated politically to (re)distribute, and therefore also to struggle for justice. As Walker concludes, "different forms and scales of space are ... a strategic resource and just as 'different groups will resort to different conceptions of justice to bolster their position' (Harvey, 1996: 398), so will different groups work with different understandings of the spatiality of the issues at hand" (2009b: 630).

2.4.2 Linking Global Ideals with Contextualized Experiences

Many theories of justice take universalist transcendent approaches, anchored in some principled idea that all human beings are equal (Schlosberg, 2004; Sen, 2009) and in an imagined "perfectly just" society. Indeed, "most dominant theories focus on what justice *should be*" (Lauderdale, 1998: 5), rather than on understanding how diverse people experience and define justice within a specific context, history and time, or on the *effects* of particular definitions on a society's distribution of wealth and authority. Amartya Sen's theorization of justice is much more pragmatic: he argues for human situatedness, capabilities, and behavior in real-life conditions when defining and understanding justice.[6] He stresses that feelings of (in)justice can have plural grounds, and be based on different reasonings, as part of an informed plea for theorizing and approaching justice on the basis of the question of how to reduce injustice and advance justice, rather than on the basis of abstract notions of idealized justice (see also Schlosberg, 2004).

Taking inspiration from Sen, we prefer not to start theorizing water justice from universal normative conceptions of what (water distribution) rights should be and how they should be related to water policies and governance.[7] Instead, we suggest to start with the empirical diversity, complexity and embeddedness of rights, to analyze relationships among multiple definitions of rights and governance frameworks produced at various levels. This suggestion recognizes that water transcends political, jurisdictional, administrative and other boundaries, and links spatial (and temporal) scales to create "places" where it is valued, used and given meaning in specific contextualized ways as part of larger cosmologies (Baviskar, 2007), and where it informs and is shaped by social relations of power (Lahiri Dutt and Wasson, 2008; Sneddon *et al.*, 2002). This makes any generalization about water property difficult, and explains why practical water governance always is and must be locally embedded (Boelens *et al.*, 2010; Mosse, 2003; Roth and Vincent, 2012). Any discussion about improving water justice starts with the realities of injustice *as experienced by* the politically oppressed, culturally discriminated and economically exploited.

Underscoring the ontological and theoretical need to recognize multiple conceptualizations of water justice or environmental justice does not sanction or romanticize local or traditional ways of life. Particular (interpretations of) localized definitions can be used for specific strategic or populist purposes, also to others' detriment (Holifield, 2001). Situated views of rights and justice cannot be presumed to be clear-cut, uncontested and fixed (see Olson and Sayer, 2009; Williams and Mawdsley, 2006). Comparing different conceptualizations of justice will yield inconsistencies and frictions that are revealing for broader debate about what is fair. This also usefully links them to wider social and political struggles over resources, places and identities (Walker, 2009b; also see Castree, 2004; Schlosberg, 2004).

Emphasis on experience, context and place does not underestimate the importance of globally defined rights discourses (human rights; indigenous rights; Castree, 2004; von Benda-Beckmann *et al.*, 2009) to make room on political agendas to discuss water equity (or justice). However, unlike what some advocates of a global human right to water would suggest, it is important not to simply assume the positive effects of such a right (Gleick, 2007) on actual on-the-ground practices of water distribution and decision-making. This will depend on how this additional level of jurisdiction further pluralizes and interacts with existing definitions of water rights. Referring to universal rights may indeed help strengthen claims to water for those whose rights tend to go unheard. It may provide marginalized groups with an additional tool to question unfair water allocations, and demand their fair share (Athukorala and Rajepakse, 2012). This is also why many activist and civil-society groups in India struggle to have water recognized as a human right and treated as a basic service to fulfill livelihood needs. They expect that this will provide legal protection and help define a minimum water assurance that must be delivered at reasonable cost and dependability. Even without immediate enforcement, many critical groups hope and think that the sheer existence of such a right may put pressure on governments to acknowledge their responsibilities to turn the right into tangible benefits.

Yet there are also dangers in relying too much on universal and global rights frameworks to justify one's claims. They may strengthen unwarranted beliefs in formalized legal solutions – a "fetishization of law" (von Benda-Beckmann *et al.*, 2009). At the end of the day, who guarantees that formulating such a right will lead to improved access for poor, disconnected people, and not at the expense of existing water rights in, for instance, rural areas? Universal rights definitions may also freeze dynamic, flexible local and place-based allocation principles by straitjacketing them into legal formats and procedures, or make solving water conflicts the professional privilege of trained legal experts. Finally, most universal water rights frameworks are operationalized as per-capita entitlements to a specific quantity (and quality) of water, rather than as socially embedded bundles of rights and obligations (see F. and K. von Benda-Beckmann, 2003). This operationalization of water is dangerously compatible with neoliberal privatization and commoditization agendas. This is why D'Souza (2008: 3) states that "this promise is informed theoretically by liberalism and is conceptually problematic."

In sum, although pleas for the human right to water have now become mainstream (Bakker, 2010), their possible consequences for place-based contents and definitions of water rights remain understudied and under-theorized (see e.g. Khadka, 2010). Moreover, as with human rights more generally, the question of how to best locally translate or "vernacularize" (Merry, 2006) universal rights definitions to make them socially acceptable and meaningful in diverse real-life societies remains unanswered. Suspicious of human rights discourses as cultural practices that involve "technologies for fixing truth in universal and legalistic forms" (Merry, 2006: 229), Merry therefore stresses that approaches must be seen as legitimate locally, as well as by audiences at other scales.

Hence, we anchor water justice theorization in everyday experiences of justice, but not as a plea to uncritically support any claim of unfair treatment. Yes, the emergence of broader forms of engagement for social justice across the boundaries of local struggles is important (Harvey, 1996, Schlosberg, 2004; also see Castree, 2004; Sneddon and Fox, 2008a, b).[8] While plurality is a basic political condition to debate and socially negotiate rights (Mouffe, 1993), basic commonalities in the language of justice (Merry, 2006) that respect plurality, difference and contestation might give justice movements more political clout.

Approaches thus should recognize specificity, diversity and difference, but also enable the emergence of a shared language of justice (Harvey, 1996; Walker, 2009b).

2.4.3 Individual Merit or Social Relations of Power?

Many theorizations of justice are informed by notions of freedom and choice that are deeply utilitarian and liberal in their assumptions of individual autonomy and rationality (Lauderdale, 1998; Mouffe, 1993; Olson and Sayer, 2009; Sen, 2009; Walby, 2011), often starting from individual notions of rights (e.g. Rawls' theory of justice). We prefer Sen's pragmatic definition of justice as contextual and relational. However, as Walby (2011) also notes, the danger in Sen's pragmatism is its compatibility with a variety of agendas and purposes, including (neo)liberal ones. Especially his concepts of choice, capabilities, and functioning matches the neoliberal emphasis on individual freedom (vis-à-vis state control) and private property rights (or entitlements). Emphasizing individuals' freedom to maximize their own interests risks reducing fairness to a matter of individual merit – attributing economic and distribution inequalities more to differences in individual effort and talent, than to the workings of power. Sen starts his analysis from the liberal assumption that all humans, at their core, are equal and share a common capacity to reason. This assumption may wrongly suggest that all people, given the same chances and capabilities, can and do compete on a level playing field and distracts analytical attention away from real-life historical inequities and contemporary social struggles. We therefore prefer a relational/contextual conceptualization of personhood that suggests that a person always exists relative to the constructed relations in which he or she lives. This is more compatible with our approach to justice as lived, which allows for plurality and embeddedness. It acknowledges that public interest – or "the common good" – will always be negotiated and often contested: there will/can never be a neutral, rationally derived, undisputed "final" conception of justice. Justice, according to Mouffe (1993: 53), is an "unresolved question" that political communities have to cope with to maintain their coherence. Forms of political association and collective participation in the public sphere, and ways of channeling and organizing dissent, are a key dimension of justice in this conceptualization.

2.5 Conclusions

Using India as the contextual backdrop, we have approached water justice starting from water's specific characteristics, and water governance access and rights dimensions. The Indian examples illustrate that current water reallocation dynamics transcend state and national boundaries, in more ways than one. They are an intrinsic dimension of globalizing markets for food and biofuels; reliable, cheap access to water drives decisions about where to produce and invest. These markets follow an actively promoted international water governance agenda, which preaches allocating water to wherever its marginal returns are highest, favoring "marketable" and "priceable" water uses over those with benefits less quantifiable. In India, this is clearest when water that was used for producing food crops or for domestic uses is reallocated to industries and agro-export companies. Water expertise and policies praise such reallocations on grounds of efficiency, effectiveness or productivity, overshadowing implications for social equity or justice. The Indian examples show that water access and control concentrate increasingly in a few privileged actors – often those who can afford the technology to access it – to the detriment of many others.

Exposing the water justice dimensions of current policies, interventions and expertise requires, first of all, repoliticizing water policies, knowledge and interventions. Naturalization, technification, and universalization obscure the political nature of water problem analyses and solutions. Underlying water resources valuations and allocation priorities become implicit, while water appropriation and allocation processes and mechanisms seem inevitable. It is difficult to ask how these processes influence access and exclusion patterns along axes of power hierarchies and social differentiation. Moreover, policy emphasis on consensus-based solutions creates pressures to conceal rather than open up to debate new laws' and interventions' allocation choices.

We approach water justice with an interdisciplinary ontological definition of water as co-produced by societal and natural processes, termed as "waterscapes." Water (re)allocation justice/injustice cannot be assessed from any neutral position, but requires identifications and engagements to be made explicit. Creative attempts to develop a collective knowledge base grounded in actual distributional dilemmas or conflicts and identification (or empathy) with those who feel they are treated unjustly can reach locally specific approaches to greater fairness. Globally defined, universal rights doctrines and discourses, e.g. recognizing water as a global human right, may lend useful, much-needed support to localized water struggles. Yet the positive effects of such globalized water-rights proclamations cannot be assumed: understanding or achieving them requires attention to how global norms and rights interact with and influence place-based definitions and practices.

Plurality in water-rights definitions and conceptions, and in experiences of (in)justice, provides a useful, pragmatic starting point for place-based appreciations and discussions about water justice. Critical theorizations of justice (e.g. Fraser, 2000, 2005) and environmental justice (e.g. Schlosberg, 2004; Walker, 2009b) show that such discussions go beyond issues of distribution, including political participation and recognition. Seeing water control issues as basically political is, in our opinion, a precondition for emphasizing the justice dimensions of distribution, participation and recognition. Taking the existence of plurality as a starting point does not make it mandatory to celebrate local definitions of justice, as Walker (2009a,b) and Olson and Sayer (2009) have convincingly argued, but it does imply that they provide the context within which to search for common ground.

Assessments of water justice are always scale- and place-sensitive, with justice on one scale sometimes implying injustice at another. The Indian cases clearly show this: more justice within cities, for instance, may be achieved by taking rural areas' water. Or more economic revenues by charging industries to use water may be achieved at the cost of food insecurity and hunger. Manipulation of scales and spaces may itself be an important domain of justice struggles. This is one reason why we caution against overly individualistic approaches to water justice, which link it to individual freedoms or capabilities or risk, reducing it to questions of merit. Instead, we suggest that justice is always deeply relational; it is about reaching agreement about how (the costs and benefits of access) water is or should be shared among different actors, or between people and the environment.

Notes

1 We approach water rights as a bundle of rights and obligations regarding water and related resources (land, infrastructure, etc.), developing and embedded in a specific hydrosocial and other context and based on a locally legitimate social institution. India's water discourse uses "the right to water" rather than "water rights" (which is associated with tradability). As explained below, this issue is very context-specific: how rights are embedded and expressed, how they interact and "work" (determining people's access to water, water security and rights) in real life – both within specific localities (e.g. rural/urban) or across scales (e.g. urban/peri-urban; urban/rural).
2 This is how entitlements and water rights are defined in the Maharashtra Water Resources Regulatory Authority Act 2010.
3 Williams and Mawdsley also stress, for an analysis of environmental justice for governance in India, the importance of "close examination of differences in the *context* in which struggles for environmental justice are located" (2006: 669).
4 A crucial difference between liberal theorists like Rawls and Miller on one hand, and Young and Fraser on the other is that the former, in their liberal search for perfect justice, assume and subsume recognition "within the distributive or procedural spheres of justice" (Schlosberg, 2004: 520).
5 For a detailed discussion of asymmetries in watershed and other ecosystem processes see Lélé (2004) and Kerr *et al.* (2002).
6 See also Mouffe's (1993) criticism of Rawls' theory of justice. Mouffe notes that Rawls' focus on individual rights over the common good leads to a theory in which rights are defined prior to the common good. Mouffe, on the contrary, stresses that rights and justice can only be defined through political participation in a community, which creates a sense of rights and a conception of justice. Hence questions of justice can be posed only by starting from specific communities and their histories, traditions, social relationships and meanings.
7 Water is often problematically categorized as, for instance, a "commons" or a "common resource." First, actually these are often normative statements about the property status water *should* have in the eyes of the user of the term; second, such statements are restricted to the broad categorical level of its property status, and not related to actual social relationships, conditions and practices of water use and control.
8 Sneddon and Fox (2008a, b) show that this fear that myriads of local and particularistic struggles will emerge is not always justified. The authors show, for anti-dam struggles in Thailand and Mozambique, that such "scale-making projects" transcend local and particularistic agendas.

References

Achterhuis, H., Boelens, R. and Zwarteveen, M. (2010). Water property relations and modern policy regimes: Neoliberal utopia and the disempowerment of collective action. In R. Boelens, A. Guevara and D. Getches (eds.), *Out of the Mainstream: Water Rights, Politics and Identity*. London: Earthscan.

Ahlers, R. and Zwarteveen, M. (2009). The water question in feminism: Water control and gender inequities in a neo-liberal era. *Gender, Place & Culture*, 16(4), 409–26.

Aiyer, A. (2007). The allure of the transnational: Notes on some aspects of the political economy of water in India. *Cultural Anthropology*, 22(4), 640–58.

Athukorala, K. and Rajapaksa, R. (2012). Water rights and gender rights: The Sri Lanka experience. In M. Z. Zwarteveen, S. Ahmed and S. R. Gautam (eds.), *Diverting the Flow: Gender Equity and Water in South Asia*. New Delhi: Zubaan, pp. 137–60.

Bakker, K. (1999). The politics of hydropower: Developing the Mekong. *Political Geography*, 18(2), 209–32.
Bakker, K. (2010). *Privatizing Water: Governance Failure and the World's Urban Water Crisis*. Ithaca, NY: Cornell University Press.
Bandyopadhyay, S. (2011). Rich states, poor states: Convergence and polarisation in India. *Scottish Journal of Political Economy*, 58(3), 414–36.
Baviskar, A. (ed.) (2007). *Waterscapes: The Cultural Politics of a Natural Resource*. New Delhi: Orient Longman.
Benda-Beckmann, F. von and Benda-Beckmann, K. von (2003). Water, human rights, and legal pluralism. *Water Nepal*, 9/10 (1/2), 63–76.
Benda-Beckmann, F. von, Benda-Beckmann, K. von and Griffiths, A. (eds.) (2009). *Spatializing Law: An Anthropological Geography of Law in Society*. Farnham: Ashgate.
Boelens, R. and Zwarteveen, M. (2005). Prices and politics in Andean water reforms. *Development and Change*, 36(4), 735–58.
Boelens, R., Getches, D. and Guevara-Gil, A. (eds.) (2010). *Out of the Mainstream: Water Rights, Politics and Identity*. Washington, DC: Earthscan.
Castree, N. (2004). Differential geographies: Place, indigenous rights and "local" resources. *Political Geography*, 23, 133–67.
Castree, N. and Brown, B. (eds.) (2001). *Social Nature: Theory, Practice, and Politics*. Oxford: Wiley-Blackwell.
Choudhury, P. R., Sahoo, B. C., Sandbhor, J., Paranjape, S., Joy, K. J. and Vispute, S. (2012). *Water Conflicts in Odisha: A Compendium of Case Studies*. Pune: Forum for Policy Dialogue on Water Conflicts in India.
Cullet, P., Gowlland-Gualtieri, A., Madhav, R. and Ramanathan, U. (eds.) (2010a). *Water Governance in Motion: Towards Socially and Environmentally Sustainable Water Laws*. Delhi: Foundation Books; Cambridge University Press India.
Cullet, P., Gowlland-Gualtieri, A., Madhav, R. and Ramanathan, U. (eds.) (2010b). *Water Law for the Twenty-first Century: National and International Aspects of Water Law Reform in India*. London and New York: Routledge.
Donahue, J. M. and Johnston, B. R. (eds.) (1998). *Water, Culture, and Power: Local Struggles in a Global Context*. Washington, DC: Island Press.
D'Souza, R. (2008). Liberal theory, human rights and water-justice: Back to square one? *Law, Social Justice and Global Development Journal* [online], 1, www.go.warwick.ac.uk/elj/lgd/2008_1/desouza.
Fraser, N. (2000). Rethinking recognition. *New Left Review*, 3, 107–20.
Fraser, N. (2005). Reframing justice in a globalizing world. *New Left Review*, 36, 69–88.
Fraser, N. (2008). *Scales of Justice: Reimagining Political Space in a Globalizing World*. Cambridge: Polity Press.
Gleick, P. H. (2007). *The Right to Water*. Oakland, CA: Pacific Institute.
Government of India (2008). *From Hariyali to Neeranchal: Report of the Technical Committee on Watershed Programmes in India*. New Delhi: Department of Land Resources, Ministry of Rural Development.
Harvey, D. (1996). *Justice, Nature and the Geography of Difference*. Cambridge and Oxford: Blackwell Publishers.
Harvey, D. (2003). *The New Imperialism*. Oxford, NY: Oxford Press.
Holifield, R. (2001). Defining environmental justice and environmental racism. *Urban Geography*, 22(1), 78–90.
Holifield, R., Porter, M. and Walker, G. (2009). Spaces of environmental justice: Frameworks for critical engagement. *Antipode*, 41(4), 591–612.

Islar, M. (2012). Struggles for recognition: Privatisation of water use rights of Turkish rivers. *Local Environment*, 17(3), 317–29.

Joy, K. J. and Paranjape, S. (2004). *Watershed Development Review: Issues and Prospects*. Bangalore: Centre for Inter-disciplinary Studies in Environment and Development (CISED), now part of ATREE.

Joy, K. J., Gujja, B., Paranjape, S., Goud, V. and Vispute, S. (eds.) (2008). *Water Conflicts in India: A Million Revolts in the Making*. New Delhi: Routledge.

Kerr, J. Pangare, G. and Pangare, V. L. (2002). *Watershed Development Projects in India: An Evaluation*. Washington, DC: International Food Policy Research Institute, Research Report No. 127.

Khadka, A. K. (2010). The emergence of water as a "human right" on the world stage: Challenges and opportunities. *Water Resources Development*, 26(1), 37–49.

Kulkarni, S. (2016). "Drought and debts: The plight of Bharat Mata in Marathwada," http://scroll.in/article/808302/drought-and-debts-the-plight-of-bharat-mata-in-marathwada

Lahiri-Dutt, K. and Wasson, R. J. (eds.) (2008). *Water First: Issues and Challenges for Nations and Communities in South Asia*. New Delhi: Sage Publications.

Lauderdale, P. (1998). Justice and equity: A critical perspective. In R. Boelens and G. Davila (eds.), *Searching for Equity: Conceptions of Justice and Equity in Peasant Irrigation*. Assen: Van Gorcum, pp. 5–10.

Lélé, S. (2004). Beyond state-community and bogus "joint"ness: Crafting institutional solutions forresource management. In M. Spoor (ed.), *Globalisation, Poverty and Conflict: A Critical "Development" Reader*. Dordrecht: Kluwer Academic, 283–303.

Li, T. M. (2007). *The Will to Improve: Governmentality, Development, and the Practice of Politics*. Durham, NC: Duke University Press.

Linton, J. and Budds, J. (2014). The hydrosocial cycle: Defining and mobilizing a relational-dialectical approach to water. *Geoforum*, 57, 170–80.

Loftus, A. (2009). Rethinking political ecologies of water. *Third World Quarterly*, 30(5), 953–68.

Mehta, L. (2003). Whose scarcity? Whose property? The case of water in western India. *Land Use Policy*, 24, 654–63.

Merry, S. E. (2006). *Human Rights and Gender Violence: Translating International Law into Local Justice*. London: University of Chicago Press.

Molden, D. (ed.) (2007). *Water for Food, Water for Life: A Comprehensive Assessment of Water Management in Agriculture*. London: IWMI/Earthscan.

Molle, F. (2008). Nirvana concepts, narratives, and policy models: Insights from the water sector. *Water Alternatives*, 1(1), 131–56.

Mollinga, P. P. (2003). *On the Waterfront: Water Distribution, Technology and Agrarian Change in a South Indian Canal Irrigation System*. Hyderabad: Orient Longman.

Mollinga, P. P. (2008). *The Rational Organisation of Dissent: Boundary Concepts, Boundary Objects and Boundary Settings in the Interdisciplinary Study of Natural Resources Management*. ZEF Working Papers # 33. Bonn: Center for Development Studies.

Mosse, D. (2003). *The Rule of Water: Statecraft, Ecology and Collective Action in South India*. New Delhi: Oxford University Press.

Mouffe, C. (1993). *The Return of the Political*. London: Verso.

Mouffe, C. (2005). *On the Political*. London: Routledge.

Olson, E. and Sayer, A. (2009). Radical geography and its critical standpoints: Embracing the normative. *Antipode*, 41(1), 180–98.

O'Connor, M. (1994). On the misadventure of capitalist nature. In M. O'Connor (ed.), *Is Capitalism Sustainable? Political Economy and the Politics of Ecology*. New York, London: The Guilford Press.

Pahl-Wostl, C., Gupta, J. and Petry, D. (2008). Governance and the global water system: A theoretical exploration. *Global Governance*, 14(4), 419–35.

Prayas. (2010). *Sinchanache Pani Udyogana va Shaharana Valavinyachya Maharashtra Rajyatil Dhoranancha va Amalbajavanicha Abhyas*. Pune: Resources and Livelihood Group Prayas.

Roth, D. and Vincent, L. (eds.) (2012). *Controlling the Water: Matching Technology and Institutions in Irrigation and Water Management in India and Nepal*. New Delhi: Oxford University Press.

Roth, D., Boelens, R. and Zwarteveen, M. (eds.) (2005). *Liquid Relations: Contested Water Rights and Legal Complexity*. New Brunswick, NJ: Rutgers University Press.

Schlosberg, D. (2004). Reconceiving environmental justice: Global movements and political theories. *Environmental Politics*, 13(3), 517–40.

Schroeder, R., St. Martin, K., Wilson, B. and Sen, D. (2008). Third World environmental justice. *Society & Natural Resources*, 21, 547–55.

Sen, A. (2009). *The Idea of Justice*. London: Allen Lane.

Sneddon, C. and Fox, C. (2008a). Struggles over dams as struggles for justice: The World Commission on Dams (WCD) and anti-dam campaigns in Thailand and Mozambique. *Society & Natural Resources*, 21(7), 625–40.

Sneddon, C. and Fox, C. (2008b). River-basin politics and the rise of ecological and transnational democracy in Southeast Asia and Southern Africa. *Water Alternatives*, 1(1), 66–88.

Sneddon, C., Harris, L., Dimitrov, L. and Ozesmi, U. (2002). Contested waters: Conflict, scale, and sustainability in aquatic socioecological systems. *Society & Natural Resources*, 15, 663–75.

SOPPECOM. (2008). *Water Rights as Women's Rights? Assessing the Scope for Women's Empowerment through Decentralised Water Governance in Maharashtra and Gujarat*. SOPPECOM-Utthan-TISS, supported by International Development Research Centre, www.soppecom.org/pdf/Water-rights-and-women's-rights-report.pdf.

SOPPECOM. (2010). *Assessing Social and Gender Equity in the Water Sector*. Pune: SOPPECOM/Gender and Water Alliance (GWA), www.soppecom.org/pdf/final_GEG1.pdf.

Srinivasan, V. and Kulkarni, S. (2014). Examining the emerging role of groundwater in water inequity in India. *Water International*, 39(2), 172–86.

Swyngedouw, E. (1999). Modernity and hybridity: Nature, *regeneracionismo*, and production of the Spanish waterscape, 1890–1930. *Annals of the Association of American Geographers*, 89(3), 443–65.

Swyngedouw, E. (2003). Modernity and the production of the Spanish waterscape 1890–1930. In K. Zimmerer and T. J. Bassett (eds.), *Political Ecology: An Integrative Approach to Geography and Environment-Development Studies*. New York: The Guildford Press, pp. 94–114.

Swyngedouw, E. and Heynen, N. (2003). Urban political ecology: Justice and the politics of scale. *Antipode*, 35(5), 898–918.

Vaidyanathan, A. (2005). Water policy in India: A brief overview. Paper prepared for the Centre for Public Policy, Indian Institute of Management, Bangalore.

Walby, S. (2011). Sen and the measurement of justice and capabilities: A problem in theory and practice. *Theory, Culture & Society*, 29(1), 99–118.

Walker, G. (2009a). Environmental justice and normative thinking. *Antipode*, 41 (1), 203–05.

Walker, G. (2009b). Beyond distribution and proximity: Exploring the multiple spatialities of environmental justice. *Antipode*, 41(4), 614–36.

Warner, J., Wester, P. and Bolding, A. (2008). Going with the flow: River basins as the natural units for water management? *Water Policy*, 10(suppl. 2), 121–38.

Wester, P. (2008). "Shedding the waters: Institutional change and water control in the Lerma-Chapala Basin, México." PhD Thesis, Wageningen University.

Williams, G. and Mawdsley, E. (2006). Postcolonial environmental justice: Government and governance in India. *Geoforum*, 37(5), 660–70.

Zwarteveen, M. (2015). "Regulating water, ordering society: Practices and politics of water governance." Inaugural Lecture, University of Amsterdam.

Zwarteveen, M., Roth, D. and Boelens, R. (2005). Water rights and legal pluralism: Beyond analysis and recognition. In D. Roth, R. Boelens and M. Zwarteveen (eds.), *Liquid Relations: Contested Water Rights and Legal Complexity*. New Brunswick, NJ: Rutgers University Press, pp. 254–68.

3
Water Grabbing: Practices of Contestation and Appropriation of Water Resources in the Context of Expanding Global Capital

Gert Jan Veldwisch, Jennifer Franco, and Lyla Mehta

3.1 Introduction

Over the past decade, much media, academic, and policy attention has focused on the rapid growth of large-scale land deals around the world (see Borras and Franco, 2010; Cotula, 2012; Cotula *et al.*, 2009; Deininger, 2011; De Schutter, 2011; GRAIN, 2008; Li, 2011; Oxfam, 2011; von Braun and Meinzen-Dick, 2009; White *et al.*, 2012; World Bank, 2010; Zoomers, 2010). The rush to acquire land as sources of alternative energy, food and feed crops, and environmental services led to the phenomenon popularly known as "land grabbing," which made global headlines and contributed to skyrocketing global food prices in 2008. Drawing on notions of "marginal," "waste," "vacant," "idle," and "unproductive" lands, powerful transnational and national actors moved into large-scale agriculture to take advantage of potential windfall gains in sub-sectors such as biofuels, "flex crops" (e.g. sugar cane, palm oil, maize, soya – see Borras *et al.*, 2012) and other major commodities (e.g. rice, wheat and other cash crops). Conservation and climate change mitigation measures also drove new demands for land (hence the notion of "green grabbing," cf. Fairhead *et al.*, 2012).

Headline attention to "land grabbing" also turned a spotlight (albeit comparatively weaker) on the implications for existing surface water and groundwater resources (hence the notion of "water grabbing," see Transnational Institute, 2014a). Early on, evidence suggested that in many cases the location of land grabbing was motivated also by the desire to capture water resources (Skinner and Cotula, 2011; Smaller and Mann, 2009; Woodhouse and Ganho, 2011). Although water is a potential constraint on large-scale agricultural projects, many land deal contracts did not explicitly mention water requirements (Woodhouse, 2012). Meanwhile, the land grabbed was rarely "marginal" but either already being used by small- and large-scale producers, or of prime quality and associated with irrigation facilities, or with the potential to acquire fresh water from river systems or aquifers; in arid areas, land is plentiful, and agricultural expansion will not create conflict until water is provided. This raised the crucial question of whether this water is truly "available," unsustainably withdrawn (ultimately undermining the land's quality), or unequally reallocated (away from existing users).

Although initial media and policy attention associated land grabbing almost exclusively with large-scale food, feed and fuel crops, later research and advocacy moved beyond this limited focus on agriculture-driven resource-grabbing. Studying water grabbing, among others, contributed to widening the analytical lens. Water grabbing is now seen in relation to a much wider range of activity that spans the food, water, energy, climate and mineral domains. In many cases, water itself is the object of grabbing, for agriculture and for purposes such as mining or hydropower development.

The literature on water grabbing shows that water's fluid nature, its fluctuating variability across time and space, and multiple scales (upstream, downstream, across the watershed) have tremendous impacts on water allocation, reallocation, distribution and quality both now and in the future (e.g. Birkenholtz, 2016; De Bont *et al.*, 2016; Duvail *et al.*, 2012; Hertzog, 2012; Islar, 2012; Matthews 2012; Mehta *et al.*, 2012; Sosa and Zwarteveen, 2012; Van Eeden *et al.*, 2016). Hydrologic complexity, in particular surface water/groundwater interactions and inter-annual variability, can obscure how reallocation happens and with what impacts on the environment and diverse social groups. The fluid properties of water interact with the "slippery" nature of grabbing processes: unequal power relations; fuzzy distinctions between legality/illegality and formal/informal rights; unclear administrative boundaries and jurisdictions, fraught with negotiation processes. All these factors combined with water's powerful material, discursive and symbolic characteristics make "water grabbing" a site for conflicts with potential drastic impacts on current and future uses and benefits of water. This chapter reiterates and updates our understanding of resource-grabbing, particularly water grabbing, its narratives and drivers, its mechanisms and facilitator, its impacts on local people and ecosystems, and how resistance is reversing some of these trends. This heavily builds on our earlier publications on the topic (Franco *et al.*, 2013 and Mehta *et al.*, 2012) but also reflects on developments and publications since then. We conclude with implications for policy and practice.

3.2 Grabbing Resources, Control and Attention

Before turning to the specific characteristics of water grabbing, we first discuss resource-grabbing more broadly. "Grabbing" is a highly political term, used to draw attention toward current injustices. If not for this word, the current phase of resource-grabbing might well have remained largely "invisible." Attention to resource-grabbing started with booming large-scale land deals in Sub-Saharan Africa observed during the 2007–08 food price crisis, when global investment money was looking for alternative high returns despite the global economic crisis. Much early attention, fueled by headline-grabbing reports and often presented and portrayed as authoritative, was later found to be limited by questionable definitions, flawed assumptions, "shaky data sets" and interpretive choices (see Scoones *et al.*, 2013). To illustrate, taking some of the most prominent studies of the time, including the World Bank's Deininger *et al.* (2011), the International Land Coalition (2011) and Oxfam (2011), close examination has shown major data discrepancies and how "[o]ften, on the basis of the same, limited sources, very different conclusions are drawn" (Scoones *et al.*, 2013: 4730). Indeed, large-scale land deals were increasingly seen as only the tip of the iceberg of a much broader, far-reaching restructuring of access and control over natural resources and the benefits from their use.

Water grabbing, green grabbing and ocean grabbing are examples of concepts that were introduced to draw attention to similar processes of resource and control capture. The work on water grabbing started by looking at the water implications of large-scale land deals, but soon also started to include analyses of extractive industries, drinking water privatizations, hydropower projects and territorial politics of basin management plans.

Originating as terms to mobilize attention and political response, "land and water-grabbing" are not always clearly defined analytically. Resource-grabbing broadly refers to appropriation of natural resources in general, including land and water, and control over their associated uses and benefits, with or without transfer of ownership, usually from poor and marginalized to powerful actors (Fairhead *et al.*, 2012). It is not surprising that most critical analysts now working on land grabbing draw on political economy and Marxist traditions, particularly David Harvey's notion of "accumulation by dispossession" (2003). Building on this, Borras *et al.* (2012) see three defining features of land grabbing that are worth considering as a backdrop to thinking about water grabbing.

First, land grabbing is ultimately "control grabbing," or capturing the power to control land and other associated resources such as water, and how they are used, in order to corner the benefits, a point that builds on Ribot and Peluso's (2003) theory of access. The ultimate aim seems to be "value capture," for which resource control is useful but not essential. The need to see beyond enclosures includes analysis of a much broader field of investment and extraction models (Vermeulen and Cotula, 2010). Beyond very visible land deals, other corporate investment models in agriculture lead to similar transformations (Hall, 2011; White *et al.*, 2012). Veldwisch (2015), for instance, shows how contract farming in the 30,000 ha Chókwè public irrigation scheme (Mozambique) led to de facto corporate control over most of the irrigated production and most of the value created, while costs and risks were left to individual producers. Control-grabbing is perhaps best viewed as a contingent process, marked variously by conflict, negotiation and friction, that can end up ratifying an existing balance of power among state and non-state actors in steps along the way, even if only temporarily. Although poor people often lose out, under certain conditions, their political action can make a difference, however small.

Second, today's land grabbing is also defined by scale – both in land area and the capital involved. Literature on land grabbing tends to view land grabbing mainly in terms of the physical size of the land (see, for example, Anseeuw *et al.*, 2011; Oxfam, 2011; World Bank, 2010), but incorporating the amount of capital makes land and other key resources such as water central to the analysis. Focusing on land can overlook the underlying logic of capital operations and diversity of biophysical requirements in capital-accumulation dynamics.

Third, current land grabbing, according to Borras *et al.* (2012), distinctly occurs within capital-accumulation strategies, largely in response to multiple converging crises: food, energy/fuel, climate change, financial crises (driving financial capital to look for new, safer investment opportunities) (McMichael, 2012), and to emerging needs for resources by newer hubs of global capital, especially Brazil, Russia, India, China and South Africa ("BRICS") and some powerful middle-income countries (MICs). Therefore, key contexts for land grabbing include: food security concerns, energy/fuel security interests, climate change mitigation strategies, demands for natural resources by new centers of capital, and demand for flex crops, i.e. crops that have multiple uses (food, feed, fuel, industrial material) that can be easily and flexibly interchanged: soya (feed, food, biodiesel), sugar cane (food, ethanol), oil palm (food, biodiesel, commercial/industrial uses), and corn (food, feed, ethanol).

As the body of empirical research on global land grabbing has grown, several other important realizations have surfaced. What was initially seen as a phenomenon driven in the global South, by only certain states in the North, has now been recognized as a much larger, broader phenomenon involving capital from the North, South, East and West, from wealthy and middle-income states, and involving many intra-regional transactions and acquisitions by national elites as well (Cotula, 2012; Kenney-Lazar, 2012; Visser *et al.*, 2012). Control is captured in diverse contexts involving varying degrees of physical violence, clientelist coercion, transparency, or formal legality. Rights range from statutory to customary land tenure rights, which may or may not be recognized by central state law, and are often very context-specific. It has also become clear that the supposed benefits of big land deals very often fail to materialize either as claimed, or at all.

Looking at these key contexts through a water lens may help both deepen and broaden the understanding of how they each operate, and make the truly far-reaching water impacts of both land- *and* water grabbing more visible. As Woodhouse (2012) explains, it is difficult (if not impossible) to grab land without grabbing water. There is a perceived general pattern that investors do not seek land that does not have water for production in the first place.

3.3 What Is Water Grabbing?

Water grabbing is a process in which powerful actors take control of, or reallocate to their own benefit, water resources used by local communities or which feed aquatic ecosystems on which communities base their livelihoods. It is one manifestation of a broader global trend involving large-scale (re)allocations of natural resources more generally. Drawing insight from the discussion on land grabbing, we understand water grabbing as capturing control not just of water itself, but also of the power to decide how it will be used – by whom, when, for how long and for what purposes – in order to control the benefits. The large body of case material on land grabbing demonstrates a broader contemporary trend to further expand large-scale capitalist control over natural resources for purposes of production, extraction and speculation.

As many analysts and observers have noted, powerful actors have been capturing land and water resources for centuries. The current cycle is what we refer to as land and water grabbing.

For water grabbing, the conventional fixation on physical size (mentioned earlier with respect to land grabbing) has a parallel in too narrow a focus on the volume of water involved, ignoring the fact that water access, impacts and implications of water grabbing concern distribution in time and space. For instance, Rulli *et al.* (2013) is an influential work that completely ignores the intricacies of local water distribution. The Hertzog *et al.* (2012) study of water grabbing in the Office du Niger demonstrates how important it is to thoroughly assess water requirements in space and time, rather than just looking at water volumes. The notion of scale is also relevant to water flows, to highlight and account more systematically for changes in water distribution and quality.

Our approach to water grabbing complements the work of Borras *et al.* (2011) and likewise combines political economy, political ecology and political sociology. It seeks to move beyond narrow, proceduralist mainstream understandings of "grabbing" as illegal by definition, with the disadvantage of emphasizing formal legality and examining grabbing only when state law is clearly contravened. This poses at least three problems. First, it glosses over the actual nature and desirability of outcomes of these "transactions" in terms of the underlying development model that the new economic arrangements usher in. Legality is not equivalent to legitimacy. Political responses to current resource-grabbing include, for example, more "corporate social responsibility"-framed moves to obtain "social licenses" for projects to minimize "reputational risk," versus more resistance-type efforts by affected villagers in targeted areas to defend traditional territory and their customary tenure, livelihood and self-provisioning practices. Many policy responses to resource-grabbing rest on the dubious assumption of basic, widespread agreement on the need and desirability of extending the dominant large-scale industrial development model.

Next, it tends to reduce the transaction itself to a technical formal-legal procedure, at times even conflating financial accounting with political accountability. This underestimates (or ignores) how natural resources are grabbed in a historical–institutional field that is pluri-legal, marked by power asymmetries and thus deeply political. Further, formal water and land management are often separated from each other – an institutional void that makes encroachment easier, because separating land and water rights can make room for water grabbing (Williams *et al.*, 2012).

Stepping back, one finds that water grabbing (like land grabbing) is diverse in appearance: 1) driven by varied forms of state–capital alliances; 2) not limited geographically; 3) happening in diverse agro-ecological contexts; 4) unfolding across various water–land property rights regimes; and 5) leading to diverse impacts. Each of these points is elaborated below using water grabbing-focused case study material.

First, the main actors behind diverse grabbing processes are varied forms of state–capital alliances, involving varied types of mechanisms enabling the grabs – among others, state law and policy reforms (Islar, 2012; Wagle *et al.*, 2012); state law and new policy interpretations (Bossio *et al.*, 2012; Bues and Theesfeld, 2012); violation of state law (Duvail *et al.*, 2012; Matthews 2012); new public–private interest business coalitions (Sosa and Zwarteveen 2012; Velez Torres, 2012; Wagle *et al.*, 2012); exploiting legal complexity (Hertzog *et al.*, 2012; Williams *et al.*, 2012); and bypassing democratic accountability processes (Matthews, 2012).

Second, water grabbing, like land grabbing, is happening around the globe. Many of the most prominent reports and studies, including those by Skinner and Cotula (2011), Woodhouse (2012) and Woodhouse and Ganho (2011), have tended to focus initially on water grabbing in Africa, reinforcing the impression (cultivated in the media) that it was mainly an African phenomenon. But empirical evidence shows it unfolded throughout Latin America (Sosa and Zwarteveen, 2012; Velez Torres, 2012); across Asia (Matthews, 2012; Wagle *et al.*, 2012), in the Middle East and in Eurasia as well (Gasteyer *et al.*, 2012; Islar, 2012).

Third, water grabbing is also happening across various agro-ecological contexts: river deltas and floodplains, inland rivers, freshwater lakes, wetlands, as well as semi-arid plains and savannahs.

Fourth, water grabbing, like land grabbing, is happening across diverse property rights regimes, including commons such as grazing corridors, as in the Tana Delta (Duvail *et al.*, 2012); communal/community tenure and resource management systems (Beekman and Veldwisch, 2012; Duvail *et al.*, 2012; Matthews, 2012; William *et al.*, 2012); land- and waterscapes understood by local communities as territory (Gasteyer *et al.*, 2012; Velez Torres, 2012); and areas under individual private property-rights regimes (Houdret, 2012; Sosa and Zwarteveen, 2012).

Finally, the impacts of water grabbing are diverse, distinguished as two broad types: exclusion and adverse incorporation (Borras *et al.*, 2013). However, water grabbing and its impacts appear to be even more diverse and "slippery" because of their dislocated, timing-relevant and quality-related effects. Interventions in the water cycle can, for instance, 1) disturb the amount of groundwater and downstream water available for existing users (exclusion from volume); 2) change peak and base flows (exclusion in timing); 3) change the agro-ecological landscape (exclusion from ecosystem benefits that require, for example, occasional flooding); and 4) affect water quality (exclusion from clean, safe water). In this last case, water grabbing does not necessarily divert water, but rather powerful upstream actors pollute water resources, externalizing problems and costs (from the polluters to local communities downstream) (Arduino *et al.*, 2012). "Watery" exclusion can also involve adverse incorporation – imposing water use and management regimes that directly or indirectly "incorporate" people into changed water regimes tied to new economic arrangements.

3.4 Governance and Resistance

Elsewhere (Franco *et al.*, 2013), we have demonstrated how disparate, seemingly isolated global processes have led to domination by neoliberal discourses and trends. In the field of water management, the Dublin Conference became a watershed moment where neoliberal economic policies of water privatization and commodification overtook principles that had been more prominent in earlier processes, such as the UN global consultation (New Delhi, 1990), which emphasized equity and universality. The Dublin Declaration and its popularity reflect the dominant Washington Consensus of the 1990s, which also influenced environmental governance. In water management, this has led to clear neoliberal tendencies, elaborated alongside policies of integration, participation, water rights formalization and basin management. After privatizing water services, water itself is now being privatized and financialized. Van Eeden *et al.* (2016) show how integrated water resource management (IWRM), promoted globally as the governance framework to manage water resources efficiently, equitably and sustainably, can also directly or indirectly facilitate water grabbing. They explore how IWRM manages competing interests and the diverse priorities of large and small water users amidst foreign direct investment. By focusing on two commercial sugar companies operating in the Wami-Ruvu River Basin in Tanzania and their impacts on surrounding villages' water and land rights, the article shows how powerful actors, seeking to satisfy their own interests, can exploit institutional and capacity weaknesses of IWRM implementation, allowing water grabbing. Like IWRM policies, with their 20-year roots, new trends, such as the food–energy–water nexus and PPPs, are reinforcing a call for integrated governance while legitimizing increased corporate involvement. Fuzzy, ambiguous global water and land governance are thus increasing local uncertainties and complexities. Usually powerful players can navigate their way through such uncertainties, which makes them into mechanisms excluding poor, marginalized people. For less powerful players, resolving ambiguities in conflicting regulatory frameworks may require tipping the balance toward the frameworks favoring them.

In another recent publication, De Bont *et al.* (2016) show that IWRM-inspired governance fora and water's fluidity can also be navigated by the less powerful and be used to defend their access to water for productive uses. Their analysis of an area along Nduruma River, part of the Pangani Basin in Tanzania, shows the continuing struggle between corporate farms and irrigating peasant communities both upstream and downstream of these export-oriented businesses. They used their bargaining power in river committees and water users' associations (WUAs) to prevent agro-export companies from abstracting surface water. In response, companies reverted to pumping groundwater, which also reduces water availability downstream. Grabbers use water's fluidity (and often invisibility) to obscure abstraction, but this can also be used to revert flows to peasant communities. Findings by De Bont *et al.* (ibid.) highlight water grabbing as an ongoing process: grabs need to be continuously sustained or they can be continuously contested.

Many dynamic initiatives have emerged around the world to resist and reverse the trends of neoliberal environmental governance and capitalist accumulation in the cases described above, including water justice movements focused on resisting and reversing privatization of urban water management and deliveries infrastructure through grassroots "re-municipalization" campaigns and grassroots driven public–public partnerships (see

TNI, 2014a). Official recognition by the UN General Assembly in July 2010 and by the UN Human Rights Council a few months later was the culmination of a major effort by water justice movement activists to secure recognition of access to clean water and sanitation as a human right. Meanwhile, anti-fracking and anti-dam movements and campaigns are fundamental to resistance, alongside local and national initiatives by small-scale and artisanal fishers and fisher communities across the globe to organize collective campaigns resisting privatization of ocean spaces and coastal and aquatic resources (TNI, 2014b; TNI, 2016). Many diverse alternatives are being proposed and constructed at the margins. In Colombia, local inhabitants, social organizations, and several scholars are calling for the need to reinforce alternative local–global linkages in order to protect their territories and work towards a Colombian "Paz-ific" (*Paz* is "peace" in Spanish) (Vélez-Torres, 2012). In India, many local communities constantly resist forced displacement by special economic zones and hydropower projects, achieving time and cost overruns, stay orders by courts and, in some cases, project termination. In Morocco, Houdret (2012) describes how the Arab Spring has allowed previously marginalized farmers new opportunities to regain control over water, their livelihoods and potentially some (political) power. Thus, movements protesting grabbing are providing us with new tools to counter some mechanisms of grabbing. Engaged scholars are also helping create new vocabularies and imaginaries that can challenge dominant narratives justifying such appropriations. Indeed, the recent flurry of scholarship on grabbing reveals that even so-called marginal land is highly productive for those who live there, so narratives of abundant, under-utilized water must be reconsidered.

3.5 Conclusions

This chapter has focused on the missing water dimensions in debates on the global land rush and demonstrated how, in many cases, water itself is an object of grabbing, to broaden resource-grabbing beyond merely agriculture. Water's fluid nature, demand and availability fluctuate in time and space, making it difficult to precisely characterize grabbing, appropriation and reallocation and their varied impacts across multiple scales and timeframes. In many cases, water is not the aim of grabbing, but rather instrumental for achieving underlying objectives of controlling revenues; water often serves a particular purpose of producing value for which also other resources are needed: land particularly, but also reordering labor/jobs, value chains and so on. These combine with blurred, obscure, fragmented negotiation, interpretation and enforcement of policies and laws, to make water grabbing a highly slippery process indeed.

While grabbing often involves disregarding or outright dismantling customary and/or even statutory land rights, violating basic human rights and generating harmful social and environmental effects, mainstream policy responses have tended to focus on minimizing these "risks" by creating mechanisms to apply international standards via a code of conduct (cf. von Braun and Meinzen-Dick, 2009), or principles of responsible agricultural investments (e.g. World Bank, 2010), or initiatives for improved transparency and information disclosure (e.g. Global Witness *et al.*, 2012). Such approaches partly assume optimistic benefits to be gained from these large-scale investments.

However, there is little evidence regarding actual social or environmental benefits or that such initiatives would work in practice. It is highly questionable whether such codes, principles or initiatives can actually work in such a charged, unequal playing field. Indeed, water and other forms of resource-grabbing question broader development and economic growth paradigms, highlighting the need to limit unfettered resource extraction, capital flows and gross injustices borne by those hardest hit when their landscapes and waterscapes are appropriated.

References

Anseeuw, W., Wily, L. A., Cotula, L. and Taylor, M. (2011). *Land Rights and the Rush for Land: Findings of the Global Commercial Pressures on Land Research Project*. Rome: The International Land Coalition.

Arduino, S., Colombo, G., Ocampo, O. M. and Panzeri, L. (2012). Contamination of community potable water from land grabbing: A case study from rural Tanzania. *Water Alternatives*, 5(2), 344–59.

Beekman, W. and Veldwisch, G. J. (2012). The evolution of the land struggle for smallholder irrigated rice production in Nante, Mozambique. *Physics and Chemistry of the Earth, Parts A/B/C*, 50, 179–84.

Birkenholtz, T. (2016). Dispossessing irrigators: Water-grabbing, supply-side growth and farmer resistance in India. *Geoforum*, 69, 94–105.

Borras, Jr., S. and Franco, J. (2010). From threat to opportunity? Problems with the idea of a "code of conduct" for land-grabbing. *Yale Human Rights and Development Law Journal*, 13(2), 507–23.

Borras, Jr., S., Fig, D. and Suárez, S. (2011). The politics of agrofuels and mega-land and water deals: Insights from the ProCana case, Mozambique. *Review of African Political Economy*, 38(128), 215–34.

Borras Jr., S. M., Franco, J. C. and Wang, C. (2013). The challenge of global governance of land-grabbing: Changing international agricultural context and competing political views and strategies. *Globalizations*, 10(1), 161–79.

Borras, Jr., S., Franco, J., Gomez, S., Kay, C. and Spoor, M. 2012. Land-grabbing in Latin America and the Caribbean. *Journal of Peasant Studies*, 39(3–4), 845–72.

Bossio, D., Erkossa, T., Dile, Y., McCartney, M., Killiches, F. and Hoff, H. (2012). Water implications of foreign direct investment in Ethiopia's agricultural sector. *Water Alternatives*, 5(2), 223–42.

Bues, A. and Theesfeld, I. (2012). Water-grabbing and the role of power: Shifting water governance in the light of agricultural foreign direct investment. *Water Alternatives*, 5(2), 266–83.

Cotula, L. 2012. The international political economy of the global land rush: A critical appraisal of trends, scale, geography and drivers. *Journal of Peasant Studies,* 39(3–4), 649–80.

Cotula, L., Vermeulen, S., Leonard, R. and Keeley, J. (2009). *Land-grab or Development Opportunity? Agricultural Investment and International Land Deals in Africa*.

London/Rome: IIED (International Institution for Environment and Development)/ FAO (United Nations Food and Agriculture Organisation)/IFAD (International Fund for Agricultural Development).

De Bont, C., Veldwisch, G. J., Komakech, H. C. and Vos, J. (2016). The fluid nature of water-grabbing: The on-going contestation of water distribution between peasants and agribusinesses in Nduruma, Tanzania. *Agriculture and Human Values*, 33(3), 641–54.

Deininger, K. 2011. Forum on global land-grabbing: Challenges posed by the new wave of farmland investment. *Journal of Peasant Studies*, 38(2), 217–47.

De Schutter, O. 2011. Forum on global land-grabbing: How not to think of land-grabbing: Three critiques of large-scale investments in farmland. *Journal of Peasant Studies* 38(2), 249–79.

Duvail, S., Médard, C., Hamerlynck, O. and Nyingi, D. W. (2012). Land and water-grabbing in an East African coastal wetland: The case of the Tana delta. *Water Alternatives*, 5(2), 322–43.

Fairhead, J., Leach, M. and Scoones, I. (2012). Green grabbing: A new appropriation of nature? *Journal of Peasant Studies*, 39(2), 237–61.

Franco, J., Mehta, L. and Veldwisch, G. J. (2013). The global politics of water-grabbing. *Third World Quarterly*, 34(9), 1651–75.

Gasteyer, S., Isaac, J., Hillal, J. and Hodali, K. (2012). Water-grabbing in colonial perspective: Land and water in Israel/Palestine. *Water Alternatives*, 5(2), 450–68.

Global Witness, International Land Coalition and Oakland Institute. (2012). *Dealing with Disclosure: Improving Transparency in Decision-Making over Large-Scale Land Acquisitions, Allocations and Investments*. London: Global Witness; Rome: ILC; Oakland: Oakland Institute.

GRAIN. (2008). *Seized: The 2008 Land-Grab for Food and Financial Security*. Barcelona: GRAIN.

Hall, R. (2011). Land-grabbing in Southern Africa: The many faces of the investor rush. *Review of African Political Economy,* 38(128), 193–214.

Harvey, D. (2003). *The New Imperialism*. Oxford: Oxford University Press.

Hertzog, T., Adamczewski, A., Molle, F., Poussin, J. C. and Jamin, J. Y. (2012). Ostrich-like strategies in sahelian sands? Land and water-grabbing in the Office du Niger, Mali. *Water Alternatives*, 5(2), 304–21.

ILC (International Land Coalition). (2011). *Land Rights and the Rush for Land: A Report*. Rome: ILC. InvestAg Savills, International farmland market bulletin [online], www.investag.co.uk/Bulletin2011.pdf.

Islar, M. (2012). Privatised hydropower development in Turkey: A case of water-grabbing? *Water Alternatives*, 5(2), 376–91.

Kenney-Lazar, M. 2012. Plantation rubber, land-grabbing and social-property transformation in southern Laos. *Journal of Peasant Studies*, 39(3–4), 1017–37.

Li, T. M. (2011). Forum on global land-grabbing: Centering labor in the land-grab debate. *Journal of Peasant Studies*, 38(2), 281–98.

Matthews, N. (2012). Water-grabbing in the Mekong basin: An analysis of the winners and losers of Thailand's hydropower development in Lao PDR. *Water Alternatives*, 5(2), 392–411.

McMichael, P. (2012). The land-grab and corporate food regime restructuring. *Journal of Peasant Studies*, 39(3–4), 681–701.

Mehta, L., Veldwisch, G. J. and Franco, J. (2012). Introduction to the Special Issue: Water-grabbing? Focus on the (re) appropriation of finite water resources. *Water Alternatives*, 5(2), 193–207.

Oxfam. (2011). *Land and Power: The Growing Scandal Surrounding the New Wave of Investments in Land.* Oxfam International Briefing Paper No. 51. Oxford: Oxfam International.

Ribot, J. and Peluso, N. (2003). A theory of access. *Rural Sociology*, 68(2), 153–81.

Rulli, M. C., Saviori, A. and D'Odorico, P. (2013). Global land and water-grabbing. *Proceedings of the National Academy of Sciences*, 110(3), 892–97.

Skinner, J. and Cotula, L. (2011). *Are Land Deals Driving "Water-Grabs"? Briefing: The Global Land Rush.* London: International Institute for Environment and Development (IIED), http://pubs.iied.org/17102IIED.

Smaller, C. and Mann, H. (2009). *A Thirst for Distant Lands: Foreign Investment in Agricultural Land and Water.* Foreign Investment for Sustainable Development Program. Winnipeg, Canada: International Institute for Sustainable Development (IISD).

Sosa, M. and Zwarteveen, M. (2012). Exploring the politics of water-grabbing: The case of large mining operations in the Peruvian Andes. *Water Alternatives*, 5(2), 360–75.

TNI. (2014a). *The Global Water-grab: A Primer.* Amsterdam: Transnational Institute.

TNI. (2014b). *The Global Ocean Grab: A Primer.* Amsterdam: Transnational Institute.

TNI. (2016). *Human Rights vs. Property Rights: Implementation and Interpretation of the SSF Guidelines.* Amsterdam: Transnational Institute.

Van Eeden, A., Mehta, L. and van Koppen, B. (2016). Whose waters? Large-scale agricultural development and water-grabbing in the Wami-Ruvu River Basin, Tanzania. *Water Alternatives*, 9(3), 608–26.

Veldwisch, G. J. (2015). Contract farming and the reorganisation of agricultural production within the Chókwè Irrigation System, Mozambique. *Journal of Peasant Studies*, 42(5), 1003–28.

Velez Torres, I. V. (2012). Water-grabbing in the Cauca basin: The capitalist exploitation of water and dispossession of afro-descendant communities. *Water Alternatives*, 5(2), 421–49.

Vermeulen, S. and Cotula, L. (2010). *Most of Agricultural Investment: A Survey of Business Models that Provide Opportunities for Smallholders.* London: IIED.

Visser, O., Mamanova, N. and Spoor, M. (2012). Oligarchs, mega-farms and land reserves: Understanding land-grabbing in Russia. *Journal of Peasant Studies*, 39(3–4), 899–931.

Von Braun, J. and Meinzen-Dick, R. (2009). *"Land-grabbing" by foreign investors in developing countries: Risks and opportunities.* IFPRI Policy Brief No. 13. Washington, DC: International Food Policy Research Institute.

Wagle, S., Warghade, S. and Sathe, M. (2012). Exploiting policy obscurity for legalising water-grabbing in the era of economic reform: The case of Maharashtra, India. *Water Alternatives*, 5(2), 412–30.

Williams, T. O., Gyampoh, B., Kizito, F. and Namara, R. (2012). Water implications of large-scale land acquisitions in Ghana. *Water Alternatives*, 5(2), 243–65.

White, B., Borras, Jr., S., Hall, R., Scoones, I. and Wolford, W. (2012). The new enclosures: Critical perspectives on corporate land deals. *Journal of Peasant Studies*, 39(3–4), 619–47.

Woodhouse, P. (2012). New investment, old challenges: Land deals and the water constraint in African agriculture. *Journal of Peasant Studies*, 39(3–4), 777–94.

Woodhouse, P. and Ganho, A.-S. (2011). Is water the hidden agenda of agricultural land acquisition in sub-Saharan Africa? International Conference on Global Land Grabbing, Institute of Development Studies and Future Agricultures Consortium, University of Sussex, UK, April 6–8, 2011.

World Bank. (2010). *Rising Global Interest in Farmland: Can It Yield Sustainable and Equitable Benefits?* Washington, DC: World Bank.

World Bank. (2011). *Rising Global Interest in Farmland; and Oxfam, Land and Power: The Growing Scandal Surrounding the New Wave of Investments in Land.* Oxford: Oxfam-International.

Zoomers, A. (2010). Globalisation and the foreignisation of space: Seven processes driving the current global land-grab. *Journal of Peasant Studies*, 37(2), 429–47.

4

De-Politicized Policy Analysis: How the Prevailing Frameworks of Analysis Slight Equity in Water Governance

Andrea K. Gerlak and Helen Ingram

4.1 Introduction: Policy Analysis for "Better" Public Policy

There is a common article of faith in the multi-disciplinary field of policy analysis: a belief that the quality of policy-making increases in proportion to available policy knowledge, and that the policy analyst's role is to generate and transmit relevant policy information and evaluation. What Alice Rivlin (1984: 18–19) wrote more than 30 years ago is even truer today, particularly when it comes to water policy: "No debate on any serious issue ... takes place without somebody citing a public policy study." Information is not neutral, however, and the kind of information gathered by policy analysts depends largely on the frameworks they adopt and the underlying perspectives and values.

The most prominent policy frameworks currently applied to water reflect the context and circumstances from when they were developed, and inventors' perspectives and values. Frameworks have evolved over time as users apply and modify them. We argue that none of the most common schemas were developed to reflect equity and public participation in decision-making. While improvements have been made, and newer critical perspectives hold promise, the kind of information provided by policy analysis continues to slight the concerns of equity and participation set out by the editors in the introductory chapter to this book.

This chapter will examine four different policy approaches and their impact on water governance: (1) efficiency-based analysis; (2) institutional analysis and development; (3) physically-based watershed and river basin approaches; and (4) discursive policy analysis. Each of these approaches developed first in the United States and spread elsewhere, often becoming greatly modified as they disseminated. In each case, we will examine the context and key influences in which each approach was invented, central concepts and underlying theoretical logic, how the approach has evolved, and an overall assessment. The conclusion will indicate how policy analysis must change to better serve equity and participation.

4.2 The Gospel of Efficiency and Moving Water to its Best Use

The progressive conservation movement of the early twentieth century placed water at the heart of the new doctrine championed by reformers. The period's authoritative historian, Samuel P. Hays (1959) wrote that the conservation movement had little to do with popular support or the people's wishes. At its core, conservation was pro-development: wise use and preventing waste. Most important, the conservation ideal depended upon experts, and in the case of water, hydrologists in the U.S. Geological Service. According to Hays, "In all Federal Programs, the Survey argued that all possible uses of water should be considered, so that rivers could produce the greatest possible benefit to man. Multiple purpose development ... arose directly from the experiences and ideas of these new hydrographers in the Geological Survey" (Hays, 1959: 9). Multi-purpose development included navigation, flood control, power production, recreation, irrigation, and other purposes. While water projects might be dedicated to regional economic development in areas such as the Columbia River Basin and the Tennessee Valley Authority, they did not target the disadvantaged in those areas. Significantly, conservation leaders, such as Fredrick Newlands, hoped that expert knowledge, notably hydrology and planning, rather than Congressional log-rolling, would determine where federal projects would be built. Newlands, who became the Irrigation Service's first Director, hoped to make the deserts bloom (Maas and Anderson, 1978). Dusty, vacant lands were to be transformed into thriving, democratic communities where people could make a good living, with time and resources to engage in public affairs. While eligibility rules restricted benefits to small farmers, those with no more than 160 acres, recipients had no voice in project selection, a job left to specialists.

While the expertise embodied in the gospel of efficiency in the progressive era was hydrology, economics became the queen of policy evaluation in the post-war period. Professor Pigou's book, *The Economics of Welfare* intellectually underpinned evaluation, calculating national economic efficiency: the benefits of a program or project (to whomsoever they accrue) must exceed its estimated cost (Caulfield, 2011). Called benefit/cost analysis, throughout the 1950s, budgeters in the White House insisted that the ratio be at least one-to-one for a project supported by the administration, although members of Congress complained bitterly that it was essentially a 'no new starts' for water project policy. In 1962, the national economic efficiency criterion was broadened to include multiple objectives, sparking argument about what to include in planning and policy-making.

In the lengthy scholarly debate over multiple, objective standards to reflect changing values in the 1960s, three additional accounts were added to national economic efficiency: environment or preservation; social wellbeing; and regional wellbeing (Michalson *et al.*, 1975). Regional wellbeing was rather quickly downgraded as an objective because many economists and others felt benefits to some regions were offset by losses in other regions and should not be counted. For instance, subsidized water development to grow cotton in Arizona came at the expense of profits for Southern cotton farmers. The social wellbeing account suffered a similar fate, mainly at the behest of the Office of Management and Budget. They argued that the data bases used were weak, and criteria were not mutually exclusive from other accounts, leading to double-counting. Further, even when water projects seemed to have positive effects on social wellbeing, critics questioned whether spending federal funds on water resources was the best way to achieve social objectives (Michalson *et al.*, 1975: 11). Ultimately, only the environment account and the national economic development account were officially endorsed, and only national economic efficiency really mattered. The bias was toward easily quantifiable consequences, and whatever the other effects of proposed projects, a favorable benefit-cost estimate was fundamental for federal agency backing.

4.2.1 Central Concepts and Theoretical Logic

Efficiency and moving water to its highest economic use assumes water is an economic commodity. Even when considering the benefits water might provide for regions or social welfare, benefits were to be measured by enhanced income, and strongly biased toward quantifiable analysis. Less tangible consequences, such as stable communities, a sense of security and opportunity, were not part of the calculation. People were clearly privileged over the rest of the ecosystem, and degradation counted only when it affected humans. From the beginning, expert information was privileged over other kinds of knowledge, increasingly from agency economists. Underlying the gospel of efficiency was the effort to keep politics out of water-resource decision-making. Progressive conservationists blamed politics, including private economic natural resource developers, for degradation and loss of precious resources. The assumption was that experts, applying broad efficiency criteria, best serve the public.

4.2.2 Evolution and New Applications

As many cost-conscious critics of conventional water development, particularly economists, increasingly questioned politically inspired and federally funded water projects, water pricing and rates became more important. Water tariffs to recover water-supply and sanitation operational and management costs were conventionally applied to consumers for generations (Lemos, 2008). Water costs did not reflect scarcity. Long-term low-cost contracts between federal agencies delivering water were set at very low rates. Similarly, municipal water utilities often charged high-volume users less for water. Rates and tariffs were re-purposed in the last quarter of the twentieth century to promote conservation and move water to higher economic value uses. Economists reasoned that, if users are forced to pay more for a commodity that they have customarily been able to access free of charge, or nearly free, they will be more likely to conserve it and make it available for others. However, the way pricing schemes are negotiated, set and implemented profoundly impacts distribution equity, access and costs to different users (Lemos, 2008). Equity and democracy in setting water tariffs and rates tend to be slighted in a policy analysis structure that portrays water as a commodity.

Life scientists and environmental economists chafe at the gospel of efficiency's narrow definitions of benefits and costs. They note that natural ecosystems provide goods and services at no immediate cost, and consequently their value is often overlooked. Further, water quantity and quality in lakes, rivers, streams and wetlands often depend on distant terrestrial ecosystems. When interconnected ecosystems are degraded, these benefits and services are threatened and are sometimes impossible to replace (Constanza *et al.*, 1997; Safriel, 2011). Methods to evaluate the broad benefits of water-related ecosystem services, including non-material benefits obtained from ecosystems (spiritual, inspirational, aesthetic, educational and recreational), have been developed (Safriel, 2011).

Ecological service analysis has considerably improved public appreciation of nature's broad contributions to a sustainable environment. Changing efficiency calculations has had less success. According to Gretchen Daily (1997: 1–10), introducing a book on natural ecosystem services, there is no absolute value of ecosystems services waiting to be discovered. However, an assessment of their magnitude is a prerequisite to incorporate them into the dominant analytical framework.

It entails real sacrifices to quantify the value of water-related ecosystem services in dollars. The bias remains towards easily quantifiable benefits. Economic measurement reinforces the view of water as a commodity that can be reallocated according to willingness and ability to pay. Monetizing nature, or what George Monbiot (2014) has termed chopping up nature and turning it into money, does not deal with how power subverts representation and equity. Allowing ecosystems not only to be priced but also bought and sold introduces yet more problems. Determining who has a legitimate right to sell means picking winners and losers. In developing countries, indigenous communities may lack the documentation or political clout to assert their ownership (Conniff, 2012).

Another recent evolution in calculating water efficiency concentrates on costs in terms of quantities. Tony Allan developed the notion of virtual water to describe how arid countries satisfy their annual food demands by importing substantial amounts of grain and other products (Allan, 1996). Water footprints were introduced somewhat later to describe direct and indirect water use by consumers and producers of particular products (Chapagain and Hoekstra, 2004; Hoekstra and Chapagain, 2007). Both virtual water and water footprints can be useful devices to describe the water burden of particular water uses. However, using these concepts for efficiency analytics is dangerous and misguided. Wichelns (2015) argues that these concepts contain too little information to be used to inform policy or trade issues. Water footprints are not appropriate to determine optimal use of land and water resources, select investment strategies, or allocate water among competing uses. For instance, water use in arid lands may have great cultural value not captured by these concepts, and there may be no other viable sources of employment besides agriculture. Trade issues can be based on the quantities and availability of water consumed by trade goods, but also on competitive advantages, opportunity costs of water and other inputs, and potential impacts of any land and water allocation changes on livelihoods.

4.2.3 Overall Assessment

Despite decades of trials and adjustments, the notion that experts can allocate water to the best uses by measurements and analyses has fallen short. Benefit/cost ratios, multiple-objective planning and analysis, water pricing to reflect scarcity value, assessment of environmental services, virtual water, and water footprints have all been attempts to substitute expert knowledge and skills for the political process. While some developments have considered impacts on equity and quality of life better than others, they are at a disadvantage because non-economic values are difficult to quantify. Economic and efficiency arguments often resonate with policy-makers who appreciate clear justifications. However, this kind of policy analysis short-changes the values that are central to this book including equity and participation. Emphasizing efficiency can deprive small users of water rights. Expert-driven water policy and notions of efficiency may interfere with local water management practices and harm livelihood and production strategies. And then water users are blamed for not achieving efficiency goals, even though their actions support other equity and participatory values (Boelens and Vos, 2012: 16).

4.3 Institutional Analysis and Development

Arguably the most prominent academic approach in the public policy field, and also widely applied to natural resources and water, is institutional analysis and development (IAD). The framework is Elinor Ostrom's brainchild, following on influential work of her husband, Vincent Ostrom. She was awarded a Nobel Prize in Economics partly for the framework's brilliance. Two contextual factors are important to understanding the emergence of IAD analytics and its relation to water. First were the adverse democratic consequences related to federal-agency dominance of water. These problems were not lost on the Ostroms, whose early work focused on water in California. Vincent Ostrom (1953) wrote that basin organizations whose decisions were based on democratic participation were better at decision-making than federal bureaucracies. Elinor Ostrom discovered in her 1965 dissertation about West Basin in Los Angeles that water users were able to collectively manage groundwater overdraft on their own. A water-user association was able to reach a collective agreement and appoint a water master to implement it. Both the Ostroms were strong supporters of self-government. Second was the cynicism among a group of social-science scholars about government's role in solving public problems. Rational choice theorists posited that agency experts would not necessarily make choices that benefit the public, but instead serve their agency's interests and their respective scholarly disciplines (rent-seeking). In contrast, rational individuals are far better than government at making decisions that serve their preferences.

4.3.1 Central Concepts and Theoretical Logic

Water, according to IAD, is a common-pool resource (CPR). CPRs are resources where one person's use subtracts from another's use, where it is often necessary, though difficult and costly, to exclude other users from the group using the resource (Ostrom, 2005). Unlike the gospel of efficiency approach to water, which sought to distribute water to the highest uses, IAD stresses the potential problem of overuse that damages the resource itself. In that way, the CPR definition is more environmentally conscious, but it is clearly utilitarian and linked to human management and control. The CPR, according to Ostrom, is generally bounded or restricted to a community who, under the best institutional designs, can manage water (Wall, 2014).

Humans make decisions on the basis of bounded rationality. Unlike the gospel of efficiency approach, ordinary humans do not need to depend upon experts to advise them. While their access to information may be limited, they have sources of information that agency officials lack or ignore. They are close to the resource and able to assess its status, its over-use, and whether they or fellow users are abusing the resource or the rules. For the Ostroms, political actors were no better or worse than the rest of us, and have good and bad motives (Wall, 2014: 194). Given appropriate institutional arrangements and opportunities to communicate, water users can and will agree on limitations to use CPRs.

The IAD framework is intended to identify the major types of structural variables present to some extent in all institutional arrangements but whose values differ from one type to another. Issues of fairness, equity and participation are among the evaluative criteria, but not reflected in other parts of the three-tiered conceptual map (Ostrom, 2007: 27). Elinor Ostrom writes that evaluative criteria can be applied to both outcome and process and may include: economic efficiency; equity through fiscal equivalence; redistributive equity; accountability; conformance to general morality; and adaptability. Trade-offs are often necessary among performance criteria, and Ostrom indicates that it is particularly difficult to choose between the goals of efficiency and redistributive equity. Evaluating how institutional arrangements compare across overall evaluative criteria can be quite a challenge, according to Ostrom, and she does not reveal her priorities (Ostrom, 2007: 35).

Water management involves negotiation among individual users in the commons, and conflict is likely to occur when one person gains more than others do. In that sense, the IAD policy approach embraces politics. Yet IAD does not cover macro forces in politics – discourses, values, biases and social constructions. It does not adequately consider how subjectivity influences choice, as well as power and politics. Psychologists have discovered that people make decisions based on emotions and values and then rationalize these non-rational decisions (Kahneman, 2011). Kahneman and Tversky (1973) discovered that – even when taught how to make so-called "rational" choices that would actually produce more goods for themselves – people still made "errors." They gave names to these cognitive limitations. Confirmation bias (seeking and paying more attention to information that already fits with what one believes) influences both ordinary people and experts. Availability bias is when people are more apt to believe something that they can easily imagine. IAD does not examine how culture, social class, prejudice and discrimination can shape such biases.

4.3.2 Evolution and New Applications

Early work concentrated on local decision-making. Later, polycentric arenas and levels of decision-making were judged preferable to top-down decisions dominated by the federal government or even concentrating decision-making authority locally. Consistent with the checks and balances Vincent Ostrom so admired in the US constitution and the Federalist papers, polycentric decision-making, including both private and public decisions at various levels, provided alternative platforms to exercise democratic voice. People who work in small groups nested in larger structures may find ways to exit some systems and join others (Wall, 2014: 195). David Harvey (2012) argues that polycentrism is less democratic than it seems. For instance, the ability to move from one arena to another depends on resources, ability to get jobs in new environs, and other factors favoring the rich and mobile over poor people and those attached to place. Moreover, Harvey (2012: 70) asserted that some higher authority may be necessary to enforce rules that mandate greater equality or protect ecologically vulnerable places. In response, Elinor Ostrom, who disagreed with public choice scholars on a number of grounds, probably would join them in their skepticism that central governments are not likely to be more egalitarian or preservationist than local or private interests (Wall, 2014: 180).

Over time, and with contributions by her many students and followers, the number of variables identified as significant to resolving CPR problems expanded greatly. Agrawal (2013) identified as many as 35 variables in the theory, likely to affect successful governance. Generalizations and consistent findings are difficult to come by with so many variables to consider. Studies that connect the different variables in causal chains or propose plausible causal mechanisms are rare. Further, as the number of IAD scholars has expanded, interest in performance criteria seems to have diminished. It seems the values of equity, participation and fairness in the IAD framework have been overwhelmed by attention to definitions, categories, and discussion of the difference between frameworks, models, and theories (for example, Ostrom *et al.*, 2013).

4.3.3 Overall Assessment

There is no question that the IAD approach is a great addition to the policy analysis toolkit. There are real doubts about how useful the framework is to water scholars interested in equity, participation and fairness in decision-making. The focus is clearly on institutions, and this comes at a cost (Agrawal, 2013) Studies of commons are relatively negligent in examining how some aspects of user group membership, such as social status, race and gender, affect institutional durability and long-term local management. While the framework's roots in rational choice have in many ways been far outgrown, the portrayal of human motivations in IAD scholarship is thin and overly utilitarian. Lejano and de Castro (2014) suggest that non-utilitarian ethics can be determinative in certain social, network or cultural settings. Payoffs to the individual can be partly a function of contributions to the public good. Further, the choice-making individual's nature is not fixed, but shaped by context and identity which, in turn, may be strongly influenced by social movements, narratives and discourses.

4.4 Physically-Based Watershed and River-Basin Approaches

Throughout much of the first half of the twentieth century, river basins served as a logical unit for economic planning, serving both multi-purpose water development projects and current bureaucratic realities (Molle, 2006: 9). Unified basin management was often proposed, under a basin authority, albeit with limited application and success (Ingram, 1973; White, 1957). While the concept went out of vogue in the 1970s as a development unit, it was revived in the 1990s to address water quality concerns. It can be seen in the Dublin Principles (1992) and through World Bank (1993) policies highlighting river basins as the appropriate unit for analysis and coordinated management and calling for institutional reform at the river-basin scale. Further, the catchment principle formed the cornerstone of the European Union's Water Framework Directive (2000), calling on all country members to realign their water management strategies at the basin level.

The river-basin concept is the foundation of integrated water resource management (IWRM), blended with ecosystem management approaches (Molle, 2009). Advanced in the 1990s as the new paradigm in water management, IWRM was promoted as an ideal approach to achieve efficient, equitable, sustainable development and manage the world's limited water resources (UN-Water, 2008, 2012). Of course, this was nothing new. Multi-objective planning had already tried to capture the variety of goals that can be served by water. What distinguishes IWRM from previous efforts of water management is an emphasis on integrating various aspects of water management, such as water uses by sectors, multiple governance levels, promoting river basins as management units, and greater stakeholder involvement in water management and decision-making (Mukhtarov, 2008). IWRM shifted water management away from territoriality and towards neoliberal management approaches, such as a greater role for the private sector, pricing mechanisms, and ecosystem valuation (Conca, 2006; Mollinga *et al.*, 2006).

4.4.1 Central Concepts and Theoretical Logic

According to Molle (2009), the river-basin notion first helped operationalize water development. It appealed to planners and engineers because water could perform multiple tasks and produce multiple benefits, including hydropower generation, flood control, irrigation, and water supply. Seen as a "natural" or "ideal" planning unit, the river basin would help rationally justify administrative and political boundaries. Molle writes: "The river basin is presented as a clear-cut concept, as uncontroversial as the physical delineation of a watershed, in other words the 'natural' unit for planning and management of water resources by societies" (Molle, 2009: 484). Proponents of IWRM argue for integrating various water users, e.g. industry and municipalities, and levels of water management governance: local, regional and national (e.g. Engle *et al.*, 2011; GWP, 2004). Molle (2009) refers to IWRM as a "nirvana concept," representing an ideal image of what the world *should* tend to.

Sustainability and quality interests also advanced the river-basin concept, which coalesced as the appropriate management unit (Molle and Wester, 2009). Environmental advocates reinforced the basin concept, transforming it from a water resource exploitation and development tool into a blueprint for holistic management (Teclaff, 1996). First, watersheds will improve ecosystem management (Montgomery *et al.*, 1995; Parkes, 2010). It is often assumed that governing water at a watershed scale will account for both upstream and downstream factors, producing better environmental outcomes (Vogel, 2012). Second, some environmental NGOs saw the river-basin scale as a vehicle for bottom-up planning and greater participation and engagement (Warner *et al.*, 2008), assuming that the public will be more likely to participate at the watershed scale (Blomquist and Schlager, 2005; Sabatier *et al.*, 2005).

4.4.2 Evolution and New Approaches

IWRM has come under considerable criticism over the past decade. Some opponents claim that such integration is doomed to fail because water governance is inherently political (Biswas *et al.*, 2005; Petit and Baron, 2009; Rahaman and Varis, 2005). Others highlight IWRM's failures. For example, Conca (2006: 157–58) points out that "malleable" IWRM language provided a broad umbrella for water users to legitimize their conflicting claims to water, without any way to sort priorities, and local water users' participation in water management is often "rhetorical," tending to privilege expert, technocratic knowledge over other forms of knowledge. In their edited volume, Huitema and Meijerink (2009) provide several international case studies of IWRM approaches that have fallen short or failed altogether due to lack of implementation and effective citizen engagement. It has been suggested that the IWRM debate is locked in a stalemate (Mukhtarov and Gerlak, 2013).

Today, critical scholars such as Alice Cohen argue that watersheds are materially, discursively and conceptually constructed, producing real winners and losers in water management. Cohen and Davidson (2011) are critical that watersheds have been conflated with governance tools, such as IWRM. They write:

The adoption of international IWRM water dialogues by regional, national, and sub-national government agencies and water policy planners appear to have been fixated on watershed boundaries. Rather than as an arm of IWRM or a technical tool (as framed by IWRM antecedents), watersheds were recast as frameworks; the watershed approach became an umbrella under which other features of IWRM, such as participation and integration, fell.

(Cohen and Davidson, 2011: 7)

As Perrault (2014: 237) writes, "to the extent that particular scales for water governance are seen as 'natural' and immutable … they run the risk of obscuring the politics underlying such scales' production." Further, territorializing water policy under an integrated watershed approach shifts responsibility from state to civil society for achieving and maintaining environmental standards (Buller, 1996). Although a community may be good at expressing values and interests, it may not be able to marshal support and competence to effectively govern water resources (Bakker, 2010, 2013). Power and resources may need to be redistributed to engage the poor and achieve substantive stakeholder participation (Wester *et al.*, 2003). Further, we know from studies of watershed councils and organizations in the US that is difficult to sustain effective public engagement over time (Sabatier *et al.*, 2005). Wilder and Ingram (forthcoming) suggest that, rather than broadening public engagement, formal and informal networks must be created and maintained as an essential part of effectively expressing public preferences.

Some scholars are working to develop concepts that can reveal both the physical and social dimensions of water governance (e.g. Bakker, 2003; Budds and Hinojosa, 2012; Linton, 2010; Loftus, 2009; Swyngedouw, 2004, 2009). Swyngedouw (2009: 56) argues for a new "hydrosocial" concept to "transcend modernist nature-society binaries" and envision water as both physical and social. Budds and Hinojosa (2012) define a waterscape as a range of moments, and integrate both social and physical properties, including physical flows, access patterns, water discourses, and governance frameworks. The chapters in Part II of this book (e.g., Swyngedouw and Boelens, and Hommes *et al.*) advance such notions by elaborating on the concept of "hydrosocial territory" (Boelens *et al.*, 2016).

4.4.3 Overall Assessment

The watershed concept has a long history and has come in and out of vogue in water management. In the early twentieth century, it appealed to US and European engineers and planners to promote multi-purpose water development. Over the past two decades, it has been revived by environmental actors to improve water quality and promote stakeholder participation. The watershed concept was tightly linked with IWRM in the 1990s, as the appropriate scale to promote integrated water resource management. Yet IWRM has recently come under considerable criticism and reached a stalemate. It has been argued that defining a community in hydrologic terms depoliticizes and technifies water decision-making (Roa-Garcia, 2014). There is growing recognition of the limits to recognizing basins as merely physical constructs. This has real consequences for water management policy, producing real winners and losers.

4.5 Discursive Policy Analysis: The Meanings of Water

Over the past several decades, discursive approaches to policy-making have been emerging in Europe and the United States to challenge positivist politics. Argumentative policy analysis subsumes a group of policy analysis approaches emphasizing language as a key feature of both policy, and policy analysis (Gottweis, 2006). These diverse theoretical approaches, including discourse analysis, frame analysis, interpretative policy analysis, and others, similarly highlight argumentation and language in policy-making and analysis, and offer new ways to analyze policy and politics. They focus on studying actors' analytical production and use of arguments, as opposed to analyzing public policies rationally.

In the early 1980s, Frank Fischer (1980) and John Forester (1985) began to map out some key features of the argumentative angle. Later in that decade, Deborah Stone's *Policy Paradox* (1988) critiqued the rationality model of policy making, emphasizing language. She challenged the notion that neutral facts could decide conflicts and shape policy decisions. According to Stone, "Facts do not exist independent of interpretative lenses, and they come clothed in words and numbers" (Stone, 1988: 307). In *Evidence, Argument, and Persuasion in the Policy Process*, Giandomenico Majone (1989) outlines the "argumentative character" of policy-making, calling for systematic attention to the role and function of words, and ways of "doing things with words" (Majone, 1989: 7).

Fischer's and Forester's *The Argumentative Turn in Policy Analysis and Planning*, published in 1993, made argumentative policy analysis a recognizable movement in policy studies, with a clearly articulated research agenda. They argued that policy analysis must place argumentation at the center of analysis and epistemology, and offered a systematic, pluralistic agenda for new, argumentative policy analysis. This publication has made the notion very influential, that argumentative policy analysis fosters political participation and deliberation (Gottweis, 2006: 472). Policy analysts' task is no longer to identify solutions for objective problems, but rather to facilitate public learning and political empowerment (Fischer, 2003).

4.5.1 Central Concepts and Theoretical Logic

The discursive paradigm, inspired by the "linguistic turn" in philosophy and social sciences, builds on constructivist perspectives in social inquiry (Durnova *et al.*, 2016: 35): attention is given to the forms of knowledge actors assemble and the multiple interpretations used to create meaning. It rejects rational assumptions about human behavior and instead "embraces an understanding of human action as intermediated and embedded in symbolically rich social and cultural contexts" (Fischer and Gottweis, 2012: 2). Durnova *et al.* (2016) write:

The best decision is frequently not the most efficient one. Instead, it is the one that has been deferred until all disagreements have either been discursively resolved or placated, at least enough to be accepted in specific circumstances.

It emphasizes the community value of water, including its organizing importance, its emotional and symbolic meaning, and the relationship between water and the community (Brown and Ingram, 1987). Poor rural Hispanics and Native Americans viewed water not just as a resource or commodity, but rather as part of their identity, solidarity and sense of opportunity (Brown and Ingram, 1987). For many desert-dwellers, giving up water means forgoing a way of life. Moreover, linkages between water scarcity and the desire to control the future has led to preoccupation with water security.

4.5.2 Evolution and New Approaches

Unlike the tightly linked researchers who use Ostrom's IAD framework, there is a diverse set of scholars, who do not necessarily fall under the public-policy umbrella, but who do share the notion that language is important. While these researchers work from a variety of disciplinary perspectives, a stream of critical geography has emerged over the past two decades, speaking to equity and justice issues in water governance. The premise is that water is neither purely "natural" or "social," but rather a hybrid; and scales are socially and materially constructed (as highlighted in our previous section on physically based watershed and river basin approaches). Karen Bakker, one of the most notable scholars working in this area, has bridged between the "more policy-oriented governance literature with the more politics-infused political ecology of water literature" (Wilder and Ingram, forthcoming). Bakker (2007) argues that viewing water as a collective good and commons resource, rather than merely an economic good, may reveal possibilities to conceptualize equity in water governance. Other critical research urges to look beyond more formal decision-making and political arenas, to conceptualize water justice within broader political, ecological and social change (Zwarteveen and Boelens, 2014). Others still argue that water governance institutions need to better address issues of democratization, human welfare and ecological conditions (Perrault, 2014).

Governance processes designed to embody water's multiple meanings, values and knowledge (Blatter and Ingram, 2001; Ingram and Lejano, 2009) and to involve a diverse "community" of human, non-human, and biophysical actors (Schmidt, 2012) can achieve water equity. Schneider and Ingram (2007) analyze how meanings produce and communicate knowing and understanding in policy-making. One way to know is how one interprets policy elements and makes sense of relationships among them (Feldman and Ingram, 2009: 3). Ways of knowing corroborate claims by political adversaries that there are multiple types of knowledge. Ingram and Lejano (2009: 68–69) argue that water can be known as the output of an engineering process, an economic good, a critical element in the environment, and a community focal point for insuring equity and sense of place. There is often little interaction or mutual understanding among these diverse perspectives, so policies inadequately embody multiple values and perspectives.

Mukhtarov and Gerlak (2013) develop a typology of three epistemic forms (prescriptive, discursive and practical) and apply it to IWRM discourse. They find that IWRM has been viewed narrowly as a universal, abstract, prescriptive concept to achieve predefined water management objectives through rigorous planning. This limited framing of IWRM does injustice to water management complexities; it ignores politics and ethics, as well as the on-the-ground context. They argue that IWRM is also a discursive concept – a point of reference to discuss water values, ethics and power, and a practical concept – in which practicing water management yields practical knowledge. By acknowledging the existence of multiple epistemic forms, it becomes possible to evaluate policy concepts such as IWRM from multiple dimensions. This can help shift and expand the debate, making space for multiple conflicting epistemologies to come together and resolve conflict. They follow a similar strategy to explore the newly emerging paradigm of water security, where they argue that water security represents a discursive way of knowing water with a greater consideration of human values, ethics and power (Gerlak and Mukhtarov, 2015).

4.5.3 Overall Assessment

There are many diverse approaches under the discursive umbrella. Regardless of the disciplinary perspective, discursive approaches all address language and argumentation, and reject the rational positivist approach to policy analysis. Many critical approaches reconceptualize or repoliticize governance, emphasizing equity and justice in water governance. Significantly, these approaches challenge the artificial boundaries between humans and nature, revealing multiple meanings, values and knowledges of water. Although many of these more discursive approaches to policy analysis highlight that existing policies fail to foster participation and equity, they only rarely offer policy proposals. So, while discursive approaches to water move us closer to issues of fairness and justice than the three earlier approaches outlined in this chapter, they remain insufficient. Ultimately, practical, implementable policies need to engage processes for making better decisions – including laws, rules, institutions, etc. – to achieve more equitable water governance.

4.6 Conclusions

Public policy analysis affects water-policy decisions. In this chapter, we examined four different policy approaches and their impact on water governance: (1) efficiency-based analysis; (2) institutional analysis and development; (3) physically based watershed and river basin approaches; and (4) discursive policy analysis. Which of these approaches– or some other analytical framework – are applied matters, because frameworks identify which factors get attention to explain policy action. This essentially then predetermines policy solutions. Further, policy analyses affect which water problems reach the agenda and how issues are framed. They also determine which sources of knowledge are privileged in policy-making. Briefly stated, policy frameworks determine winners and losers, and for the most part they slight equity and fairness. In assessing policy frameworks, it is crucial to ask: what and who is left out?

Measured by the impact that frameworks have on policy, efficiency-based frameworks clearly have the greatest influence, and are the least sensitive to equity and participation. While no longer so persuasive as they once were to academic scholars, efficiency analytics have a firm hold on bureaucratic analysis, at least in part because they favor experts. Efficiency ideas have influenced institutional analysis and development, as have integrated water resources management notions within river basin approaches. All three approve moving water to its highest economic use and market-based institutions and incentives. IAD and river basin approaches favor grass-roots participation, but both slight the importance of language and narratives, group dynamics, and the role of networks. In multi-objective planning and analysis, equity and participation are supposed, on paper, to be evaluated together with economic efficiency and environmental quality. However, in practice, only economic efficiency really matters, and only occasionally can the dictates of efficiency be blocked by environmental and social grass-roots mobilizations.

Policy analysis frameworks tend to treat water too narrowly, weighing most heavily its benefits to people rather than ecosystems, and separating water from its social setting. Discursive policy approaches go the farthest in embracing water's multiple, complex meanings. Further, discursive approaches are most sensitive to issues of equity and participation. However, discursive approaches have been the least influential in actually affecting policy choices. While discursive analyses are exceptionally good at identifying weaknesses in other frameworks and in water policies, they tend to avoid policy prescriptions. We believe that discursive policy analysts must play a larger role in crafting contextually relevant, equitable, participatory governance solutions. Further, users of other frameworks must address their shortcomings by adding factors addressing equity and participation, and giving these values priority in policy evaluation.

References

Agrawal, A. (2013). Studying the commons, governing common-pool resources outcomes: Some concluding thoughts. *Environmental Sciences and Policy*, 36, 86–91.

Allan, J. A. (1996). Water use and development in arid regions: Environment, economic development and water resource politics and policy. *Review of European Community International Environmental Law*, 5(2), 107–15.

Bakker, K. (2003). *An Uncooperative Commodity: Privatizing Water in England and Wales*. Oxford: Oxford University Press.

Bakker, K. (2007). The "commons" versus the "commodity": Alter-globalization, anti-privatization and the human right to water in the global South. *Antipode*, 39, 430–55.

Bakker, K. (2008). The ambiguity of community: Debating alternatives to private sector provision of urban water supply. *Water Alternatives*, 1(2), 236–52.

Bakker, K. (2010). *Privatizing Water: Governance Failure and the World's Urban Water Crisis*. Ithaca, NY: Cornell University Press.

Bakker, K. (2013). Constructing "public" water: The World Bank and water as an object of development. *Society and Space Environment and Planning D*, 31(2), 280–300.

Biswas, A., Varis, O. and Tortajada, C. (2005). *Integrated Water Resources Management in South and Southeast Asia*. New Delhi: Oxford University Press.

Blatter, J. and Ingram, H. (2001). *Reflections on Water: New Approaches to Transboundary Conflicts and Cooperation*. Cambridge, MA: MIT Press.

Blomquist, W. and Schlager, E. (2005). Political pitfalls of integrated watershed management. *Society & Natural Resources*, 18(2), 101–17.

Boelens, R. and Vos, J. (2012). The danger of naturalizing water policy concepts: Water productivity and efficiency discourses from field irrigation to virtual water trade. *Agricultural Water Management*, 108, 16–26.

Boelens, R., Hoogesteger, J., Swyngedouw, E., Vos, J. and Wester, P. (2016). Hydrosocial territories: A political ecology perspective. *Water International*, 41(1), 1–14.

Brown, F. L. and Ingram, H. M. (1987). *Water and Poverty in the Southwest*. Tucson: University of Arizona Press.

Budds J. and Hinojosa, L. (2012). Restructuring and rescaling water governance in mining contexts: The co-production of waterscapes in Peru. *Water Alternatives*, 5(1), 119–37.

Buller, H. (1996). Towards sustainable water management: Catchment planning in France and Britain. *Land Use Policy*, 13, 289–302.

Caulfield, H. P., Jr. (2011). Early federal guidelines for water resource evaluation. *Journal of Contemporary Water Research & Education*, 116, 1.

Chapagain, A. K. and Hoekstra, A. Y. (2004). *Water Footprints of Nations*, Vol. 2: Appendix. Value of Water Research Report Series No. 16. Delft, the Netherlands: UNESCO-IHE Institute for Water Education.

Cohen, A. (2015). Nature's scales? Watersheds as a link between water governance and the politics of scale. In E. S. Norman and C. Cook (eds.), *Negotiating Water Governance: Why the Politics of Scale Matter*. Surrey: Routledge, pp. 25–40.

Cohen, A. and Davidson, S. (2011). The watershed approach: Challenges, antecedents, and the transition from technical tool to governance unit. *Water Alternatives*, 4(1), 1–14.

Conca, K. (2006). *Governing Water: Contentious Transnational Politics and Global Institution Building*. Cambridge, MA: MIT Press.

Costanza, R., d'Arge, R., de Groot, R., Farber, S., Grasso, M., Hannon, B., Limburg, K., Naeem, S., O'Neill, R. V., Paruelo, J., Raskin, R. G., Sutton, P., and van den Belt, M. (1997). The value of the world's ecosystem services and natural capital. *Nature*, 387, 253–60.

Coniff, R. (2012). What's wrong with putting a price on nature? *Yale Environment*, 360. October 18.

Daily, G. C. (1997). *Nature's Services: Societal Dependence on Natural Ecosystems*. Washington, DC: Island Press.

Durnova, A., Fischer, F. and Zittoun, P. (2016). Discursive approaches to public policy: Politics, argumentation, and deliberation. In B. G. Peters and P. Zittoun (eds.), *Contemporary Approaches to Public Policy Theories, Controversies and Perspectives*. New York: Palgrave Macmillan, pp. 35–58.

Engle, N., Johns, O., Lemos, M. and Nelson, D. R. (2011). Integrated and adaptive management of water resources: Tensions, legacies, and the next best thing. *Ecology and Society* 16(1), 19.

Feldman, D. L. and Ingram, H. (2009). Multiple ways of knowing water resources: Enhancing the status of water ethics. *Santa Clara Journal of International Law*, 7, 1.

Fischer, F. (1980). *Politics, Values, and Public Policy: The Problem of Methodology*. Boulder, CO: Westview Press.

Fischer, F. (2003). *Reframing Public Policy: Discursive Politics and Deliberative Practices*. Oxford: Oxford University Press.

Fischer, F. and Forester, J. (1993). *The Argumentative Turn in Policy Analysis and Planning*. Durham, NC: Duke University Press.

Fischer, F. and Gottweis, H. (eds.) (2012). *The Argumentative Turn Revisited: Public Policy as Communicative Practice*. Durham, NC: Duke University Press.

Forester, J. (ed.) (1985). *Critical Theory and Public Life*. Cambridge, MA: MIT Press.

Gerlak, A. K. and Mukhtarov, F. (2015). "Ways of knowing" water: Integrated water resources management and water security as complementary discourses. *International Environmental Agreements: Politics Law and Economics*, 15(3), 257–72.

Global Water Partnership (GWP). (2004). *IWRM Plans and Water Efficiency Plans by 2005: Why, What and How?* Stockholm: Global Water Partnership.

Gottweis, H. (2006). Argumentative Policy analysis. In B. Guy Peters and J. Pierre (eds.), *Handbook of Public Policy*. London: SAGE Publications, pp. 461–80.

Harvey, D. (2012). *Rebel Cities: From the Right to the City to the Urban Revolution*. London: Verso.
Hays, S. P. (1959). *Conservation and the Gospel of Efficiency: The Progressive Conservation Movement 1890–1920*. Cambridge, MA: Harvard University Press.
Hoekstra, A. Y. and Chapagain, A. K. (2007). Water footprints of nations: Water use by people as a function of their consumption pattern. *Water Resources Management* 21(1), 35–48.
Huitema, D. and Meijerink, S. (eds.). (2009). *Water Policy Entrepreneurs: A Research Companion to Water Transitions around the Globe*. Cheltenham: Edward Elgar.
Huitema, D. and Meijerink, S. (eds.). (2014). *The Politics of River Basin Organisations: Coalitions, Institutional Design Choices and Consequences*. Cheltenham: Edward Elgar.
Ingram, H. (1973). The political economy of regional water institutions. *American Journal of Agricultural Economics* 35(1), 10–18.
Ingram, H. and Lejano, R. (2009). Transitions: Transcending multiple ways of knowing water resources in the United States. In D. Huitema and S. Meijerink (eds.), *Water Policy Entrepreneurs: A Research Companion to Water Transitions around the Globe*. Cheltenham: Edward Elgar, pp. 61–79.
Kahneman, D. (2011). *Thinking, Fast and Slow*. New York: Macmillan.
Kahneman, D. and Tversky, A. (1973). On the psychology of prediction. *Psychological Review*, 80(4), 237–51.
Lejano, R. and de Castro, F. (2014). Social dimensions of the environment: The invisible hand of community. *Environmental Science and Policy*, 36, 73–85.
Lemos, M. C. (2008). Whose water is it anyway? Water management, knowledge and equity in Northeast Brazil. In J. Whitely, H. Ingram, and R. Perry (eds.), *Water, Place and Equity*. Cambridge, MA: MIT Press, pp. 249–70.
Linton, J. (2010). *What is Water? The History of a Modern Abstraction*. Vancouver: University of British Columbia Press.
Loftus, A. (2009). Rethinking political ecologies of water. *Third World Quarterly*, 30, 953–68.
Maas, A. and Anderson, R. (1978) *... and the Desert Shall Rejoice: Conflict, Growth, and Justice in Arid Environments*. Cambridge, MA: MIT Press.
Majone, G. (1989). *Evidence, Argument and Persuasion in the Policy Process*. New Haven, CT: Yale University Press.
Michalson, E. L. (1975). What objectives for multiple objective planning. In E. Michalson, E. Engelbert and A. Wade (eds.), *Multiple Objectives Planning Water Resources*. Vol. 2(5). Moscow: Idaho Research Foundation Inc., pp. 9–14.
Molle, F. (2006). *Planning and Managing Water Resources at the River-Basin Level: Emergence and Evolution of a Concept*. Colombo: International Water Management Institute.
Molle, F. (2009). River-Basin Planning and Management: The Social Life of a Concept. *Geoforum*, 40(3), 484–94.
Molle, F. and Wester, P. (eds.) (2009). *River Basin Trajectories: Societies, Environments and Development*. Wallingford: CABI.
Mollinga, P., Dixit, A. and Athukorala, K. (2006). *Integrated Water Resources Management Global Theory, Emerging Practice, and Local Needs*. New Delhi: Sage.
Monbiot, G. (2014). Put a price on nature? We must stop this neoliberal road to ruin. *The Guardian*, July 24, www.theguardian.com/environment/georgemonbiot/2014/jul/24/price-nature-neoliberal-capital-road-ruin.

Montgomery, D. R., Grant, G. E. and Sullivan, K. (1995). Watershed analysis as a framework for implementing ecosystem management. *Journal of the American Water Resources Association*, 31(3), 369–86.

Mukhtarov, F. (2008). Intellectual history and current status of integrated water resources management: A global perspective. In C. Pahl-Wostl, P. Kabat and J. Moltgen (eds.), *Adaptive and Integrated Water Management: Coping with Complexity and Uncertainty*. Heidelberg: Springer.

Mukhtarov, F. and Gerlak, A. K. (2013). Epistemic forms of integrated water resources management: Towards knowledge versatility. *Policy Sciences*, 47, 101–20.

Ostrom, E. (1965). "Public entrepreneurship: A case study in ground water basin management." A dissertation submitted in partial satisfaction of the requirements for the degree Doctor of Philosophy in Political Science. University of California at Los Angeles.

Ostrom, E. (2005). *Understanding Institutional Diversity*. Princeton, NJ: Princeton University Press.

Ostrom, E. (2007). Institutional rational choice: An assessment of the institutional analysis and development framework. In Paul Sabatier (eds.), *Theories of the Policy Process*. Denver: Westview Press, p. 27.

Ostrom, E., Cox, M. and Schlager, E. (2013). An assessment of the institutional analysis and development framework and introduction to the social ecological systems framework. In P. Sabatier and C. Weible (eds.), *Theories of the Policy Process*, 3rd edn. Boulder, CO: Westview Press, pp. 267–306.

Ostrom, V. (1953). *Water and Politics*. Los Angeles: John and Dora Haynes Foundation.

Parkes, M. W., Morrison, K., Bunch, M., Hallström, L., Neudoerffer, R., Venema, H., and Waltner-Toews, D. (2010). Towards integrated governance for water, health and social-ecological systems: The watershed governance prism. *Global Environmental Change*, 20(4), 693–704.

Perreault, T. (2014). What kind of governance for what kind of equity? Towards a theorization of justice in water governance. *Water International* 39(2), 233–45.

Petit, O. and Baron, C. (2009). Integrated water resources management: From general principles to its implementation by the state. The case of Burkina Faso. *Natural Resources Forum*, 33, 49–59.

Rahaman, M. and Varis, O. (2005). IWRM: Evolution, prospects and future challenges. *Sustainability: Science, Practice, and Policy*, 1(1), 15–21.

Rivlin, A. M. (1984). A public policy paradox. *Journal of Policy Analysis Review* 4(1), 17–22.

Roa-Garcia, M. (2014). Equity, efficiency and sustainability in water allocation in the Andes: Trade-offs in a full world. *Water Alternatives*, 7(2), 298–319.

Sabatier, P. A., Focht, W., Lubell, M., Trachtenberg, Z., Vedlitz, A. and Matlock, M. (eds.). (2005). *Swimming Upstream: Collaborative Approaches to Watershed Management*. Cambridge, MA: MIT Press.

Safriel, U. (2011). Balancing water for food and nature. In A. Garrido and H. Ingram (eds.), *Water for Food in a Changing World*. London: Routledge, pp. 135–70.

Schmidt, J. (2012). Scarce or insecure? The right to water and the changing ethics of global water governance. In F. Sultana and A. Loftus (eds.), *The Right to Water: Politics, Governance and Social Struggles*. London: Routledge, pp. 94–109.

Schneider, A. and Ingram, H. (2007). Ways of knowing: Implications for public policy. Annual meeting of the American Political Science Association, Chicago.

Stone, D. (1997). *Policy Paradox: The Art of Political Decision Making*. New York: W. W. Norton & Company.

Stone, D. A. (1998). *Policy Paradox and Political Reason.* Glenville, IL: Scott Foresman & Co.

Swyngedouw, E. (2004). *Social Power and the Urbanization of Water: Flows of Power.* Oxford: Oxford University Press.

Swyngedouw, E. (2009). The political economy and political ecology of the hydro-social cycle. *Journal of Contemporary Water Research & Education* 142, 55–60.

Teclaff, L. A. (1996). Evolution of the river basin concept in national and international water law. *Natural Resource Journal,* 36, 359.

United Nations Water (UN-Water). (2008). *Report on Integrated Water Resources: Management and Water Efficiency Plans.* New York: The Commission on Sustainable Development.

United Nations Water (UN-Water). (2012). *The UN-water Status Report on the Application of Integrated Approaches to Water Resource Management.* Stockholm: UNEP, UNDP, GWP.

Vogel, E. (2012). Parceling out the watershed: The recurring consequences of Columbia River management within a basin-based territory. *Water Alternatives* 5(1), 161–90.

Wall, D. (2014). *The Sustainable Economics of Elinor Ostrom: Commons, Contestation and Craft.* London: Routledge.

Warner, J., Wester, P. and Bolding, A. (2008). Going with the flow: River basins as the natural units for water management? *Water Policy* 10 Supplement 2, 121–38.

Wester, P., Merrey, D. J. and De Lange, M. (2003). Boundaries of consent: Stakeholder representation in river basin management in Mexico and South Africa. *World Development* 31(5), 797–812.

White, G. F. (1957). A perspective of river basin development. *Law and Contemporary Problems,* 22, 157–87.

Wichelns, D. (2015). Virtual water and water footprints: Overreaching into the discourse on sustainability, efficiency, and equity. *Water Alternatives* 8(3), 396–414.

Wilder, M. and Ingram, H. (forthcoming). Knowing equity when we see it: Water equity in contemporary global contexts. In K. Conca and E. Weinthal (eds.), *The Oxford Handbook on Water Policy and Politics.* Oxford: Oxford University Press.

World Bank. (1993). *Water Resources Management.* Washington, DC: IBRD.

Zwarteveen, M. and Boelens, R. (2014). Defining, researching and struggling for water justice: Some conceptual building blocks for research and action. *Water International* 39(2), 143–58.

5

Urban Water and Sanitation Injustice: An Analytic Framework

Ben Crow

5.1 Introduction

Most of the world's population now lives in urban areas. Getting water and using sanitation in cities can reproduce a range and intensity of inequalities, and opportunities for collective action that may not exist in the countryside. The urban poor are entangled with their cities' economy and culture (Srivastava, 2014). Many work as domestic servants and guards for wealthier households, cleaners for retail spaces and offices, street hawkers and electricians, construction workers and more. The urban poor are a large part of a city's labor force. They perform jobs no one else is willing to do, work long hours for low wages, and lower the living costs of those living outside low-income neighborhoods (Perlman, 1976: Ch. 8). They seek to survive and to improve their lives through these interactions, seeking education, employment, shelter, security and more. In Delhi and many other cities, the urban poor's contributions to the formal city occur in the context of "constant, frequently enforced threats of displacement (through 'slum demolitions' for example), [and] their never-ending efforts to secure foothold within it (say, by purchasing fake identity cards)" (Srivastava 2014: xxxiv).

In the wide range of capabilities sought by the urban poor, and the contexts of displacement, uncertainty and exclusion that they face, water and sanitation may appear to be just two among many exclusions. There is some evidence (Arputham, 2016; Devoto, *et al.*, 2012; Swallow, 2005), nonetheless, that enhanced access to water and sanitation may be a high priority for the urban poor. I explore some possible reasons for this priority below.

Some important aspects of urban water and sanitation include:

– Providing infrastructure in some neighborhoods, and not others, leading to particular coping mechanisms and power inequalities.
– Household water and sanitation replace irrigation as a principal focus of government attention and collective action.
– The presence, absence and governance of municipal infrastructure are central to urban water and sanitation.
– Water-borne disease transmission may be intensified in dense urban areas.

This chapter has a particular focus on informal settlements, such as shantytowns, *favelas*, *bustees* and homeless encampments, the unregulated and sometimes illegal, often peri-urban and poorly provisioned settlements where many live and many dimensions of inequality and injustice are pronounced.

In Sub-Saharan Africa, more than 60 percent of urban residents live in informal settlements. In other regions of the global South, the figure ranges from 20 percent to 35 percent (UN Habitat, 2012: Figure 3). In the global North, refugee camps, homeless encampments and migrant populations are a smaller but not insignificant part of the urban population, facing comparable deprivations. There are between 2.3 and 3.5 million people experiencing homelessness each year in the United States, with approximately 33 percent of them living outside (HUD, 2016; National Coalition for Homelessness, 2009). France has nearly 4 million with inadequate housing or without personal housing (Fondation Abbé-Pierre, 2017), and Paris had an 84 percent increase in homelessness between 2001 and 2012 (France Bleu, 2015).

In most cities across the world, ideas of gender, and the division of labor between men and women, assign women the unpaid reproductive work of childcare and maintaining the home. Women suffer the indignities, risks, time expenditure and bodily exertion of collecting household water (Truelove, 2011). At the same time, social norms put upon the female body make sanitation demands particularly oppressive for women (Joshi and Zwarteveen, 2012; Ray, 2016; see Chapter 11 by Rusca *et al.*, this volume). In this chapter, I argue that inequalities in accessing water and sanitation have profound consequences for deprivation, dignity and gender subordination, which are magnified in informal settlements.

The chapter is organized as follows. Section 5.2 sketches a framework to illuminate how everyday practices of getting water and using sanitation in urban informal settlements may constitute poverty and subordination. Section 5.3 describes some historical background of informal settlements and the resulting water and sanitation characteristics. The ways gender relations and deprivation are constituted by everyday water and sanitation practices are described in Section 5.4. Indications of what can happen when collecting household water is made easier are assembled in Section 5.5. Some areas of struggle and contestation that can enable household capabilities, including water and sanitation, to improve are outlined in Section 5.6. The final section asks what this might mean for urban water and sanitation justice.

5.2 A Framework Illuminating Capabilities and Change: Water and Sanitation Injustice

This section begins to sketch a framework to illuminate urban water and sanitation justice. It starts by describing key dimensions of deprivation. Then it introduces a focus on capabilities to facilitate disaggregated assessment of the empirical circumstances and injustices of urban deprivation. The third sub-section introduces a relational analysis of inequality reproduction to illuminate the origins of capability deprivation in historical and political change. The final sub-section suggests that gender relations and domestic work are key elements in analyzing urban water and sanitation justice.

5.2.1 Interlocking Dimensions of Powerlessness and Ill-Being

Narayan *et al.* (2000) sought to listen to 'the voices of the poor' in a study of 20,000 poor men and women in the cities and countryside of 23 countries. They distilled what they learned into ten "interlocking dimensions of powerlessness and ill-being" (Narayan *et al.*, 2000: 2):

1	Livelihoods and assets are precarious, seasonal and inadequate.
2	Places of the poor are isolated, risky, unserviced and stigmatized.
3	The body is hungry, exhausted, sick and poor in appearance.
4	Gender relations are troubled and unequal.
5	Social relations are discriminating and isolating.
6	Security is lacking in the sense of both protection and peace of mind.
7	Behaviors of those more powerful are marked by disregard and abuse.
8	Institutions are disempowering and excluding.
9	Organizations of the poor are weak and disconnected.
10	Capabilities are weak because of the lack of information, education, skills and confidence.

These ten dimensions of ill-being and powerlessness illuminate the breadth of pressing concerns and begin to suggest how they are embedded in social relations that shape the agency, dignity, security and experience of being poor. The circumstances and daily tasks of getting household water and finding sanitation contribute to almost all dimensions. In the following section, I describe one approach to justice that can shed light on the role of water and sanitation in the making of ill-being and powerlessness.

5.2.2 The Capabilities Approach to Poverty, Development and Social Justice

What idea of justice sheds light on water and sanitation inequalities? Most obviously, it is grounded in the everyday realities of people's lives, rather than abstract, transcendental ideas of justice and notions of a perfect society. It also goes beyond calculating gross domestic product and, instead, reaches toward those capabilities and freedoms that enable people to live lives that they have reason to value (Sen, 1999, 2008).

Amartya Sen's concepts of poverty, development and social justice can provide a starting-point to analyze water and sanitation justice. Sen (1999) suggests that poverty can be thought of as deprivation from capabilities and freedoms. He builds on this insight to formulate an influential theory of development and social justice focusing on people's freedoms and functioning: "the ends of well-being, justice and development should be conceptualized in terms of people's capabilities to function; that is, their effective opportunities to undertake the actions and activities that they want to engage in, and be whom they want to be" (Robeyns, 2005: 95).

The idea of poverty as capability deprivations facilitates disaggregated analysis of poverty's multiple dimensions. The social dynamics that reproduce or enable advances in collecting water may not be the same as for improving sanitation, or for shelter or work. Sen's capability approach thus suggests that enhanced access to water and sanitation can be analyzed as both means and ends of development and social justice. Adequate household water, for example, can sustain the reproductive activities of childcare, cleaning, and laundry that may represent both means to a better life in the future and the ends of a better life in the present. Adequate household water can also sustain small home enterprises offering a means to reduce poverty in terms of both income and consumption. Gaining access to adequate sanitation facilities is also both a means to a better life in the future (reduced illness and risks) and, in itself, one element of a valued life in the present. The perspective of capability deprivation, however, is not strong on the historical and political contexts behind deprivation.

5.2.3 Capability Deprivation, Relational Poverty and Domestic Labor

Sen's focus on people's capabilities to function emphasizes individual and household endowments and entitlements. The historical, political and collective dimensions reproducing capability deprivations need to be integrated into this framework. With the idea of relational poverty, Mosse (2010) suggests how history and power can be included. Mosse abstracts three general processes that reproduce structural, durable poverty. First, the logic of capitalist transformation and economic growth creates opportunities for some while reproducing poverty for others. Dispossession from land, livelihood and opportunities, for example, is part of economic development. Second, categories of social distinction, such as caste, race, gender, class, age and citizenship, can be used to enable particular groups to hoard opportunities and to segment access to jobs, assets, finance, education and skill acquisition. Third, skewed political representation accords agenda-setting power to some while excluding others.

These three processes, drawing from Tilly's (1998, 2003) work on durable poverty, suggest a backdrop for understanding some of the collective and historical processes leading to lack of capabilities and increase in injustice. They illuminate, for example, the creation and reproduction of informal settlements. Accumulation and dispossession, particularly through land control, give opportunities to some and exclude others. Divisions built on categorical lines of ethnicity, race and class shape the making, continuation and dynamics of informal settlements and their attendant ranges of capability exclusions.

5.2.4 Gender Relations and Domestic Labor

Largely missing from analysis of both capability deprivations and relational poverty is unpaid domestic labor and its consequences for women, children and whole households in the global South. Robeyns (2003) describes a general approach to gender in the capabilities tradition, with a focus on post-industrial societies. Robeyns does not, however, analyze gender relations, unpaid domestic labor and deprivation in the global South. Childcare, laundry, washing, cleaning, cooking, collecting water, and using sanitation are all activities necessary for social reproduction. Constructions of gender and the gender division of labor, in many societies, give women primary responsibility for domestic labor. These activities shape women's and children's daily lives, including their work, time, dignity and opportunities.

The quality of domestic work and household activities has wide consequences for the whole household. Men are able to attend job interviews when they have clean, presentable clothes. Children can go to school and play. The whole household's health and livelihoods depend upon domestic work and its quality. Understanding capability deprivation and the social dynamics of poverty must integrate gender relations and household work.

The preliminary framework introduced in this section seeks to connect three perspectives on social life and social relations. The capabilities approach enables disaggregated analysis of change in particular capabilities, including capabilities to get water and use sanitation. Then, the notion of relational poverty suggests a broad set of structural, or durable, processes shaping capability deprivation. And, in the last sub-section, an introductory discussion of gender relations and domestic work draws our attention to their centrality in reproducing deprivation and subordination and to possibilities for change.

5.3 Urban Informal Settlements and Water and Sanitation Justice

This section first describes some history of informal settlements in the post-colonial South. Then, the section notes some key elements of what Anand (2012) calls abject water access. Anand (2012) describes how water system engineering and operation in Mumbai renders Muslim settlers abject residents of the city, denied social and political entitlements and subject to a bodily response of revulsion. He quotes Murphy (2006: 152), suggesting that abjection "designates unlivable, uninhabitable zones of social life which are nonetheless densely populated by those who are not enjoying the status of subject but whose living, under the sign of unlivable, is required to circumscribe the domain of subject." Abjection is, I suggest, an appropriate term to describe informal settlements in general and their water and sanitation access.

In the global South, many of these uninhabitable, unlivable spaces have their origins in colonial rule. Many big cities and their spatial and social divisions are shaped by colonial rulers' actions and their categories of race and ethnicity. In East Africa and much of Sub-Saharan Africa, for example, urban spatial divisions, and informal settlements, have roots in colonial racial and ethnic segregation. Colonial rulers in Kenya established separate residential areas for white settlers, Asians and Africans. Then, colonial policy established planning and service requirements for each area, with little provision for the "native reserves." K'Akumu and Olima (2007: 87) describe how, in Kenya, "Racial segregation

[was] transformed into socio-economic and legal-tenurial residential segregation upon attainment of independence."

In early twentieth-century colonial India, similar divisions were deepened as the colonial government saw the urban poor as a political threat to order, public health and moral fabric. Gooptu (2001: 420) describes how "social constructions of the poor as dangerous and unstable ... and measures to exclude and discipline the poor served to intensify class divisions."

After World War II, historically unprecedented rural–urban migration took place across the global South. The segregated, fragmented cities initiated by colonial rule expanded rapidly. This rapid expansion of areas segregated by race, ethnicity and class was rarely accompanied by a commensurate increase in service provision to informal settlements, or by efforts to reformulate policies that discriminated against these neighborhoods.

The consequence of this history is to generate some of the most egregious injustices of urban living, including those regarding water and sanitation in informal settlements. Modes of access to water and levels of inequality, as Ranganathan (2014) describes for Bangalore and Anand (2012) describes for Mumbai, can be differentiated by categories of occupation, class, caste, ethnicity, and religious affiliation.

The absence of water and sanitation access in these areas leads to the construction of a range of informal or illicit arrangements. Networks of often-leaky pipes may be delivering water to land or structure owners (the de facto building owners, when tenure is unclear), who then sell it to households. Small and medium enterprises may emerge to deliver water in tankers, carts and other manually powered vehicles (Crow and Odaba, 2010; Ranganathan *et al.*, 2009; Swyngedouw, 2004). Households may interact with many modes, actors and technologies for getting water: several different types of water (groundwater for bathing, rainwater for laundry, bottled water for drinking), a range of different water providers (municipal water supply, tanker truck drivers, local water vendors and bottled water sellers), and a variety of technologies from industrialized to artisanal. These interactions require "intimate knowledge of the political ecology of the city's water, where it flows at times of year, how cost and quality vary across time and space" (Bakker, 2013: 286).

Even with that knowledge, informal arrangements often deliver expensive water that may be contaminated. Water availability may be unreliable and restricted to certain days of the week and/or times of the day, constrained in quantity available, and at some significant distance from the home. McFarlane (2014) identifies a paradox of water and sanitation provision. Historically, he writes, sanitation "has been a bringer of life and health ... but in practice it symbolizes the threat of violence, and is intimately linked to crushing forms of torturous poverty." Desai *et al.* (2014) note how unequal sanitation provision in urban areas produces "profoundly unequal opportunities for fulfilling bodily needs." The authors describe how, in the absence of sanitation provision, improvised coping mechanisms often reproduce and deepen inequalities.

The consequences of abject water and sanitation access fall most heavily upon women and children. Women are most frequently tasked with collecting water. As Ahmed and

Zwarteveen (2012: 8) write, "Water problems, therefore, cannot be understood – nor can effective solutions to such problems be found – without an explicit realization that water realities are deeply gendered." Women may suffer most from sanitation indignities and risks, as the next section describes.

5.4 Gender, Water and Sanitation

Abject water and sanitation access falls on women and girls through different, but related, social processes. Women's water collection in informal settlements, with its attendant constraints on time, opportunities and dignity, arises from the gender division of domestic labor. Women's experience of sanitation use, by contrast, arises substantially from widespread, but varying, rules about the female body. Nauges and Strand (2013) write that "women face disproportionate challenges in accessing water supply and sanitation services." The allocation of domestic labor, including the collection of water, to women and girls has profound consequences for their time and opportunities. Joshi *et al.* (2012: 200) describe how "a higher level of shame and secrecy is levied on the female body across many societies." Gendered demands "levied on the female body" make defecation, urination and managing menstrual hygiene a constant challenge for women and girls' dignity, privacy, security, and daily lives. Ray (2016) provides a powerful description of the implications of inadequate sanitation for women and girls.

The heterogeneity of households, and of domestic labor within households, must be recognized (Nightingale 2011). Some households, even within informal settlements, are able to hire paid domestic labor. In other households, women, men and children sell their labor power to middle-class households. Some individuals and households, in the global North and South, have no homes and live on the streets. For such homeless people, it is not clear that gendering water and sanitation will work as described above.

An analysis of the origins of gender subordination is beyond the scope of this chapter. Nevertheless, Elson's (1991) description of underlying support for gender subordination and gender division of labor is pertinent. Elson (1991: 320) suggests that male dominance, or gender subordination, arise because women do not have independent ways of making a living while they are giving birth and breast-feeding a child:

lack of an independent entitlement forces women into dependence for these [biological] phases of child-rearing, phases which are particularly difficult to combine with income-earning, then women are likely to get locked into all the other phases [of child-rearing and domestic work].

Lack of an independent income, Elson suggests, sustains male dominance. Then, given the gender division of labor, women have primary responsibility for the household's social reproduction, i.e. domestic work. Women have to deal with their own personal hygiene needs plus other household members' needs: "washing clothes for the entire household, keeping children clean, sweeping and scrubbing floors, washing utensils and cleaning the immediate home surroundings. In the absence of adequate water and waste disposal services, these tasks become hugely challenging" (Joshi *et al.*, 2012: 185).

This yields at least two sets of outcomes for women and girls. First, women having to devote time and energy to collecting water at uncertain times of day pre-empts a wide range of opportunities. Second, the indignities and risks required to get water and use sanitation may constitute a significant part of what it means to be poor and subordinate. I outline these outcomes, in turn, below.

5.4.1 Time and Opportunities

Time taken collecting water, waiting in line and waiting for water to be available is widely noted in studies of many countries' rural and urban areas in the global South. The findings of two studies are particularly illuminating. One situates the significance of time saving in the overall economic benefits of closer access to water and sanitation. The second notes that inadequate sanitation also consumes significant time and constrains opportunities.

In a study for the WHO, Hutton and Haller (2004) estimated the value of time savings at 65 percent of the substantial economic benefits of closer water and sanitation access, including reduced health-care costs and schooling. To arrive at this conclusion, Hutton and Haller estimated the time savings in rural and urban areas and valued time saved using the minimum wage for unskilled labor. Whittington *et al.* (1990) had shown that households in Kenya valued water collection time at about this rate. While this is only a crude estimate, it indicates the significance of the time taken, or opportunity costs, primarily for women and girls, of getting water.

The time devoted to the household's sanitation is also significant. Mothers in a rural area reported to O'Reilly (2012: 8–9) that "a toilet afforded them time savings and stress reduction because they did not need to take their children to the village outskirts when either the children or they themselves needed to defecate." This is no less a challenge in urban areas. We shall see in Section 5.7 that reductions in this time constraint open a range of opportunities for women and girls. Further, collecting water and using sanitation can impose a second tier of harm for women and girls.

5.4.2 Everyday Practices, Gender Subordination and Class Position

The social repercussions of women's water collection and use of sanitation, beyond the opportunity costs, are both deep and wide. Women's and girls' subjectivities, on the one hand, are formed by the devaluation of their work and lives. Then, on the other hand, abject water and sanitation access make, and are made by, a wide series of social disadvantages, from malnutrition and violence to ethnic, educational and religious disadvantage. Everyday practices of collecting water and using sanitation, in themselves, reproduce and perform gender subordination and deprivation. These gendered processes are not limited to informal settlements and the global South (Gershenson and Penner, 2009).

Drawing on feminist political ecology, Truelove (2011: 146, 150) writes that, "women's bodies encounter gendered hardships, physical labor and public shame shaped by their situated position within families, communities and class groups ... gender and class formations, patterns of risk, criminality, social exclusion are tied to and reproduced through daily water practices." In Delhi, the hardships embodied in everyday water collection practices lead to girls being kept out of school to look after younger children, contribute to devaluing women's labor and rights at work, and curtail women's leverage in bargaining within their household (Truelove, 2011). Early-morning travel to seek places to defecate incurs risks and may impose extra pressure when the woman has diarrhea. These and other experiences accumulate to devalue women's lives. Undie *et al.* (2006) describe the prevalence of similar practices in a Kenyan informal settlement. Similarly, Joshi *et al.* (2006) outline comparable processes arising from sanitation access in informal settlements in Dhaka, Bangladesh.

Jackson (1998: 323) describes how such experiences constitute "embodied subjectivities" of collecting water. She describes how the work of collecting water extends from women's experiences of work to the making of gender relations. The "physical work intensity," she writes of women's work collecting domestic water, "is a characteristic of the bodily experience of work that influences well-being, both physical and perceptual, generates working experiences that pattern subjectivities, preferences and perceptions of individual women, and feeds into the social relations of gender in intra-household bargaining, producing divisions of labor, and extra-household gendered social relations of work." In her paper, Jackson is focusing primarily on water collection. Experiences of abject sanitation add another dimension.

McFarlane (2014: 1) argues that gendered sanitation practices have wider origins and consequences than these noted above. He describes the rape and death of two teenage Dalit women as they walked through fields seeking a place for defecation. They are vulnerable, he argues, because of "caste oppression, the normalization of horrific violence against poor women, and the structural and systematic failure of the Indian state to provide the minimum of everyday rights: a clean, functional toilet." He goes on to describe a range of urban and rural examples and to argue that abject sanitation conditions are submerged in, and coproduced by, a broad, changing range of social deprivations and degradations:

The tragedy of sanitation is that it is fundamentally not just about toilets and pipes. Sanitation is far more than the staples of daily life, the unglamorous backstage of everyday life, as Erving Goffman once put it. Sanitation is an incessantly morphing object: it is immersed in and enrolled through malnutrition, violence, caste cultural politics, sexism, ethno-religious hatred, political patronage, educational disadvantage, illness and disease, precarious livelihoods, unemployment, environmental pollution, and more.

Although abject access to water and sanitation reproduces elements of deprivation and gender subordination, many contemporary projects to improve water and sanitation provision do little to ameliorate male dominance and impoverishment. O'Reilly (2012: 8–9) notes that women's inability to use public spaces, as a consequence of social rules on women's seclusion, creates a need for toilets in the home. Sanitation projects that target women may reduce childcare work, nonetheless, "leave unchallenged gendered inequalities as they relate to access of public and private space." Ahmed and Zwarteveen (2012:162), summarizing a set of empirical chapters on "Gender in Drinking Water and Sanitation" in Zwarteveen, Ahmed and Gautam (2012), show how "the incorporation of gender into drinking water and sanitation projects and policies risks 'watering down' the objective of redressing gender inequalities." Sanitation projects in the slums of Bangladesh (Joshi *et al.*, 2012), for example, seek to address only women's need for privacy in defecation without recognizing the broad, interrelated set of women's gendered responsibilities and varied poverty levels between the housed and those living on the streets.

While enhanced provision of water and sanitation may not address the contexts of gender subordination and class position, there is evidence that reducing time taken to collect water can lead to new opportunities for women. The next section turns to that question.

5.5 What Can Happen When Getting Water Takes Less Time?

As mentioned above, enhanced access to water and sanitation may offer possibilities for change, including these elements of household activities:

- Providing water and time for small-scale commodity production and in peri-urban household agriculture.
- Making improvements possible in domestic work quality, such as more effective laundry, house cleaning, bathing and child care.
- Reducing time constraints that cause tension and conflict in the household.
- Providing time for sleeping and leisure.

The breadth of these changes may illuminate Devoto *et al.*'s (2012) striking finding from a controlled study in Tangiers, Morocco, that women in households gaining in-home water supply report reduced household conflict and increased happiness. While a household connection freed time for household members and increased quality of life, in this case, it did not increase income or labor-market participation. The study recognized (Devoto *et al.*, 2012: 69) that lack of a household water connection could foreclose "an important source [for women] of opportunities to socialize, and possibly reduce their well-being." It did not investigate the effects on domestic work quality or on other household activities.

We do not have quantitative evidence describing improvement in all these dimensions. Studies of improved water access, nonetheless, describe a diverse set of opportunities enabled by reduced time collecting water and increased quantity of water (Table 5.1). Apart from Devoto *et al.* (2011) and Nauges and Strand (2013), studies report enhanced access to water in rural areas. Similar possibilities are likely to result in urban areas.

Of the eight studies in Table 5.1, seven mention time savings. The diversity of consequences, however, is striking. Enhanced access to water can bring these elements of emancipation: improving the quality of domestic work, doing it in less time, enabling a

Table 5.1: *Consequences of enhanced access to water*

Study	Consequences of enhanced access to household water
Rural Senegal (van Houweling et al., 2012)	Time savings and increased activities, including horticulture, livestock, car-washing, ice- and brick-making.
Rural Vietnam (Noel et al., 2010)	Accessible, reliable, sufficient water enabled livelihood diversification and increased income in treatment vs. control villages.
Rural Kenya (Crow et al., 2012)	Community-organized, piped household water enabled more diverse livelihoods, increased income and allowed women more sleep than in villages without household water, but also increased men's work.
Rural India (James et al., 2002)	Time released from water collection enables small enterprise and higher incomes.
Distance to water and health, Sub-Saharan Africa (Pickering and Davis, 2012)	Reduced collection time cuts diarrhea and infant mortality because long walk times restrict the volume of water collected leading to less effective hygiene, water stored for longer periods becomes contaminated and because there is more time for childcare and health care.
Urban and rural Ghana (Nauges and Strand, 2013)	15-minute decrease in water collection time increases the proportion of girls attending school by 8–12 percent.
Piped household water in Tangiers, Morocco (Devoto et al., 2012)	Reduced conflict over water, increased leisure time, and generated higher levels of happiness.
Global South – rural (Koolwal and van de Walle, 2013)	More leisure for women and less work on family farm but no evidence of increased market participation by women.

diversity of livelihoods, increasing household incomes, women getting more sleep and leisure, reducing household conflict, increasing opportunities for girls to go to school, reducing indignity and gender subordination, reducing women's bodily taxing tasks, reducing risk of attack and increasing privacy, reducing mortality and increasing longevity. In sum, significant opportunities arise when households can get more water and devote less time to collecting water. Then, in what contexts can that happen?

5.6 Justice and Contestation in Relation to Water, Sanitation and Poverty

Under what circumstances can those facing deprivation and subordination prevail against capability deprivation and the forces of relational poverty, outlined in Section 5.2? What social processes could empower the claims of the urban poor for water and sanitation and other valued capabilities? In this section, I describe four spaces of contestation where the agency of women and the urban poor may enable claims for justice in capabilities, including water and sanitation: (i) women's organizing around water, (ii) encroachments by the urban poor, (iii) movements for insurgent citizenship, and (iv) interaction in the many and varied entanglements of urban poor with other elements of the city.

There is a significant record of women's organizing related to water (Crow and McPike, 2009; James *et al.*, 2002). This collective action can be situated on a continuum of women's agency, as suggested by Molyneux (1998), from women's practical needs to their strategic interests. In this continuum, women's practical interests and action relate to reproductive labor and daily household needs, while strategic interests and organizing seek to challenge social relations that reproduce gender subordination. In Latin America during the 1970s and 1980s, women's organizing around (gender-specific) practical interests (including water and energy) sometimes raised consciousness and led, over time, to a rise in independent women's organizations focused on larger strategic questions of gender and class relations (Johnson, 1992). So, one possible connection between gender and domestic labor and connections to the world of work involves rising women's collective action from practical interests to strategic gender needs.

Social *non-movements* of the poor (Bayat, 2013) also may have significant influence on unpaid domestic labor, women's work and household capabilities (see Chapter 13 by Cleaver, this volume). Bayat describes a wide set of *encroachments*: "dispersed and disparate struggles in the immediate domains of their everyday life – in the neighborhoods, places of work, street corners, court houses, communities, and in the private realms of taste, personal freedom, and preserving dignity." He argues that these encroachments go beyond everyday acts of resistance, because the poor "engage in surreptitious and incremental encroachments to further their claims" (Bayat, 2013: 17–18).

In contrast to these social non-movements of the poor, Holston (2009) has studied and documented social movements by urban poor in Brazil seeking legitimacy for illegal residence, house building, and land conflict. This third form of agency (see, for example, Ranganathan 2013) seeks *urban citizenship*, a right to the city (Harvey, 2003; 939). This may be seen as an organized action related to, and perhaps relatively rarely, growing out of, encroachments and entanglements.

A more encompassing analysis of agency of the urban poor, a fourth space of contestation, focuses on *entanglements* where the poor engage with the city's middle class, wealthy and powerful (Srivastava, 2015). These sites, crossing borders between the poor and the rest, include domestic service, interaction with street-level officials and brokers over shelter and land, work in shopping malls and retail stores, street-level hawking, seeking city services and more. Such entanglements offer survival and opportunities and, Srivastava suggests, they also constitute sites where negotiation and change may take place.

Srivastava's picture of many and varied entanglements opens up the ways that the urban poor are crucial to a city's metabolism. Social non-movements of the poor, insurgent citizenship and women's organizing around practical needs can be seen as key elements in this varied set of relations and their combined opportunities. Action for water and sanitation justice fits into this larger picture, along with action for the many capabilities sought by the urban poor. Can agency by the deprived and subordinated, exercised in these four spaces of contention, prevail against the social forces described in Section 5.2?

5.7 Conclusion: An Encompassing Analytic Framework and What This Could Mean for Justice

Concerns for urban water and sanitation justice are, at one level, simple to describe. There are striking differences between urban neighborhoods in access to safe water and sanitation. This is the level of analysis at which the WHO/UNICEF Joint Monitoring Committee estimates the number of people without access to improved drinking water and improved sanitation (e.g. UNICEF and WHO, 2017). That level of analysis also enters into government and international discourse, and shapes much policy, about water and sanitation.

That understanding of inequality and injustice, however, leaves out the much broader and deeper sociological, political and historical picture of abject access to water and sanitation. As I have suggested in this chapter, the indignities and risks, bodily exertion, material constraints and performance of everyday practice produce and reproduce gender- and class-subordinated lives. This reframing of abject water and sanitation access suggests that a focus on collection times, safe drinking water and improved sanitation overlooks the social significance of the issues and the limits of technological approaches, seeking somehow to circumvent politics and power.

In this reframing, some limits of the relational poverty analytic framework, outlined in Section 5.3, also become clear. That framework sketches a political economy of water and sanitation access, but it is not sufficient to engage the broader, gendered experience at household and bodily levels. A framework needs to include gendered divisions of labor and gender relations. Both everyday practices and their consequences for a water collector and her household constitute poverty and gender subordination. It then becomes clear that depicting poverty as *capability deprivation* is insufficient. Capability-deprivation is too atomized and mechanical. It leaves out many dimensions of experiencing poverty, the embodied subjectivities of subordination.

Is the framework of relational poverty still useful? Yes. That analytic framework suggests how the experience of abject access to water and sanitation may be located in the three processes reproducing durable poverty described by Mosse (2010): Logic of capitalist transformation and economic growth, deployment of categorical distinctions to enable segmentation and hoarding of opportunities/resources, and skewed political representation denying agenda-setting power to the deprived and subordinated. Reframing abject access to illuminate gendered, embodied subjectivities opens new opportunities to advance theoretical and ethnographic investigation of interactions with capitalism and economic growth, and intersections with social categories enabling segmentation, hoarding and more.

Reframing abject access to recognize the gendered experience of deprivation and subordination also sheds light on the four spaces of contestation described in Section 5.6. Those contestations articulate broader dimensions of experiencing poverty and subordination. They may speak, for example, to women's practical interests and needs. Then, the extent to which such contestations engage gender relations and women's strategic interests needs further exploration.

We need a framework for analysis that goes beyond the idea of poverty as capability deprivations, to explore the embodied subjectivities arising from abject access to water and sanitation, and to examine how the "tragedy" of water and sanitation is "immersed in and enrolled through" (McFarlane, 2014) malnutrition and violence, illness and disease and the unemployment and precarious livelihoods segmented by social divisions of class, ethnicity, religion and more. Such a framework will begin to illuminate the sociological breadth and depth of water and sanitation injustice, the limits on global policy discourse and the significance of grassroots forms of contestation and struggle. Such an analytic framework may also illuminate the priority given by the poor, as noted in Section 1, to relief from water and sanitation injustice.

Acknowledgements

This chapter has been shaped by discussions with, and the work of, two teams of which the author is a member. One team is embarking on collaborative research in four cities on "Capabilities and domestic work, dignity and deprivation in entangled urbanization." The second is producing an online undergraduate course on "Water and sanitation justice" for the University of California. None bear responsibility for errors and omissions. Those remain the sole responsibility of the author. Research team: Bernard Barraque, Abigail Brown, Chris Butler, Owiti K'Akumu, Meenoo Kohli, Frank Poupeau and Sanjay Srivastava. University of California online Water and Sanitation Justice course team: Carolina Balazs, Abigail Brown, Isha Ray and Kirsten Rudestam. In addition, the author wishes to acknowledge the advice and ideas of Peter Mollinga, Rutgerd Boelens and Margreet Zwarteveen.

References

Ahmed, S. and Zwarteveen, M. (2012). Gender and water in South Asia: Revisiting perspectives, policies and practice. In M. Zwarteveen, S. Ahmed, and S. R. Gautam (eds.), *Diverting the Flow: Gender Equity and Water in South Asia*. New Delhi: Zubaan Books.

Anand, N. (2012). Municipal disconnect: On abject water and its urban infrastructures. *Ethnography*, 13(4), 487–509.

Arputham, J. (2016). Ward diaries: Crucial evidence for planning in Mumbai's slums. Guest blog for International Institute for Environment and Development, www.iied.org/ward-diaries-crucial-evidence-for-planning-mumbais-slums.

Bakker, K. (2013). Constructing "public" water: The World Bank, urban water supply, and the biopolitics of development. *Environment and Planning D: Society and Space*, 31(2), 280–300.

Bayat, A. (2013). *Life as Politics: How Ordinary People Change the Middle East*. Standard, CA: Stanford University Press.

Brown, A. (2016). Personal communication based on extensive research in Paris.

Burt, Z., Nelson, K. and Ray, I. (2016). Towards gender equality through sanitation access. Discussion Paper, UN Women.

Crow B. and McPike, J. (2009). How the drudgery of getting water shapes women's lives in low-income urban communities. *Gender, Technology and Development*, 13(1), 43–68.

Crow, B. and Odaba, E. (2010). Access to water in a Nairobi slum: Women's work and institutional learning. *Water International*, 35(6), 733–47.

Crow, B. and Swallow, B. (2017). Poverty and water: Pathways of escape and descent. In K. Conca, and E. Weinthal (eds.), *Oxford Handbook of Water Politics and Policy*. Oxford: Oxford University Press.

Crow, B., Swallow, B. and Asamba, I. (2012). Community-organized water increases household incomes but also men's work. *World Development*, 40 (3), 528–41.

Desai, R., McFarlane, C. and Graham, S. (2014). The politics of open defecation: Informality, body, and infrastructure in Mumbai. *Antipode,* 47 (1), 98–120.

Devoto, F., Duflo, E., Dupas, P., Parienté, W. and Pons, V. (2011). Happiness on tap: Piped water adoption in urban Morocco. *American Economic Journal: Economic Policy*, 4(4), 68–99.

Dill, B. and Crow, B. (2014). The colonial roots of inequality: Access to water in urban East Africa. *Water International*, 39 (2), 187–200.

Elson, D. (1991). *Male Bias in the Development Process*. Manchester: Manchester University Press.

Fondation Abbé-Pierre (2017). L'état du mal-logement en France 2017. Rapport Annuel #22.

France Bleu. 2015. "Le nombre de SDF monte en fleche Paris," www.francebleu.fr/infos/societe/le-nombre-de-sdf-monte-en-fleche-paris-1422971532.

Gershenson, O. and Penner, B. (eds.) (2009). *Ladies and Gents: Public Toilets and Gender*. Philadelphia, PA: Temple University Press.

Gooptu, N. (2001). *The Politics of the Urban Poor in Early Twentieth-Century India*, Cambridge: Cambridge University Press.

Harvey, D. (2003). The right to the city. *International Journal of Urban and Regional Research* 27(4): 939–41.

Holston, J. (2009). *Insurgent Citizenship: Disjunctions of Democracy and Modernity in Brazil*. Princeton, NJ: Princeton University Press.

HUD. (2016) *The 2016 Annual Homeless Assessment Report to Congress*. US Department of Housing and Urban Development Office of Community Planning and Development, www.hudexchange.info/resources/documents/2016-AHAR-Part-1.pdf.

Hutton, G. and Haller, L. (2004). *Evaluation of the Costs and Benefits of Water and Sanitation Improvements at the Global Level*. Geneva: World Health Organization.

Jackson, C. (1998). Gender, irrigation and environment: Arguing for agency. *Agriculture and Human Values*, 15(3), 313–24.

James, A. J., Verhagen, J. and van Wijk, C., Nanavaty, R., Parikh, M., and Bhatt, M. (2002). Transforming time into money using water: A participatory study of economics and gender in rural India. *Natural Resources Forum*, 26, 205–17.

Johnson, H. (1992). Women's empowerment and public action: Experiences from Latin America. In M. Wuyts (ed.), *Development Policy and Public Action.* Oxford: Oxford University Press.

Joshi, D. and Zwarteveen, M. (2012). Gender in drinking water and sanitation: An introduction. In M. Zwarteveen, S. Ahmed, S. R. Gautam (eds.), *Diverting the Flow: Gender Equity and Water in South Asia*. New Delhi: Zubaan Books.

K'Akumu, O. and Olima, W. (2007). The dynamics and implications of residential segregation in Nairobi. *Habitat International*, 31, 87–99.

Koolwal, G. and van de Walle, D. (2013). Access to water, women's work, and child outcomes. *Economic Development and Cultural Change*, 61(2), 369–405.

McFarlane, C. (2014). "The everywhere of sanitation: Violence, oppression and the body," www.opendemocracy.net.

Mollinga, P. (2008). Water, politics and development: Framing a political sociology of water resources management. *Water Alternatives*, 1(1), 7–23.

Molyneux, M. (1998). Analysing women's movements. *Development and Change*, 29(2), 219–45.

Mosse, D. (2010). A relational approach to durable poverty: Poverty, inequality and power. *Journal of Development Studies*, 46(7), 1156–78.

Murphy M. (2006). *Sick Building Syndrome and the Problem of Uncertainty: Environmental Politics, Technoscience, and Women Workers*. Durham, NC: Duke University Press.

Narayan, D., Chambers, R. and Shah, M.K. (2000). *Voices of the Poor: Crying Out for Change*. Oxford: Oxford University Press for the World Bank.

National Coalition for the Homeless. (2009). "How many people experience homelessness?" Fact Sheet, www.nationalhomeless.org/factsheets/How_Many.html.

Nauges, C. and Strand, J. (2013). *Water Hauling and Girls' School Attendance: Some New Evidence from Ghana*. Washington, DC: World Bank. Policy Research Working Paper 6443.

Nightingale, A. J. (2011). Bounding difference: Intersectionality and the material production of gender, caste, class and environment in Nepal. *Geoforum*, 42, 153–62.

Noel, S., Phuong, H.T., Soussan, J. and Lovett, J.C. (2010). The impact of domestic water on household enterprises: Evidence from Vietnam. *Water Policy*, 12, 237–47.

O'Reilly, K. (2012). Geography matters: The importance of land, water, and space in sanitation studies. *WH$_2$O: Journal of Gender and Water*, 1(1), 8–9.

Perlman, J. (1976). *The Myth of Marginality: Urban Poverty and Politics in Rio de Janeiro*. Berkeley, CA: University of California Press.

Pickering, A. J. and Davis, J. (2012). Freshwater availability and water fetching distance affect child health in sub-Saharan Africa. *Environmental Science and Technology*, 46, 2391–97.

Ranganathan, M. (2013). Paying for pipes, claiming citizenship: Political agency and water reforms at the urban periphery. *International Journal of Urban and Regional Research*, 38(2), 590–608.

Ranganathan, M. and Balazs, C. (2015). Water marginalization at the urban fringe: Environmental justice and urban political ecology across the North–South divide. *Urban Geography*, 36(3), 403–23.

Ranganathan, M., Kamath, L. and Baindur, V. (2009). Piped water supply to Greater Bangalore: Putting the cart before the horse? *Economic and Political Weekly*, 44, 33, 53–62.

Ray, I. (2016) "Gender equality: A view from the loo." TEDx Berkeley talk, www.youtube.com/watch?v=CvE8AWl8Wb4.

Ray, I. (2007). Women, water, and development. *Annual Review of Environmental Resources*, 32, 421–49.

Robeyns, I. (2003). Sen's capability approach and gender inequality: Relevant approaches. *Feminist Economics*, 9(2–3), 61–92.

Robeyns, I. (2005). The capability approach: A theoretical survey. *Journal of Human Development*, 6(1), 93–117.

Sen, A. (1999). *Development as Freedom*. Oxford: Oxford University Press.

Sen, A. (2008). The idea of justice. *Journal of Human Development and Capabilities*, 9(3), 331–42.

Srivastava, S. (2015). *Entangled Urbanism: Slum, Gated Community, and Shopping Mall in Delhi and Gurgaon*. New Delhi: Oxford University Press.

Swallow, B. (2005). Potential for poverty reduction strategies to address community priorities: Case study of Kenya. *World Development*, 33(2), 301–21.

Swyngedouw, E. (2004). *Social Power and the Urbanization of Water: Flows of Power*. Oxford: Oxford University Press.

Tilly, C. (1998). *Durable Inequality*. Berkeley, CA: University of California Press.
Tilly, C. (2003). Changing forms of inequality. *Sociological Theory*, 21(1), 31–6.
Truelove, Y. (2011). (Re-)Conceptualizing water inequality in Delhi, India through a feminist political ecology framework. *Geoforum*, 42, 143–52.
UN Habitat. (2012). *Streets as Tools for Urban Transformation in Slums: A Street-Led Approach to Citywide Slum Upgrading*. A UN-Habitat Working Paper, Nairobi, Kenya.
UNICEF and WHO. (2017). *Safely Managed Drinking Water*. New York: UNICEF and Geneva: WHO: Joint Monitoring Program.
US Department of Housing and Urban Development, Office of Community Planning and Development. (2016). "The 2016 Annual Homeless Assessment Report to Congress," www.hudexchange.info/resources/documents/2016-AHAR-Part-1.pdf.
Van Houweling, E., Hall R. P. and Diop, A. S. (2012). The role of productive water use in women's livelihoods: Evidence from rural Senegal. *Water Alternatives*, 5(3), 658–77.
Whittington, D., Mu, X. and Roche, R. (1990). Calculating the value of time spent collecting water: Some estimates for Ukunda, Kenya. *World Development*, 18(2), 269–90.
Zwarteveen, M. and Boelens, R. (2014). Defining, researching and struggling for water justice: Some conceptual building blocks for research and action. *Water International*, 39(2), 143–58.
Zwarteveen, M., Ahmed, S. and Guatam, S. R. (eds.) (2012). *Diverting the Flow: Gender Equity and Water in South Asia*. New Delhi: Zubaan books and University of Chicago Press.

Part II
Hydrosocial De-Patterning and Re-Composition

Grapes for export in Ica, Peru.
Photo by Jeroen Vos.

Introduction: Hydrosocial De-Patterning and Re-Composition

Rutgerd Boelens, Tom Perreault, and Jeroen Vos

Water governance fundamentally deals with the question of how to organize decision-making about water access, use and management in contexts of diverging interests, conflicting normative repertoires, and unequal power relations. It aims to produce particular socio-natural orders by controlling water resources, infrastructure, investments, knowledge, truth, and ultimately, water users and authorities (Boelens, 2014; Bridge and Perreault, 2009). To achieve this, as the chapters in Part I have illustrated, water governance reforms and interventions commonly emphasize strongly de-politicized common wellbeing, shared progress, and efficient, rational resource management. This naturalizing discourse of sustainable, progressive, clean development, achieved by egalitarian "stakeholders," obscures the fact that water reforms and interventions entail competing claims and conflicts over water, territorial ordering, and reconfiguring socio-economic and politico-cultural realities (Harris and Alatout, 2010; Hommes *et al.*, 2016; Kaika, 2006). These issues are directly related to disputes over problem definitions, knowledge frameworks, ontological meanings, decision-making powers and preferred solutions.

The chapters in Part II will focus on this field of water (in)justice, the dynamics of contested imaginaries and materializing socio-natural, techno-political networks: de-patterning and re-patterning multi-scalar hydrosocial territories (Baviskar, 2007; Boelens *et al.*, 2016; Linton and Budds, 2014; Swyngedouw, 2004, 2009). A governmentality perspective toward disputed, overlapping, hybridizing hydrosocial territories may help us to understand how water control is embedded in the broader political context of governance over and through socio-natures. Governmentality refers to the ways societies are governed, not only through direct application of laws and military force, but also through subtle and invisible "capillary" working of power to control, at once, the conduct of people and their socio-environment (Foucault, 1991).

The concept of hydrosocial territory provides a lens to analyze water flows, water infrastructures, and water control as simultaneously, interactively constituted compositions of physical, social, political and symbolic entities and dimensions. Together, these domains form multi-scaled, networked mosaics, which produce techno-political and socio-ecological territoriality (Boelens *et al.*, 2016; cf. Hommes *et al.*, 2016; Hoogesteger *et al.*, 2016; Swyngedouw, 2007). This analytical framework views humans and nature – social, technical and natural – not as separate entities that interact, but as mutually influencing, co-producing and constituting each other, in complex ways (see also Harris and Alatout, 2010; Latour, 1994).

Thus, territorial spaces are sites of contested control over socio-natural composition. Examining these conflicting hydrosocial imaginaries offers insight in how different social groups envision the distribution of water benefits and burdens (Boelens *et al.*, 2016). Thereby, scale and scalar assemblages do not refer to an ontologically, empirically given hierarchical framework but constitute the contingent outcome of contestations and struggles to order these hydrosocial territories (cf. Marston, 2000; Smith, 1984; Swyngedouw, 1997).

Hydraulic interventions and institutional reforms aim to re-pattern existing hydrosocial territories, changing the constituents and their respective functions, interrelations and boundaries, and thereby also the political order, world views, and ways by which water actors identify and represent themselves (Hommes *et al.*, 2016). This includes the ways in which hydraulics themselves transform/are transformed – how social-political norms, morals and hydro-cultural relations are "re-embedded" and "concretized" in artifacts and technological network relationships – "moralizing" hydro-territorial infrastructure (e.g. Hommes and Boelens, 2017; Latour, 1994; Obertreis *et al.*, 2016; Winners, 1986).

The chapters in this section make it clear that dominant hydrosocial territorial arrangements interweave scientific knowledge, technological, and state-administrative networks to enhance local-global commodity transfers, resource extraction, water development and conservation to suit non-local political-economic interests. Thereby, localized water authorities and governance are commonly bypassed or co-opted, creating a techno-political order to render these spaces legible, comprehensible, exploitable and controllable.

An illustrative example is the way many community-managed irrigation or drinking water systems are being (or are targeted to be) re-patterned in many places of the world. Particularly in the global South but equally in the West, there is a huge variety of local water uses, rights and governance systems, which often go largely unnoticed by formal water administrations (e.g. Lynch, 2012; Molle, 2004; Romano, 2017; Ruiz-Ballesteros and Gálvez-García, 2014). The mission to socially engineer "rational, efficient" water society by installing modern systems blinds policy-makers to the enormous variety of locally developed hydro-territorial solutions. Beyond misunderstanding and neglecting local water societies, this often results in interventions that reconfigure fundamental socio-hydraulic grids, while causing water distributive inequality and social differentiation (Meehan, 2013; Oberborbeck-Andersen, 2016; Paerregaard, 2016; Perreault *et al.*, 2011).

As we can witness in many cases, increasingly, the production of water knowledge, disciplines and truths – and the ways they inform the shaping of particular water technologies, rules, and systems – concentrates on the effort to align local people, mind-sets, identities and resources with interests and water governance as imagined by water-power hierarchies. Efforts to de-pattern and re-pattern locally existing water control systems and establish so-called rational frames of "water order" are at the heart of water struggles. In the end, state- and capital-driven reconfiguration deprives local water user communities of control over their hydrosocial territories. If local communities protest, the powers that be tend to dismiss, ignore or violently oppress them.

Again, we suggest viewing hydrosocial transformations as forms of governmentality, involving politics to construct new meanings and social-natural-technological connections. Regimes of water injustice tend to build hydraulic and administrative power grids that link nation-state authority, markets and companies in such a way as to externalize knowledge, rule-making and authority from local water realities. Historically, hydrosocial re-patterning has often been imposed through coercive force by governments and powerful water actors combining legal, military and extra-economic compulsion – in Foucault's (2008: 313) terms, "government according to sovereign power." Or they were devised and constructed by deploying authoritarian mythical-religious representations, "government according to truth" (ibid.; see, for instance, Boelens, 2014). While top-down state imposition is still common in the hydro-social re-patterning and "governmentalization" projects that rule the water world, modernist government-rationalities usually transform territories and control subject populations by applying subtler governance techniques – such as moralizing-scientific "disciplinary governmentality" and market rationality-based "neoliberal governmentality" (Foucault, 2008: 313). These aim for "subjectification" (Foucault, 1975) whereby water actors come to adopt the dominant discourse and turn into self-disciplining, obedient citizens of the ruling hydro-territorial system.

Rather than just aiming to manage water as a resource, ruling groups seek to deploy, and convince subjugated groups to adopt, discourses that define and position social and material issues in a human-material-natural network that leaves the political order unchallenged. Thus, "the effort is to simultaneously govern water-through-mentality and mentality-through-water" (Hommes *et al.*, 2016:12). Water actors are made to self-correct in accordance with the ruling group's hydro-territorial imaginaries (e.g. Duarte-Abadía and Boelens, 2016; Ferguson and Gupta, 2002; Perreault *et al.*, 2011). More than imposed legal or military power, this 'capillary Foucauldian power' to include, rather than exclude, is fundamental to comprehend the modernist re-patterning of hydrosocial territories. Dominant water-extractivist power coalitions aim to erase existing or alternative water territorial relationships and solutions. By deploying different governmentalities, including outright oppressive and subtle capillary forms of power, they legitimize particular extraction, distribution and control practices, as if these were entirely natural.

As a consequence, shaping dominant hydro-political order as "imagined communities" (Anderson, 1983) takes place through both "un-imagining" existing communities (Nixon, 2010) and actively "re-imagining water communities": reshaping and resignifying hydrosocial territories to produce and rule through "communities of convenience" (i.e. government through micro-societies that, in the eyes of the ruling groups, are conveniently ordered and linked to meso and macro scales: see Valladares and Boelens, 2017; cf. Li, 2011; Perreault, 2014; Rodríguez-de-Francisco and Boelens, 2016). Governmentality endeavors do not always necessarily attempt to eradicate vernacular or opposing territorialities. Commonly, subtler territorialization strategies seek to "recognize" and discipline them. They include and encapsulate local norms, resources, practices and water actors in the spatial/political organization of dominant governmentality schemes. "Through 'inclusive' strategies it recognizes the 'convenient' and sidelines 'problematic' water cultures and identities" (Boelens *et al.*, 2016: 7). In general, particular local normative and technological infrastructures are embraced (often, suffocatingly), while the more contentious, defiant, disloyal hydro-territorial norms and institutions are repressed.

Despite ruling groups' efforts to build techno-political hegemony, it is common to find multiple competing imaginaries and ways to materialize hydrosocial territoriality within each geographical-political space (Agnew and Oslender, 2013; Hoogesteger *et al.*, 2016). Dominant territorial imaginaries are contested by counter-imaginaries and alliances seeking alternative frames of meanings, problems, solutions and possibilities. Hydrosocial territories compete and superimpose. "These overlapping hydro-political projects generate 'territorial pluralism' and continuously transform the water arena's hydraulic grid, cultural reference frames, economic base structures, and political relationships" (Boelens *et al.*, 2016: 8).

An illustrative African case is given in *Seeing like an Oil Company*, where Ferguson (2005) disputes the idea framed by Scott (1998) that in the current period of neoliberal globalization, global capitalist companies have taken over the role of nation-states (*Seeing like a State*): as homogenizers and standardizers of local territories within a national-international, coherent grid of power that would encompass contiguous geographical space. Instead, oil companies have concentrated on intensive exploitation and techno-political and military organization of spatially separated but networked enclaves throughout Africa. They generate "usable" and "unusable" Africa; the first to be controlled by private security enterprises, the second left to the "transnational governmentalities" of international NGOs and weak nation-states (Ferguson and Gupta, 2002). Within the enclaves, state rules and agents are converted and co-opted to generate company-convenient territories. As Bebbington *et al.* (2010) show, the same happens in Latin America; mining and oil companies generate their own extractivist territories in which entirely different water rules and government techniques apply as compared to other spaces and places in the nation-states. To achieve this, they deploy a mix of territorial government techniques (creatively assembling customary, state and private norms) entirely convenient to the companies' extractivist objectives (see Valladares and Boelens, 2017). Van Teijlingen (2017) presents a detailed, fascinating study of how a large-scale copper mine in the Ecuadorian Amazon introduces yet another territorial layer on top of the historically created pluralist territorial configuration. The contentious outcome shows how a diversity of indigenous, colonist, state-centric and oil-company hydrosocial territories encounter, interact and overlap.

Influenced by transnational extraction, new technological developments, global policies, and the wish to combat the lingering "threat of overall water scarcity," new forms of "governmentality through water" are deployed that entail the profound re-patterning of water space and territorial decision-making (e.g. Vos and Hinojosa, 2016). This de-patterning of local water control systems in the name of rationality, decentralization, participation, and aligning of local water actors, identities and resources is often an everyday, subtle, bottom-up process – far less visible than large-scale water grabbing – but may still result in large-scale water injustices. Transnationalizing dominant hydro-territoriality means that locally based efforts to defend hydro-territorial rules and resources must upscale their "territorial reach." Hence, from the globalized local water war that started in Cochabamba, to the territorial defense networks formed in the Chevron-Texaco affected Amazon, the Shell-polluted Niger Delta, or around the Dakota-Access pipeline that threatens water flowing through the Standing Rock reservation, local communities and territories coordinate with translocal alliances, including actors not linked directly through water flows.

Part II starts with the chapter by Erik Swyngedouw and Rutgerd Boelens, examining interaction among the evolving political ideologies and socio-cultural imaginaries in twentieth-century Spain, the country's hydro-techno-social configurations and transformations, and its changing visions and practices of water justice. Starting with a late nineteenth-century naturalistic vision that considered unjust the physical distribution of water and rainfall, a broad-based progressive, hydraulic revival movement (*regeneracionismo*) emerged. It would rectify nature's mistakes through human intervention, seeking to improve the "degenerated quality of man and nature." In the following decades, between 1939 and 1975, correcting "unjust" water distribution through intensified, interconnected mega-hydraulic dam development was spearheaded as one of the principal ideological discourses and programs of Franco's fascist regime. The authors explore how "water justice" was reframed in the transition to democracy, in a context of both social differentiation and intense neoliberalization.

Chapter 7, by Renata Moreno and Theresa Selfa, explores how different conceptualizations and uses of wetlands clash in a flood-control project to improve the Cauca River's regime in Colombia. The chapter examines how water governance conflict in the region was triggered by strongly divergent visions and interests in wetland uses, expressed in encounters between Afro-Colombian smallholders and large industrial sugar cane growers. These two groups produce contradictory narratives about the flood-management conflict, the definition of wetlands and the services they provide, and who makes decisions about the river and the region's wetlands. Dominant proposals for wetland management correlate strongly with large agri-business interests, but neglect small traditional farmers' uses and meanings for wetlands. They also sideline alternative solutions to cope with flooding events, proposals that would benefit ecological and community interests.

Chapter 8 is a multi-author, multi-case paper in which Lena Hommes, Rutgerd Boelens, Bibiana Duarte-Abadía, Juan Pablo Hidalgo-Bastidas and Jaime Hoogesteger use the hydrosocial territories conceptualization to comprehend territorial politics regarding conflicts over water governance, development and distribution. The authors apply political ecology to diverse cases in Peru, Colombia, Ecuador and Turkey. In all of them, they scrutinize how powerful territorial reconfiguration projects generate and stabilize water injustices. They also explain how civil-society movements engage in important political discursive battles to challenge dominant hydro-territorialization projects and propose alternative configurations.

Chapter 9, by Barbara Rose Johnston, focuses on how large-scale hydraulic interventions, particularly large dams, are defended by discourses of national interest and the common good, but profoundly de-pattern local territories and generate huge local and regional inequality and suffering. Large dams generate benefits but commonly only for industry and people in adjacent and distant cities and towns. With an entry point provided by the lens of historical events that shaped a specific case, Guatemala's Chixoy Dam, this chapter explores how and why the relative perception of hydro-development and the enclosure of the water commons fuels movements for water justice.

References

Agnew, J. and Oslender, U. (2013). Overlapping territories, sovereignty in dispute. In W. Nicholl and B. Miller (eds.), *Spaces of Contention: Spatialities and Social Movements*. London: Ashgate, pp. 121–40.

Anderson, B. (1983). *Imagined Communities: Reflections on the Origin and Spread of Nationalism*. London: Verso.

Baviskar, A. (2007). *Waterscapes: The Cultural Politics of a Natural Resource*. Delhi: Permanent Black.

Bebbington, A., Humphreys-Bebbington, D. and Bury, J. (2010). Federating and defending: Water, territory and extraction in the Andes. In R. Boelens, D. Getches and Guevara-Gil (eds.), *Out of the Mainstream: Water Rights, Politics and Identity*. London: Earthscan, pp. 307–27.

Boelens, R. (2014). Cultural politics and the hydrosocial cycle: Water, power and identity in the Andean highlands. *Geoforum*, 57, 234–47.

Boelens, R., Hoogesteger, J. and Baud, M. (2015). Water reform governmentality in Ecuador: Neoliberalism, centralization and the restraining of polycentric authority and community rule-making. *Geoforum*, 64, 281–91.

Boelens, R., J. Hoogesteger, J., Swyngedouw, E., Vos, J. and Wester, P. (2016). Hydrosocial territories: A political ecology perspective. *Water International*, 41(1), 1–14.

Bridge, G. and Perreault, T. (2009). Environmental governance. In N. Castree, D. Demeritt, D. Liverman, and B. Rhoads (eds.), *Companion to Environmental Geography*. Oxford: Blackwell, pp. 399–475.

Duarte-Abadía, B. and Boelens, R. (2016). Disputes over territorial boundaries and diverging valuation languages: The Santurban hydrosocial highlands territory in Colombia. *Water International*, 41(1), 15–36.

Ferguson, J. (2005). Seeing like an oil company. *American Anthropologist*, 107(3), 377–82.

Ferguson, J. and Gupta, A. (2002). Spatializing states: Toward an ethnography of neoliberal governmentality. *American Ethnologist*, 29(4), 981–1002.

Foucault, M. (1975). *Discipline and Punish: The Birth of the Prison*. New York: Vintage Books.

Foucault, M. (1991). Governmentality. In G. Burchell, C. Gordon, and P. Miller (eds.), *The Foucault Effect: Studies in Governmentality*. Chicago: University of Chicago Press, pp. 87–104.

Foucault, M. (2008). *The Birth of Biopolitics*. New York: Palgrave Macmillan.

Harris, L. M. and Alatout, S. (2010). Negotiating hydro-scales, forging states: Comparison of the upper Tigris/Euphrates and Jordan River basins. *Political Geography*, 29, 148–56.

Hommes, L. and Boelens, R. (2017). Urbanizing rural waters: Rural-urban water transfers and the reconfiguration of hydrosocial territories in Lima. *Political Geography*, 57, 71–80.

Hommes, L., Boelens, R. and Maat, H. (2016). Contested hydrosocial territories and disputed water governance: Struggles and competing claims over the Ilisu Dam development in southeastern Turkey. *Geoforum*, 71, 9–20.

Hoogesteger, J., Boelens, R. and Baud, M. (2016). Territorial pluralism: Water users' multiscalar struggles against state ordering in Ecuador's highlands. *Water International*, 41(1), 91–106.

Kaika, M. (2006). Dams as symbols of modernization: The urbanization of nature between geographical imagination and materiality. *Annals of the Association of American Geographers*, 96(2), 276–30.

Latour, B. (1994). *We Have Never Been Modern*. Cambridge, MA: Harvard University Press.

Li, T. (2011). Rendering society technical: Government through community and the ethnographic turn at the World Bank in Indonesia. In Mosse, D. (ed.), *Adventures in Aidland*. Oxford: Berghahn.

Linton, J. and Budds, J. (2014). The hydrosocial cycle: Defining and mobilizing a relational-dialectical approach to water. *Geoforum*, 54, 170–80.

Lynch B. D. (2012). Vulnerabilities, competition and rights in a context of climate change toward equitable water governance in Peru's Rio Santa Valley. *Global Environmental Change*, 22, 364–73.

Marston, S. (2000). The social construction of scale. *Progress in Human Geography*, 24(2), 219–42.

Meehan, K. (2013). Disciplining de facto development: Water theft and hydrosocial order in Tijuana. *Environment and Planning*, 31, 319–36.

Molle, F. (2004). Defining water rights: By prescription or negotiation? *Water Policy*, 6, 207–27.

Nixon, R. (2010). Unimagined communities: Developmental refugees, megadams and monumental modernity. *New Formations*, 69, 62–80.

Oberborbeck Andersen, A. (2016). Infrastructures of progress and dispossession: Collective responses to shrinking water access among farmers in Arequipa, Peru. *Focaal*, 74, 28–41.

Obertreis, J., Moss, T., Mollinga, P. and Bichsel, C. (2016). Water, infrastructure and political rule: Introduction to the Special Issue. *Water Alternatives*, 9(2), 168–81.

Paerregaard, K., Bredholt Stensrud, A. and Oberborbeck Andersen, A. (2016). Water citizenship: Negotiating water rights and contesting water culture in the Peruvian Andes. *Latin American Research Review*, 51(1), 198–217.

Perreault T. (2014). What kind of governance for what kind of equity? Towards a theorization of justice in water governance. *Water International*, 39(2), 233–45.

Perreault, T., Wraight, S. and Perreault, M. (2011). "The social life of water: Histories and geographies of environmental injustice in the Onondaga Lake watershed," www.justiciahidrica.org.

Rodriguez-de-Francisco, J. C. and Boelens, R. (2016). PES hydrosocial territories: De-territorialization and re-patterning of water control arenas in the Andean highlands. *Water International*, 41(1), 140–56.

Romano, S. (2017). Building capacities for sustainable water governance at the grassroots: "Organic empowerment" and its policy implications in Nicaragua. *Society & Natural Resources*, DOI: 10.1080/08941920.2016.1273413.

Ruiz-Ballesteros, E. and Gálvez-García, C. (2014). Community, common-pool resources and socio-ecological systems: Water management and community building in southern Spain. *Human Ecology*, 42, 847–56.

Scott, J. (1998). *Seeing Like a State*. New Haven, CT and London: Yale University Press.

Smith, N. (1984). *Uneven Development: Nature, Capital and the Production of Space*. Oxford and Cambridge, MA: Blackwell.

Swyngedouw, E. (1997). Excluding the other: The production of scale and scaled politics. In Lee, R. and Wills, J. (eds.), *Geographies of Economies*. London: Arnold, pp.167–76.

Swyngedouw, E. (2004). Globalisation or "glocalisation"? Networks, territories and rescaling. *Cambridge Review of International Affairs*, 17(1), 25–48.

Swyngedouw, E. (2007). Technonatural revolutions: The scalar politics of Franco's hydro-social dream for Spain, 1939–1975. *Transactions of the Institute of British Geographers*, 32(1), 9–28.

Swyngedouw, E. (2009). The political economy and political ecology of the hydrosocial cycle. *Journal of Contemporary Water Research & Education*, 142, 56–60.

Valladares, C. and Boelens, R. (2017). Extractivism and the rights of nature: Governmentality, "convenient communities" and epistemic pacts in Ecuador. *Environmental Politics*, 26(6), 1015–34.

Van Teijlingen, K. van (2017). Large-scale mining and territorial pluralism: Mapping conflicts and convergence in the Ecuadorian Amazon. *Revista de Estudios Atacameños*.

Vos, J. and Hinojosa, L. (2016). Virtual water trade and the contestation of hydrosocial territories. *Water International*, 41(1), 37–53.

Winner, L. (1986). Do artifacts have politics? In L. Winner (ed.), *The Whale and the Reactor*. Chicago: University of Chicago Press.

Zwarteveen, M. and Boelens, R. (2014). Defining, researching and struggling for water justice: Some conceptual building blocks for research and action. *Water International*, 39(2), 143–58.

6

"… And Not a Single Injustice Remains": Hydro-Territorial Colonization and Techno-Political Transformations in Spain

Erik Swyngedouw and Rutgerd Boelens

6.1 Introduction

In this chapter, we explore how changing political visions, socio-cultural imaginaries and hydro-territorial configurations interact with shifting practices of water justice. In Plato's *Republic*, Socrates comments that justice is what those in power consider just. Over the centuries, this statement has haunted any discussions and efforts to create a fairer society. Recent social-justice debate has extended to include the physical world as an integral component in structuring just/unjust socio-ecological relations. This chapter examines how hydro-territorial politics finds expression in the diverse actors' confluences and encounters with spatial and political-geographical projects that compete, superimpose and align their territorialization strategies to strengthen their governance positions, ideologies and water-control claims. This continuously transforms the territory's hydraulic grid, cultural reference frames, economic base structures and political relationships.

Territorial struggles go beyond battles over natural resources per se, as they also involve conflicts over meaning, norms, knowledge, decision-making authority, representations and discourses. Policy actors commonly tend to present socio-natural, geopolitical territories as mere biophysical "nature" or legal-administrative "governance units," portraying water problems and solutions as politically neutral, technical and managerial issues to be objectively managed through "rational water use" and "good governance" – a conscious or unconscious veil to legitimize deeply political choices sustaining specific political orders (Harris, 2009; Hommes *et al.*, 2016; Perreault, 2014). Challenging such powerful conventions, we examine the contradictions, conflicts and societal responses generated by the configuration of hydrosocial territories (see Boelens *et al.*, 2016; Swyngedouw and Williams, 2016); how water politics are ingrained in such socio-natural and techno-political arrangements, enhancing or challenging unequal distribution of resources and decision-making power in water governance (Boelens, 2015; Swyngedouw, 2015). Therefore, hydrosocial territories (imagined, planned or materialized) have contested functions, values and meanings as they define processes of inclusion and exclusion, development and marginalization, and the distribution of benefits and burdens that affect different groups of people in distinct but often deeply unequal manners.

Taking the co-production of "nature and society" in twentieth-century Spain as our entry point, we seek to elucidate the relationship among transformations in and of "hydrosocial territory," the state, and the contested modernization, and to tease out the multiple power relationships that enroll, transform and distribute water. In doing so, we seek to excavate how nature becomes political and, through this, how environmental reconfiguration parallels ongoing state transformation (Swyngedouw, 2014; cf. Carroll, 2012; Perreault *et al.*, 2015).

The chapter illustrates how there is indeed a strong relationship between the changing nature of the state and modes of producing nature, and both are intrinsically bound up with often radical, contested imaginaries, cultural practices, and political-ecological power relations. In this interaction among nature, technology and society, the materiality of the physical environment and its dynamics – more than merely an external given – provide a historically constituted, materially produced environment that becomes enrolled in shifting frames of reference, new cultural and social imaginaries, and new discursive registers (see Swyngedouw, 1999, 2007, 2015). It is precisely in and through the contested production of new hydrosocial territories that new forms of state and societal organization are forged (see also Duarte-Abadía and Boelens, 2016; Hoogesteger *et al.*, 2016; Kaika, 2006).

This chapter[1] illustrates the notion of hydrosocial territories with Spain's tumultuous twentieth century, as a political-ecological project marked by profound transformations, punctuated by periods of great hope and expectations, intense social and political conflict, democratic reform as well as political closure, brutal civil war and dictatorship, deep crises and remarkable socio-spatial and cultural change.

6.2 Hydrosocial Territories

A political ecology perspective – to conceptualize and understand hydrosocial territories, their constitution and reconfiguration – must go beyond dichotomies separating nature from society (Latour, 1993; Lefebvre, 1991; Smith, 1984). As hybrids, these territories are simultaneously biophysical and cultural; hydrological and hydraulic; material and political. Different, divergent human interest groups inscribe their life worlds, particularly their biophysical environments, by using, inhabiting and/or managing them according to their ideologies, knowledge and socio-economic and political power, thereby generating territory. Creating hydrosocial territories involves socializing nature based on social, political, and cultural visions of the world-that-is and the world-that-should-be (Boelens, 2015; Swyngedouw, 2015). Water flows and hydraulic technologies connect places, spaces and people to each other, while human and/or non-human induced variations in its flow create, transform or destroy social linkages, lived spaces and boundaries as they produce new configurations (Linton and Budds, 2014; Mosse, 2008). These, in turn, create and transform social/political hierarchies, conflicts, and forms of collaboration (Barnes and Alatout, 2012; Brenner, 1998).

Conceptualizing hydrosocial territories as "socialized nature" or "socionatures" means insisting that they are not fixed, bounded, spatially coherent territorial entities. Rather, territory and territorialization are – and should be examined as – spatially bound, subject-built, socio-natural networks produced by actors who collaborate and compete in defining, composing and ordering this networked space (Boelens *et al.*, 2106; Swyngedouw and Williams, 2016). Politically speaking, *territory is the socio-materially constituted and geographically delineated organization and expression of and for the exercise of political power.* Territories are actively constructed and historically produced through the power-laden interfaces among society, technology and nature. Consequently, hydrosocial territories can be conceptualized as "the contested imaginary and socio-environmental materialization of a spatially bound multi-scalar network in which humans, water flows, ecological relations, hydraulic infrastructure, financial means, legal-administrative arrangements and cultural institutions and practices are interactively defined, aligned and mobilized through epistemological belief systems, political hierarchies and naturalizing discourses" (Boelens *et al.*, 2016:2).[2] Hydrosocial territory, therefore, is socionature deeply embodying its constituting societies' contradictions, conflicts and struggles (e.g. Agnew, 1994; Baletti, 2012; Elden, 2010).

A hydrosocial-territory perspective also highlights how local human actors and non-human actants connect to broader political-economic, cultural, and ecological scales (e.g. Swyngedouw, 2009). Spatial scales – geographically constituted levels of social interactions and interconnectedness, from households to global networks – are neither natural nor fixed, but produced, contested and reconfigured through myriad state, market, civil-society and individual actions (Bridge and Perreault, 2009; Heynen and Swyngedouw, 2003; Kaika, 2005). All hydro-territorial configurations relate to and are embedded within other territories that operate at broader, overlapping, counterpoised and/or hierarchically organized scales. In the (trans)formation of hydrosocial territories, scales and the ways they connect require continual reproduction and are therefore subject to negotiation and struggle (e.g. Ferguson and Gupta, 2002; Hommes and Boelens, 2017). Divergent territorial interest groups struggle to define, influence and command particular scales of resource governance (Swyngedouw, 2004).

Projections of how these territories, their water and people are and ought to be organized may commonly empower certain groups of actors while disempowering others, and offer arenas for claim-making and contestation (Duarte-Abadía *et al.*, 2015; Hoogesteger *et al.*, 2016; Hommes *et al.*, 2016; Rodríguez-de-Francisco and Boelens, 2016; Seemann, 2016; Vos and Hinojosa, 2016). Therefore, though the impacts of de-territorialization and re-patterning hydrosocial territories may be felt mostly by individuals and organizations at the local level, the processes dynamically interconnect various scales (Brenner, 1998; Hoogesteger and Verzijl, 2015; Swyngedouw, 2009).

In sum, using the illustrative case of twentieth-century Spain, this chapter's core argument is that hydrosocial territories, constituted and reconfigured on different interrelated scales, are sites of political contestation that provide deep insight into defending existing socio-natural relationships and producing alternative ones. It argues that re-patterning territories' scale, composition and control crucially depends on support and power from an interlocked multi-scale coalition of heterogeneous actors that provides technical-scientific, political-economic and discursive support for this reconfiguration.

6.3 Regenerating Spain's Hydrosocial Territory: Imagining a Modern Hydraulic State

half the reconstruction work involves hydraulic policy, to civilize our land; the other half falls to pedagogical policy, to civilize the populace: the two are complementary and either, without the other, would prove sterile.

(Ricardo Macías Picavea, 1896. *La Tierra de los Campos*)

When, in 1898, Spain's once invincible Armada was sent to the bottom of the ocean by the American Navy in the Philippines and in Cuba, the Spanish empire came to an end. Modernization, progress, economic power and political glory could no longer be secured through geographical expansion, imperial conquest and colonial robbery. There was no alternative but to turn their gaze to the lands and people of Spain itself. A new hydro-territorial imaginary took over, centered on internal geographical conditions and reworking the nation's environment (Swyngedouw, 2014). A new modernizing discourse, articulated around water, and nurtured by heterogeneous social and political actors emerged, who shared a disdain for the imperial discourse of a "great" Spain and a concern for the need for a national vision, in the early twentieth century. This imaginary gradually imposed itself, but the traditional power choreography kept its implementation in check, boosting antagonistic political and social tensions.

After *el Disastre Colonial,* and faced with a mounting economic crisis, rising social tensions, and an antiquated, largely feudal social order, Spain's modernizing political and intellectual elites were desperately searching for a way to revive or to "regenerate" the nation's socio-economic base (Ortega Cantero, 1995). This drive to revive the nation's "spirit" became known as *"el regeneracionismo,"* associated with a movement of intellectuals, writers, technocratic modernizers (particularly the Corps of Engineers), enlightened politicians, journalists, smallholder farmer leaders and some industrialists (see e.g. Costa, 1981; Macías Picavea, 1977; Mallada, 1890). Their key themes center on sensitivity to central and southern Spain's arid, austere lands, the nation's disintegrated character, political critique, intellectual crisis, social malaise and the need to revive Spain both physically and spiritually (Maeztu, 1997). In their attempt to carve out a new socio-ecological destiny for the nation, the regenerationists pursued a mythical vision of an integrated, cohesive Spain, one that would cut through the more radical imaginaries pursued by communists or anarchists, while attenuating the more socially reactionary ideologies of the liberal and conservative forces (Swyngedouw, 1999). This section summarizes the key threads of this emerging hydro-territorial discourse and the hesitant, ultimately failed, attempts at its material realization.

Geographical isolation and the disappearance of its external geographical expansion (within which the traditional myth of a glorious Spain was framed) was to be overcome by materializing a new discourse of radically transforming Spain's *internal* geography and, particularly, its water resources (Gómez-Mendoza and Ortega-Cantero, 1987). To achieve this, regenerationism advocated reorganizing the country's physical and social geography through intervention directed by a strong state and wise men, moral revival and regional and municipal autonomy (Boelens and Post Uiterweer, 2013). This new geographical imaginary would have to lead to a new socio-physical configuration: through "internal colonization." The utopian project was to colonize Spain internally instead of distant lands, incorporating all regions and people into modernity.

The key protagonist of this revival-through-modernization was Joaquín Costa (1846–1911). He coined the term "hydraulic regenerationism" (Ortí, 1984) and argued that state-organized hydraulic politics should be a national objective "capable of reworking the Fatherland's geography and solving its complex agricultural and social problems" (Costa, 1911:90). He diagnosed the core of the nation's problem in strictly post-imperial terms and identified the foundations for a remedial therapy, a mission he formulated as a military-geographic project (Costa, 1981:13). The remaking of the fatherland would require urgently implementing a "veritable surgical politics," one that demanded "an iron surgeon" (Costa, 1998:15). Long before Neil Smith coined the term "Production of Nature" (Smith, 1984), Costa already insisted on the need to literally produce a new geography: "there is no land in Europe that less resembles a paradise than the Spanish one ... [I]f in other countries it is enough for man to help nature, here more must be done; it is necessary to create her" (Costa, 1911:3).

Many contemporaries of Costa voiced this concern for a re-engineered geography (Ayala-Carcedo and Driever, 1998). The regenerationist project became formulated as a "hydrological correction of the nation's geographical problem" (Gómez Mendoza, 1992: 236). Spain's presumably unfavorable natural conditions had to be corrected: God had made a mistake here and it was for humans to rectify the defects etched into the hydro-fluvial landscape. Hydraulic politics sublimated the totality of the nation's socio-economic program; "utopian hydraulism" (Boelens and Post Uiterweer, 2013; Ortí, 1984) became the political-ideological driving force for the great modernizing transformation Spain would have to embark on – the material and symbolic kernel around which the possibility for a national rebirth was articulated (Hernández, 1994). Just, inclusive development would depend, so the argument went, on rectifying the injustices inscribed in the fluvial regime.

The hydraulic argument was grounded in both historical facts and idealization of the Arab past: "If you wish to leave traces of your passage through power, irrigate fields; the Arabs passed through Spain: their race, their religion, their codes, their temples, their tombs have all vanished, but their memory remains alive, because their irrigation has persisted" (Costa, 1911: 1). In addition, utopian hydraulism was simultaneously built on the uneasy fusion of geographical facts, arguments and dreams: the uneven distribution of rainfall, and the torrential, intermittent nature of Spain's fluvial system. Costa and allies argued: "there are immense deposits in the crests and bowels of the mountains, and we can, with mathematical regularity, distribute it over the land, crisscrossing the country with a hydraulic arterial system that mitigates its heat and quenches its thirst" (Costa, cited in Gil Olcina, 2001:10). Ricardo Macías-Picavea summarized the hydraulic mission as a necessary strategy for national development:

There are countries which ... can solely and exclusively become civilized with such a hydraulic policy, planned and developed by means of designated grand works. Spain is among them ... Therefore, hydraulic politics imposes itself; this requires changing all the national forces in the direction of this gigantic enterprise ... We have to dare to restore great lakes, create real interior seas of fresh water, multiply vast marshes, erect many great dams, and mine, exploit and withhold the drops of water that fall over the peninsula without returning, if possible, a single drop to the sea.

(Macías Picavea, 1977: 318–20)

Ortí shows the symbolic force of this material intervention by producing a new hydraulic geography: "hydraulic regeneration" constituted "a mythical power, a collective illusion and the imagined reconciliation of diverse ideologies" (1984: 12). For the regenerationists, producing a new geography while revolutionizing the state's internal operation would help mitigate social tensions. This, in turn, enabled a strong political bloc with a modernist vision of Spain's future to form – an alliance aiming to defeat the traditionalists *and* keep revolutionary socialists and anarchists at bay. Rather than following contemporary Socialist and Marxist thinking (which sought to reform basic economic and political structures) regenerationists aimed to "change men to change structures ... this creation of new men will necessarily save Spain from its slump – cultural, economic and political – largely caused by uncultured governance" (Maurice and Serrano, 1977: 55). Therefore, far beyond a mere material hydro-technological project, utopian hydraulism sought a radical change in nature and humans simultaneously; as such, Joaquín Costa proposed hydraulic policy as a means to "combat the misfortunes of geography and our breed, a work of art to remedy our inferiority in both respects" (quoted in Ortí, 1984: 93) – civilizing nature and people at once.

For the regenerationists, the national state was the vehicle for this reform and was called upon to take charge of a national plan for large dams and irrigation canals. Moreover, productivist modernization through the hydraulic engine would consolidate the liberal state in Spain: a free-market-based national economy driven by an alliance of small owners, industrialists, and modernizing engineers, supported by the state – grand hydraulic works would lay the foundations for modernizing Spain. Strongly driven by regenerationist imaginaries, the Corps of Engineers advocated state-based dam and irrigation projects to counter the agricultural crisis and to reverse the dismal results of private initiative in water works. They emphatically stated that "we [The Corps] have begun the war of peace, the war of labor, the fight for progress that, instead of devastating, restores; instead of destroying, builds; instead of draining, enriches" (Cuerpo de Ingenieros, 1899: 131).

The liberal bourgeoisie also embraced the state as the key protagonist of hydraulic revival: regenerationists expected harmony and collective progress from the "natural pacts" that (should) prevail between the state (represented by the Head of State) and "the people" (considered a collaborative network among classes). They assumed that hydraulic policy would be broadly accepted, because of its collective benefits and "intrinsic, unquestionable logic."

Once in government positions, though, the hydraulic reformers encountered what they felt as "broken pacts," blaming it on "the nature of the Spanish people" and deeply entrenched traditional elite practices. They concluded that the reforms would require force and "guardian dictatorship." Evidently, regenerationist mythology had always contained the seeds of strong, totalitarian, "enlightened" leadership. Costa had already called for "surgical policy" depicting a compassionate dictator as "an iron-hearted surgeon, familiar with the Spanish people's anatomy and feeling infinite compassion for them" (Costa, 1901: 86).

In practice, very few of the proposed projects were realized. While regenerationist policy-makers and engineers put their efforts into pushing through their techno-natural dreams, the social and political edifice around them was crumbling. Social agitation by anarchists, socialists and communists intensified, often leading to violent uprising and bloody repression; regionalist demands for autonomy grew. Strong traditional forces fought to maintain control over key state functions and stalled the nascent modernizers' rise to political power. All this resulted in social antagonisms that would eventually pave the way for dictatorial regimes from the 1920s onwards: the "iron surgeons" who would assemble heterogeneous forces to enable Spain's multi-scale hydraulic reordering (Swyngedouw, 2014). While a discourse of hydraulic national modernization was now firmly in place, it also pointed to the need of a "benevolent" authoritarian state to turn this watery dream into concrete, steel and institutional infrastructure.

Box 6.1 The Politics of Scalar Territorial Reconfiguration

Spain's recent history exemplifies the notion of contested hydrosocial territories, highlighting elites' material, political and discursive strategies to position and align humans, nature and thought within a network that aims to transform diverse socio-natural water worlds into a dominant governance system (Boelens, 2014). In general terms, "governmentalizing" territory (Foucault, 1991) and reshaping territorial configurations entwine technological, industrial, state-administrative, and scientific knowledge networks that enhance local-global commodity transfers, resource extraction, and development/conservation driven by non-local economic and political interests (e.g. Carrol, 2012; Hommes et al., 2016; Kaika, 2005). To achieve this, they curtail local sovereignty and create a political order that makes these local spaces comprehensible, exploitable and controllable (e.g. Ferguson and Gupta, 2002; Rodriguez-de-Francisco and Boelens, 2016).

Territorial governmentalization projects like Spain's seek to fundamentally alter local water users' identification with community, neighborhood, kinship or federative solidarity organization, to change water users' ways of belonging and behaving, according to new identity categories and hierarchies (Boelens and Post Uiterweer, 2013). Making such new subjects requires these water users to frame their world views, needs, strategies and relationships differently, building and believing in new models of agency, causality, identity and responsibility. Simultaneously, such frames exclude other options and thus "delimit the universe of further scientific inquiry, political discourse, and possible policy options" (Jasanoff and Wynne, 1998: 5). Governmentalizing territories, indeed, means creating particular forms of consciousness that are called upon – presumably in a self-evident manner – to defend particular water policies, authorities, hierarchies, and management practices (e.g. Duarte-Abadía et al., 2015; Harris, 2009; Swyngedouw and Williams, 2016).

Subtle imposition (or less-subtle indoctrination) of particular perspectives toward hydrosocial territories constitutes a politics of truth, legitimizing certain water knowledge, practices and governance forms and discrediting others. They separate "legitimate" forms of water knowledge, rights and organization from "illegitimate" forms (Forsyth, 2003; Foucault et al., 2007). As a result, water knowledge and truth production – and the ways they inform and shape particular water artifacts, rules, rights and organizational structures – concentrate on how to align local users and livelihoods to the imagined multi-scale water-power hierarchies (Boelens, 2015). Discourses about "hydrosocial territory" weld power and knowledge together (Foucault, 1980) to ensure a specific political order as if it were a naturalized system, by making fixed linkages and logical relations among a specified set of actors, objects, categories and concepts that define the nature of problems as well as the solutions to overcome them (Linton and Budds, 2014; Swyngedouw, 2004; Zwarteveen and Boelens, 2014).

As Boelens et al. (2016) have explained, hydrosocial territoriality, as a battle of divergent (dominant/non-dominant) discourses or narratives, has the consolidation of a particular order of things as its central stake. Though thoroughly mediated in everyday praxis, ruling groups strategically deploy discourses that define and position the social and the material in a human-material-natural network to leave the political order unchallenged and stabilize their ways of "conducting subject populations' conduct" (Foucault, 1980, 1991). Spanish history shows how these disputed socio-natural networking efforts entwine material-physical, political-structural and symbolic-discursive forms of violence, while always being challenged by "counter-conducts."

6.4 Erasing and Reconstructing Spain's Hydrosocial Territory: Imposing a Modern Hydraulic Nation

Early in the last century, despite practical failures, regenerationist discourse and policy efforts had infused the imaginary of an integrated hydraulic politics, and solidified in the minds of many Spanish leaders. Relentless campaigning, the endless flow of reports, plans, speeches, analyses, and proposals had produced a new imaginary around water and territory: as a quilting point around which a particular metonymic string of signifiers was woven: modernization, development, regeneration, irrigation, engineering technologies, steel and concrete, dams, integration, social cohesion, national pride (Swyngedouw, 2014). This progressive, utopian-inspired hydro-territorial mindset set the stage for right-wing dictatorships to show that, unlike left-wing dreamer politicians' failures, the multi-scale technopolitical water society was achievable. This also required scalar reorganization of territorialities for hydraulic organization and governance, enabling hydro-modernizers to take command of the country's hydraulic modernization.

6.4.1 Enlightened Hydraulic Despotism

Spain's September 13, 1923 coup-d'état by General Primo de Rivera set up a dictatorship to "save the nation from professional politicians," abolished all political parties, depoliticized government and technified governance. Direct action replaced political debate, with particular focus on agrarian production, hydraulic development, transport and improving the country's domestic economy. What had proven impossible to achieve during the first decades of the century was finally set in motion, although it would prove too little, too late to save Spain from slipping into the abyss of Civil War. Many regenerationists welcomed the dictatorship as a means to unblock the political system's inertia and deadlock. The Corps of Engineers cautiously endorsed the dictatorship, combining military tradition with Costa's mythical "iron surgeon." The colonial disaster figured prominently in this ideological edifice; only after a major effort of national reconstruction could "the pulse of Spain" be restored (Swyngedouw, 2014).

The authoritarian regime's regenerationist credentials were secured by appointing a hydraulic engineer with impeccable conservative credentials, Rafael Benjumea-Burín, as Minister of Development in 1925. Benjumea, famous for his construction of impressive large dams, had always enthusiastically admired Costa and his utopian hydraulism (Boelens and Post Uiterweer, 2013; Martín Gaite, 2003); both shared their idealism for improving society through concrete works, avoiding political debates and abstract, bureaucratic solutions; they shared an ideology of a positivistic, plannable society based on scientific technical-managerial rationality, firmly rooted in natural sciences. They also shared an "admiration for the policy of enlightened despotism" (Martín-Gaite, 2003). On March 5, 1926, shortly after taking office, Benjumea changed the nation's water policy administration decisively by creating the River Basin Confederations. He appointed engineer Manuel Lorenzo Pardo, founder of the Ebro River Basin Confederation, to organize the Confederations throughout Spain. Lorenzo Pardo, an enigmatic figure in the unfolding of Spain's hydrosocial edifice and author of the First National Plan of Hydraulic Works in 1933, had coordinated some of the largest hydraulic projects at the time.

The new vision of integrated basin governance extended the hydraulic community's aspirations by insisting on the need to up-scale from a focus on individual projects to integrated, interconnected management of entire river basins. Based on the exemplary Ebro model (Lorenzo Pardo, 1930, 1931), Benjumea and Lorenzo Pardo materialized the internationally acclaimed concept of River Basin Confederations, to manage Spain's national water system. Benjumea later called this national process "the splendor of my loves, integrating river management by organizing industry, agriculture and society as a whole." The Royal Decree praised its supposed political neutrality, its technical and ecological superiority and its inherent "justice": "This undertaking entails justice, a great moral value, as a significant example of social solidarity and patriotic exaltation ... free of all parties and factions, creating a meeting-ground for Spaniards' regenerating drive" (Martín Gaite, 2003: 79).

In 1930, by the end of the dictatorship (overthrown because of scandalous squandering of resources and large-scale protests), several river basin authorities had been established as quasi-autonomous techno-administrative bodies to plan and implement the nation's hydraulic policy. However, the entrenched opposition to the new territorial power base, configured around river basins, signaled profound distrust of multiple, mutually conflicting political and socio-economic interests. Nonetheless, throughout this tumultuous period that would culminate in the horror of the Civil War (1936–39), the scalar geometry of hydraulic management was reordered to strategically align the national scale and the river-basin scale. This, in turn, would set the framework that the Franco dictatorship later mobilized effectively (Swyngedouw, 2014). With river-basin authorities in place, the hydraulic-modernization discourse began to integrate gradually with emerging new water-governance configurations.

6.4.2 The Republican Interval

With the first constitutional government of the Republic in 1931, socialist Minister Prieto consolidated the river-basin authorities. Prieto appointed Lorenzo Pardo as head of the Hydraulic Planning Section and instructed him to create a national institute to plan and manage the nation's water resources. The First National Water Plan was developed, which set the parameters for water governance for the next 60 years. This plan radically "re-imagineered" the hydrosocial cycle and mobilized H_2O in a national framework, a fundamental idea that later would be embraced fully by the Franco regime. The Plan envisaged constructing 215 new dams and canals, extending irrigation by an additional 1.3 million ha, thereby doubling the surface area of irrigated land. This, in turn, would make it possible to repopulate deserted areas and improve Spanish food production self-sufficiency, while strengthening its international competitive position. The most significant contribution is the view of integrated river basin management under a national assessment of regional water availabilities and requirements. This would transform the imaginary of water, from something to be considered and engineered within the unit of each basin, to a national vision. What Pardo defined as the "hydrological imbalance" between the Atlantic and Mediterranean basins would now be "rectified" by inter-basin water transfers. Water transfers from the Ebro and the Tajo River to the basins of the Levant and the south would solve the problem of systematic under-supply. A national hydraulic imaginary was brewing, one that linked national development with a national vision of how to organize water flowing on the surface.

Rectifying unequal distribution of abundant waters, soon to be redressed as rectifying socio-spatial inequalities and injustices, would become the leitmotiv of the state's hydraulic intervention. Of course, turmoil in the thirties and forties would prevent early realization of Pardo's national hydraulic dream. The fantasy, however, would galvanize the imaginaries around which Franco and his allies would cement a hegemonic vision of autarchic national development and spatial integration (Swyngedouw, 2014). Indeed, after the Civil War, General Franco's iron hand, with the co-operation of most strata of Spanish society, would implement the long-cherished dream of national hydro-territorial transformation as the lynchpin for Spain's modernization.

6.4.3 Paco el Rana's Watery Dream for Spain

General Francisco Bahamonde Franco constructed more than 600 dams, small and large, as well as the first major water transfer (Vallarino, 1992), leading to a complete re-engineering of mainland Spain's ten continental river basins. As Gómez-De Pablos (1972: 242) puts it, "during the two decades after the Plan of 1940, Spanish rivers were really 'created'." Under

Franco's rule, Spain's hydraulic development indeed reached its apogee and its logic continued after the transition to democracy. The construction of a large inter-basin water transfer scheme, from the Tajo to the Segura basin, would become the pivotal project, which physically consolidated the national scale as the central arena for hydrological planning. Throughout the Franco years, water infrastructure and transforming Spain's techno-natural edifice was mobilized with relentless zeal by the propaganda machinery, to such an extent that the popular nickname for General Franco is *Paco El Rana* (Frankie the Frog). The omnipresent image of Franco during this period was "on the water," while inaugurating yet another hydro-infrastructure (Swyngedouw, 2007).

The fascist national project would seamlessly weld together the national engineering dream of "balancing" Spain's uneven hydrology with a discourse and vision of justice, equality and organic national development (Ortega Cantero, 1975; Ortí, 1984). If problems of scarcity and hydrosocial injustice remained, this was simply because the state was unable to perform its functions adequately. This vision of state water management generated a sense of unlimited potential availability. The older notion of water "scarcity" became rescripted and "scientifically" defined as "deficit," "imbalance" or "disequilibrium" between the regionally desired volumes and the nationally available quantities. The uneven distribution of rainfall and water availability became resymbolized as an imbalance requiring "rectification." Of course, imagining abundant "national" waters but regional "imbalances" required a scalar reconfiguration that shifted the gaze from considering the hydraulic balances at the river-basin scale to the national territorial scale. The latter, in turn, was predicated upon disavowing regional(ist) demands and regional hydraulic autonomy, rearticulating an integrated space by transferring water from "surplus" to "deficit" river basins in a vision of national and just solidarity. Indeed, this particular staging and mobilization of water was captured effectively by Franco himself: "We are prepared to make sure that not a single drop of water is lost and that not a single injustice remains" (Franco, 1959b:1).

The argument that water infrastructures would pave the way to a wet, fertile, balanced future was staged as one of the vital and central projects for realizing the Fascist utopia. The extract below is just one among dozens in which Franco mobilizes water as an integral part of his politics:

Spain aches because of its drought, its misery, the needs of our villages and hamlets; and all Spain's pain is redeemed by these grand national hydraulic works, with this Reservoir of the Ebro and all the others that will be created in all the basins of our rivers, embellishing the landscape and producing this golden liquid that is the basis of our independence.

(Franco, cited in Río Cisneros, 1964: 122–23)

Towards the end of Franco's life, he was seen as the great dam-builder. *Paco El Rana* had indeed directed a complete socio-hydraulic revolution. Achieving this water "activism" depended crucially on the loyal support of a series of powerful (inter)national networks of interests (Swyngedouw, 2007) and on a new relationship between the authoritarian state and the international community.

However, despite significant support from a variety of actors, the early autarchic fascist model (1939–55) did not generate enough capital and investment to move the rocks and build the dams. To obtain the necessary capital, Spain's elites began to recognize that Spain's geo-political insertion into the Western Alliance was vital to modernize Spain, and to prop up the dictatorial regime. The US also started to turn its geo-political gaze toward Spain as a possible ally in the strategic post-war geometry. A most significant moment was undoubtedly the secret Pact of Madrid, signed in September 1953, in which Spain agreed to let the US use parts of Spain's territory for military bases in exchange for economic, military and technical aid. This capital inflow enabled rapid infrastructure development. Between 1951 and 1963, more than US$1.3 billion were granted to Spain as economic aid, a substantial part for agricultural machinery, steel, electrical equipment, and infrastructure. Most of the Spanish counterpart funding was invested in agricultural irrigation projects, railroads, and hydraulic works. The dictatorship's scalar rearrangement of networks proved vital for modernization. The great hydraulic leap forward happened after 1955, with 276 dams built by 1970. The total volume of water reservoirs capacity skyrocketed exponentially from 8.3 to 36.9 billion cubic meters (Swyngedouw, 2014). The Tajo-Segura water transfer project was approved and construction commenced. Franco himself insisted that his "great hydraulic and irrigation works are changing the geography of Spain" (Franco, 1959:1).

Indeed, Spain modernized extraordinarily quickly during the 1960s and early 1970s, a process that further ensured the regime's survival. Franco's death in 1975 ended one of Europe's most repressive and long-lasting dictatorial-fascist regimes (Swyngedouw, 2014). The Franco era witnessed how utopian hydraulism necessarily became a large-scale dystopian drama, leaving a bitter legacy. The ideology and its materialization in practice are full of intrinsic contradictions and paradoxes (Boelens and Post Uiterweer, 2013), and the results achieved were the exact opposite of early regenerationism's progressive objectives. Under Franco, a system of inter-basin transfers was established as the backbone of this hydro-political territory, integrating the whole country under centralistic despotism – the sad, contradictory legacy of the regenerationists' dreams of autonomy and decentralization.

6.4.4 Political and Socio-Technical Transformation

Long after the democratic transition, hydraulic administration continued with its mission to build large-scale state-led hydraulic interventions, militating to complete the transformation they had spearheaded during the Franco era. Estevan typifies this as the country's hydrological and cultural anomaly: "no country in Europe has gone as far as Spain, where exclusive identification of water policy with large hydraulic works was not simply a main feature of Franco's dictatorship, but had already existed for a long time, and after *Franquismo* survived in the political and administrative institutions of democracy" (Estevan, 2008: 22).

But the 1980s and 1990s also marked an era of heated controversies over water, that would gradually, and amidst great dispute and proliferating inter-regional conflict, move the water agenda – together with state transformation – into new, largely uncharted terrain. In the transition years, policy discourse insisted on Spain's deficit and surplus river basins. Julián Campo, Public Works Minister in the first Socialist government, exclaimed with great pride: "I am going to build more dams than Franco" (Llamas Madurga, 1984: 18). Democratization, believed most hydro-experts, would finally complete Franco's watery dream for Spain.

Soon after the restoration of democracy, debate started over the need to replace the 1879 water law. In 1985, a new Water Law was passed. Key ingredients again emphasized the state's centrality and water as a public good. River basins were defined as the central management unit. Apparently significant power over water was devolved to the Autonomous Regions. However, trans-communitarian river basins stayed under central state control. Subsequent preparations for a new National Hydraulic Plan (NHP) would become a hornet's nest of intense social and political passions, sharpening social and spatial conflicts that had lain dormant for years. The draft NHP (published in 1993) hit the water community like a bombshell. It proposed an "Integrated System of National Hydraulic Equilibrium," seeking to upscale management of Spain's mainland waters to the national scale. Its doctrinal core was the thesis of national hydraulic disequilibrium. Hundreds of new dams and 14 major water transfers would rectify "nature's imbalance" and finally achieve a just distribution of the available resources. The draft plan proposed to interconnect all river basins, with an eye to transferring significant volumes of water from the Ebro, Duero and Tajo (with their presumed "excess" water) to the "deficit" basins of the South, the Levant and Catalonia. At the same time, a State Organization for National Hydraulic Equilibrium, a kind of super-basin management organization, would be established to organize and police the national water grid (Gil Olcina, 1995).

Soon after the draft plan was published, major controversies began to arise. Even from within the hydraulic engineering community, voices of dissent were raised against the single-minded supply-side measures based on constructing large hydro-infrastructures. Manuel Llamas felt that the nationally integrated hydraulic system was created "to reinforce even more the Orwellian figure of Big Hydraulic Brother who will decide all Spanish waters' destiny" (Llamas-Madurga, 1996: 101). Moreover, having the lever on water transfers enabled the national government to use water as quid pro quo in negotiating agreements with regionalist parties and with the newly established autonomous regions. The plan indeed became a central political controversy and political deadlock well into the early twenty-first century. The cacophony of new voices articulating their visions produced a stalemate. It also signaled the demise of the traditional hydraulic policy community. Greater regional identification cut through traditional nationally organized party cleavages, and water soon became a pivotal axis in choreographing inter-regional conflict (Lopez-Gunn, 2009). The proposed large-scale water transfers turned into a veritable Gordian knot in managing territorial political tensions, particularly between water-ceding and water-receiving regions (Sauri and Moral, 2001).

A final plan was approved in 2001; still reveling in constructing 200 more anticipated dams and continuing transfers, although its ambitions had been scaled down significantly, while water pricing and environmental concerns took a more prominent place. Immediately after approval, a highly active Platform to Defend the Ebro River had been established. The

Platform succeeded in staging an extraordinarily successful resistance campaign, bringing together 400,000 people in Barcelona and mass demonstrations in Zaragoza and Madrid in 2002. A "Blue March" was organized from Aragón to Brussels to bring the cause of the Ebro water and its peoples to broader European attention (Lopez-Gunn, 2009). The fusion of ecological arguments, the rights of humans and non-humans to water and defense of the intimate relationship between water practices and local/regional identity all forged a loose alliance of activists into a formidable oppositional force. Of course, the receiving regions, Valencia and Murcia in particular, defended with similar zeal "Agua para Todos" (Water for All) to make a case for "an equitable and just distribution" of Spain's water.

After intensive struggles, the Ebro transfer was suspended, while simultaneously bringing forward a new hydro-political configuration and discursive construction project: a total of 21 newly constructed or upgraded desalination plants were programmed along the Mediterranean coast. Extending hydrosocial cycle management into the sea, as a new geographical "fix" for the country's uneven distribution of water, had been contemplated since the dying days of Fascism, but was now rapidly emerging as the new panacea (Swyngedouw and Williams, 2016). While terrestrial waters are increasingly marred by complex property rights, inserted in dense regulatory and institutional arrangements, subjected to all manner of social, cultural and ecological conflicts, and integral parts of multi-scale tensions and inter-regional rivalries, seawater is seemingly free of these highly charged meanings, practices and claims. And while water transfers and other big infrastructure projects relayed an imaginary associated with top-down bureaucratic politics, desalination was staged as "local," democratic, decentralized, market-efficient and ecologically sustainable. The techno-natural configuration of "desal" made it possible to align the water conflict with some stakeholders' desire to reinforce regional autonomy, defending it from what many consider the invasive powers of centralized Spain (Swyngedouw, 2015). In addition, some stakeholders discerned the kernel for "green" or "ecological" modernization in this new socio-technical edifice. So, again, the Spanish state's transformation was significantly paralleled by a new hydrosocial assemblage of social, technical, and physical relations, sustained by new imaginaries and symbolic framings.

6.5 Concluding Remarks

Investigating the hydrosocial territories concept, contested territorialization, and notions of water justice, in this chapter we have shown how Spain's hydraulic history expresses different forms of socio-ecological justice and equality. This was originally formulated around a naturalistic, utopian vision considering physical distribution of water and rainfall in Spain as unjust, requiring rectification through decentralized human intervention. The relative failure of this idealistic endeavor during the early twentieth century shifted their gaze to consider the national state as the conduit through which to deliver water justice. Rectifying "unjust" water distribution became one of the key ideological support structures of Spain's long fascist regime between 1939 and 1975. Water and justice became tightly knitted together in the fascist regime's national development vision. In the process, the discursive framing shifted from considering localized and absolute scarcity as the object-cause of an unjust hydrosocial configuration to a focus on redistributing an unequally distributed resource throughout the nation-state, if necessary by outright force and violence. In the final part, we have examined the reworking of water justice in the transition to democracy, in a context of both social differentiation and intense neoliberalization. The Spanish case illustrates how justice is indeed in the eye of the beholder, and makes a case for considering hydrosocial equality as a politically more performative perspective, one that Spain is currently experimenting with, too.

We argue that water governance and production of new hydrosocial territories simultaneously steers and results from the intersection and confrontation of divergent territorial perspectives and the realization of contested political-economic, socio-environmental imaginaries and discourses, under unequal power relations. More concretely, we have shown how production of particular socio-technical configurations (such as dams and inter-basin water transfers) depends on assembling/enrolling particular social groups, cultural discourses, technical expertise, material conditions, the variegated actions involving water, and shifting political-economic power relations within ongoing state reconfiguration.

This way, a hydrosocial-territories perspective enables us to understand how everyday territorial politics finds expression in encounters among diverse, divergent political and geographical projects, such as: state organization, spatial control over water, and power relations among local, regional, national and global political and economic alliances. All of these compete, superimpose, and foster their territorial projects to strengthen their water control. Additionally, focusing on hydrosocial territories reveals the dynamics of water control, alternative ways of conceptualizing and building nature-society-power relations, and divergent claims and political strategies to foster "water justice." This argues for the need to politicize the water access and distribution mechanisms that are built into hydro-territorial planning, the relationships that shape norms, rights, and rules regarding water decision-making, and the discursive "regimes of truth" that underpin water policies and hydrosocial territorial reform.

Struggles for water justice therefore necessarily entail the effort to "redesign" and reshape the hydraulic grids, units and artifacts underlying dominant hydrosocial territories' structure and logic. They involve transforming technology-embedded cultural and distributional norms and political relations, including the corresponding definitions of proper functioning, social suitability, and technical efficiency. They also involve building and engaging in new multi-scale networks. Whether, to what extent, and in what ways, the dominant or opposing alliances are successful in producing, reinforcing, or reordering the hydrosocial territories and associated water justice paradigms they envision, depends on their capacity to network, mobilize, and exercise power.

Notes

1 Various parts of this chapter are based on Swyngedouw (1999, 2007, 2014, 2015); Boelens *et al.* (2016), and Boelens and Post Uiterweer (2013). See also Swyngedouw and Williams (2016).
2 "Hydrosocial territory" presents a further elaboration compared to related concepts (e.g. hydrosocial networks (waterscape and cycle), on the one hand, and literature on territories, on the other). Waterscape, socionature, and imagined geography literature provide important building blocks.

Most territory literature does not fundamentally theorize the particularities and fluid properties of technology (materialized and embedded norms) in shaping geographies (e.g. the (de)politicization and "moralization" of technology). Most waterscape literature focuses particularly on the hegemonic structures and discourses that drive (and follow from) socio-natural reconfiguration. While maintaining the essence of how water flows, technologies, institutions and power structures are arts of statecraft and shape multi-scalar political geography; hydrosocial territory conceptual innovation stresses the resulting diversity in terms of overlapping, simultaneously existing hydro-territorial regimes and imaginaries (in one and the same geo-political location, and with unequal power). Further, actor-oriented approaches and interface concepts have inherently stronger presence. Moreover, incorporating the governmentality focus turns attention to the domination-resistance web's normalizing subtleties. The concept also explicitly deals with modernist territorial recognition politics: the tactics of "recognizing" and "incorporating" local territorialities (integrating local norms, practices, and discourses into mainstream government rationality and its spatial/political organization) make state/formal and local/vernacular modes of territorial ordering depend on each other in complicated (and often confrontational) ways.

References

Agnew, J. (1994). The territorial trap: The geographical assumptions of international relations theory. *Review of International Political Economy*, 1(1), 53–80.

Ayala-Carcedo, F. J. and Driever, S. L. (eds.) (1998). *Lucas Mallada – La Futura Revolución Española y otros Escritos Regeneracionistas*. Madrid: Editorial Bibliotheca Nueva.

Baletti, B. (2012). Ordenamento territorial: Neo-developmentalism and the struggle for territory in the lower Brazilian Amazon. *The Journal of Peasant Studies*, 39(2), 573–98.

Barnes, J. and Alatout, S. (2012). Water worlds. *Social Studies of Science*, 42(4), 483–88.

Boelens, R. (2014). Cultural politics and the hydrosocial cycle: Water, power and identity in the Andean highlands. *Geoforum*, 57, 234–47.

Boelens, R. (2015). *Water, Power and Identity: The Cultural Politics of Water in the Andes*. London: Earthscan, Routledge.

Boelens, R., Hoogesteger, J., Swyngedouw, E., Vos, J. and Wester, P. (2016). Hydrosocial territories: A political ecology perspective. *Water International*, 41(1), 1–14.

Boelens, R. and Post Uiterweer, N. C. (2013). Hydraulic heroes: The ironies of utopian hydraulism and its politics of autonomy in the Guadalhorce Valley, Spain. *Journal of Historical Geography*, 41, 44–58.

Brenner, N. (1998). Between fixity and motion: Accumulation, territorial organization and the historical geography of spatial scales. *Environment and Planning D*, 16, 459–81.

Bridge, G. and Perreault, T. (2009). Environmental governance. In N. Castree, D. Demeritt, D. Liverman, and B. Rhoads (eds.), *Companion to Environmental Geography*. Oxford: Blackwell, pp. 475–497.

Carroll, P. (2012). Water and technoscientific state formation in California. *Social Studies of Science*, 42(4), 489–516.

Costa, J. (1911). *Política Hidráulica (Misión Social de los Riegos en España)*. Madrid: Biblioteca J. Costa.

Costa, J. (1981(1900)). *Reconstitución y Europeización de España*. Madrid: Instituto de Estudios de Administración Local.

Costa, J. (1998(1901)). *Oligarquía y Caciquismo*. Madrid: Editorial Biblioteca Nueva.

Cuerpo de Ingenieros. (1899). Pantanos y Canales de Riego. *Revista de Obras Publicas*, 46, 1229.

Duarte-Abadía, B. and Boelens, R. (2016). Disputes over territorial boundaries and diverging valuation languages: The Santurban hydrosocial highlands territory in Colombia. *Water International*, 41(1), 15–36.

Duarte-Abadía, B., Boelens, R. and Roa-Avendaño, T. (2015). Hydropower, encroachment and the repatterning of hydrosocial territory: The case of Hidrosogamoso in Colombia. *Human Organization*, 74(3), 243–54.

Elden, S. (2010). Land, terrain, territory. *Progress in Human Geography*, 34(6), 799–817.

Estevan, A. (2008). *Herencias y problemas de la política hidráulica española*. Bilbao: Bakeaz/Fundación Nueva Cultura del Agua.

Ferguson, J. and Gupta, A. (2002). Spatializing state: Toward an ethnography of neoliberal governmentality. *American Ethnologist*, 29(4), 981–1002.

Forsyth, T. (2003). *Critical Political Ecology: The Politics of Environmental Sciences*. London and New York: Routledge.

Foucault, M. (1980). *Power/Knowledge: Selected Interviews and Other Writings 1972–1978*, C. Gordon, ed. New York: Pantheon Books.

Foucault, M. (1991[1978]). Governmentality. In G. Burchell, C. Gordon and P. Miller (eds.), *The Foucault Effect: Studies in Governmentality*. Chicago: University of Chicago Press, pp. 87–104.

Foucault, M., Sellenart, M. and Burchell, G. (2007). *Security, Territory, Population*. New York: Palgrave Macmillan.

Franco, F. B. (1959a). Speech of chief of state in Medina del Campo, *Diario ABC*, nr. 16734, October 30, 1959.

Franco, F. B. (1959b) Discourse of the head of state. *Diario ABC*, November 3, 1959.

Gil Olcina, A. (1995) Conflictos autonómicos sobre trasvases de agua en España. *Investigaciones Geográficas*, 13, 17–28.

Gil Olcina, A. (2001). Del plan general de 1902 a la planificación hidrológica. *Investigaciones Geográficas*, 25, 5–31.

Gómez de Pablos-González, M. (1972). El Centro de Estudios Hidrográficos y la Planificación Hidráulica Española. *Revista de Obras Públicas*, 120, 3096.

Gómez-Mendoza, J. (1992). Regeneracionismo y regadíos. In A. Gil Olcina and A. Morales Gil (eds.), *Hitos Históricos de los Regadíos Españoles*. Madrid: Ministerio de Agricultura, Pesca y Alimentación.

Gómez-Mendoza, J. and Ortega-Cantero, N. (1987). Geografía y regeneracionismo en España. *Sistema*, 77, 77–89.

Harris, L. (2009). States at the limit: Tracing contemporary state–society relations in the Borderlands of Southeastern Turkey. *European Journal of Turkish Studies*, 10, 2–19.

Hernández, J. M. (1994). La planificación hidrológica en España. *Revista de Estudios Agro-Sociales* 167, 13–25.

Heynen, N. and Swyngedouw, E. (2003). Urban political ecology, justice and the politics of scale. *Antipode*, 34 (4), 898–918.

Hommes, L. and R. Boelens (2017). Urbanizing rural waters: Rural-urban water transfers and the reconfiguration of hydrosocial territories in Lima. *Political Geography*, 57, 71–80.

Hommes, L., Boelens, R. and Maat, H. (2016). Contested hydrosocial territories and disputed water governance: Struggles and competing claims over the Ilisu Dam development in southeastern Turkey. *Geoforum*, 71, 9–20.

Hoogesteger, J. and Verzijl, A. (2015). Grassroots scalar politics: Insights from peasant water struggles in the Ecuadorian and Peruvian Andes. *Geoforum*, 62, 13–23.

Hoogesteger, J., Boelens, R. and Baud, M. (2016). Territorial pluralism: Water users' multi-scalar struggles against state ordering in Ecuador's highlands. *Water International*, 41(1), 91–106.

Jasanoff, S. and Brian, W. (1998). Science and decision making. In S. Rayner and E. L. Malone (eds.), *Human Choice and Climate Change*. Columbus, OH: Battelle Press, pp. 1–87.

Kaika, M. (2005). *City of Flows: Modernity, Nature and the City*. London: Routledge.

Kaika, M. (2006). Dams as symbols of modernization: The urbanization of nature between geographical imagination and materiality. *Annals Association of American Geographers*, 96(2), 276–301.

Latour, B. (1993). *We Have Never Been Modern*. Cambridge, MA: Harvard University Press.

Lefebvre, H. (1991). *The Production of Space*. Oxford: Blackwell.

Linton, J. and Budds, J. (2014). The hydro-social cycle: Defining and mobilizing a relational-dialectical approach to water. *Geoforum*, 57, 170–80.

Llamas Madurga, M. R. (1984). Política hidráulica y genesis de mitos hidráulicos en España. *Cimbra*, 218, 16–25.

Llamas Madurga, M. R. (1996). *El Agua en España: Problemas Principales y Posibles Soluciones*. Madrid: Real Academia de Ciencias. Instituto de Ecología y Mercado.

Lopez-Gunn, E. (2009). Agua para todos: A new regionalist hydraulic paradigm in Spain. *Water Alternatives*, 2/3, 370–94.

Lorenzo Pardo, M. (1930). *La Confederación del Ebro: Nueva Política Hidráulica*. Madrid: Compañía Ibero-Americana de Publicaciones.

Lorenzo Pardo, M. (1931). *La Conquista del Ebro*. Zaragoza: Editorial Heraldo de Aragón.

Macías-Picavea, R. (1896). *La Tierra de los Campos*. Madrid: Suárez.

Macías-Picavea, R. (1977 (1899)). *El Problema Nacional*. Madrid: Instituto de Estudios de Administración Local.

Mallada, L. (1890). *Los Males de la Patria y la Futura Revolución Española*. Madrid: Fundación Banco Exterior.

Maeztu, R. de (1997 (1899)). *Hacia otra España*. Madrid: Editorial Biblioteca Nueva

Martín Gaite, C. (2003). *El Conde del Guadalhorce. Su Epoca y su Labor*. Madrid: Tabla Rasa.

Maurice J. and Serrano, C. (1977). *J. Costa: Crisis de la Restauración y Populismo (1875–1911)*. Madrid: Siglo XXI.

Mosse, D. (2008). Epilogue: The cultural politics of water. *Journal of Southern African Studies*, 34(4), 939–48.

Ortega Cantero, N. (1975). *Política Agraria y Dominación del Espacio – Orígenes, Caracterización y Resultados de la Política de Colonización planteada en la España posterior a la Guerra Civil*. Madrid: Editorial Ayuso.

Ortega Cantero, N. (1995). El Plan general de Canales de Riego y Pantanos de 1902. In A. Gil Olcina and A. Morales-Gil (eds.), *Planificación Hidráulica en España*. Murcia: Fundación Caja del Mediterráneo, pp. 107–36.

Ortí, Alfonso (1984). Política hidráulica y cuestión social: Orígenes, etapas y significados del Regeneracionismo Hidráulico de Joaquín Costa. *Revista Agricultura y Sociedad*, 32, 11–107.

Perreault, T. (2014). What kind of governance for what kind of equity? Towards a theorization of justice in water governance. *Water International*, 39(2), 233–45.

Perreault, T., Bridge, G. and McCarthy, J. (eds.) (2015). *The Handbook of Political Ecology*. London: Routledge.

Río Cisneros, A. del (1964). *Pensamiento Político de Franco – Antología*. Madrid: Servicio Informativo Español.

Rodríguez-de-Francisco, J. C. and Boelens, R. (2016). PES hydrosocial territories: De-territorialization and re-patterning of water control arenas in the Andean highlands. *Water International*, 41(1), 140–56.

Sauri, D. and del Moral, L. (2001). Recent developments in Spanish water policy: Alternatives and conflicts at the end of the hydraulic age. *Geoforum* 32, 351–62

Seemann, M. (2016). Inclusive recognition politics and the struggle over hydrosocial territories in two Bolivian highland communities. *Water International*, 41(1), 157–72.

Smith, N. (1984). *Uneven Development*. Oxford: Blackwell.

Swyngedouw, E. (1999). Modernity and hybridity: Nature, regeneracionismo, and the production of the Spanish waterscape, 1890–1930. *Annals of the American Association of Geographers*, 89(3), 443–65.

Swyngedouw, E. (2004). Globalisation or "glocalisation"? Networks, territories and rescaling. *Cambridge Review of International Affairs*, 17(1), 25–48.

Swyngedouw, E. (2007). Technonatural revolutions: The scalar politics of Franco's hydro-social dream for Spain, 1939–1975. *Transactions of the Institute of British Geographers*, 32(1), 9–28.

Swyngedouw, E. (2009). The political economy and political ecology of the hydrosocial cycle. *Journal of Contemporary Water Research & Education*, 142, 56–60.

Swyngedouw, E. (2014). "Not a drop of water …": State, modernity and the production of nature in Spain, 1898–2010. *Environment and History*, 20(1), 67–92.

Swyngedouw, E. (2015). *Liquid Power: Contested Hydro-Modernities in 20th Century Spain*. Cambridge, MA: MIT Press.

Swyngedouw, E. and Williams, J. (2016). From Spain's hydro-deadlock to the desalination fix. *Water International*, 41(1), 54–73.

Vallarino Canovas del Castillo, E. (1992). Política hidráulica. *Revista de Obras Publicas*, 139, 3310.

Vos, J. and Hinojosa, L. (2016). Virtual water trade and the contestation of hydrosocial territories. *Water International*, 41(1), 37–53.

Zwarteveen, M. and Boelens, R. (2014). Defining, researching and struggling for water justice: Some conceptual building blocks for research and action. *Water International*, 39(2), 143–58.

7

Making Space for the Cauca River in Colombia: Inequalities and Environmental Citizenship

Renata Moreno-Quintero and Theresa Selfa

7.1 Introduction

During 2010 and 2011, severe floods inundated Colombia, affecting 2.8 million people, causing 519 deaths, and destroying 7,403 houses (Campos *et al.*, 2012). In the aftermath, the Colombian Government initiated a partnership with the Netherlands to prevent new flood disasters. As part of this partnership in the Cauca River Valley region of Colombia, Dutch consultants began working with the regional environmental authority CVC (Autonomous Regional Corporation of the Cauca Valley) to devise a master plan for the Cauca River aimed at limiting flooding and improving land use planning in the Cauca River watershed, under the "Room for the River" approach (RRA). Cauca River Valley is a region in southwest Colombia spanning an area of 440,000 ha, and was home to a population of 4.5 million inhabitants in 2012, primarily mestizo but also Afro-Colombians (30 percent) and less than 1 percent of indigenous population.

From the outset, the plan to make room for the river was controversial among local actors with competing values and interests regarding the river, and with different power resources to influence the ensuing negotiations. Various narratives reveal different meanings of floods and wetlands surfaced when discussing how to best manage the territory and its water resources. Floods are considered a source of life for some social groups, while for others they are regarded as destructive forces that need to be controlled. In the Cauca River Valley, Afro-Colombian communities' special relationship with the wetlands and flooding is both historical-cultural and related to their present livelihoods. Competing narratives vie to influence the definition of and the solution to the "flooding problem." In this chapter, we analyze the different narratives around the flooding issue that surfaced when proposing and debating the RRA in Colombia and how certain narratives became dominant during the process, altering the initial objectives of the envisioned interventions, and paradoxically reinforcing the use of megastructures to manage flood risk.

7.2 The Room-For-The-River Approach

The RRA falls under the category of "building with nature" (Vörösmarty *et al.*, 2010). This approach seeks to adapt to flooding in a way that harmonizes with natural processes, distinct from other approaches that emphasize "taming the river" by building floodwalls, diverting watercourses and raising dikes, which may enhance flood security, but ignore ecological conditions and increasing river discharges related to climate change (Begg *et al.*, 2015; Deeming *et al.*, 2012; Vörösmarty *et al.*, 2010). The "room for the river" approach has taken different names such as "*Ruimte voor de Rivier*" (Room for the River) in the Netherlands, "Living River" in the US, "Making space for water" in the UK, as it has developed since the 1990s to address concerns about the destruction of valuable landscapes, cultural heritage and ecosystems protection that were overlooked when implementing infrastructure and engineering solutions aimed at risk control.

However, we have recently witnessed the return of mega-structure projects such as dams around the world, despite the development and availability of new "soft" approaches such as RRA.[1] How can this resurgence be explained? Is it simply the predominance of special-interest politics and/or the developmentalist paradigm? The Cauca River Valley case contributes to answering these questions by analyzing why traditional infrastructure projects were chosen as the preferred alternative, after a negotiation process that originally intended to "make room" for the river.

Indeed, implementing RRA for flood-risk management has not proceeded without controversy. As argued by Warner *et al.* (2012), "making space for the river" means removing space for something else, since floodplains are usually occupied either by crops or by urban development. Making space for the river, then, means dealing with multiple stakeholders with different values and interests that occupy riverbanks and use and value the river and its floods in different ways. This poses many challenges for these new landscape interventions' governance, spurring research to understand how this approach is expressed in different implementation contexts and how emerging conflicts are managed when seeking to recover space for rivers. Examining this approach in the US and Europe (Warner *et al.*, 2011, 2012), and in the Netherlands (De Bruijn *et al.*, 2015; Edelenbos *et al.*, 2017; Roth and Warner, 2007; Van Buuren *et al.*, 2016; Wolsink, 2005), scholars have shown how the more horizontal governance approach entailed by "Room for the River" initiatives has actively involved actors in influencing policy framing and choosing alternatives that incorporate their local values and interests. Discursive coalitions are formed to defend or oppose flood-control projects, and actors mobilize scientific information to support their claims. Stakeholders mobilize political support through lobbying and media attention, and propose their own interventions to influence governmental plans.

In the Colombian case, there were similarities in the way local stakeholders organized, to oppose some alternatives entailing land-use changes and to develop their own alternative proposals. However, in Colombia, unequal environmental citizenship rights to implement RRA, along with dominant elites' particular imaginaries of the territory, led to a return to traditional engineering, constructing dams and raising dikes for flood risk control, sidelining alternatives and narratives from less-powerful actors more aligned with RRA. We analyze environmental citizenship, the exclusion/inclusion dynamics in defining who makes environmental decisions and large-scale reorganizations of space, identifying which groups qualify as political actors, decision makers and therefore environmental citizens (Sundberg, 2012).

To analyze these different narratives around flooding and the resulting RRA implementation plans, we draw on qualitative data gathered through case-study research on the Cauca River Corridor Project (CRCP).[2] The CRCP aimed to design a master plan for the Cauca River under RRA to be completed by the end of 2015 (although not publicly released as of July 2016), which was to be approved by the CVC board of directors and regional authorities between 2016 and 2019, providing the final alternatives for flood-risk management. To build our case study, we conducted 35 formal in-depth interviews and many informal conversations with CRCP participants and other key respondents, such as community leaders, CVC officials, sugarcane growers, and local and international project consultants.[3] Interview questions revolved around meanings and uses of wetlands, relation of wetlands to flooding, and citizen participation in CRCP. In addition, we extensively analyzed reports, workshop proceedings and bulletins produced by policy actors, sugarcane associations, and grassroots organizations. We also draw on observations at CRCP meetings, wetland-management committees, and public CRCP events. All interviews were transcribed. Interviews, field notes, and secondary documents were coded to identify commonalities and differences among actors in the values they attach to wetlands and flooding, and differential water-resource rights, use, and decision-making participation.

In the sections below, we first describe different narratives about flooding in the region. Next, we analyze how some narratives were privileged over others in devising solutions to the "flooding issue." Finally, we conclude by analyzing how the alternatives selected reflect and reinforce social inequalities in the region.

7.3 Floods are Disastrous Events, but also a Source of Life

As in the Netherlands, a flood crisis prompted the shift to a more resilient approach to flood safety in Colombia. In 2010 and 2011, extreme weather conditions resulted in devastating flooding events that inundated about 41,000 ha in 2010, and 37,000 ha in 2011, in the Cauca River Valley region, with approximate losses of US$522 million (Univalle, 2011). In December 2010, the Ministry of the Environment, Housing, and Territorial Development declared the situation a "disaster" and issued Decree 4579 to attend to the emergency. In 2011, the Colombian Government established a partnership with the Netherlands Government to design flood control plans in the affected regions (*La Mojana* and *el Canal del Dique* in the Atlantic Coast; *la Sabana de Bogotá* and *Cauca River Valley*). The CRCP included mathematical river-dynamics modeling and spatial analysis of past flooding events, after which the expert team designed alternatives. The project was intended to be participatory, including stakeholder workshops to discuss different alternatives designed by the expert team, so that decisions would benefit different interests and needs involving the river and wetlands.

In 1993 and 1995, near-floods had sparked the agenda for change in the Netherlands. However, unlike the Netherlands, in Colombia this flooding crisis was not accompanied by societal resistance against traditional engineering approaches such as dike enforcements, as described by Warner *et al.* (2012) in the Dutch case. On the contrary, most

Figure 7.1: Sugar mills in Cauca River Valley

landowners in Cauca River Valley asked for engineering solutions during and after the flooding crisis. Twenty-one dike reinforcements, levees, and protection-ring projects, and several dams to regulate river flow, were submitted by regional authorities to the Colombian Government's Adaptation Fund in 2011 for the Cauca River Valley region. Some 20,000 ha of the region affected by flooding were sugarcane areas (Vejarano-Alvarez and Saavedra-Rodríguez, 2013).

Cauca River Valley is the country's main sugar-producing region, with 13 sugar mills and six ethanol plants along the Cauca River floodplain (Figure 7.1), on 220,000 ha, 81.9 percent of the region's agricultural land, other crops being mainly fruits, maize and vegetables (Departamento Administrativo de Planeación, 2015). The Colombian sugar industry is the world's most productive, producing 15.5 tons of sugar per hectare per year. It also produced 456 million liters of ethanol in 2015.

The sugar industry has thrived in the region as a result of large engineering projects constructed to prevent periodic flooding from the Cauca River. After a World Bank study on

Colombia, the Currie mission (Currie, 1951) recommended river-basin development, similar to the Tennessee Valley Authority (TVA) model in the U.S., to increase agricultural output and energy generation. Sugar elites took the lead in implementing the TVA model in 1954 by persuading the central Government to create the CVC. This agency had the mandate to manage the Cauca River for electricity generation, irrigation, and flood control. To accomplish these outcomes, the CVC, with international support from the World Bank, implemented the Lilienthal Plan,[4] which included the Salvajina dam, put in operation in 1985, along with large drainage systems and dikes, facilitating agricultural expansion on the floodplain and dislocating Afro-Colombian farmers, especially from areas flooded by the dam. The TVA model enacted the ideal of modern river-basin development, by controlling "unruly" rivers to maximize benefits for society (Ekbladh, 2002); however, due to governance problems, only 60 percent of the projected dikes, 20 percent of the drainage channels and six pumping stations, out of 27 planned, were ever built (Sandoval, 2012).

At the time of the Lilienthal Plan, the dominant agricultural entrepreneurs' view was that wetlands located in the Cauca River floodplain were, and (according to interviews) continue to be unproductive features of the landscape, and an obstacle to progress (Tobasura, 2006). In addition, the dominant view was that floods were dramatic events that regularly wrought devastation and hence required great efforts to prevent and control them. Implementing the TVA model in Colombia after 1954 remains a source of pride and a central element of the sugar sector's identity, as they see themselves as visionaries who succeeded in taming natural forces and forging a prosperous industry in the region. The elites who promoted the Lilienthal Plan are usually portrayed in historical accounts using adjectives such as: "eminent," "outstanding intelligence," "illustrious vallecaucano" and "pioneer" (e.g. CVC, 2004).

Ideas of floods as "catastrophic forces to be tamed" were foundational for the sugarcane industry's growth and expansion in the region. The type of measures proposed by sugar growers to control flood risk at CRCP workshops demonstrated that these assumptions remain strong for many members of this economic sector. Those measures go back to the initial dream of the never-fully-implemented TVA model in the region. Workshop attendees repeatedly raised the need to build the complementary works (drainage and dikes) proposed when the Salvajina dam was built. Dike enhancement along the riverbanks and dredging the river to improve its carrying capacity were the most popular solutions envisioned by this sector (CVC, 2014). Agricultural encroachment into the floodplains and wetlands is not identified as a cause of disastrous consequences of flooding.

In contrast, alternatives under the RRA provoked great discontent among members of this sector. The proposal to move dikes off the riverbanks and relocate them, to incorporate the wetlands into the river's dynamics (Figure 7.2) received much criticism due to the costs associated with moving dikes that had just been built during the last rainy season, and because this intervention would reduce the sugarcane cultivation area.

Another modality to make space for the river proposed under RRA that met with fierce opposition from sugarcane landowners was to assign selected tracts of land adjacent to the river for temporary or incidental storage of floodwaters, preventing further damage downstream; these were called *"Lagunas de laminación"* (known in the Netherlands as "overflow polders"). CRCP workshop attendees were presented with 15 of these proposals, as shown in Table 7.1, through graphic simulations, as shown in Figure 7.3.

In most of these areas, sugarcane crops are predominant. Sugarcane growers see the floodplain as a big factory in which industrial agriculture and technology have changed the land's natural characteristics to make it more productive. Their narratives

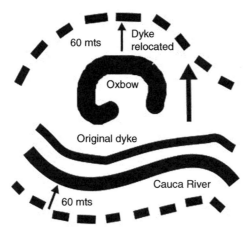

Figure 7.2: Proposed relocation of dikes

Table 7.1: *Lagunas de laminación, preliminary proposal*

	Name	Area (ha)	Storage capacity (Mm³)	Municipality
1	La Dolores	1290	39	Palmira
2	Guachal	258	9	Palmira
3	Caucaseco	1176	35	Cali
4	Yumbo	95	2.9	Yumbo and Vijes
5	Vidal	87	2.6	Vijes
6	Zabaletas	88	2.6	El Cerito
7	Videles-Cocal	540	16	Guacari, Riofrío and Yotoco
8	Yocambo	422	13	Yotoco
9	Sonso	2150	104	Buga and Guacari
10	Burriga	1645	50	Buga, San Pedro and Tulua
11	Charco de Oro	182	5.5	Andalucía
12	San Antonio	170	5	Bugalagrande
13	La Cañada	682	20	Bugalagrande
14	Guare-Zarzal	610	18	Zarzal
15	Ansermanuevo	386	12	Ansermanuevo

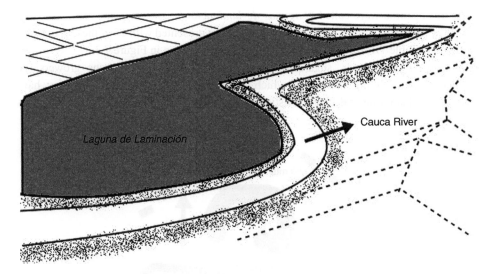

Figure 7.3: Laguna de laminación

divide the territory into two parts, with the upper watershed for water production and lower parts for economic production and water consumption (see also Rodríguez de Francisco and Boelens, 2015). Upstream, the sugar industry promotes reforestation and payment-for-ecosystem-services programs to ensure water provision. This division of the territory also marks land prices. Sugarcane growers usually refer to the elevated land costs in the valley (ten times more expensive than on the hillsides), to highlight why land in this area is so valuable and worthy of protection. Consequently, the visual impact of hundreds of hectares of sugarcane covered with water, as shown in Figure 7.3, ignited heated responses from participants in the second (July 2014) project workshop.

Sugarcane landowners also opposed wetland-restoration proposals, questioning the flood-control functions attributed to wetlands. Some argued against protective riparian forest strips, insisting that the river should be managed as a gigantic drainage or irrigation channel, and therefore without trees (CVC, 2014).

Interviews and sugarcane growers' participation in these workshops revealed that they conceptualize water simply as an input to the production process, a simple economic resource, as in many irrigation projects in Peru discussed by Martínez-Alier (2002).

This concept of water helps explain why members of this sector showed a marked preference for traditional engineering solutions to flood risk prevention. The ideas put forward by RRA – the need to live *with* water, conceptualizing floods as a consequence of undue agricultural expansion on the floodplain, and wetlands as important elements for flood control, were alien to the sugarcane growers. In addition, the RRA narrative conflicted with their identity as the group that had brought progress to the region. The RRA tacitly portrayed them as wrongdoers who had encroached on the floodplain, altered the ecosystems' regulatory functions, and exacerbated floods' negative effects.

New ideas brought by the Netherlands consultants to discuss flood-risk management under RRA portrayed floods as not bad per se but rather as important to maintain ecosystem balance by recharging wetlands and enriching floodplain soils. The RRA perspective suggested that, in order to reduce flood risk, it would work to give the river back part of the land taken from it, so water could spread slowly across the floodplain. In addition, restoring riparian vegetation and wetlands would enable them to perform their flood-control functions by absorbing excess water during rainy seasons.

The new RRA ideas resonated with some actors in the region, such as some engineers and biologists, who were participating in the project, and who agreed that the TVA model was a mistake; they believed that unregulated agricultural encroachment on the floodplain had reduced the natural capacity for floodwater retention, which needed to be restored.[5] We found that RRA narratives on flood-risk management were also more compatible with visions held by traditional fishermen and Afro-Colombian traditional farmers.

When asked about the floods, Afro-Colombian communities located in the south of this region[6] told a very different story about the Salvajina dam, which sugarcane growers see as a landmark for progress, but they see as a turning point that led to their decline. They point to the Salvajina dam as an elite-imposed project that brought poverty, dispossession, and isolation to their villages. Many farms, crops and roads were flooded, dislocating communities and eliminating cultural traditions and practices. In interviews, they called the Salvajina dam the "*Salvajada*" which means "savagery." According to Guzmán and Rodríguez (2015), the FARC (Revolutionary Armed Forces of Colombia) had entered this territory by 1982–84 and tried to persuade the people to oppose state policies and projects such as the Salvajina dam, capitalizing on people's discontent with this intervention to gain these communities' sympathy. Afro-Colombian actors in community organizations told how they had adapted to periodic flooding before the Salvajina dam through their traditional farming system, as an example of living with water.

Small landholders in the area usually had two small properties, one by the river or wetlands, where we grew crops such as plantain, maize and citrus on a small scale, and our homestead where we lived, with a large backyard. During flood season, we retreated to our homesteads to grow the same crops in our backyards while the floods fertilized the riparian lands. This way, we had very good harvests by the time the floods passed.[7]

They also used flooding periods to fish, thus complementing their agricultural production. Floods brought nutrients for their crops, and fish for these communities that had adapted to constantly varying water regimes. The riverine communities learned to deal with periodic flooding by reading the river's signs and acting accordingly. When the river started to rise, they moved away from shorelines and retreated to higher places to avoid harm.[8]

For these communities, the Lilienthal Plan both took away floods' positive effects and dispossessed them of much of their land for sugarcane expansion and eliminating the wetlands (oxbow lakes) created by the meandering river before the Salvajina dam was built.

Besides their importance for livelihoods, wetlands are also very important cultural and historical symbols for these communities. For example, the first inhabitants of the

La Guinea wetland were escaped slaves from Guinea (Africa) who established a *Palenque* (an area where they lived) around the wetland. The escaped slaves could survive by fishing in the oxbow lakes and cultivating small plots in the fertile surroundings. Around these wetlands, various traditions were practiced by these communities, some of which still remain.[9] Many myths and legends related to magical wetlands characters also populate these communities' oral histories.

Reforesting riparian vegetation, restoring wetlands, and making flooding areas available, as proposed by RRA, were consonant with the solutions envisioned by Afro-Colombian farmers and fishers, who perceive recovering their traditional livelihoods as a political affirmation of their territorial rights, which are threatened by modernizing agricultural development. They see restoring river dynamics and wetlands as a way to restore an imagined past in which traditional farmers were better off.

Sugarcane growers' and Afro-Colombian narratives about flooding and its solutions are not only dissimilar, but reflect the region's power hierarchy, underpinning differing claims about the river system and wetlands. Sugar-industry expansion in the region drove most peasantry from the floodplain to the hillsides. Traditional farmers, once scattered all over the region, were pushed to the mountain foothills and southward. This was reinforced by national economic policies such as open-market policies, which since 1991 left these peasants' grain and oilseed production unprotected. In addition, violence against rural people by paramilitary groups, especially during 1999–2003, weakened their social organizations (CNMH, 2014) and capacity to act collectively to advance their interests and exercise participatory rights in regional environmental governance.

7.4 Choosing Alternatives for Flood Risk Management in the Context of Unequal Environmental Citizenship Rights

This section analyzes how local social inequity and power imbalances had profound consequences in shaping the technologies chosen to manage flood risk in Cauca River Valley. We also point to some differences with the Netherlands case regarding characteristics of the social and public actors involved.

Originating in the Netherlands in 1996, the Room for the River Policy (*Ruimte voor de Rivier*) contained regulations regarding the use of riverbeds and set the goal of removing vulnerable land uses from the floodplains (Wolsink, 2005). To achieve resilience and legitimacy for this new paradigm, it proved essential to incorporate diversity and pluralism into planning, which is contrary to technocratic and high-modernist planning. Consulting with stakeholders and engaging actors locally provides more horizontal governance, which acknowledges the river system's multiple functions and multiple values.

Notwithstanding the long history of participation in water governance in the Netherlands, several authors (De Bruijne *et al.*, 2015; Edelenbos *et al.*, 2017; Roth and Warner, 2007; Van Buuren *et al.*, 2016; Warner and Van Buuren, 2011; Warner *et al.*, 2012; Wolsink, 2005) have shown how making space for the river in the Netherlands proved to be controversial, ridden with conflicts and local opposition. Proposals to build flood zones, calamity polders (areas designated for controlled flooding during extreme discharges to prevent damage in more densely populated downstream areas), and by-passes (deliberate overflow and discharge areas to increase control over peak flows) have been the most controversial, sparking social unrest, resistance and criticisms, especially from inhabitants of candidate areas. Opposition to the *Lagunas de Laminación* in the Cauca River Valley region is similar to these types of conflicts, although in a context of greater social inequity and power imbalances among local stakeholders.

In all these cases, uncertainty regarding the outcome of interventions has prompted local actors who will be affected by these measures to seek what Naratan and Venot (2009) call "social shaping" of the technology, through diverse strategies, such as lobbying and political interventions. In the Colombian case, socially and politically influential groups such as sugarcane growers have greater chances to shape technologies at the expense of less-influential groups such as Afro-Colombian and hillside farmers.

A quick comparison to the Netherlands case points to several differences that help to explain why it has been more difficult in Colombia to adopt alternatives under RRA such as the *Lagunas de Laminación*. In the Netherlands, areas have been designated for flooding in combination with goals that create incentives for the proposed land-use changes. For instance, the need for water retention in some areas has been promoted not only as a flood protection measure, but also attached to the concept of "spatial quality" that includes developing tourism, new forms of housing, and new economic activities in the areas to be set aside for flooding. By contrast, the *Lagunas de Laminación* proposal was presented to sugarcane growers without any reference to complementary goals.

The Vice-Minister's role, steering the process, and a strong state in the Netherlands were important to deter resistance, by allocating substantial compensation to the affected parties[10] and because of the perception of greater state power. According to Edelenbos *et al.* (2017: 57–58), for dike relocation in the city of Lent in the Netherlands, opposing groups "estimated that NIMBY behavior would not be a successful strategy, because the Government will win in the end." De Bruijn *et al.* (2015) also show how the Vice Minister and the status of the institute that made calculations for the Overdiepse Polder (a place where the river was given more space) certainly contributed to widespread agreement. By contrast, alternatives in the Colombian case were negotiated through the regional environmental authority (CVC), which lacks "practical authority"[11] (Abers and Keck, 2013) in the territory and is biased towards large landowners' interests.

Finally, Warner *et al.* (2012) argue that nearly experiencing floods (1993, 1995) in the Netherlands made it possible to analyze the river basin as an interconnected physical-ecological system, which was important to understand water and the river as more than just an economic production input and to adopt a whole-landscape approach, changing old paradigms to flood-risk management and accommodating new concepts, such as giving water more room. In Colombia, such a paradigm change was not locally initiated or actively pursued. Moreover, it seems unlikely to occur, since the project has not involved broad debate about the region's developmental path, nor have any actors other than the economically powerful been involved in discussing alternatives. Unlike the Netherlands, where environmental NGOs play an important role in local flood-risk management governance, and actively participate in designing interventions under RRA, such actors are almost non-existent and/or too weak in Colombia, making it challenging to bring ecological concerns to the forefront.

One of the most striking differences between the Netherlands and Colombian cases is the power differentials among local social groups, reflected in unequal environmental citizenship rights. On the one hand, the sugar industry is one of the country's most-organized economic sectors, with strong vertical and horizontal linkages through various organizations such as: Asocaña, which represents 12 of the 13 sugar mills; an international sugar trading company (CIAMSA) whose shareholders are mainly sugar mills; another company for industrial development and commercialization (DICSA) whose shareholders are also 11 sugar mills; a sugarcane research center (Cenicaña); an association of sugarcane technicians (Tecnicaña), and Procaña, which represents sugarcane suppliers. Privileged environmental citizenship rights for sugarcane growers have been guaranteed by the institutional set-up in which they dominate rule-making inside the CVC. Though Law 99 mandated inclusion of new actors on the CVC board of directors, such as representatives of Afro-Colombian and Indigenous organizations and NGOs, those groups were, and continue to be very weak, while industrialists and sugarcane-grower representatives remain on the CVC board and exert a great deal of influence on decisions affecting them. Sugarcane grower representatives on the board of directors have been able to block approval of wetland management plans as well as certain projects that would restrict agricultural activities.

Most wetlands grassroots associations involve low-income population groups, mostly small farmers, with more formal organizations located in Jamundí, constituted by ethnic associations that were only recently recognized by the state as community councils. None of these organizations is professional: they rely on voluntary work and donations from their own members and have no operating budgets. Marginalizing rural and environmental actors from environmental management has a long history within the CVC. Sánchez-Triana and Ortolano (2001) describe how the CVC has prevented citizen involvement in environmental decision-making processes since its creation. The authors explain this opposition and prevention because some of Cauca River Valley's elite have seats on the CVC board of directors, and own industrial facilities that need CVC permits, which could be threatened by social contradiction in a more-open process.

In Cauca River Valley, proposals to give more space to the Cauca River encountered resistance, especially by sugarcane growers, and motivated them to develop strategies to defend their values and interests. Discussing the Netherlands case, Edelenbos *et al.* (2017) argue for supporting local stakeholders' initiatives to incorporate local values, priorities and interests. In the Colombian case, however, since local stakeholders are at a great power imbalances and institutional participation has serious democratic deficits, interventions often reinforce social water use and access inequities.

Sugar associations formally rejected proposals such as the *Lagunas de Laminación*, not only at the workshops, but also through appeals lodged against this proposal at the CVC and pressure on the agency's board of directors. Due to sugar growers' opposition, *Lagunas de Laminación* were eliminated from the alternatives to manage flood risk before they were thoroughly discussed, and many other alternatives, such as dike relocation and land-use changes on the floodplain, were sidelined. After getting these alternatives rejected, the sugar associations were invited to present their own proposals. No other sector was given this opportunity, reinforcing the sugar sector's influence on project decision-making.

As in the case of the calamity polders in the Netherlands (Roth and Warner, 2007), sugar associations mobilized their own experts to question the knowledge and technical assumptions underlying controlled flooding. To develop their own alternatives, Asocaña hired an expert team led by one of the engineers who designed the Salvajina dam, to provide expertise and evidence to counter the plans that could hinder productive interests, because "they feared the project vision could be too ecological or environmentalist."[12] The proposal under construction is inspired by the Lilienthal Plan, and seeks to complete the dikes and drainage channels on the floodplain that were left unfinished in the 1980s, and add dams on the tributaries. These actions made sugarcane narratives predominant and allowed their expression and representation in the project, while excluding ecological or community narratives from the discussion and alternatives.

As of July 2016, the master plan for flood-risk management in the region has not been released, but the final alternatives have been narrowed down to a few: hydraulic recovery of only one wetland, the Sonso Lagoon; a protection ring in the municipality of La Victoria; reinforcing the dike surrounding the city of Cali; an early warning system; review of the Salvajina dam operating rules; three pilot vegetation enrichment projects; and four multi-purpose dams (see Table 7.2 and Figure 7.4).

The purpose of the dams is to regulate water entering the Cauca River from tributaries during flooding periods, while serving as reservoirs for irrigation water. In line with sugarcane growers' images of the territory, these works will avoid any changes in their current land and displace intervention from the floodplain toward the hillsides, affecting peasant smallholders' land and water rights.

These dam projects are now justified on the grounds of climate-change preparedness (see Materón, 2016) and have Inter-American Development Bank financing, but are currently being contested by municipalities located in the region's hillsides. They contest the

Table 7.2: *Proposed dams*

Dams	Height (m)	Reservoir (Mm^3)	Regulated flow[a] (m^3/s)
Río Claro	115	28.8	3.09
Río Jamundí	45	19.3	1.95
Quebrada Chambimbal-La Honda	69	14.25	3.79
Quebrada El Buey	65	74.1	5.88

[a] This is the discharge that will flow out continuously.
Source: CVC, 2016.

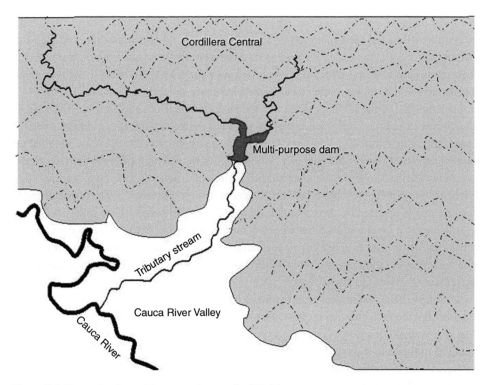

Figure 7.4: Example of a multipurpose dam on the hillside

climate-change justifications and argue that these dams are devised mainly to provide irrigation water to downstream sugarcane crops, affecting local availability of water as well as local fisheries and small farming. In addition, they complain about their municipalities' non-involvement in designing those projects (e.g. letter of complaint of two mayors: Perez and Eliecer, 2016).

While sugarcane growers refused to move off the floodplain, 8,777 low-income families who had settled on the dike and had been living on the river banks, some of them for over 30 years – now defined as "population at risk" – were evicted to reinforce 26.1 km of the dike protecting Cali. Spokespersons from these communities have complained repeatedly about this process, arguing they had not been consulted, and that the people living on the river bank were well adapted to living with the floods. They are now obliged to move to tiny apartments, which will break the tight relationship they had between the river and their community, and they will be unable to carry on their livelihood activities associated with riverine habitats. Alternative design options for this dike reinforcement were not considered, although "overflow polders" could have reduced the city's flooding risks and the need to reinforce the dike. Subsidies for the forced resettlement could have been used to compensate sugarcane growers, in flooding years, for "overflow polders."

7.5 Conclusions

As in the Netherlands, the Room for the River project in the Cauca River Valley initially sought to encourage public participation in planning, to acknowledge multifunctionality and multi-stakeholder diversity in river and wetlands values. However, the Netherlands consultants underestimated the biased institutional context in which participation took place and failed to devise strategies to identify and include less powerful actors in discussing and selecting alternatives.

Since local stakeholders were identified by floodplain land ownership, sugarcane growers' narratives became the main input to design alternatives for flood risk management. In contrast to the Netherlands, RRA implementation in the Cauca River Valley involved more unequal local actors, with weaker organizations representing community and environmental values. The lack of trust in the state, and the lack of clear compensation schemes and incentives for the proposed measures, put sugarcane growers on the defensive, retreating to protect their interests by mobilizing their resources to legitimize their values and narratives in selecting alternatives.

Sugar interests were able to emphasize traditional engineering approaches, such as dike enhancement and dams, due to the sector's privileged environmental citizenship, and imaginaries dividing the floodplain and the hillsides according to the conceptualization of water as an input for production. Paradoxically, the major alternatives selected to control flood risk were reinforcing the dike protecting Cali and building multi-purpose dams on tributaries, which would affect Afro-Colombian traditional farmers, and peasant hillside inhabitants who were not involved in decision-making, yet would bear the costs of these interventions.

Notes

1 As pointed out by Jeroen Warner of Wageningen University at the International Conference: Political Ecologies of Conflict, Capitalism and Contestation. July 7–9, 2016. Wageningen, the Netherlands.
2 The actual name of the project is Conservation Corridor and Sustainable Use of the Cauca River System Project.
3 All interviewees were offered anonymity, to encourage open and free dialogue, and we elicited individuals' permission to quote their responses.
4 David Lilienthal was one of the first directors of the Tennessee Valley Authority (TVA) of the US and founder of the consultant group Development and Resources, which advised Colombia, Puerto Rico, Iran, and South Vietnam on modernization projects inspired on the TVA model.
5 See the diagnostic document of the project: Vejarano-Alvarez and Saavedra-Rodriguez (2013).
6 These communities have approximately 21,089 inhabitants living in an area of 195.8 km^2, divided into 10 townships, with 12 wetlands in their territory. The most representative grassroots organizations are the Afro-Colombian Community Councils, the Funecorobles Foundation and Palenque 5.
7 Interview with an Afro-Colombian leader of Robles, Jamundí, February 5, 2015.
8 Interview with Benjamín, a fisherman from Buga, Buga, March 11, 2015 and an interview with José Luis, an Afro-Colombian leader from Quinamayó, Jamundí, April 19, 2015.
9 *La Varita* consists of cooking rice and potatoes while the men fish, *La Topa* consists of gathering around 15 canoes to alternately fish a certain wetland, and *El Copón* is women gathering to fish with *atarraya* (a special net for shallow waters) in a wetland.

10 The Room for the River program in the Netherlands had a budget of 2.3 billion euros over 20 years (CVC, 2015).
11 It is the kind of power (in) practice generated when particular actors (individual or organizations) develop capabilities and win recognition within a particular policy area, enabling them to influence other actors' behavior (Abers and Keck, 2013: 2).
12 Interview with a sugar growers' consultant, Cali, February 16, 2015.

References

Abers, R. N. and Keck, M. E. (2013). *Practical Authority: Agency and Institutional Change in Brazilian Water Politics*. New York: Oxford University Press.

Asocaña. (2016). *Aspectos generales del sector azucarero colombiano 2015–2016*. Annual Report, www.asocana.org/modules/documentos/11992.aspx.

Begg, C., Walker, G. and Kuhlicke, C. (2015). Localism and flood risk management in England: The creation of new inequalities? *Environment and Planning C: Politics and Space*, 33(4), 685–702.

Centro Nacional de Memoria Histórica (CNMH). (2014). *Patrones y campesinos: tierra, poder y violencia en el Valle del Cauca (1960–2012)*. Bogotá: CNMH.

Campos, A., Holm-Nielsen, N., Díaz, C., Rubiano, D., Costa, C., Ramírez, F., and Dickson, E. (2012). *Análisis de la gestión del riesgo de desastres en Colombia: un aporte para la construcción de políticas públicas*. Bogotá: World Bank.

Corporación Autónoma Regional del Valle del Cauca (CVC). (2004). *Génesis y Desarrollo de una visión de progreso, CVC cincuenta años*. Cali: CVC.

CVC. (2014). *Análisis de las jornadas de trabajo participativo con el sector azucarero (Asocaña-Procaña). Análisis de medidas preliminares propuestas*. Cauca River Corridor Project internal report.

CVC. (2015). *La protección contra inundaciones en el Valle Alto del Río Cauca. Nuevo Paradigma Corredor de conservación y uso sostenible del sistema Río Cauca*, www.cvc.gov.co/images/CVC/microsite/corredor_rio_cauca/documentos/gobernanza/seminario_taller_rio_cauca/20150907_Proyecto percent20Corredor_Maria percent20C percent20Sandoval.pdf.

CVC. (2016). "*Agua en el Valle del Cauca. El Aporte Técnico a las Soluciones.*" Powerpoint presentation by Ruben Darío Materón Muñoz, Director General CVC, www.infraestructura.org.co/filef.php?IDe=2237.

Corporación Autónoma Regional del Valle del Cauca (CVC) and Universidad del Valle (Univalle). (2007). *El Río Cauca en su Valle Alto: Un aporte al conocimiento de uno de los ríos más importantes de Colombia*. Cali: CVC, Universidad del Valle.

Currie, L. (1951). *Bases de un programa de Fomento para Colombia. Mission Report*, 2nd edn. Bogotá: Banco de la República.

De Bruijn, H., de Bruijne, M. and ten Heuvelhof, E. (2015). The politics of resilience in the Dutch "Room for the River" project. *Procedia Computer Science*, 44, 659–68.

Deeming, H., Whittle, R. and Medd, W. (2012). Investigating resilience, through "before and after" perspectives on residual risk. In S. Bennett (ed.), *Innovative Thinking in Risk, Crisis and Disaster Management*. Farnham: Ashgate, pp. 173–200.

Departamento Administrativo de Planeación. (2015). *Anuario Estadístico del Valle del Cauca*. Cali: Gobernación del Valle del Cauca, http://anuarioestadisticovalle.comli.com/#!/page_Home.

Edelenbos, J., Van Buuren, A., Roth, D. and Winnubst, M. (2017). Stakeholder initiatives in flood-risk management: Exploring the role and impact of bottom-up initiatives in three

"Room for the River" projects in the Netherlands. *Journal of Environmental Planning and Management*, 60(1), 47–66.

Ekbladh, D. (2002). "Mr. TVA": Grass-roots development, david lilienthal, and the rise and fall of the Tennessee Valley Authority as a symbol for U.S. overseas development, 1933–1973. *Diplomatic History*, 26(3), 335–74.

Guzmán-Barney, A. and Rodríguez-Pizarro, A. N. (2015). *Orden social y conflicto armado: El Norte del Cauca 1990–2010*. Cali: Programa Editorial Universidad del Valle.

Materón-Muñoz, R. D. (2016). "Comunicado a la opinión pública: La situación de los embalses en el Valle." June 2, 2016, www.cvc.gov.co/index.php/carousel/2319-comunicado-a-la-opinion-publica-2.

Martinez-Alier, J. (2002). *The Environmentalism of the Poor: A Study of Ecological Conflicts and Valuation*. Cheltenham, UK; Northampton, MA: Edward Elgar Publishing

Perez, W. and Eliecer, J. (2016). "Comunicado conjunto suscrito por los alcaldes de los municipios de Andalucía y Bugalagrande." Letter to CVC, April 20, 2016, http://andalucia-valle.gov.co/apc-aa-files/39353635313532356464353264386533/comunicadoconjuntosuscritoporlosalcaldesdeandaluciaybugalagrande.pdf.

Rodriguez de Francisco, J. C. and Boelens, R. (2015). Payment for environmental services: Mobilising an epistemic community to construct dominant policy. *Environmental Politics*, 24, 481–500.

Roth, D. and Warner, J. (2007). Flood risk, uncertainty and changing river protection policy in the Netherlands: The case of "calamity polders." *Tijdschrift voor Economische en Sociale Geografie,* 98(4), 519–525.

Sánchez-Triana, E. and Ortolano, L. (2001). Organizational learning and environmental impact assessment at Colombia's Cauca Valley Corporation. *Environmental Impact Assessment Review*, 21(3), 223–39.

Sandoval, M. C. (2012). *La protección contra inundaciones en el Valle del Cauca: Historia y Nuevo Paradigma*. Cali: CVC.

Sundberg, J. (2012). Negotiating citizenship in the Maya Biosphere Reserve, Guatemala. In A. Latta and H. Wittman (eds.), *Environmental and Citizenship in Latin America: Natures, Subjects, and Struggles*. CEDLA Latin America Studies 101. New York: Berghahn Books, pp. 97–111.

Tobasura, I. (2006). La Laguna de Sonso Valle del Cauca, Colombia: Más de tres décadas de lucha ambiental. Un caso de historia ambiental. *Gestión y Ambiente*, 9(2), 13–26.

Univalle (Universidad del Valle). (2011). *Valoración económica de la infraestructura existente para el control de inundaciones y estimación de los costos de las inundaciones 2010–2011 en el Corredor de Conservación del Río Cauca*. Partnership agreement CVC-Univalle-Asocars.

Univalle. (2014). *Planteamiento de alternativas estructurales para la gestión de inundaciones en el Valle Alto del Río Cauca*. Partnership Agreement No.001 of 2013 ASOCARS – Universidad del Valle. Escuela de Ingeniería de Recursos Naturales y del Ambiente, grupo de investigación en hidráulica fluvial y marítima.

Van Buuren, A., Vink, M. and Warner, J. (2016). Constructing authoritative answers to a latent crisis? Strategies of puzzling, powering and framing in Dutch climate adaptation practices compared. *Journal of Comparative Policy Analysis: Research and Practice*, 18(1), 70–87.

Vejarano-Alvarez, P. and Saavedra-Rodríguez, C. A. (2013). *Diagnóstico ambiental y de las inundaciones del 2010–2011 sobre los ecosistemas del Corredor de conservación y uso sostenible del valle alto del Río Cauca en el Valle del Cauca: un análisis con miras a la conservación y la restauración ecológica.* Internal report Partnership agreement CVC-Asocars.

Vörösmarty, C. J., McIntyre, P., Gessner, M., Dudgeon, D., Prusevich, A., Green, P., Glidden, S., Bunn, S., Sullivan, C., Liermann, C., and Davies, P. (2010). Global threats to human water security and river biodiversity. *Nature*, 467, 555–61.

Warner, J. and Van Buuren, A. (2011). Implementing room for the river: Narratives of success and failure in Kampen, the Netherlands. *International Review of Administrative Sciences*, 77(4): 779–801.

Warner, J., van Buuren, A. and Edelenbos, J. (eds.) (2013). *Making Space for the River: Governance Experiences with Multifunctional River Flood Management in the US and Europe*. London: IWA Publishing.

Wolsink, M. (2005). River basin approach and integrated water management: Governance pitfalls for the Dutch Space-Water-Adjustment Management Principle. *Geoforum*, 37, 473–87.

8
Reconfiguration of Hydrosocial Territories and Struggles for Water Justice

Lena Hommes, Rutgerd Boelens, Bibiana Duarte-Abadía, Juan Pablo Hidalgo-Bastidas, and Jaime Hoogesteger

8.1 Introduction

A vast and growing body of scholarly studies has shown how large-scale hydraulic and hydro-managerial projects, such as large dam and irrigation developments or market-environmentalist ecosystem payment schemes, have diverse socio-cultural and political-economic implications beyond merely altering water flows and raising socio-economic productivity. Concepts such as the hydrosocial cycle (Boelens, 2014; Linton and Budds, 2014), waterscapes (Baviskar, 2007; Budds and Hinojosa-Valencia, 2012; Swyngedouw, 1999) and water as socio-nature (Barnes and Alatout, 2012; Perreault, 2014) express connected insights about water being coproduced by social relations and in turn shaping these relations. The hydrosocial cycle, for instance, is described as a socio-natural process in which "water and society make and remake each other over space and time" cyclically (Linton and Budds, 2014: 170). Such coproduction of water and society is also reflected in the notion of waterscapes, conceptualized as socio-spatial configurations of water flows, artifacts, institutions and imaginaries embodying a particular world view (Budds and Hinojosa-Valencia, 2012; Zwarteveen 2015). However, these notions have so far largely focused on established hegemonic structures and discourses that drive and succeed from waterscape configurations. Less attention has been given to the multiplicity of diverging and overlapping hydrosocial territories that exist within one and the same space.

To address this, we employ the hydrosocial territories approach, analyzing water territories not merely as materializations of dominant discourses and interests, but as multi-scalar networks in which water flows, hydraulic infrastructure, legal-administrative and financial systems, and socio-cultural institutions and practices are interactively produced, aligned, negotiated and contested (Boelens *et al.*, 2016). Furthermore, combining the hydrosocial territories notion with Foucault's governmentality approach highlights different forms of "government rationalities" and how they are entwined with hydraulic and hydro-managerial projects. We focus specifically on analyzing how ruling groups' efforts to "conduct the conduct" of the governed (Foucault, 2008: 313) penetrate, operate through, and assimilate the rationality of the governed to advance neoliberal projects (Fletcher, 2010; Hommes *et al.*, 2016; Zwarteveen and Boelens, 2014).

Building on the work of Agnew (1994), Gupta and Ferguson (1992) and Elden (2010), we understand territories not as fixed spaces, but as spatially entrenched multi-scalar networks evolving from social interactions and practices, and materializations of these practices (see also Baletti, 2012; Brighenti, 2010). Social encounters and acts, including legal-administrative arrangements, technical reconfigurations and symbolic, cultural and political mechanisms of boundary- and place-making, actively produce territories. However, territories are not a stable set of rules and materialized relations. Instead, they are continuously contested and renegotiated by diverse actors engaged in particular acts of territory making within the same time and space (Baletti, 2012; Brighenti, 2010; Escobar, 2001). Therefore, there is always multiplicity of territories and territorial imaginaries; resulting from the diversity of people, communities and actor coalitions, trying to materialize their diverging interests, discourses and knowledge (Baletti, 2012; Duarte-Abadía and Boelens, 2016; Hommes et al., 2016). Therefore, differing from the notion that there exists only a single territory, we explore territorial pluralism (Boelens et al., 2016; Hoogesteger et al., 2016).

Our focus on hydrosocial territories particularly addresses the role of water resources, flows, artifacts, water rules, knowledge and imaginaries in territories. Hydrosocial territories are time- and space-bound networks entwining physical, hydraulic, political-economic and socio-cultural actors, acts, artifacts and institutions. They evolve through contestations between and materialization of particular imaginaries about how water and people are and should be managed. Water-user groups and other actors hold specific imaginaries about how water should be managed by whom, pursuing particular hydrosocial-ecological projects and "hydro-political dream schemes" (Boelens, 2015a; cf. Perreault, 2014; Pfaffenberger, 1988, Swyngedouw, 2015). Given that different actors uphold diverging imaginaries, shaped by their respective interests, territory-making by realizing imaginaries becomes a constant struggle and negotiation, mobilizing political, economic, discursive and physical powers. At the same time, in any hydrosocial territory it is not only human actors who act, encounter, and entwine, but also water flowing and hydraulic infrastructure operating in a network between "the human" and the "non-human," interlacing as "actants" (Latour, 1993). The notion of hydrosocial cycles builds on this understanding that "the natural" and "the social" mutually (re)produce each other.

Lastly, the fact that creating hydrosocial territories entails reconfiguring physical, social, and symbolic features to realize a certain socio-territorial imaginary about what is and what ought-to-be, means that many hydraulic projects can be seen as governmentality projects. Through the norms, knowledge and government mentalities inscribed in hydraulic projects, these projects become means to create a reality according to dominant ideas. They materialize government mentality ("governmentality," Foucault, 1991), attempting to form subjects who behave according to what is considered "appropriate" and "normal" behavior and "good" governance. It is, however, not (only) traditional coercive forms of power at work here, but subtle forms of Foucauldian "capillary/inclusive power" that induce norms and create self-measuring, self-correcting subjects by invoking guilt, morality, conformity, and compliance (Boelens, 2014, 2015a). This "disciplinary governmentality" tends to combine and overlap in subtle ways with other governmentalities, in particular "neoliberal governmentality" and top-down hierarchical modes of power based on "sovereignty" (e.g. Cadman, 2010; Fletcher, 2010). Territories are thus governmentalized through discourses and ideologies that establish and defend specific water and power hierarchies, water policies, institutions and management practices – triggering also counter-discourses and alternative territorial designs (Hoogesteger *et al.*, 2016; Hommes *et al.*, 2016; cf. Cadman, 2010).

Hydrosocial governmentality projects become materialized through several mechanisms and government techniques. Using specific valuation languages (Duarte-Abadía and Boelens, 2016; Martínez-Alier, 2004; Sullivan, 2009), establishing knowledge systems that regard some knowledge as valid and true while discrediting others (Zwarteveen and Boelens, 2014), and policies of inclusion and managed multiculturalism are just a few amongst many (Assies, 2010; Häkli, 2009; Hale, 2002). The following case studies will further illustrate and analyze such mechanisms, unraveling the politics and governmentalities embedded in socio-hydraulic and territorial reconfigurations.

As the title of this book already suggests, deeply connected to the notions of governmentality and hydrosocial territories are concerns for water justice. Insights into the multidimensionality and networked, processional nature of hydrosocial territories and contestations surrounding territorial projects call for a similarly encompassing and multidimensional understanding of water justice. Correspondingly, we see water (in)justice in terms of access to and distribution of water resources, and as a matter of cultural, political and socio-economic justices. This is to say that individuals, communities or societal movements contesting governmentality projects struggle to be able to use and control water, but also – and not less importantly – for recognition (of water-based identities, values, beliefs, knowledge, rights and situated justice understandings), participation, political co-decision making, and the right to socio-ecological integrity (cf. Boelens, 2015b; Fraser, 2000; Schlosberg, 2004; Zwarteveen and Boelens, 2014).

8.2 Territorial Reconfigurations and Stabilizing Hydrosocial Injustices

8.2.1 Hydrosocial Territorial Reconfiguration as a Governmentalizing Endeavor: The Chone Multipurpose Dam in Coastal Ecuador

As the Governor of Manabí accurately explained during the inaugural ceremony of the Chone multipurpose dam, this work is much more than hydraulic infrastructure damming water; it is also an undertaking that is intended to profoundly reconfigure hydrosocial territories beyond their biophysical materiality:

Today we have to thank life, when our Citizen's Revolution Government hands over this work; because this work has soul and body. The body is the majestic dam, but the soul is sown through educative unity and the *comunidad del milenio* [see below]. This is what will transcend ... This work is not simply infrastructure.

(November 24, 2015)

As will be shown, Chone's material reconfigurations convey a profound transformation of the way local people experience, practice and reproduce their social, cultural, and agro-productive lives. Through the Chone scheme and associated resettlement villages, the state and its allies seek to introduce particular meanings, values and practices into people's lives and minds, in accordance with what they consider progressive thinking, good water governance, advanced agro-productive practices and dignified ways of living. Thus, the Chone multipurpose dam is not neutral technical hydraulic infrastructure, but a hydro-political governmentality endeavor that stabilizes water injustices.

The Chone dam, which was inaugurated in 2015, is located in a low mountainous tropical zone of the Río Grande watershed in northwest coastal Ecuador. The Chone dam was built to prevent major flooding events in the lower plains, distribute irrigation water and provide drinking water to the city of Chone and its surroundings. The reservoir behind the 50-meter dam inundates over 1,600 ha and directly affects over 240 peasant families, most of whom relied on small-scale traditional farming. For centuries, these peasant families have shaped their livelihoods and related territories according to their own notions of well-being, farming and labor relations. The recent construction of the Chone multipurpose dam has, however, significantly reconfigured these local territories and related lifestyles.

Despite intense social resistance by Río Grande communities against the project, which was designed in the late 1970s, the self-identified progressive and participatory (and broadly characterized as authoritarian) Government of Rafael Correa continued building the dam, using diverse government techniques, for example economic compensation programs and resettlement of affected populations to a so-called *comunidad del milenio* ("millennium community"). These compensation and resettlement programs reconfigured socio-cultural relations and agro-productive practices according to government rationality in subtle, long-lasting ways.

What was central to Chone project governmentality was the Citizen's Revolution Government's strategic deployment of a discourse that associated Chone with economic development, inclusion, participation, and modernization. Established according to official notions of progress, dignity and *buen vivir* ("living well"), the discourse simultaneously denounces any opponents as backward and anti-modern people.

We should be united by our search for development, but there are still those who oppose everything; they would like us to practically live in caves ... they are backward villages that do not want development for us.

(Vice-President of Ecuador, November 2, 2013)

Alternative understandings of development and progress are actively rejected and dismissed. To strengthen, institutionalize and propagate officialdom's notions of development, the Government created a public enterprise called *Ecuador Estratégico* (Strategic Ecuador, EEEP in its Spanish acronyms). Although EEEP was set up to provide socio-economic compensation to local communities negatively affected by strategic national projects regarding water, oil, gas and mining, in practice EEEP's investments are also consciously geared to breaking down social opposition to government projects. Introducing official development discourses to local communities, they subtly seek to change the ways in which communities perceive their own positioning in Ecuador's (under)development. More so, by including communities in government-conditioned development and evoking guilt amongst those opposing it, EEEP's work intends to create conformity and compliance among communities. Its tactics work as a capillary power force.

EEEP's efforts materialized powerfully in the *comunidad del milenio* resettlement site and the associated high-tech *escuela del milenio* school, both built as compensatory measures 15 kilometers away from the Chone dam. Besides transforming "abnormal," "backward" livelihoods into "modern," "progressive" lifestyles, socio-physical changes were deployed to disarticulate social resistance at Río Grande. As the President of the new *comunidad del milenio* remembers: "We opposed the project, when it was signed, one wanted to cry. But when they started to give us houses the people started to become divided. At this moment, I left the opposition to join the Government's side. We gave in, and others of the opposition also now live here in these houses" (Personal communication, December 10, 2014).

However, only 81 affected families were resettled to the *comunidad del milenio* while the rest were left out. This points to the new exclusionary relations created among Río Grande's peasants. On the one hand, there are the resettled inhabitants of the *comunidad del milenio* who believe that the Chone dam has brought them progress. They now view their previous lives as underdeveloped, thereby reproducing government discourse and behaving as the self-conducting citizens envisaged by EEEP. On the other hand, there are those peasants that were excluded from compensation and who feel offended when being qualified as backward and undignified by government representatives simply because they were opposed to the project.

Yet, those families who were included in the new community also confront profound changes and challenges regarding control over their lives. Contrary to their livelihoods before the resettlement, the new community is designed and ruled as if it were an urban gated community. For example, farmland is not provided next to the houses, but has been relocated to two farms acquired by the government, 20 minutes by car from the resettlement site. Regardless of local realities, they are collective monoculture farms, which materialize governmental blueprint norms and ideas of efficiency and modernity. Still, peasants who claim that the resettlement has significantly increased their dependence, contest the new arrangements:

> Where we lived before, we sowed green beans, vegetables, and everything one can sow in the field. Here we cannot do anything, not even have chickens or pigs. Now we buy almost everything.
> (Resident of the *comunidad del milenio*. Personal commication, January 10, 2015)

This indicates that even though the Chone project has been implemented, people have been resettled and state development discourses may dominate among some, the governmentality endeavor and indented reconfigurations of territory are neither complete nor stable, but will remain under constant renegotiation by the multiple actors engaged in territory-making in Río Grande.

8.2.2 Hydrosocial Control, Developmentalism and Injustices in Guanajuato's Groundwater Territories, Mexico

The State of Guanajuato in Mexico's central highlands is one of the areas with the most intensive groundwater use in the country. While most cities, rural communities and industries depend exclusively on groundwater to meet their needs, the lion's share (82 percent) of groundwater is used by the agricultural sector (CNA, 2016). The advent of tube-well technologies in the 1940s marked the start of state-supported groundwater-based socio-economic development and reconfiguration of Guanajuato's hydrosocial territories. The number of tube wells rose from fewer than 100 in the 1940s to around 2,000 in the 1960s, 6,000 in the 1980s (CEAG, 2006, 2010) and over 17,300 (registered wells) in 2016 (CNA, 2016). This translates into an increase of groundwater-irrigated areas from 24,000 ha in the 1960s to over 250,000 ha by the end of the 1990s (CEASG, 1999) and an estimated 265,000 ha at present (personal communication, 2016). This reconfiguration of the rural landscape, extending irrigated areas, has dropped the water table by an average of two to three meters yearly (Wester *et al.*, 2011) and several pockets of fluoride and arsenic contamination have appeared (Ortega *et al.*, 2009; Gevaert *et al.*, 2012).

Groundwater over-extraction in Guanajuato results from the Government's political strategy that establishes legal mechanisms for groundwater regulation and control (Hoogesteger and Wester, 2017), while at the same time facilitating and stimulating groundwater-based economic growth. The first groundwater drilling ban (*veda*) to regulate groundwater use around some of the bigger cities, was established in Guanajuato in 1948, to control state territories under and above ground (Wester *et al.*, 2009). Since then, additional decrees were issued and, by 1983, the entire state of Guanajuato was placed under *veda*. These regulations portray an imaginary of state control over social and natural aspects of groundwater use in a specific territory, by administrative procedures, rules and penalties. Accordingly, groundwater – officially state property and responsibility – could be used only under a temporary water title concession by the state, specifying where, how and how much water could be extracted. The National Water Commission (CONAGUA) is now responsible for issuing and administrating all water rights (both surface and groundwater), granting the state, at least on paper, monopoly over groundwater use in its territory. Nevertheless, the law has never been strictly enforced, and the number of wells and irrigated areas keeps on expanding. As CONAGUA officials recognize: "the *veda* decrees and regulations stayed on bureaucrats' tables" (personal communication, 2016).

Moreover, the state has continued regularizing new illegal wells through legal loopholes and laissez-faire policies that favor drilling companies and let them drill new wells. For new users to acquire water rights or to transfer water rights among users, a water rights transmission system was introduced in the late 1990s. As put by Reis (2014: 546): "This means that someone who wants to extract water and does not have a concession, can do so only via a transfer of rights from another user who wants to cede his/her rights. A change of use type is then possible and the transferred volume can be extracted from a well at another location in the aquifer, as long as it meets CONAGUA's technical requirements." This new legal provision opened the door to regularize many more wells, and to issue new drilling permits when rights are transferred.

The still-increasing number of registered (and thus legalized) wells in the state of Guanajuato (most with no water meters) shows how administrative territorialization of groundwater control has been unable to limit water extraction or ensure equity (Hoogesteger and Wester, 2017). On the contrary, it has administratively supported sustained increase in groundwater extractions, facilitating the governmentality project of advancing unevenly distributed economic growth and elite-based water governance and resource capture in its territory. Through government techniques (e.g. subsidies for irrigated agriculture), the industrial sector and a vibrant groundwater-based economy have developed in Guanajuato in the last half century. In its strengthened political economy, a small elite controls most land, water and industrial capital; while holding very close ties to Guanajuato's governing elite. Accordingly, the territorial development project triggers important and critical water justice and equity issues.

Indeed, while urban growth, industrialization and advancing large-scale commercial agriculture have all been enabled by secure groundwater access (see also Wester and Hoogesteger, 2011), in other areas of the state poor rural communities still haul water from ponds and rivers with buckets. For others, it takes over ten years of constant pressure on local authorities to get a communal well installed, because funds are short for drinking water projects in rural communities. In the communities that have a well, families commonly have access to water only two or three days a week, for a restricted number of hours, forcing individuals to invest in water storage systems and restrict their domestic water use. As groundwater levels continue dropping, many community wells have declining yield or have dried up. The alarming levels of fluoride and/or arsenic in groundwater pumped in many areas of the state further aggravate the situation for the poor who depend on this water.

Yet, relatively little open protest and confrontation has taken place; part of which can be attributed to the fact that groundwater is underground and invisible, making it difficult to point to a single culprit (Hoogesteger and Wester, 2015). Nonetheless, to defend their hydrosocial territories and secure future access to water in terms of both quality and quantity, in 2012 a broad coalition of communities and NGOs of northern Guanajuato took the state to the international Permanent Tribunal of the People (PTP) to hold it accountable for severe groundwater over-extraction and the serious problems this brings for communities. They won their case and held the state accountable; yet no meaningful changes have happened, as the PTP is only morally binding. Similarly, in 2016, a broad group of communities and organizations launched a petition, on the international website platform, change.org, to demand that the Government take responsibility for groundwater management. Locally, communities' water problems are reshaping hydrosocial relations, as many households are now installing water harvesting systems with NGOs' support, making them less dependent on groundwater. These are some of the ways different strategies contest or deal with existing water injustices under the dominant hydrosocial imaginary and territorial constellation.

Thus, the situation in Guanajuato is a clear example of how administrative systems and selective developmentalism shape particular groundwater hydrosocial territories that manifest and stabilize water injustices, creating territories of abundance and capital accumulation for industries and agribusiness; territories of scarcity and everyday struggle for many rural communities; and territories of state control to be economically developed by capital by using and abusing existing groundwater regulations.

8.3 Territorial Pluralism and Contestations of Transformation Projects: Struggles for Territories and Hydrosocial Justice

8.3.1 Negotiating Transformation of Hydrosocial Territories through Territorial Representations and Valuation Languages: The Case of the Santurban Páramo in Colombia

In Colombia, the páramos (Andean highland wetlands) are strategic hydrosocial territories encompassing agricultural production systems, biodiversity conservation practices, water supply for urban centers, rural livelihoods, and multi-sectorial activities. Values and meanings attached to these hydrosocial territories have important implications for their management, shaping different users' rights. In this context of multiple diverging values, páramos are increasingly becoming objects of disputes about how to manage, use and conserve them. The case of Santurban páramo illustrates how – beyond myths of "good governance" based on singular truths or notions of objectivity about how to conduct socio-environmental management – pluriform interests, values, and rights systems are key to comprehend territorial control practices, obstacles and opportunities. They involve contestations among discordant languages of valuation. These languages of valuation, expressed through concepts, discourses and normative frames, represent actors' world views and knowledge systems, manifest actors' regimes of representation, and embody socio-economic interests and cultural and political relationships (cf. Escobar, 2001; Martínez-Alier, 2004).

Since pre-Hispanic times, the páramos have been represented as sacred places where different divinities dwelled, controlling territorial space and the origins and continuation of life. Human-natural-divine relationships integrated and ensured socio-productive exchange among diverse altitudinal and climatic zones. These referents survive in present-day rural livelihoods, influencing communities' self-organization and páramo lake protection to ensure water supply to community aqueducts. As Doña Aura, inhabitant of Vetas, Santurban páramo, explains, "The people who live close to the lakes have looked after them. The lakes provide water ... and when they are not protected the water sources may get angry. Therefore, indigenous communities have always tended to worship these sources, making ritual offerings" (quoted in Buitagro, 2012: 89–91).

Such local practices and cultural identity constructions have coexisted with intensifying capitalist modernization. As a result of the 1960s and 1970s Green Revolution, potato, livestock and scallion-growing have transformed the páramo territories into areas of agricultural and economic importance. "The páramo is home to farmers who, by sowing potatoes, provide food to the people who live in the towns and cities" (Villamizar, farmer, cited in Franco, 2013: 127). Likewise, and already since colonial times, small-scale mining activities have been part of the livelihoods of some páramo inhabitants for whom gold represents wealth, income, cultural tradition and the basis to organize water and territorial management. Páramos have become represented through both socio-cultural valuation languages – as a rooted dwelling place endowed with site-specific cultural identity – and agro-economic valuation languages – as a space for local economy, livelihoods and food production (Duarte-Abadía and Boelens, 2016).

Páramos are also of great importance for various large-scale mining operations, for example the Canadian Eco Oro company (formerly Greystart), which exploits mineral resources from páramo territories. While these operations are officially framed as enhancing national development, environmental policies have recently become stricter, arguing to protect the páramos from any economic activity. These policies are underpinned by natural-scientific language that depicts páramos as natural spaces with unique biodiversity and important functions for hydrological regulation and drinking water provision. Using scientific language de-politicizes contestations around páramos, and also seemingly implies a retreat from earlier government policies strongly promoting a national development model based on mining, energy and multinational activities.

Citizens from nearby metropolitan areas saw their drinking water security was jeopardized by the mining activities already underway in Santurban. They joined forces with the academic sector, environmentalist non-governmental organizations (NGOs) and other societal movements to unite against gold extraction in the Santurban páramo, basing their demands on their human right to water. They reached a consensus to deny a social license for the Eco Oro Company in February 2011, thereby preventing open-pit gold mining. What was central in their struggle was a natural-scientific valuation language that established the páramo as natural space that guarantees water quality and supply; and mining and other economic activities as damaging this natural space. The languages and claims were enforced by climate change adaptation and mitigation discourses materialized in the new protectionist policies:

Páramo is an ecological unit with high importance for water regulation ... Páramos are also ecosystems with high capacity to capture carbon; these two functions can contribute to mitigate the effect of climate change. Hence páramos need to be protected from economic activities.

(Ministry of the Environment, 2001)

While the natural-scientific valuation language proved effective for urban water consumers to press their demands concerning páramo protection, it also constrains local livelihoods (farmers and small-scale miners) and polarizes páramo inhabitants against páramo conservationists from urban areas. In fact, environmental arguments particularly marginalize smallholder residents of highland areas, portraying them as water polluters who threaten the health of other citizens and ecosystems (cf. Duarte-Abadía and Boelens, 2016). Páramo inhabitants had no alternative way to defend their hydro-territorial rights, and had to ally with multinational companies, who are likewise opposed to páramos becoming a protected area, although they have several disagreements. Ironically, multinational companies assumed socio-cultural and agro-economic valuation languages, defending the territory and its inhabitants' rights, claiming continuation of historical access.

To deal with the multiple divergent values about how to manage páramos and to reconcile state policies' contradictions in Santurban, the Minister of the Environment organized a discussion group with different stakeholder representatives in early 2014. Neo-institutionalist scholarship based on game theory and experimental economics, was strategically adopted to introduce the key notions of collective rationality to manage shared resources. To mediate stakeholder meetings, this approach would argue that when incentives are correct, individuals' motives to maximize their profits will ensure that opposing groups automatically collaborate, to find the most efficient way to organize distribution of water, funds and other relevant resources. Neo-institutionalism attempts to solve divergent interests through rational arrangements for cooperation and agreements for trust, which built on presumably universal values and, in practice, the universalist language of monetary profit-making (Cárdenas, 2009 in Duarte-Abadía and Boelens, 2016). This introduced neoliberal governmentality in Santurban páramo to reconcile incommensurable valuation languages, simplifying complex realities and behaviors, while subtly steering thinking and acting in very particular directions.

Yet these approaches ignore fundamental causes for non-cooperation, disguising social groups' opposing interests and profoundly unequal economic powers in a discriminatory, exclusionary political structure. They also obscure the existence and juxtaposition of multiple cultures, world views, knowledge systems and livelihood and conservation practices. Questions arise whether a neo-institutional approach to reach "consensus" actually furthers social justice (as it acclaims), or whether it supports arrangements and relationships that clearly favor winners while sidelining losers in this game with water and life.

In 2014, contestations and claims around the Santurban páramos resulted in two political decisions to achieve a balance among different valuation languages. First, a zoning regime was established that delimitates areas for restoration (25,227 ha), sustainable agriculture (5,502 ha), and preservation (98,993 ha). Nevertheless, mining activities can continue within "restoration areas" if the mining titles were acquired before 2010. In the "preservation areas" and in "sustainable agriculture zones," Payment for Environmental Services (PES) and other market-environmentalist instruments are being planned to promote conservation as an economic activity. In practice, these instruments of a green capitalism model tend to worsen the situation for peasant families who have no official land titles, and for smallholders who have no other income than what is generated from (now illegalized) homestead production (e.g. Rodríguez-de Francisco and Boelens 2015, 2016). Besides, PES schemes end up bolstering actors who can afford to pay for conservation and water supply, while local Santurban páramo dwellers lose control over their hydrosocial territories (cf. Rodriguez-de-Francisco *et al.*, 2013).

Second, to contest a ruling of the Colombian Constitutional Court that bans mining activities in páramos (see Judgment C-035, 2016), the Canadian Eco Oro company announced they would sue the Colombian Government for not adhering to international commercial and free trade agreements between Colombia and Canada. This means that decisions on what and how to conserve páramos are beyond PES agreements, but subject to international trade law. Foreign capital has thus blurred borders and reorganized scales, building strategic multi-scale compositions in response to multinationals' commercial flows and geopolitical interests (Garay, 2013; Swyngedouw, 2009). Colombian páramos are subject to globalization's fluidity, free of territorial constrictions in a "liquid modern world" (Bauman, 2000).

To conclude, the case of the Santurban páramos demonstrates how making and shaping hydrosocial territories is a constant struggle and negotiation among multiple actors and their territorial values, imaginaries and projects. Particular valuation languages play a central role in territorial projects, revealing how struggles are not just about land and water resources but also intrinsically connect to recognition of identities, values, knowledges and situated justice understandings.

8.3.2 Contestations around the Ilisu Dam Project in Turkey and Evolving Territorial Pluralism

Water-dam construction for electricity generation and irrigation in Turkey dates back to the beginnings of the Turkish Republic. The most notorious project has been the Southeastern Anatolia Project (GAP, in its Turkish acronym – Güneydogu Anadolu Projesi), in planning since the 1950s and now in its final stage of completion. GAP will build 22 dams, 19 hydroelectric power plants and 1.7 million ha of irrigation areas in the Euphrates and Tigris river basins, to become the "breadbasket of the Middle East" (Warner, 2012). The project area stretches over nine provinces, covering approximately 10 percent of Turkey's total land area (Ministry of Development, n.d.), making it an impressive mega-project. While originally envisioned as a purely hydro-technical project, GAP planners soon added socio-cultural development objectives such as empowering women and youth, promoting birth control and family planning, settling nomadic communities and resettling people from dispersed rural settlements into state-designed, centralized villages (Özok-Gündogan, 2005).

While official state reasoning justifies GAP by Southeastern Anatolia's low development status and Turkey's need for nationally produced energy (Bagis, 1997), GAP must be understood in the context of "the Kurdish question," the decades-old, blood-stained conflict between the Turkish state and Kurds living within Turkish national boundaries. Topics in the conflict include recognition of Kurdish cultural identity, language rights, equal status before the law and greater political autonomy for Kurdish regions.

Accordingly, Özok-Gündogan (2005) and Hommes et al. (2016) observe how GAP is clearly geared to reconfigure Southeastern Anatolian territory physically and hydrologically, and also socio-culturally, economically and politically (cf. Çarkoglu and Eder, 2001; Jongerden, 2010), a governmentality project materializing state presence and mentality in "disobedient," "insecure" areas. The project's website, for example, explicitly states that GAP will bring "civilization back to Upper Mesopotamia" (GAP Regional Development Administration, 2013: History of GAP) and that "local sub-culture elements may form a positive synthesis with the national culture" (GAP Regional Development Administration, 2013). This shows how Southeastern Anatolia is depicted as a place in need of civilizing, where existing culture is regarded as below national Turkish culture. The region's ethnic make-up is also left ambiguous: while pointing out its backwardness, subordination, and need for integration. Such politics of inclusion and managed multiculturalism clearly establish which cultural elements are "right" and "wrong," including only those aspects that are convenient to the territorial state project (cf. Assies, 2010). The state establishes development conditions and the imperative of aligning sub-ordinated, uncivilized cultural elements with the superior national culture. Turkish culture, materialized in GAP, is equated to modernity and development, and Kurdishness to backwardness, implying that those who want to become modern need to become Turkish. This "capillary/inclusive power" mechanism as described by Foucault makes opposition to GAP extremely difficult because of its subtle productive discourse. Indeed, as Harris (2009) notes, many welcome GAP as state attention for long-neglected, poverty-stricken areas.

However, GAP does not simply create a homogenous territory in which people, institutions and rivers are aligned according to state imaginaries. Instead, GAP triggers contestations and constant (re)negotiations of territory-making and imagining by multiple actors, resulting in territorial pluralism. This can be illustrated with the case of the Ilisu Dam, the last centerpiece to be completed within GAP. Situated on the Tigris River, 65 km upstream from the Turkish-Syrian border, and expected to affect around 78,000 people, the Ilisu Dam has been the subject of heated discussions since the 2000s: possible negative downstream effects, ecological damages, resettlement issues, and flooding of numerous archaeological and cultural heritage sites (Eberlein *et al.*, 2010). Turkish state institutions try to materialize specific imaginaries about what the dam hydrosocial territory should look like, whereas Kurdish communities, environmental organizations, archaeologists and European NGOs contest such imaginary and pursue different territorial projects (Hommes *et al.* 2016).

The Turkish government has shown great determination to build the dam, presenting it as a matter of national pride and a necessity for national development, given Turkey's growing energy demands and heavy dependence on energy imports (Çarkoglu and Eder, 2001). A selective development discourse prioritizes national development over the fate of towns and villagers to be inundated and resettled (cf. Nixon, 2010). Whereas GAP is employing an inclusive development discourse, claiming to integrate the region socio-economically e.g. by developing massive irrigation schemes, the Ilisu Dam is constructed solely to generate electricity for cities and industries. Some local employment is generated during dam construction, but local benefits do not seem to be a government concern. Nevertheless, the dam is imagined to reconfigure local territories desirably, by resettling population groups from currently dispersed rural settlements to state-designed, centralized villages, replacing traditional social structures with "modern organizations and institutions" (GAP Regional Development Administration, 2013). This is imagined to be beneficial for state service delivery, while also increasing people's dependence on those very same services and eroding local support for the Kurdish guerrilla group (Özok-Gündogan, 2005). Furthermore, resettlement patterns envisaged according to state planners' design principles are making the area and its population more legible and thus controllable (see similarities with the above Ecuadorian case), and more accessible by building dam-associated infrastructure such as roads. Lastly, the dam is imagined as a matter of security: a massive physical barrier against insurgent activity, flooding major guerrilla hideouts (Jongerden, 2010).

At the same time, Kurdish activists describe flooding as deliberately erasing part of Kurdish history and culture, drowning heritage sites such as the famous town of Hasankeyf, and suspected gravesites from the Turkish-Kurdish conflict, to exclude them from any possible future justice research and action (Ronayne, 2005). As Izady (1996:1) denounces: "dams built to benefit non-Kurds are drowning Kurdish heritage." The Ilisu Dam accordingly sparked protests from Kurdish communities in project areas and outside of Turkey, the Kurdish diaspora in the UK, Germany and the US playing an important role in attracting international attention and media coverage of the case of the Ilisu Dam. The Ilisu Dam was interpreted as an intrusion into Kurdish territory, which then turned anti-dam struggles into struggles for recognition and protection of Kurdish identity, cultural heritage, and participation in decision-making.

Environmental NGOs, in turn, imagine the dam project in terms of its potential destructive environmental effects. Reframing the dam as an environmental issue rather than a Kurdish issue reinterprets the dam's hydrosocial imaginary, up-scaling the anti-dam movement to the international level, supported by particular knowledge creation, labeling the dam's biophysical environment as "universally valuable" (Ahunbay and Balkiz, 2009; see also the Colombian case), and incorporating Ilisu as an example of destructive large-scale dams in global anti-dam campaigns (Damocracy, 2016). Historical-cultural arguments and discourses add to the environmental imaginary: portraying the project area as "Mesopotamia, the cradle of Man" has helped involve and politicize archaeologists, who call the project "a tool for cultural and ... ethnic cleansing" (Ronayne, 2007: 255).

Finally, it is important to note that these state, Kurdish and environmental imaginaries do not match the local population's territorial projects. Accounts of local opinions are contradictory: while government reports offer villagers' quotes such as "Let [the Turkish government] save us from here so that we will have civilization" (DSI and Ilisu Consortium, 2005: 44), anti-dam campaigners agree among each other that most local people oppose the dam (Ronayne, 2005; Setton and Drillisch, 2006; Eichelmann, 2009). Different actors construct diverging on-the-ground realities, according to own interests, mentalities and interpretation frames. At the same time, there is certainly no single local idea about the Ilisu Dam. Some might be more critical while others, such as large estate owners, may profit from the project driving land prices up (cf. Evren, 2014). Furthermore, local involvement in the anti-dam campaign seems to have died down over time, due to delays and seesaw changes in project-related decision. A journalist notes "the people here are tired. For decades, they heard there was going to be a dam ... They lost patience. Many of them have told me, 'If we have to go, if we are going to lose this place, let's do it as quickly as possible'." (Today's Zaman, 2013).

Analyzing the GAP project as a governmentality endeavor has shown how a supposedly technical hydraulic project is in fact designed to comprehensively reconfigure territories physically and socio-politically. At the same time, the ways in which coalitions opposing the Ilisu Dam deploy diverging imaginaries about the dam's hydrosocial territories has illustrated how territory-making is a constant multi-scalar, multi-dimensional struggle mobilizing political, discursive and physical powers. Territorial pluralism and continuous (re)negotiations are thus part and parcel of transformation projects and water justice struggles.

8.4 Conclusions

Examples from Latin America and Turkey have illustrated how water management endeavors as diverse as multi-purpose and hydropower dams, payment for environmental services schemes, conservation projects and groundwater governance regimes profoundly reconfigure local territories. In this context, territories and their reconfigurations not only alter water flows but conjointly also socio-ecological and political relations. For example, resettlement to government-designed villages within the Chone and Ilisu Dam projects led to new ways of living and new citizen-state relations according to government mentality. Similarly, PES and conservation languages, increasingly employed in the context of Colombian páramos, also aim to materialize certain ideas of how land and water resources should be used, and how people should be governed. Central in all analyzed contexts is the way territory-making involves a range of physical, socio-political and discursive strategies, such as discourses of development, modernity and progress that define desirable uses of territory.

At the same time, territorial projects pursued by politically and economically powerful actor alliances of governments, large-scale farmers, urban elites and mining companies are contested by counter-imaginaries and alternative projects. Contestations take different forms, ranging from simple discontent to up-scaling anti-dam struggles, employing different valuation languages and searching for complementary or independent water resources. These different contestations and strategies for territory-making, according to actors' interests and understandings, demonstrate how hydrosocial territories are plural, diverse and continuously renegotiated.

Thus, even though territorial reconfiguration projects may generate and stabilize profound water injustices, movements' and water users' struggles for techno-political recomposition challenge dominant hydro-territorialization projects. These power plays and contestations continuously transform or question the territory's hydraulic grid, cultural reference frameworks, economic base structures, and political relationships; being struggles that go beyond battles over natural resources to also involve meaning, norms, knowledge, decision-making authority and discourses.

References

Agnew, J. (1994). The territorial trap: The geographical assumptions of international relations theory. *Review of International Political Economy*, 1(1), 53–80.

Ahunbay, Z. and Balkiz, Ö. (2009). Outstanding universal value of Hasankeyf and the Tigris Valley, Doğa Derneği, Ankara (leaflet).

Assies, W. (2010). The limits of state reform and multiculturalism in Latin America: Contemporary illustrations. In R. Boelens, D. Getches, and A. Guevara-gil (eds.) *Out of the Mainstream. Water Rights, Politics and Identity*. London: Earthscan.

Bagis, A. I. (1997). Turkey's hydropolitics of the Euphrates-Tigris Basin. *International Journal of Water Resources Development*, 13(4), 567–82.

Baletti, B. (2012). Ordenamento territorial: Neo-developmentalism and the struggle for territory in the lower Brazilian Amazon. *Journal of Peasant Studies*, 39, 573–98.

BanCo2. (2016). "Servicios Ambientales Comunitarios," www.banco2.com/v2/index.php.
Barnes, J. and Alatout, S. (2012). Water worlds: Introduction to the special issue of Social Studies of Science. *Social Studies of Science*, 42(4), 483–88.
Bauman, Z. (2000). *Liquid Modernity*. Cambridge: Polity Press.
Baviskar, A. (2007). *Waterscapes: The Cultural Politics of a Natural Resource*. Ranikhet: Permanent Black.
Boelens, R. (2014). Cultural politics and the hydrosocial cycle: Water, power and identity in the Andean highlands. *Geoforum*, 234–47.
Boelens, R. (2015a). *Water, Power and Identity: The Cultural Politics of Water in the Andes*. London: Earthscan, Routledge.
Boelens, R. (2015b). *Water Justice in Latin America: The Politics of Difference, Equality and Indifference*. Amsterdam: University of Amsterdam.
Boelens, R., Hoogesteger, J., Swyngedouw, E., Vos, J. and Wester, P. (2016). Hydrosocial territories: A political ecology perspective. *Water International*, 41(1), 1–14.
Brighenti, A. M. (2010). On territorology: Towards a general science of territory. *Theory, Culture & Society*, 27, 52–72.
Budds, J. and Hinojosa-Valencia, L. (2012). Restructuring and rescaling water governance in mining contexts: The co-production of waterscapes in Peru. *Water Alternatives*, 5, 119–37.
Buitrago, E. (2012). *Entre el Agua y el Oro: Tensiones y Reconfiguraciones Territoriales en el Municipio de Vetas*. Santander, Colombia: Universidad Nacional de Colombia.
Cadman, L. (2010). How (not) to be governed: Foucault, critique, and the political. *Environment and Planning D*, 28(3), 539–56.
Çarkoglu, A. and Eder, M. (2001). Domestic concerns and the water conflict over the Euphrates–Tigris River Basin. *Middle Eastern Studies*, 37(1), 41–71.
Comisión Estatal del Agua de Guanajuato (CEAG). (2001). *Actualización de los Balances de los Estudios Hidrológicos y Modelos Matemáticos de los Acuíferos del Estado de Guanajuato*. Mexico: Guanajuato City.
CEAG. (2006). *Memoria Institucional 2000–2006 de la Comisión Estatal del Agua de Guanajuato (CEAG)*. Guanajuato City: CEAG.
CEAG. (2010). "Planes de manejo de acuíferos," www.guanajuato.gob.mx/ceag/planes.
CEASG. (1999). *Plan Estatal Hidráulico de Guanajuato 2000–2025, Fase 1: Diagnóstico base de la situación hidráulica del Estado de Guanajuato*. Guanajuato City: CEASG.
Comisión Nacional del Agua (CNA). (2016). *El Registro Público de Derechos de Agua (REPDA)*.
Damocracy. (2016). "Damocracy: Global campaign to protect rivers from destructive dams," www.damocracy.org/about/.
DSI and Ilisu Consortium. (2005). *Ilisu Dam and HEPP Project Update of Resettlement Action Plan – Final Report*. Ankara: Environmental Consultancy Co.
Duarte-Abadía, B. and Boelens, R. (2016). Disputes over territorial boundaries and diverging valuation languages: Santurban hydro-social highlands territory in Colombia. *Water International*, 41, 15–36.
Eberlein, C., Drillisch, H., Ayboga, E. and Wenidoppler, T. (2010). The Ilisu Dam in Turkey and the role of export credit agencies and NGO networks. *Water Alternatives*, 3(2), 291–312.
Eichelmann, U. (2009). *Stop Ilisu – Controversial Project on the Tigris River*. Vienna: ECA Watch Austria.
Elden, S. (2010). Land, terrain, territory. *Progress in Human Geography*, 34(6), 799–817.
Escobar, A. (2001). Culture sits in places: Reflections on globalism and subaltern strategies of localization. *Political Geography*, 20, 139–74.

Evren, E. (2014). The rise and decline of an anti-dam campaign: Yusufeli Dam project and the temporal politics of development. *Water History*, 6, 405–19.

Fletcher, R. (2010). Neoliberal environmentality: Towards a poststructuralist political ecology of the conservation debate. *Conservation & Society*, 8, 171–81.

Foucault, M. (1991). Governmentality. In G. Burchell, C. Gordon and P. Miller (eds.), *The Foucault Effect: Studies in Governmentality*. Chicago: University of Chicago Press, pp. 87–104.

Foucault, M. (2008). *The Birth of Biopolitics*. New York: Palgrave Macmillan.

Franco, B. M. (2013). "Characterization and analysis of production systems in Guerrero, Rabanal and Santurban páramos: 'Páramos and life system project'." Alexander von Humboldt Institute, Bogotá. Unpublished manuscript.

Fraser, N. (2000) Rethinking recognition. *New Left Review*, 3, 107–20.

GAP Regional Development Administration. (2013). About GAP, www.gap.gov.tr/english.

GAP Regional Development Administration. (2013). About GAP: Objectives, www.gap.gov.tr/aboutgap/objectives-of-gap.

Garay, L. (2013). Globalización/ Glocalización soberanía y gobernanza. A propósito del cambio climático y extractivismo minero. In V. Saldarriaga, O. Alarcon, and R. Medina (eds.), *Minería en Colombia: Fundamentos Para Superar el Modelo Extractivista*. Bogotá: Imprenta Nacional, pp. 9–19.

Gevaert, A. I., Hoogesteger, J. and Stoof, C. R. (2012). *Suitability of Using Groundwater Temperature and Geology to Predict Arsenic Contamination in Drinking Water: A Case Study in Central Mexico*. Internet-First University Press.

Gupta, A. and Ferguson, J. (1992). Beyond "culture": Space, identity, and the politics of difference. *Cultural Anthropology*, 7(1), 6–23.

Häkli, J. (2009). Governmentality. In R. K. Thrift (ed.), *International Encyclopedia of Human Geography*. Oxford: Elsevier.

Hale, C. R. (2002). Does multiculturalism menace? Governance, cultural rights and the politics of identity in Guatemala. *Journal of Latin America Studies*, 34(3), 485–524.

Harris, L. M. (2009). States at the limit: Tracing contemporary state-society relations in the Borderlands of Southeastern Turkey. *European Journal of Turkish Studies*, 10, 2–19.

Hommes, L., Boelens, R. and Maat, H. (2016). Contested hydrosocial territories and disputed water governance: Struggles and competing claims over the Ilisu Dam development in southeastern Turkey. *Geoforum*, 71, 9–20.

Hoogesteger, J. and Wester, P. (2015). Intensive groundwater use and (in)equity: Processes and governance challenges. *Environmental Science and Policy*, 51, 117–24.

Hoogesteger, J. and Wester, P. (2017). Regulating groundwater use: The challenges of policy implementation in Guanajuato, Central Mexico. *Environmental Science and Policy*, 77, 107–13.

Hoogesteger, J., Boelens, R. and Baud, M. (2016). Territorial pluralism: Water users' multiscalar struggles against state ordering in Ecuador's highlands. *Water International*, 41, 91–106.

Izady, M. R. (1996). The drowning of the Kurdish historical and artistic heritage. *The Kurdish Life*, 19.

Jongerden, J. (2010). Dams and politics in Turkey: Utilizing water, developing conflict. *Middle East Policy*, XVII(1), 137–43.

Latour, B. (1993). *We Have Never Been Modern*. Cambridge, MA: Harvard University Press.

Linton, J. and Budds, J. (2014). The hydrosocial cycle: Defining and mobilizing a relational-dialectical approach to water. *Geoforum*, 57, 170–80.

Martínez-Alier, J. (2004). Los conflictos ecológicos distributivos y los indicadores de sustentabilidad. *Revista Iberoamericana de Economía Ecológica*, 1, 21–30.

Ministry of Development. (n.d.). Information Booklet. Ankara.

Ministry of Environment (MMA). (2001). *Program for the Sustainable Management and Restoration of Colombian High Land Ecosystems: Páramos*. Bogotá: General Direction of Ecosystems.

Nixon, R. (2010). Unimagined communities: Developmental refugees, megadams and monumental modernity. *New Formations*, 69, 62–80.

Ortega-Guerrero, M. A. (2009). Presencia, distribución, hidrogeoquímica y origen de arsénico, fluoruro y otros elementos traza disueltos en agua subterránea, a escala de cuenca hidrológica tributaria de Lerma-Chapala, México. *Revista Mexicana de Ciencias Geológicas*, 26(1), 143–61.

Özok-Gündogan, N. (2005). "Social development" as a governmental strategy in the southeastern Anatolia project. *New Perspective on Turkey*, 32, 93–111.

Perreault, T. (2014). What kind of governance for what kind of equity? Towards a theorization of justice in water governance. *Water International*, 39(2), 233–45.

Pfaffenberger, B. (1988) Fetishised objects and humanised nature: Towards an anthropology of technology. *Man (N.S.)*, 23, 236–52.

Reis, N. (2014). Coyotes, concessions and construction companies: Illegal water markets and legally constructed water scarcity in central Mexico. *Water Alternatives*, 7(3), 542–60.

Rodriguez-de-Francisco, J. C. and Boelens, R. (2015). Payment for environmental services: Mobilising an epistemic community to construct dominant policy. *Environmental Politics*, 24(3), 481–500.

Rodriguez-de-Francisco, J. C. and Boelens, R. (2016). PES hydrosocial territories: De-territorialization and re-patterning of water control arenas in the Andean highlands. *Water International*, 41(1), 140–56.

Rodriguez-de-Francisco, J. C., Budds, J. and Boelens, R. (2013). Payment for environmental services and unequal resource control in Pimampiro, Ecuador. *Society & Natural Resources* 26(10), 1217–33.

Ronayne, M. (2005). The cultural and environmental impact of large dams in southeast Turkey, fact-finding mission report. National University of Ireland, Galway and Kurdish Human Rights Project.

Ronayne, M. (2007). The culture of caring and its destruction in the Middle East: Women's work, water, war and archaeology. In Y. Hamilakis and P. Duke (eds.), *Archaeology and Capitalism: From Ethics to Politics*. Walnut Creek, CA: Left Coast Press.

Schlosberg, D. (2004). Reconceiving environmental justice: Global movements and political theories. *Environmental Politics*, 13(3), 517–40.

Setton, D. and Drillisch, H. (2006). *Zum Scheitern Verurteilt – Der Ilisu Staudamm im Südosten der Türkei*. Berlin: World Economy, Ecology & Development.

Sullivan, S. (2009). Green capitalism, and the cultural poverty of constructing nature as service provider. *Radical Anthropology*, 3, 18–27.

Swyngedouw, E. (1999). Modernity and hybridity: Nature, regeneracionismo, and the production of the Spanish waterscape, 1890–1930. *Annals of the American Academy of Political and Social Science*, 89, 443–65.

Swyngedouw, E. (2009). The political economy and political ecology of the hydro-social cycle. *Journal of Contemporary Water Research & Education*, 142, 56–60.

Swyngedouw, E. (2015). *Liquid Power: Contested Hydro-Modernities in Twentieth Century Spain*. Cambridge, MA: MIT Press.

Today's Zaman. (2013). "Hasankeyf locals struck in limbo after Ilisu Dam decision," www.todayszaman.com/national_hasankeyf-locals-stuck-in-limbo-after-ilisu-dam-decision_304540.html.

Warner, J. (2012). The struggle over Turkey's Ilısu Dam: Domestic and international security linkages. *International Environmental Agreements*, 12, 231–50.

Wester, P. and Hoogesteger, J. (2011). Uso intensivo y despojo del agua subterránea: hacia una conceptualización de los conflictos y la concentración del acceso al agua subterránea. In R. Boelens, L. Cremers, and M. Z. Zwarteveen (eds.), *Justicia Hídrica: Acumulación, Conflicto y Acción Social*. Lima: IEP, pp. 111–33.

Wester, P., Hoogesteger, J. and Vincent, L. (2009). Local IWRM organizations for groundwater regulation: The experiences of the Aquifer Management Councils (COTAS) in Guanajuato, Mexico. *Natural Resources Forum*, 33(1), 29–38.

Wester, P., Sandoval-Minero, R. and Hoogesteger, J. (2011). Assessment of the development of aquifer management councils (COTAS) for sustainable groundwater management in Guanajuato, Mexico. *Hydrogeology Journal*, 19(4), 889–99.

Zwarteveen, M. (2015). Regulating water, ordering society: Practices and politics of water governance. Inaugural lecture, University of Amsterdam.

Zwarteveen, M. and Boelens, R. (2014). Defining, researching and struggling for water justice: Some conceptual building blocks for research and action. *Water International*, 39(2), 143–58.

9

Large-Scale Dam Development and Counter Movements: Water Justice Struggles around Guatemala's Chixoy Dam

Barbara Rose Johnston

9.1 Introduction

In this chapter I explore the case of Guatemala's Chixoy Dam, highlighting first, the role of large dam development as a proxy-weapon in the Cold War arsenal, next, the varied injustices accompanying the commodification and destruction of the ancestral commons, and finally, the emergence of struggles to achieve meaningful, restorative justice.

World War II introduced new and expansive forms of environmental warfare, including targeted bombings to unleash the massive force of water as a weapon, generating an array of consequential damage. To halt the advance of Japanese troops in 1938, China destroyed its Huang He Yellow River levee system, drowning several thousand Japanese soldiers but also destroying 11 Chinese cities, more than 4,000 villages, millions of hectares of farmland and causing the loss of life of an estimated 800,000 Chinese civilians (Dutch, 2009: 287; Schofield, 1998: 634). In 1943, the Royal Air Force used powerful dam-buster bombs in their raid on German dams and hydroelectric facilities, generating concussive earthquake impacts and a tsunami-force wave that wiped out power and industrial facilities, swept away bridges, flooded mine shafts and agricultural fields, and drowned some 1,600 civilians and 1,000 Soviet forced-laborers. In 1944, Germany targeted the seawalls and dyke systems of the Netherlands, causing a saltwater flood that destroyed the productive capability of some 200,000 ha of arable land. In 1952, the United States and its allies engaged in a massive series of airstrikes against North Korea with the aim of destroying that nation's ability to produce its staple food, rice. Their goal was accomplished through the strategic bombing of the Sui-ho Dam, levees, and the destruction of 13 hydroelectric dams (Westing, 1980: 56). Water has been used as a weapon in war throughout the ages, but never before at such a scale.

In the aftermath of war, reconstruction of hydro-development (dams, aqueducts, hydroelectric works, transmission grids, bridges, and related roadways) became a high priority, and for good reason. Hydro-development (using water as a force in economic development by building large dams, generating and transmitting hydroelectricity, and impounding and diverting water) is an immensely productive means to provide water and power for domestic, agricultural, extractive, and manufacturing industries. Only recently have we come to understand the cumulative, synergistic impacts of this immensely destructive force. Over the past century, large dams have been built on more than 60 percent of the world's major river drainages, resulting in significant releases of methane, carbon and other atmospheric emissions, changing the nature of river and estuary ecosystems to the point of extinction for some 50 percent of the world's freshwater fisheries, and playing a major role in the growth of near-shore hypoxic zones worldwide (McAllister *et al.*, 2001; Postel and Richter, 2012; Renaud *et al.*, 2013; Richter *et al.*, 2010; Vörösmarty *et al.*, 2010). If current planned projects on the free-flowing systems of the Amazon, Southeast Asia, and Africa are achieved, hydro-development will have impacted river flow and altered ecosystemic conditions on 90 percent of world river drainages, anthropogenic change that poses a major threat to the biocultural diversity that has long sustained planetary life (Johnston and Fiske, 2014; Van Cappelen and Maavara, 2016). In short, as a tool and transformative force to enclose and commodify the commons, large dams drive industry and economies, while simultaneously degrading the ecosystems in which these dams are introduced. In this regard, large-scale hydro-development can be considered one of the major driving forces in this Anthropocene age: achieving short-lived gains at the cost of biocultural diversity and planetary life.

To my mind, no case in recent years illustrates these points in stronger terms than that of Guatemala's Chixoy Dam. Planned as a means to expand agriculture, extract mineral and energy resources, and fuel industry, Chixoy development opened up previously isolated regions, disenfranchised indigenous communities, and its flow of international funds financed a military dictatorship waging genocidal war against its indigenous civilian population. In this chapter, I situate the case-specific experience of proposing, financing, building and contesting the consequential damages of the Chixoy hydroelectric dam within the larger backdrop of political economic agendas, power struggles, and the consequential damages of Cold War development. This focus on the driving forces and controlling processes that shaped and nurtured development allows a place-based case of profit and pain that is emblematic of the larger development enterprise. It is my argument that troubling outcomes create their own momentum and force, forging – in the moments of injustice – the ideal and notion that justice might, some day, in some way, be reclaimed in the tentative spaces created by the collective recognition of fundamental rights. Such spaces represent a potentially transformative shift in loci of power over critical resources and the relative rights of indigenous peoples and local communities, and in this recognition, backlash may predictably occur.

9.2 Development as a Power and Force in the Cold-War Toolkit

In the 1950s, the human and environmental horrors of World War II gave way to an era of development: public infrastructure works to increase access, expand productive capability, and generate the critical resources essential to an increasingly globalized economy. As Cold-War power struggles took shape, large infrastructure development became a significant way to wage the battle to control the form and function of national/international political and economic systems. The template for this era had its roots in the early years of World War II.

Beginning in 1941, the architecture for a new international financial system was proposed in top-secret meetings by Special Assistant to the US Secretary of the Treasury Harry Dexter White and John Maynard Keynes, advisor to the British Treasury. In July 1944, diplomats from the 44 Allied nations attending the United Nations Monetary and Financial Conference in Bretton Woods, New Hampshire met to consider the resulting US/UK plan and come to an agreement.[1] Some of the key concerns were: how can we increase employment and productivity to levels that support a global free-market economy when national economies were severely depressed before the war and are in even worse straits because of the war? And, how can we finance an economic stimulus that puts people to work, to repair and expand infrastructure systems that will support and expand productivity and trade, when most of the world's nations lack even the minimal resources to provide basic social needs? The proposed solution: create a global monetary system with the US dollar replacing the British pound to give birth to an international financial system to invest in opportunities to expand export production and enable free trade between all nations. Following a three-week negotiation, the 1944 Bretton Woods Agreement was approved and signed, creating a system of fixed exchange rates tied to gold and the US dollar, the International Monetary Fund (IMF), and the International Bank for Reconstruction and Development (IBRD).[2] The IMF would manage the system of fixed exchange rates and help expand world trade by providing short-term loans or modifying exchange rates to countries struggling with temporary deficits in their balance of payments. The IBRD, first in what would eventually grow to be five separate banks forming the World Bank Group, would provide financial assistance to reconstruct war-ravaged nations and support economic development of less-developed countries.

The stated goals of this endeavor were laudatory, and urgently needed in the aftermath of global war. Bank staff would assess country needs, develop project proposals and, with contributions from member nations, finance projects to build roads, bridges, buildings, mines and metal works, dams, public and agricultural water supplies, sewage treatment systems, railroads, electrical generation and transmission grids, and telephone and telegraph networks. Infrastructure would be rebuilt, and development in previously isolated regions would allow access to commercially exploit timber, minerals, oil, gas, hydroelectric energy, and agricultural products for an export market. Taxes on production and sale of services and goods would generate revenue for governments to provide social services for their civilian population, open up local economies to global capital, and enable eventual repayment of loans with interest. Interest paid from these development investments would return to the World Bank's capital fund to further finance modernity, where economies around the world are stabilized and thrive.[3]

Operations began in 1947 with Bank staff working with member nations on project-specific proposals to rebuild or add new infrastructure, expanding capacity to engage in the international economy. World Bank staff also conducted country assessments in resource-rich, under-developed nations to generate country-wide development plans. The goals of such initiatives: outline opportunities and needs; identify necessary political and economic reforms; propose projects that strengthen infrastructure; improve social conditions; and set in place the political means to insure the growth of a free-market rather than socialist economy. What was funded, and where, when, and why such loans were approved, reflected the political and economic interests of the World Bank's major shareholders: the US with its initial 35.07 percent of the shares and effective right of veto, and Britain, second largest shareholder with 14.52 percent (Toussaint, 2014). Case in point, when France submitted their proposal for a post-war reconstruction loan for US$250 million in October 1946, financing was delayed by the World Bank Board until the political priorities of the French Government reflected those held by the major shareholders. This loan was not approved until the French Communist Party, a leading party in the elections of 1945 and 1946, was removed from power in 1947 (Gwin, 2007).

9.3 Guatemala Development: A Plan for Plunder with Hydropower as the Driving Force

Guatemala became, in 1950, one of the first countries to host a World Bank mission, with staff and consultants evaluating the national economy's relative strengths and weaknesses, identifying underexploited resources, and generating a National Economic Development Plan.[4] Made public in 1951 in support of a loan request from President Jacobo Árbenz, the National Economic Development Plan describes primary barriers to economic development in Guatemala as "primitive" agriculture, inequitable distribution of land, and widespread underemployment in rural areas due to the seasonal nature of subsistence agriculture, or the relatively unproductive cultivation of marginal lands. Economic transformation would build public infrastructure, beginning with a highway system to link population centers and shipping ports to the mountainous rural regions. Road construction would also support expansion of communication and power lines; movement of goods; access to and exploitation of timber, mineral reserves (lead, silver, copper, zinc, chromium, bismuth, iron, antimony, quartz, coal); and the opportunity to explore for petroleum in northeastern Guatemala. Since extractive industry requires a reliable, accessible energy source, and mineral resources were located in mountainous sub-tropical regions where major rainfed rivers flowed, hydroelectric energy was identified as the key means to power a national export-oriented economy. Thus, once built, Guatemala would enjoy "a better division of labor ... the present highland population would be brought to depend less on the growing of corn, which can be grown elsewhere" and, "at a later stage, with ample labor and cheap supply of hydroelectric power, it is possible to visualize small manufacturing plants located in this area" to transform the way of life and productive output of Guatemala's indigenous and *ladino*[5] communities. They would earn wages laboring in mines, corporate agriculture, manufacturing and industry, all fueled by an autonomous National Power Authority to plan, build, and operate publicly owned power facilities (World Bank, 1951: 82, 121, 233).

President Jacobo Árbenz's 1951 request for funds to implement the plan developed by World Bank staff was denied by the Bank's Board. This rejection occurred at the behest of the US and their concerns that, in the years since Guatemala's 1944 revolution (a popular uprising replacing a military dictatorship with a new constitution and democratically elected President and Congress), socialist sentiments and policies were being promoted. Specifically, the US reacted to President Árbenz's legislative agenda on land and labor that abolished slavery, outlawed unpaid labor, and restored communal title to ancestral lands through a redistribution program that nationalized, with compensation, the uncultivated lands of large landowners. These reforms cut into the profits of United Fruit Company, a US-owned major banana producer, which lobbied for US intervention in Guatemala through the efforts of its legal representatives: then US Secretary of State John Foster Dulles, head of the CIA and director of the United Fruit Company Board Allen Dulles, and major shareholders such as the US representative to the United Nations, Henry Cabot Lodge (Cohen, 2012: 186). In 1954, the US Central Intelligence Agency engineered a coup to topple the government of President Jacobo Árbenz (Cullather, 2006; Doyle and Kornblugh, 2013). Thirty-six years of Guatemalan dictatorships and civil war would follow.

In 1955, with the restoration of a political economic climate in Guatemala favoring US companies, the first project to launch the Guatemalan National Development Plan was financed with a highway project loan (World Bank Archives, 2001). Road construction opened up access to the mountainous regions and in 1961, a year after Guatemala's Civil War started, a loan from the new Inter-American Development Bank (IADB) was secured to finance agriculture infrastructure projects in Alta Verapaz. Also that year, the United Nations Special Fund, Government of Guatemala, and World Bank signed an agreement for a comprehensive study of the region and its potential for electric power and irrigated agriculture.[6] In 1963, IADB issued a loan in support of the operations of Instituto Nacional de Electrificación (INDE, an autonomous self-financing state/military government entity established in 1959 to modernize electricity, expand service into rural areas, and establish energy development policy). The World Bank began funding INDE directly, dispersing funds for energy surveys and feasibility studies in 1967 and 1968.[7] The resulting National Energy Plan identified 22 sites where large dam construction to support hydroelectric energy appeared feasible, including the Pueblo Viejo-Quixal site at the confluence of the Río Negro and Chixoy Rivers. Additional loans and technical assistance agreements between INDE and the World Bank, IADB, Central American Bank of Economic Integration, and other partners (the governments of West Germany, Venezuela, Canada, and the United States) financed a hydroelectric feasibility study and plan, engineering studies, site-specific construction and development plans, design, production, and delivery of the generating plant, turbines, pipelines, and other materials and equipment, and the construction of the Chixoy Dam with site-specific work beginning December 1976.[8]

The people who lived at the confluence of the Rio Negro and Chixoy Rivers were completely unaware of these plans. In the fall of 1976, visitors arrived via helicopter to deliver the news that road and other construction in their river valleys would begin soon, and they must move. When construction of access roads and worker camps began in December 1976, a census of all people whose lives and livelihoods would be adversely affected by the project had not yet been contracted, let alone begun. In November 1977, when INDE submitted a hastily prepared three-page resettlement plan to the World Bank (to demonstrate to US Congress that the loan was for humanitarian, not military purposes, and thus not a violation of the US Foreign Financial Services Act of 1977), no census had yet been commissioned. Prepared with no input from affected communities, INDE proposed resettlement to Finca La Primavera, a plantation in the forests of Alta Verapaz.[9]

In October 1978, a census and household survey was begun and INDE created an Office of Human Resettlements. Formal negotiations began and some families took up INDE's offer to move to Finca La Primavera. There, they found wood shacks instead of cement homes, no reliable supply of water, small plots of infertile land, and the distance from their old homes to the new made it impossible to maintain the family and community relationships that sustained their way of life. They returned to their old homes and rejected the resettlement offer as it failed to reflect what was promised. Resettlement negotiations in all affected communities became increasingly tense, with meetings held under the hostile eye of an armed military guard. Tensions were further escalated when, in two instances in 1979, dam-affected communities saw their leaders leave to bring to the Resettlement Office requested community records of resettlement promises and communal land title, and fail to return (presumably assassinated en route).

Tensions became incendiary following the January 31, 1980 massacre of 39 people when government security forces fire-bombed the Spanish Embassy in Guatemala City, killing staff and visiting officials, members of the *Campesino* Unity Community, and students and indigenous leaders who had joined the march to the city to occupy the Embassy and deliver a petition the Spanish Ambassador. Their petition requested the Spanish Ambassador to act as a mediator with the Guatemalan state, to denounce the military repression they had experienced to the international community, and to demand the creation of commissions to investigate their allegations of repression (CEH, 1999: 163–82). One of the people who died that day was Francisco Chen, a Maya Achi leader who organized the economic cooperative in Río Negro, and hoped to inform the Spanish Ambassador on the specific incidents of violence and repression accompanying Chixoy Dam development and request international attention to these concerns (Achi, Museo Comunitario Rabinal, 2003: 47).[10] For the next 18 months, state-sponsored violence in and around the Chixoy dam construction site, river basins and the larger region was applied systematically and selectively, targeting community leaders with threats, kidnapping, torture, and assassination; and, when these tactics failed, rounding up and killing dam-affected communities.

9.4 Development Loans and Fungibility: Hydro-Development as a Means to other Ends

When construction was complete and the reservoir waters rose, in January 1983, ten communities in the Chixoy River Basin had been massacred. In Río Negro alone, 444 of the original 791 inhabitants had been killed. Survivors were hunted in the surrounding hills, and beginning in 1984, forcibly resettled at gunpoint in guarded concentration camps built with development funding, with designs provided by the United States military to control the civilian population. The use of this US military template allowed the Rios Montt Government to establish a new phase in their campaign against Mayan communities through

Reencuentro Institucional 84, a "model villages" program involving forced relocation, re-education, forced labor, and military control of peasants (Manz, 2004: 156–82). In the Rabinal and Chixoy River Basin areas, the army concentrated captured civilians in "military colonies" on the sites of previously burned villages of Xesiguan and of Chichupac; survivors lived in these camps for six months, until the army permitted a local NGO to help rebuild their homes. Survivors of massacres whose villages were now inundated by Chixoy Dam, however, were concentrated in the "development poles" of San Pablo (1983–96) and Pacux (1983–present) in and near Rabinal; designated "potentially salvageable subversives," they were kept under constant surveillance and guard by the army.[11]

Some of the violent incidents associated with the Chixoy Dam construction and forced evictions were reported in the national and international press and investigated by the Inter-American Human Rights Commission. Shortly after the first Río Negro massacre occurred, on March 4, 1980, a formal complaint describing the incident and its relationship to dam construction was filed with the Inter-American Human Rights Commission and in October 1981 the massacre was discussed in the Court's preliminary and final country report on Guatemala. Their conclusion: events that led to the massacre are a result of Chixoy Dam construction in the area. Despite this knowledge, World Bank and Inter-American Development Bank financiers continued their disbursements to INDE, and expanded their support in years to come.

An important facet of bank operations is that project-specific development loans are granted and oversight mechanisms implemented within a broader context of *fungibility* (meaning, money, once allocated to the donor party, can be used for any purpose). While oversight mechanisms have emerged over the years in an effort to ensure that financed projects achieve stated goals, the fungible nature of foreign aid means that oversight serves more as a means to reduce donor liability than to honestly protect the health and wellbeing of vulnerable groups and the environment in which they live. Internationally financed hydro-development in Guatemala illustrates the point. Major financing for Guatemala's Chixoy Dam was twice provided by the World Bank (in 1978 and 1985) and three times by the Inter-American Development Bank (1976, 1981, and 1985) with loan agreements that required the Government of Guatemala to provide proof of legal title to development project land and obligated the Bank to ensure legal title before disseminating funds. Review of title record in 2004 demonstrated that full title was never acquired and loans were disbursed in violation of contractual obligations.

In 1981, with IADB evaluating a second loan for the project and still no viable mechanism to address social impacts, anthropologist William Partridge was contracted to do an independent evaluation. He found, among other things, a document indicating that the IADB and World Bank were aware that legal title to all the lands required for the construction of the Chixoy Dam had not been secured, and that title to Río Negro communal lands (soon to be submerged by the reservoir) had been granted in the previous century and thus was protected from seizure by a clause in Guatemala's National Constitution (1983). This and other compliance issues led to a second contract for Partridge to develop a remedial resettlement plan, with implementation funded by line item allocation in new loans (1984).[12]

Shortly after the completed dam began operations, severe water losses from the headrace tunnel prompted a shutdown of the power plant. Inspection revealed considerable damage to tunnel areas, and extensive repairs were needed to reroute water around karst deposits. A full assessment of geological conditions had not been conducted during the site evaluation phase. The largest dam in this part of the world was built on soft, porous, limestone. Later audits, interviews and testimony demonstrated INDE had received funds designated for resettlement repairs, deposited into the general operating fund, and used them for other purposes including "site security" operations. Site security workers were in fact "G-2" military intelligence officers whose work included gathering information, orchestrating disappearances, torture, massacres, hunting survivors, and if found, massacring other dam-affected communities that sheltered survivors (Schirmer, 1998: 292).

In his June 29, 1983 testimony to the US House of Representatives on the *Environmental Impact of a Multilateral Development Bank-Funded Project*, anthropologist Shelton Davis reported that from 1977 to 1981, Chixoy Dam funding was one means by which the Government of the United States could circumvent section 701 of the International Financial Institutions Act of 1977 (which mandates US opposition to lending to governments that consistently engage in gross violations of human rights except when a loan expressly meets basic human needs). Davis commented: "it appears as if hydroelectric development in Guatemala was related to the modernization of the Guatemala Army and its concern to turn the northern lowlands into a vast cattle ranching, petroleum, mining, and timber frontier. By carrying out this frontier-development program, with international assistance, the Guatemalan Army hopes to consolidate its own political and economic power" (Davis, 1983). Despite human-rights concerns raised by some in Congress, President Ronald Reagan's open support for the newest dictator, General Efraín Ríos Montt, carried the day and Congress approved the US financing of additional loans to Guatemala.

9.5 Struggles to Secure Justice

In 1992, Río Negro massacre survivors helped guide a team of forensic archaeologists to the killing grounds and, in 1993, assist in Guatemala's first exhumation of a massacre site, uncovering the remains of 107 Maya Achi children and 70 women outside their rural village. In 1994, Río Negro's survivors formed the Association for the Integral Development of the Victims of the Violence of the Verapaces, Maya Achi (ADIVIMA) to encourage exhumations of other massacre sites in the surrounding communities and the prosecution of those responsible. Their work generated evidence that was central to adopting the 1994 Oslo Peace Accords.

In 1996, Río Negro and other dam-affected community experiences received international attention with the Witness for Peace report documenting this history (Pacenza, 1996). Following a two-week World Bank Mission to Guatemala City where staff reviewed development records and confirmed that resettlement financing had not been spent in ways agreed, an aide-memoire between the Bank and the Guatemalan Government was negotiated to provide a series of modest corrective actions (an agreement with no mechanism to monitor compliance). At the same time, World Bank staff recommended restructuring INDE to privatize the energy delivery system, and in this sale, develop the means to repay all Chixoy loans in full. To implement the privatization plan, a World Bank technical assistance grant was issued in 1998, assets were sold and Bank loans were repaid in full. Loan repayment resulted in the closure of the Chixoy dam resettlement office and a local complaint mechanism. Initially projected to cost US$340 million, corruption, mismanagement, and incompetence during the era of military dictatorship resulted in a total financing of US$955 million; an amount that represented 45 percent of Guatemala's foreign debt by the mid-1990s.[13]

Massacre survivors presented the social impacts of Chixoy Dam construction to the World Commission on Dams regional consultation in 1999. In its final report, the WCD drew attention to the serious human rights abuse associated with internationally financed hydro-development in general and the Chixoy case in particular, and recognized that international financial institutions have specific obligations regarding reparations and the right to remedy in this and other cases of grossly flawed, inept development. WCD conclusions recommended best practices that prioritized environmental and social sustainability as decisive indicators in determining proposed projects' feasibility, operational management plans, and distribution of benefits from existing facilities. The WCD also called for governments, industry, and financial institutions to accept responsibility for social issues associated with existing large dams and develop remedies, including reparation, restitution, restoration of livelihoods, and land compensation for relocated communities (WCD, 2000).[14] The evidence in the Río Negro massacre case was also examined in 1999 by the UN-sponsored Historical Verification Commission and determined to be an exemplary case of genocide (CEH, 1999).

After the 1999 WCD consultation, Río Negro massacre survivors presented a claim to the World Bank's primary accountability mechanism, the Inspection Panel. Their petition was rejected, as the World Bank no longer had any investment in the Chixoy Dam, their loans had been repaid in full and, with no current bank investment, the petitioners lacked standing.

These disappointments gave way to renewed commitments to secure justice. Río Negro leaders contacted and held community meetings with upstream, adjacent and displaced dam-affected communities. As outreach expanded, it became clear that many, many villages had been affected adversely by dam construction and hydro-management and continued to suffer greatly from the lack of compensation or remedial attention. In 2003, men and women from villages across Baja Verapaz, Alta Verapaz and Quiche met to form an assembly of dam-affected communities, elect leadership, and articulate their goal of pursuing just compensation and reparation. They agreed to participate in an independent truth commission process involving an audit of the development record, a consequential damage assessment of social impacts, and a reparations-planning process.[15] Among other things, the resulting study demonstrated that legal title had never been fully secured: one community was still paying taxes on lands submerged beneath the reservoir, another still held title to a portion of the land supporting the dam. Documentation of these facts led to civil protests in 2004, with indigenous communities occupying their land at the dam site, demanding a reparations negotiation, and threatening to shut down the nation's primary source of electricity. The three-day occupation was resolved by the Government's agreement to establish a negotiation process. Meetings facilitated by the Inter-American Human Rights Commission began in 2006 with representatives from dam-affected communities and Guatemala Government joined by World Bank and IADB observers to review the evidentiary record and generate a verification of damages signed by all parties. In 2010, a reparations plan was finalized and signed by all parties; terms included development assistance, social and cultural support, and environmental restoration of the river basin.

Given the lack of a viable judicial system in Guatemala, justice for the violence accompanying hydro-development was sought in international courts. In 2012, the Inter-American Human Rights Court case issued their findings in *Río Negro massacres v. the Government of Guatemala,* concluding that hydroelectric development of the Chixoy river basin was the background event that led to the Río Negro massacres and an inhibiting factor to achieving full reparation. Members of the Río Negro community cannot perform their funeral rituals because some disappeared villagers have yet to be identified; construction of the dam and its reservoir destroyed indigenous sacred sites; and inundation physically and permanently hinders the Río Negro communities' return to their ancestral lands. The Court also found that living conditions in the resettlement community have not allowed the inhabitants to resume traditional economic activities; basic health, education, electricity and water needs have not been fully met; and these conditions have caused the disintegration of the social structure and cultural and spiritual life of the community. Observing that the Río Negro community massacres were part of a systematic context of grave, massive human rights violations, the Court found that, in addition to historical damages, "the surviving victims of the Río Negro massacres experience deep suffering and pain … which fell within a state policy of 'scorched earth' intended to fully destroy the community" (IAHRC, 2012).

By 2013, when General Efraín Ríos Montt's first genocide trial was underway, the Government of Guatemala had taken no action to implement the Inter-American Human Rights Court order for reparations in the Río Negro massacre case, nor had it implemented the 2010 Chixoy Dam reparations agreement. The history of violence and massacres accompanying Chixoy Dam development, however, received international media attention, specifically with attention to the record of international financing for the Chixoy Dam that, at the height of the genocide, were the sole source of funding for Guatemala's military government.[16] These concerns prompted several US Senators to explore corrective action in a Foreign Appropriations bill adopted by the Senate in spring 2013, and signed into law with the January 2014 Congressional Appropriations Bill (modified language also appeared in the 2015 and 2016 legislation). US representatives on the World Bank and other Bank Boards were instructed to report to Congress on Guatemala's progress in implementing the Chixoy reparations agreement, and instructed to vote against new loans for Guatemala until such evidence has been presented (Consolidation Appropriations Act, 2014).

Implementation of US law disrupted the flow of money into Guatemala, essential funds that supported Government operation. After three decades of struggle to demonstrate the depth and breadth of development-induced disaster in Guatemala, the political will to implement a reparations plan and negotiated agreement was finally achieved. In November 2014, President Otto Pérez-Molina officially apologized to 33 dam-affected communities for the massacres, violence, and varied failures associated with development of the Chixoy Dam. A month later, Guatemala's Congress adopted legislation adding Chixoy Reparations as a national budget line item for the next 15 years, creating a means to implement the US$155 million social and economic development commitment. Reparations payments to families of those massacred began in 2015. The Inter-American Development Bank supported remedial actions with technical assistance grants and repurposing existing loans to address dam-affected communities' social infrastructure needs: repairing schools, supplying medical clinics, building a bridge across the reservoir, repairing resettlement homes, providing electricity and potable water (Albertos, 2015).

9.6 The Distance between Negotiated Settlements and Justice[17]

If success is measured by the achievement of initial goals, the long struggle to secure reparation for an immense injustice in Guatemala has achieved some success. If success is measured by on-the-ground reality, justice has different meanings for different people. In this case, a win for Chixoy communities may be a small price to pay to allow "business as usual" in other parts of Guatemala. Recall that Chixoy Dam was first in a series of 22 dams proposed by the World Bank in their 1951 country assessment; refined versions of this template still guide investment attention and energies. These days, a new hydro-development boom in Guatemala is being financed through public and private capital in partnership with international corporate entities. Hydro-development fuels state-sanctioned extractive industry (especially gold, silver, nickel and other mining) on indigenous lands. Guatemala's Natural Resources Law recognizes indigenous rights, including the right to free and prior informed consent, and hundreds of indigenous community consultations have been held to implement free and prior informed consent requirements. In 99 percent of these cases, community votes rejecting development have been ignored, precipitating violent confrontations with armed troops assigned to support private investment in a complicit effort that uses violence and targeted assassination of community leadership.

The timing and events associated with Santa Rita Dam development on the upper reaches of the Chixoy River Basin is illustrative. At the same time that the Guatemalan Government worked with affected communities to negotiate reparations in the Chixoy case, they sought financing for the Santa Rita dam (proposed in 2008), received construction approval (in 2010), rejected the applicability of the laws requiring free and prior consent of indigenous communities, and utilized a familiar strategy of violence and fear to stifle community protest (Oxfam Issue Briefing, 2015). Development financing includes private and public equity funds from European development banks channeled as Green Energy investment through the World Bank's International Financial Corporation. When preliminary site work began in 2012, river basin communities protested. President Pérez-Molina declared them subversive, suspended civil rights, and sent 1,500 national police in to protect development interests. By 2014, when President Molina issued his apology to Chixoy massacre survivors, the 22 indigenous communities that had expressed opposition to the Santa Rita Dam had endured targeted violence, forced evictions, burning of property and crops, and other abuses. As of this writing, five indigenous leaders who oppose the Santa Rita dam and two young children of another leader have been assassinated. Once completed, Santa Rita dam will generate carbon credits tradable under the European Union's emissions trading system. Much of the energy produced will be exported to Mexico and the United States. The dam and its reservoir will displace thousands and adversely impact an estimated 200,000 people, most of whom are indigenous peoples, whose livelihoods are casualties in the climate-change opportunism accompanying efforts to build global "clean, green energy" systems.

Thanks to cell phones, Internet access, and media coverage, indigenous communities in Guatemala today are keenly aware that, despite an official apology and plan to repair harms from Chixoy development, rights-abusive development continues with callous disregard for the rule of law. Yet the experience of Chixoy communities also serves as a model that inspires and fosters the belief that someday, in some contexts, some measure of justice can be realized. Thus, in the Santa Rita case, community members have developed partnerships with national and international advocacy groups, documenting abuses, filing complaints with international institutions and courts, and demanding their rights. Media attention to this case and other repression and violence in Guatemala's highlands drove thousands of rural workers into Guatemala City streets in early February 2016 to call for an end to development projects that displace communities and exploit natural resources. The pendulum continues to swing. Reaction to that protest was swift, with Vice-Minister of Sustainable Development Roberto Velásquez announcing, a few days later, that an existing moratorium on new mining licenses, prompted by findings of human-rights abuse and an order from the Inter-American Human Rights Court, will no longer be enforced. Instead, the Ministry of Energy and Mining will expedite the license granting because such projects "help to reduce the high levels of poverty within the country" (Guatemala Human Rights Center, 2016).

9.7 Conclusion

In presenting these brief stories of Chixoy and Santa Rita hydro-development, I am arguing this point: when you move beyond the place-based scenario and consider the cumulative, synergistic impacts of large-scale hydro-development, building dams and related infrastructure is, in essence, a form of environmental warfare. In the overt/covert global war between political and economic actors to ensure continued access to critical resources, the consequential damages to host communities – human and otherwise – are dismissed as trivial casualties, necessary tradeoffs for a greater good. Yet the place-based and planetary consequences are profound.

The dam-building industry's political-economic power, coupled with international agendas to ensure access to water, land, food, minerals, and energy in a climate-changed world, has launched a new era of dam building. New clothes dress the embedded model and it is animated by a diverse array of global actors (China, corporations, venture capital firms) but the same shortsighted aims prop up and contort the outcomes from this architecture of power. Thanks to efforts by the hydropower industry lobby and their counterparts in the international institutions and agencies, hydroelectricity was declared a climate-friendly source of energy during the political negotiations of the 1990s that resulted in the UN Climate Change Treaty and the related creation of carbon credit/green energy markets. Thus, at this writing, some 17 out of the world's 64 remaining large free-flowing rivers are sites of proposed or approved large dam projects that, assuming no reversal in the permit or financing process, will be dammed by 2020 (World Wildlife Federation, 2015).

Water justice refers to the struggle to expose that which has been historically denied: there are life-threatening conditions that are the unjust, inequitable result of corrupt, dysfunctional governance. Such struggles aim to stop that which generates harm, secure acknowledgment by culpable parties that injustice has occurred, receive apologies and remedial actions that repair harm, and ensure no re-occurrence. Recent experience demonstrates that water justice is often elusive, a point sadly illustrated by the simultaneous apology and reparation for Chixoy Dam communities, and the militarization and assassinations in upstream communities protesting the Santa Rita Dam. Within these contexts, the struggle to secure physical, political, and temporal rights-protective space is essential, as such space allows linkage of ulcerating crises to larger regional/global factors and forces.

In any context, achieving environmental justice necessarily requires confronting and transforming the architecture and flow of power. The push for justice requires tackling inequities head-on, resulting in actions that can reinvigorate or reinforce the community's meaning, power, and integrity. At the same time, reinvigoration of community-based power implicitly threatens entrenched power and their exploitative agendas. This fear drives backlash and repression when fundamental human rights violations are exposed and demands for accountability capture the public imagination. Backlash is inevitable, yet backlash also prompts further progressive change.

Notes

1 The USSR signed but did not ratify the agreements; Russia joined in 1992. Australia attended as an observer, but did not sign until years later. New Zealand did not formally participate, but later joined the IMF and World Bank.
2 For the sake of brevity, "World Bank" or "Bank" is used to refer to the institution and staff of the IBRD and other members of the World Bank Group.
3 This discussion of the World Bank's launching and laudatory goals is revised from Johnston (2012: 296–97).
4 Except where further noted, details and discussion of the Chixoy Dam case reflect research and findings reported in Johnston (2005).
5 "Ladino" is a recognized ethnic group in Guatemala; Spanish-speaking communities that can include people of Hispanic origin, mixed parentage, as well as westernized Amerindians.
6 This produced a study of the Petén (one-third of Guatemala's total area), Alta Verapaz, and Izabal; mountainous areas populated by indigenous Mayan and Ladino communities where, in years to come, much of the state-sponsored genocidal violence was waged against a civilian population.

7 INDE was created under military rule. A new constitution and elected government was put in place following the 1994 Oslo Peace Accords, but INDE remains an independent public entity run by current or retired military officers. For an analysis of Guatemalan military government operations during the civil war that sheds light on its deeply embedded role in the current architecture of power, see Schirmer (1998).
8 Agreements, contractual requirements, and compliance failures are summarized in Johnston (2005: 7–9, 12–21). Initial IADB financing contracts include no mention of resettlement or compensatory programs, but do stipulate that before construction in 1976, INDE must provide evidence of legally secured title to lands where project works are to be constructed.
9 Site preparation, financed by the loan, reportedly allowed an under-the-table opportunity to profit from the harvesting of valuable tropical hardwoods.
10 The Historical Verification Commission concluded that the Spanish Embassy fire was started by the police with the use of a flame or gas thrower, that Guatemalan government forces were responsible for the arbitrary execution of those who died in the attack, and that "the highest [Guatemalan] authorities were the intellectual authors of this extremely grave violation of human rights" (CEH, 1999: 163–82). These findings were cited as part of the evidence in the case against Pedro García-Arrendondo, the architect of this operation and chief of the national police unit for Guatemala City at the time of the Spanish Embassy burning. Arrendondo was convicted by unanimous verdict in 2015 for orchestrating the massacre and sentenced to 90 years in prison for his crimes against humanity, homicide, and attempted homicide.
11 With the signing of the 1996 Peace Accords, model villages were demilitarized. Yet, in 2003, when I visited Pacux to meet with survivors of Río Negro and other massacres in the Chixoy Dam region, I passed through the grounds of a small military base, stopping at a gatehouse barrier to answer questions from the armed guard and present identification before being allowed entry into the fenced-in compound. In 2004, Fundación de Antropología Forense de Guatemala exhumed a clandestine grave located in the well on this military base, recovering the remains of some 73 bodies.
12 Partridge's audit and resettlement plan was accomplished through a review of the records submitted to or prepared by financiers. No visit by World Bank staff or consultants to the dam site or discussion with affected communities occurred.
13 By 1991, 51 percent of INDE's revenues were used to service this debt. Privatization of INDE in 1996 allowed all of the World Bank and most of the IADB loans to be repaid in full. While the World Bank portfolio statement includes interest rates, maturation dates and payment status on these loans, it does not report total income earned on this debt. IADB however does report their income; privatization of INDE allowed IADB to collect some US$140 million in interest income from their loans to INDE (Johnston, 2005: 7).
14 See also the substantive discussion of World Bank safeguard failures in many projects financed around the world, and the critique of that institution's failure to strengthen praxis by implementing the World Commission on Dams' recommendations (an entity it helped to create) (Scudder, 2005).
15 For a detailed, reflexive review of this research, methods, ethical issues, and key outcomes, see Johnston (2010).
16 For media coverage exploring who profited while others suffered, and the role of international financiers in supporting a military government engaged in genocide, see Mychalejko (2013) and Deardon (2012).
17 This discussion includes ideas and detail revised from Johnston (2016).

References

Albertos, C. (2015). "Guatemala: Chixoy Dam-affected communities: Current status and challenges," paper presented to the Society for Applied Anthropology Conference, Pittsburgh, March 24–28.

Cohen, R. (2012). *The Fish that Ate the Whale: The Life and Times of America's Banana King*. New York: Random House.

Commission for Historical Clarification (Comisión para el Esclarecimiento Histórico, CEH). (1999). *Memory of Silence: The Guatemalan Truth Commission Report*. Basingstoke: Palgrave Macmillan.

Consolidated Appropriations Act. (2014). Public Law No. 113–76. January 17, 2014. *Congress. Gov.* Guatemala City.

Cullather, N. (2006). *Secret History: The CIA's Classified Account of its Operations in Guatemala, 1952–1954*. Stanford, CA: Stanford University Press.

Davis. S. (1983). Testimony on the environmental impact of a multilateral development bank-funded project: Hearings before the Subcommittee on International Development Institutions and Finance of the Committee on Banking, Finance, and Urban Affairs. House of Representatives, Ninety-eighth Congress, first session, Washington, DC, June 28 and 29.

Deardon, N. (2012). "Guatemala's Chixoy dam: Where development and terror intersect." *The Guardian*, December 10, 2012.

Doyle, K. and Kornbluh, P. (1997). CIA and assassinations: The Guatemala 1954 documents. *National Security Archive*, http://nsarchive.gwu.edu/NSAEBB/NSAEBB4/.

Dutch, S. (2009). The largest act of environmental warfare in history. *Environmental & Engineering Geoscience*, 15(4), 287–97.

Guatemala Human Rights Center (GHRC). (2016). "Guatemala news update, February 6–12," *Monitoring Guatemala: GHRC's Human Rights Blog*. February 12, 2016.

Gwin, C. (1997). US relations with the World Bank, 1945–1992. In D. Kapur, J. Lewis, and R. Webb (eds.), *The World Bank: Its First Half Century*, Vol. 2. Washington, DC: Brookings Institution Press.

Inter-American Human Rights Court. (2012). *Rio Negro Massacres v. Guatemala*, www.corteidh.or.cr/docs/casos/articulos/seriec_250_ing.pdf.

International Financial Institutions Act (1977) (22 U. S. C. 262(d)), Public Law 95–118, October 3, 1977, www.cfr.org/human-rights/international-financial-institutions-act-1977/p27148.

Johnston, B. (2000). "Reparations and the right to remedy." Briefing paper prepared for the World Commission on Dams, http://siteresources.worldbank.org/INTINVRES/214578-1112885441548/20480101/ReparationsandtheRighttoRemedysoc221.pdf.

Johnston, B. (2005). *Chixoy Dam Legacy Issues Study*, vol. 1: *Executive Summary Consequential Damages and Reparation: Recommendations for Remedy*; vol. 2: *Document Review and Chronology of Relevant Actions and Events*; vol. 3: *Consequential Damage Assessment of Chixoy River Basin*. New York: American Association for the Advancement of Science. Center for Political Ecology.

Johnston, B. (2010). Chixoy dam legacies: The struggle to secure reparation and the right to remedy in Guatemala. *Water Alternatives*, 3(2), 341–61.

Johnston, B. (ed.) (2011). *Life and Death Matters: Human Rights, Environment, and Social Justice*. London: Routledge.

Johnston, B. (2012). Water, culture, power: Hydro-development dynamics. In B. Johnston, L. Hiwasaki, I. Klaver, A. Ramos-Castillo, and V. Strang (eds.), *Water, Cultural Diversity, and Global Environmental Change*. Dordrecht: Springer, pp. 295–318.

Johnston, B. (2013). Human needs and environmental rights to water: A biocultural systems approach to hydro-development and management. *Ecosphere*, 4(3), 1–15.

Johnston, B. (2016). Action-research and environmental justice: Lessons from Guatemala's Chixoy Dam. In U. Heiss, J. Christensen and M. Neimann (eds.), *Routledge Companion to the Environmental Humanities*. London: Routledge.

Johnston, B. and Fiske, S. (2014). The precarious state of the hydrosphere. *WIREs Water* 1, 1–9.

Manz, B. (2005). *Paradise in Ashes: A Guatemalan Journey of Courage, Terror, and Hope.* Vol. 8. Berkeley: University of California Press.

McAllister, D. E., Craig, J. F., Davidson, N., Delany, S. and Seddon, M. (2001). Biodiversity impacts of large dams. *World Commission on Dams Background paper, 1.* New York: IUCN.

Museo Comunitario Rabinal Achi. (2003). *Oj K'aslik, Estamos Vivos: Recuperación de la Memoria Histórica de Rabinal (1944–1996).* Baja Verapaz: Rabinal.

Mychalejko, C. (2013). Profiting from genocide: The World Bank's bloody history in Guatemala. *Truthout,* March 8, 2013.

Oxfam Issue Briefing. (2015). The suffering of others: The human cost of the International Financial Corporation's lending through financial intermediaries. *Oxfam Issue Briefing,* April 2015, pp. 11–13. Oxford: Oxfam.

Pacenza, M. (1996). A people damned: The Chixoy dam, Guatemalan massacres and the World Bank. *Multinational Monitor,* 17(7–8), 8–12.

Partridge, W. (1983). *Comparative Analysis of Bid Experience with Resettlement: Based on Evaluations of the Arenal and Chixoy Projects.* New York: Inter-American Development Bank.

Partridge, W. (1984). *Recommendations for Human Resettlement and Community Reconstruction Components of the Chixoy Project.* New York: World Bank.

Postel, S. and Richter, B. (2012). *Rivers for Life: Managing Water for People and Nature.* Washington, DC: Island Press.

Renaud, F., Syvitski, J., Sebesvari, Z., Werners, S. E., Kremer, H., Kuenzer, C., Ramesh, R., Jeuken, A., and Friedrich, J. (2013). Tipping from the Holocene to the Anthropocene: How threatened are major world deltas? *Current Opinion in Environmental Sustainability,* 5, 644–54.

Richter B., Postel, S., Revenga, C., Scudder, T., Lehner, B., Churchill, A. and Chow, M. (2010). Lost in development's shadow: The downstream human consequences of dam. *Water Alternatives* 3, 14–42.

Schofield, T. (1998). Environment as an ideological weapon: A proposal to criminalize environmental terrorism. *Boston College Environmental Affairs Law Review,* 26(3), 619–47.

Schirmer, J. (1998). *The Guatemalan Military Project: A Violence Called Democracy.* Philadelphia: University of Pennsylvania Press.

Scudder, T. (2005). *The Future of Large Dams: Dealing with Social. Environmental, Institutional and Political Costs.* London: Earthscan.

Toussaint, E. (2014). "Domination of the United States on the World Bank. Bretton Woods, the World Bank and the IMF: 70th anniversary (Part 7). Committee for the Abolition of Illegitimate Debt," www.cadtm.org/.

Van Cappellen, P. and Maavara, T. (2016). Rivers in the Anthropocene: Global scale modifications of riverine nutrient fluxes by damming. *Ecohydrology & Hydrobiology,* 16(2), 106–11.

Vörösmarty, C., McIntyre, P. B., Gessner, M. O., Dudgeon, D., Prusevich, A., Green, P., Glidden, S., Bunn, S., Sullivan, C., Liermann, C., and Davies, P. (2010). Global threats to human water security and river biodiversity. *Nature,* 467(7315), 555–61.

Westing, A. (1980). *Warfare in a Fragile World: Military Impact on the Human Environment.* Stockholm: SIPRI.

World Bank. (1951). "Guatemala Economic Development Plan," http://documents.worldbank.org/curated/en/527361468251181949/text/multi0page.txt.
World Bank Group Archives. (2001). "World Bank Group historical chronology," http://siteresources.worldbank.org/EXTARCHIVES/Resources/Bank% 20chronology.pdf.
World Commission on Dams. (2000). *Dams and Development: A New Framework for Decision-making: the Report of the World Commission on Dams.* London: Earthscan.
World Wildlife Federation. (2015). "Free-flowing rivers: Economic luxury or ecological necessity?" Appendix 1:12, www.wwf.se/source.php/1120326/free.

Part III
Exclusion and Struggles for Co-Decision

Indigenous water control, Licto, Ecuador.
Photo by Rutgerd Boelens.

Introduction: Exclusion and Struggles for Co-Decision

Jeroen Vos, Tom Perreault, and Rutgerd Boelens

Water justice is often sought in "good water governance." Yet, what "good governance" means is not something that can be straightforwardly decided or linearly implemented: different stakeholders hold different power positions, have conflicting interests and deploy different valuation languages regarding water, land and livelihoods. Deliberative policy-making processes, including local communities' participation in decision-making, are often presented as the tool to help craft inclusive, democratic water governance arrangements. However, here, a fundamental but commonly neglected or actively suppressed question is, "who participates in whose project"? Although water governance is about institutional configurations, regulations and policy-making and implementation, it is also about capabilities, powers and social struggle over access to resources, setting the agenda and discursively framing problems. Water justice, then, is not something that can easily be crafted through tinkering with governance arrangements, but requires struggles and continuous renegotiation as part of larger battles for justice and democracy.

Water injustices often imply exclusion of vulnerable groups from access to clean water and affordable services, but also from representation in water-control decision-making. This exclusion can be based on gender, race, caste, class, ethnicity, religion, or political affiliation. Maria Rusca, Cecilia Alda-Vidal, and Michelle Kooy (Chapter 11) provide clear examples of this in their chapter on drinking water in Kampala. Joyeeta Gupta (Chapter 14) contends that privatizing irrigation water services may often exclude smallholders.

Climate Justice

Climate change also causes major distributive injustices. Droughts and floods tend to affect the poor more severely than the relatively rich (Adger, 2001; Ikeme, 2003; IPCC 2014; Ribot, 2010; Schneider and Lane, 2006). Skewed vulnerabilities in relation to effects of climate change lead to asymmetrical impacts (Gardiner and Hartzell-Nichols, 2012). This is even more unfair and imbalanced considering that the poor's share in emission of greenhouse gases is much less than the gigantic emissions by the rich. A report commissioned by the World Bank (2008) estimates the impacted populations killed or left homeless per region by seven common chronic and sudden disasters that are increasingly related to climate change: droughts, extreme temperatures, floods, landslides, tidal surges and wind storms. Already millions of people are affected by floods in Southern and Eastern Asia and droughts in South America, South Asia and East Africa. Likewise, health effects of climate change affect the poor disproportionally (Costello *et al.*, 2009).

Nancy Fraser (2001, 2005, 2007) has defined three interrelated domains of social struggle: redistribution, recognition and representation. These can be seen as irreducible dimensions of empowerment (see also Chapter 1). Excluded groups might gain access to resources, income and security by struggling for representation in policy- and decision-making domains and gaining recognition of their needs and values in these domains. Clearly, in many cases, deprived groups show no signs of overt struggle. Cleaver (Chapter 13) draws attention to the reasons for this "massive non-struggle," and populations' apparent acceptance of and accommodation to unjust circumstances. She argues that overt struggle might have too high a cost, and accommodation to unjust situations is a rational strategy for many marginalized groups.

In the case of climate change, marginalized groups struggle for recognition of their knowledge and perceptions of climate change. Farmers, for instance, have specific local knowledge about their changing environments, such as detecting shifting crop flowering dates well before scientists do. However, international forums (such as IPCC) and national policy arenas are dominated by scientific discourses that hardly allow for voices of others (Adger *et al.*, 2011; Vanderheiden, 2008). And what is worse, increasingly, global water policy and expert institutes use climate change as a pretext and justification to dismiss and reconfigure functional local water governance forms and knowledge, to replace them with "modern" expert-, state- and market-biased rules, rights and institutions. For instance, Lynch (2012) critiques the 2010 World Development Report argument sustaining that traditional agricultural and water management practices are no longer able to counteract the rapid, unpredictable impacts of climate change (World Bank, 2010: 137). Climate change is (ab)used to delegitimize local communities' water management knowledge and practices. Community-managed water-use systems will indeed be substantially affected by climate change, but there is no historical or contemporary evidence to expect that universal expert-based, World Bank-advocated private water rights markets and "full-value" water service pricing will improve local water governance and generate more equitable water allocation. On the contrary, as Lynch and many authors of this book prove, it has shown to dramatically increase vulnerability of smallholder food producing communities and their agroecosystems (Lynch, 2012).

Adger *et al.* (2011) argue that vulnerability to environmental hazards should be context-specific, taking into account the values, perceptions, and practices of involved water user families and communities, rather than apply generic vulnerability indexes. In a broader sense of recognizing knowledge and values of subalterns, Bakker, Simms, Joe and Harris (Chapter 10) describe the difficulties in gaining recognition for indigenous knowledge and spiritual values among Canadian First Nations in the recent major overhaul and modernization of the British Columbia Water Sustainability Act. Rusca, Alda-Vidal and Kooy (Chapter 11) emphasize the problem with non-recognition of social and emotional dimensions of sanitation programs. In both these cases, the struggle for recognition is closely linked to the struggle for representation and co-decision.

In the case of climate change catastrophes, clearly the people who decide on targets to reduce greenhouse gas emissions and implement mitigation and adaptation measures are mainly male, and mainly from the Global North (cf. Terry, 2009). Thus, the struggle for "climate justice" is also a struggle for representation of climate-change victims on national and international decision-making forums (Parks and Roberts, 2010; Vanderheiden, 2008), and for recognition that representatives of vernacular, non-scientific natural resource governance alternatives, throughout the world, have very much to add to expert-biased national and international debates. Since governments, elites and scientific expert-knowledge centers will not be prepared to automatically and structurally involve marginalized water user groups' and grassroots federations' visions and interests in decision-making, gaining representation requires social mobilization with a multi-scalar strategy. This also holds true for all other fields of water allocation, management and governance. Van den Berge, Vos and Boelens (Chapter 12) show how the "Human Right to Water" movement in Europe employed a multi-scalar strategy to collect 1.9 million signatures from people opposed to privatizing public water utilities in Europe. Rusca, Alda-Vidal and Kooy (Chapter 11) show that public participation in defining "adequate sanitation" proved crucial to attain "sanitation justice."

The Chapters of Part III

Part III of this book includes five chapters on exclusion and struggles for co-decision. Karen Bakker, Rosie Simms, Nadia Joe and Leila Harris (Chapter 10, "Indigenous People and Water Governance in Canada: Regulatory Injustice and Prospects for Reform") explore the regulatory injustices affecting indigenous people in Canadian water governance. In northern and western Canada, high resource extraction rates are creating intense socio-environmental pressures on indigenous peoples' traditional territories. Fresh water systems are particularly affected by mining, oil and gas extraction, and logging. This, in turn, has significant impacts on indigenous communities, including compromised access to safe drinking water, threats to environmental and water quality, and related livelihood and health issues. This chapter analyzes several instances of regulatory injustice within Canada's colonial water governance framework and recent reforms. The authors first provide an overview of Canada's legal and regulatory architecture for water governance, with specific examples of the disjuncture between colonial (Western) law and indigenous water laws. Next, the chapter describes a regulatory injustice in the province of British Columbia around the "First in Time, First in Right" water-rights regime. The authors then explore current responses, focusing on the potential for indigenous water co-governance – concluding with some concrete suggestions for reform.

Maria Rusca, Cecilia Alda-Vidal and Michelle Kooy (Chapter 11, "Sanitation Justice? Addressing the Multiple Dimensions of Urban Sanitation Inequalities") reflect on sanitation justice in current development approaches to increase access to sanitation services. The authors define three dimensions of sanitation (in)justice – distributive, procedural, and recognitional – and review how these dimensions are related to, but distinct from, existing water and environmental justice frameworks. Using a case study of sanitation development projects in Kampala, Uganda, the authors illustrate the concept's utility to broaden current conceptualizations of inequalities regarding sanitation. The authors suggest that moving current development terminology about improved sanitation toward defining "just" sanitation can help shift development approaches from a singular focus on sanitation infrastructure to a multidimensional understanding of sanitation services.

Jerry van den Berge, Jeroen Vos, and Rutgerd Boelens (Chapter 12, "Uniting Diversity to Build Europe's Water Movement Right2Water") describe the European Citizens' Initiative (ECI) "Right2Water." This massive campaign ran from April 2012 to September 2013 and collected nearly two million signatures against water utility privatization. Right2Water proposed to implement the human right to water and sanitation in European legislation, as a strategic-political tool to fight privatization. The authors examine how the ECI succeeded. With illustrations from Germany, Spain, Italy, and Greece, the authors show the national campaigns' diverse contexts, actors involved, and dynamics. Despite enormous support for the campaign in most European Union counties, the European Commission's official response was that many of the suggestions were already part of Europe's policy and that it would not change any existing legislation. Notwithstanding this reaction, the authors consider the campaign a success because it brought together many highly diverse organizations and forged unity in the message against privatizing water utilities in Europe.

In Chapter 13 ("Everyday Water Injustice and the Politics of Accommodation"), Frances Cleaver analyzes the politics of accommodation to water-injustice conditions. Scholars who focus on water justice issues tend to emphasize the struggles involved in water access, variously emphasizing contestations over rights and distribution; rules; authority and representation; discourses and knowledge. However, Cleaver argues that such a perspective can distort the analytical gaze towards conflictual events and processes and towards relationships of domination and resistance. Contestations over water are ubiquitous and the project of problematizing technical and managerial water-distribution models is an important contribution to water justice thinking. However, the author suggests that we also need to focus on how and why everyday water-access relations are so often characterized by acceptance, compromise and adjustment, and overlooking injustices. This draws us into the realm of explaining the nature of human agency, social life and the ways in which water arrangements become institutionalized. The chapter explores four notions to help understand the politics of accommodation: hegemony, pragmatism, connectedness and moral ecological rationality. The chapter further reflects on the challenges of studying non-contestation rather than more obvious manifestations of conflict and struggle.

Finally, Joyeeta Gupta (Chapter 14, "Sharing Our Water: Inclusive Development and Glocal Water Justice in the Anthropocene") explores the notion of inclusive development and water justice in the Anthropocene. An inclusive development perspective toward water justice has three key implications. First, ecological inclusiveness implies recognizing limits on water use, water flows, and their ecosystem services or "water ecospace." Second, relational inclusiveness requires acknowledging that such limits must be democratically and discursively defined. Third, water governance is a "glocal" issue, as causes and effects of water use and abuse occur at local to global levels. Actors tend to monopolize control over water through discourses of liberalization, privatization, technocratization and securitization; and de jure and de facto rules of water ownership and control through permits, infrastructure, finance, land ownership and "grabbing," combined with market principles for water pricing. Gupta argues that a socially inclusive model is required, acknowledging that water ecospace is a glocal public good, that access and allocation of rights, responsibilities and risks should be undertaken in a socially just manner. Accountable collectivities – whether communities, states or inter-governmental entities – should govern water by creatively linking the local, national, international, and global scales.

References

Adger, N. (2001). *Social Capital and Climate Change*. Tyndall Centre for Climate Change Research Working Paper 8. Norwich: University of East Anglia.

Adger, W. N., Barnett, J., Chapin III, F. S. and Ellemor, H. (2011). This must be the place: Underrepresentation of identity and meaning in climate change decision-making. *Global Environmental Politics*, 11(2), 1–25.

Barnett, J., Lambert, S. and Fry, I. (2008). The hazards of indicators: Insights from the environmental vulnerability index. *Annals of the Association of American Geographers*, 98(1), 102–19.

Costello, A., Abbas, M., Allen, A., Ball, S., Bell, S., Bellamy, R., Friel, S., Groce, N., Johnson, A., Kett, M. and Lee, M. (2009). Managing the health effects of climate change. *The Lancet*, 373(9676), 1693–1733.

Fraser, N. (2001). *Social Justice in the Knowledge Society: Redistribution, Recognition, and Participation*. Berlin: Heinrich Boll Stiftung.

Fraser, N. (2005). Mapping the feminist imagination: From redistribution to recognition to representation. *Constellations*, 12(3), 295–307.

Fraser, N. (2007). Feminist politics in the age of recognition: A two-dimensional approach to gender justice. *Studies in Social Justice*, 1(1), 23–35.

Gardiner, S. M. and Hartzell-Nichols, L. (2012). Ethics and global climate change. *Nature Education Knowledge*, 3(10), 5.

Ikeme, J. (2003). Equity, environmental justice and sustainability: Incomplete approaches in climate change politics. *Global Environmental Change*, 13(3), 195–206, http://dx.doi.org/10.1016/S0959-3780(03)00047-5.

IPCC. (2014). *Climate Change 2014: Impacts, Adaptation, and Vulnerability. Part A: Global and Sectoral Aspects. Contribution of Working Group II to the Fifth Assessment Report of the Intergovernmental Panel on Climate Change*, C. B. Field, et al. (eds.). Cambridge: Cambridge University Press.

Lynch, B. D. (2012). Vulnerabilities, competition and rights in a context of climate change toward equitable water governance in Peru's Rio Santa Valley. *Global Environmental Change*, 22(2), 364–73.

Parks, B. C. and Roberts, J. T. (2010). Climate change, social theory and justice. *Theory, Culture & Society*, 27(2–3), 134–66.

Ribot, J. (2010). Vulnerability does not fall from the sky: Toward multiscale, pro-poor climate policy. *Social Dimensions of Climate Change: Equity and Vulnerability in a Warming World*, 2, pp. 47–74.

Schneider, S. H. and Lane, J. (2006). Dangers and thresholds in climate change and the implications for justice. In W. Neil Adger, J. Paavola, S. Saleemul and M. J. Mace (eds.), *Fairness in Adaptation to Climate Change*. Cambridge, MA: MIT Press, pp. 23–52.

Terry, G. (2009). No climate justice without gender justice: An overview of the issues. *Gender & Development*, 17(1), 5–18. DOI: 10.1080/13552070802696839.

Vanderheiden, S. (2008). *Atmospheric Justice: A Political Theory of Climate Change*. Oxford: Oxford University Press.

World Bank. (2008). *Assessing the Impact of Climate Change on Migration and Conflict*. Paper commissioned by the World Bank Group to C. Raleigh, L. Jordan, I. Salehyan for the "Social Dimensions of Climate Change" workshop. The Social Development Department/World Bank Group, Washington, DC.

World Bank. (2010). *Development and Climate Change: World Development Report 2010*. Washington, DC: World Bank.

10

Indigenous Peoples and Water Governance in Canada: Regulatory Injustice and Prospects for Reform

Karen Bakker, Rosie Simms, Nadia Joe, and Leila Harris

10.1 Introduction

This chapter explores two interrelated examples of injustice in access to water for Indigenous peoples in Canada. Injustice, from our perspective, has two main dimensions: limited access to safe water (water security); and exclusion from water governance and management. In Canada, the roots of this injustice run deep. Indigenous[1] peoples in Canada have historically been excluded from water governance by the colonial settler state. To date, the question of whether water is included within Aboriginal title (legal land rights) has not yet been settled by the Canadian courts (Laidlaw and Passelac-Ross, 2010; Phare, 2009).[2] As a result, Indigenous water rights in Canada, with few exceptions, have been treated implicitly within land-focused legal claims. Moreover, historical inequalities have often constrained Indigenous communities' access to water and exercise of Indigenous rights (Phare, 2009; Simms, 2014; Von der Porton, 2012; Von der Porten and De Loë, 2013a, 2013b). High-risk water systems – systems with major deficiencies, which pose a high risk to water quality – pose a threat to the health of one-third of First Nations people living on reserves (McLaren, 2016; Neegan Burnside Ltd., 2011). In short, stark injustices exist with respect to water for Indigenous communities in Canada.

To some degree, this situation is a direct result of Canada's fragmented system of governance: municipal, provincial and federal governments hold responsibility for different aspects of water management. Provincial governments are largely responsible for fresh water (and delegate drinking water to municipalities), whereas federal responsibilities include a range of issues related to water rights for Indigenous communities.[3] Competing jurisdictional priorities, lack of clarity on roles and responsibilities, and a failure to cooperate have resulted in systemic governance gaps leading to increased risk in water supply systems and widespread underfunding, which some scholars have characterized as institutionalized racism (Mascarenhas, 2007; Murdocca, 2010).

These issues are particularly acute in Canada's westernmost province of British Columbia (BC). In most of the province, formal treaties were never signed between the Crown and Indigenous communities. This is a significant issue for many reasons, not least of which is the fact that the potential for legal recognition of water rights creates the possibility for extensive Indigenous control of water governance. This, however, is currently far from the reality in practice.

This chapter documents and analyzes examples of regulatory injustice within British Columbia's water governance framework by exploring two short case studies. The first example (Section 2) is a critical analysis of the FITFIR (First in Time, First in Right) water rights regime; although First Nations are undeniably "First in Time," they are often not the first rights-holders under BC water-licensing practices (BCAFN, 2010; Simms, 2014). In Section 3, we explore a second example: the approach taken to consultation between First Nations and the provincial government under the recent major overhaul and modernization of the BC Water Sustainability Act. In Section 4, we reflect on the implications for Indigenous water governance in Canada. In Section 5, we conclude the chapter with possible responses, focusing on the potential for water co-governance – concluding with some concrete suggestions for reforms that might further water justice for Indigenous communities.

10.2 First in Time, But Not First in Right

The introduction of the water-licensing system by the Province does not change the fact that Aboriginal peoples of BC, and indeed across Canada, were the first users of the water, and continue to use water pursuant to exercising their constitutionally protected Aboriginal and Treaty rights.

(Union of British Columbia Indian Chiefs, 2011: 3)

One of the most controversial topics in British Columbia's water governance framework is the system for allocating water rights: First in Time, First in Right (FITFIR). Similar to other provinces in western Canada and states in the western United States, FITFIR was originally implemented with the onset of colonial settlement. In British Columbia (BC), the process of designating Indian reserves and water rights also became closely tied to FITFIR. When reserve lands were delineated, the struggle over reserve water rights and allocation also began. Water allocation systems "provide the rules and procedures for assigning rights and establishing the processes used to decide how water should be shared among various users" (Brandes *et al.*, 2008: 6). As its name suggests, the FITFIR allocation scheme operates on a first-come, first-served logic:

Water rights in BC may be exercised under a system of priorities according to their respective priority dates. During times of scarcity, water licenses with the earlier priority dates are entitled to take their full water allocation over the junior licenses, regardless of the purpose for which the water is used.

(MOE, 2013: 17)

The irony of the title "First in Time, First in Right" could hardly be more blatant: despite the fact that Indigenous peoples undeniably are the *first* peoples in the land of BC and were clearly first in time for all water in the Province, the water rights defined and assigned to reserve lands were not consistently registered as such – an outstanding issue that to this day Indigenous communities continue to raise as an injustice.[4]

Creating reserves was, at its core, a strategy to remove Indigenous peoples from the landscape – to "keep them quiet" (Carstens, 1991) – in the face of increasing Euro-Canadian settlement pressures in the 1800s. Historian Cole Harris (2004) is explicit on this point: "[Indigenous] people were in the way, their land was coveted, and settlers took it" (ibid.: 167). Although some 140 reserves had been established prior to 1871 when BC joined the Canadian Confederation,[5] most were mapped throughout the late 1800s and early 1900s (Harris, 2001). As a consequence of colonization and the reserve system, Indigenous peoples were denied access to much of their territories and were confined to tiny tracts of fragmented land. The ratio of reserve land area to total provincial land area demonstrates the scale of dispossession: there are 1500 small reserves in what is today the province of BC, comprising a mere one-third of 1 percent of the provincial land base (Harris, 2004). While it is beyond the scope of this chapter to consider the full breadth of impacts that the reserve system had on physical, socioeconomic and cultural wellness, it is clear that it affected almost every aspect of Indigenous lives and livelihoods (Harris, 1997, 2001, 2004; Kelm, 1997; Miller, 2000; Thom, 2009). As Armstrong and Sam (2013: 245) describe:

Forced removal and displacement from land-imposed restrictions on water and other resources that disallowed freedom of movement to vital subsistence procurement sites and inhibited the ability of Indigenous peoples to continue ancient customary relationships and responsibilities within their ecosystems.

Mirroring the provincial–federal wrangling over the so-called "Indian land question" and delineation of reserve boundaries, jurisdictional strife was pronounced in the debate over how to allocate water to the reserve-land parcels (BCAFN, 2010; Bartlett, 1998; Harris, 2004; Matsui, 2005, 2009; Richard, 1999). British Columbia began to grant water licenses to settlers in 1865. Significantly, these licenses were registered *prior to* the establishment of reserves (BCAFN, 2010). In 1871, Indigenous peoples and their lands became a federal responsibility, while jurisdiction over land and water was transferred to the Province (Harris, 2001). With this division, a prolonged dispute was initiated between the two governments over whether or not water rights should be assigned to reserve lands and, further, both governments were determined to maintain exclusive authority for defining water rights in the Province (Bartlett, 1998). Indigenous peoples were completely barred from applying for water licences in their own name until 1888 (BCAFN, 2010). This dispossession in the 1800s did not proceed in the absence of resistance from Indigenous communities: "Indigenous people were not quiescent in the face of deteriorating conditions on their reserves ... Repeatedly [the Indigenous leaders] petitioned the Government for assistance in getting adequate water" (Kelm, 1997: 48). As reserve water rights were designated, however, colonial governments did not consider these well-articulated demands and desires. Rather, these communities were barred from applying for water rights; as Matsui (2009: 63) writes:

Despite their differing opinions on the question of [Indigenous] water rights jurisdiction, both provincial and federal officials shared the belief that whatever rights [Indigenous peoples] had, they were held at the "pleasure of the Crown." A number of [Indigenous] testimonies and petitions, which asserted inherent Aboriginal rights to water, did not sway either federal or provincial officials.

In 1876, a joint federal-provincial Reserve Commission was established with the aim of settling reserve-land boundary disputes, and by extension, resolving reserve water allocations.

Federal commission officials generally recognized that reserve water rights were essential to fulfill the objectives with which the reserve lands were designated (Bartlett, 1998). Arguing that water was necessary for irrigated agriculture, which in turn was a critical element in the project of assimilating Indigenous peoples, Federal Reserve commissioners took measures to register water licenses to accompany reserve lands in BC. The federal government held these licenses on behalf of Indigenous communities (Matsui, 2005, 2009; Walkem, 2004). Provincial government officials, on the other hand, were reluctant to recognize the federal government's jurisdiction over reserve water allocation, or to recognize Indigenous water rights to the detriment of settlers (Bartlett, 1998). Provincial authorities maintained that the reserve water rights granted by federal officials were not legitimate grants but merely unauthorized records (Matsui, 2009). In 1920, for example, a provincial board of investigation ruled that the reserve water records entered by federal Indian Commissioners were legally invalid (Matsui, 2009). This federal–provincial divide over reserve water allocation is evident in the case of the Lower Similkameen Indian Band, for instance, where water rights data indicate that the provincial government still does not formally recognize two reserve water allotments that were granted by federal officials in the 1880s. Overall, as a result of the back-and-forth between federal and provincial officials, there was a great deal of ambiguity and inconsistency in how water allocations were being documented and assigned to reserves.

The ultimate outcome of the provincial–federal clash over reserve water licenses, and of the fact that Indigenous peoples could not apply for their own licenses, was that reserve water rights were often left far down the priority list within the provincial system and were frequently cancelled or overridden by settler water licenses (Bartlett, 1998; Richard, 1999). Bartlett (1998) notes that the allotments of water made by the Reserve Commissioners were, "invariably made subject to the prior recorded rights of white settlers, though it was noted at the time that in many instances water had been recorded by white settlers which was really necessary for the Indians" (ibid.: 48).

As an illustrative example, Simms (2014) documents the case of the Lower Similkameen Indian Band (LSIB). According to historical records, in 1969, the LSIB applied for the right to use 2,000 gallons of water a day from Nahumcheen Brook, but the application was refused on the basis that there was insufficient water to grant this allocation. However, 17 years later, the Province issued a license to a private landowner for 150 acre-feet of water annually (1 acre-foot is approximately 893 gallons per day, or 326,000 gallons per year) in addition to 1,000 gallons a day from Everden Spring and Nahumcheen Brook (Ministry of Environment, 1997a, 1997b). Also highlighting the low priority given to a number of First Nations communities, the LSIB has 61st priority out of 105 licenses on the Simalkameen River (Simms, 2014: 49). Sam (2013a, 2013b) has further documented how this inconsistency unfolded in the case of the Penticton Indian Band, where, "settler water licenses were given priority on all streams that flowed through reserve lands, despite the Commission's recognition of existing water rights."

Indigenous communities in BC have actively resisted the water licensing scheme from the outset and continue to voice opposition to the system today. Our review of the recent BC

Water Act Modernization process, discussed in the next section, shows that issues of water allocation and licensing continue to be a key concern for many Indigenous communities, as well as for others in the Province (see also Jollymore *et al.*, 2017). FITFIR is likely to be increasingly contentious moving forward, particularly for those Indigenous communities that do not have priority water licenses within the ranking scheme. It is important to recall that under FITFIR, the earliest recorded license has priority access to water *in times of shortage*. Water scarcity is becoming a tangible, pressing reality across the Province: many water bodies are nearing full or over-allocation; 25 percent of water sources in BC now have restricted licensing (Brandes and Curran, 2008). In the presence of water scarcity, the FITFIR system will come into force substantially and those license holders lower down the priority list, including many Indigenous communities, will face a great deal of uncertainty in water access.

10.3 Constrained Consultation: Modernizing BC's Water Act

The province's approach maintains the fundamental flaw of assuming that the Province has sole jurisdiction over water and thus the authority to delegate water resources where there is a reasonable legal basis for Aboriginal jurisdiction.

(BCAFN, 2013: 23)

Indigenous peoples in British Columbia have long contested the basic principles of the First in Time, First in Right system. Despite opposition, the doctrine was enshrined within BC's new Water Sustainability Act (WSA) when it was passed in 2014 (MOE, 2013) and entered into law in 2016. The decision to uphold FITFIR was justified on the basis that this allocation system is easy to understand and administer, and does not require a re-ranking of users which "could change with time and be highly subjective" (MOE, 2013: 116). Although the WSA does not provide substantive changes to the provincial water allocation mechanism, it does enable decision-makers to make allowances for essential household use regardless of the priority of other licenses. This is deemed a sufficient amendment to satisfy concerns with the FITFIR system: "These modifications ... to allow for essential household use should address many of its perceived shortcomings" (MOE, 2013: 116). Further, although the WSA includes a provision for decision-makers to consider environmental flow needs[6] in new water-license applications, the adopted definition of environmental flow needs does not include cultural or spiritual flows and only mentions that there will be "continued mechanisms to reserve water for First Nations" (MOE, 2013: 6). The licensing measures adopted in the WSA fall short of responding to the many issues and questions that Indigenous communities and organizations have raised about FITFIR, and do not speak to Aboriginal rights and title to water more generally.

The decision to strategically disregard Indigenous peoples' input is evidenced by a review of the consultation process for the Water Sustainability Act (Joe *et al.*, 2016, see also Jollymore *et al.*, 2017). To elicit feedback on proposed changes to the Act, the Province initiated a public engagement process and invited members of the public, interest groups, stakeholders and Indigenous peoples to participate. This process proved to be challenging, as Indigenous peoples are not an "interest group" or even a stakeholder under the law, yet they were treated as such in the consultation. However, the consultation did not recognize that Indigenous peoples have constitutionally protected rights that have been "recognized and affirmed" in the Constitution of Canada and thus require an appropriate degree[7] of consultation directly with First Nations governments.[8] Aside from a handful of public information and consultation workshops, no additional resources were provided to meaningfully engage Indigenous peoples in the review and discussion of the proposed changes to the legislation.

The result was that Indigenous participation in the consultation process was limited. A total of 34 Indigenous organizations, governments and individuals throughout the province of British Columbia participated in the public engagement (while others actively resisted the process, based on the concerns suggested above, namely treating Indigenous people and communities as "stakeholders"). We can take these generally low numbers as indication that the process was not designed and implemented in a way that solicited meaningful Indigenous input (cf. Jollymore *et al.*, 2017).

In their submissions, Indigenous individuals, organizations and governments expressed the view that the public consultation process set out by the Province did not meet their expectations or legal obligations to engage in meaningful consultation over changes to the legislative framework governing water use in BC. Many submissions argued that Indigenous peoples must be involved in any and all decisions that impact, or have the potential to impact, their lands and resources, and by extension this includes their full participation in developing legislation and regulations pertaining to water resources (Joe *et al.*, 2017: 24).

Furthermore, many respondents clearly stated that their submission did not constitute consultation. Many described consultation as a longer period of engagement, compared to the relatively short timeframe set by the Province for submissions. A meaningful consultation was also understood to require mutual negotiation and agreement by all parties prior to engaging in the consultation process – which did not happen – raising concerns about the likelihood that the Province had failed in its legal obligation to uphold the Crown's honor in its duties to consult with Indigenous peoples (*Ratcliff & Company, LLP on behalf of the Coldwater Indian Band*, 2010).

The Courts have also been clear in their position on what constitutes meaningful consultation. In a decision on *Halfway River First Nation* v. *British Columbia* (Minister of Forests, 1999), the BC Supreme Court declared that the process for adequate and meaningful consultation must "ensure that Aboriginal peoples are provided with all necessary information in a timely way, so that they have an opportunity to express their interests and concerns, and to ensure that their representations are seriously considered and, wherever possible, demonstrably integrated into the proposed plan of action."

The process undertaken by the Province to solicit feedback from First Nations also created barriers to participation. Most submissions by BC First Nations expressed interest in participating in the discussion due to the significance of water and its management on their reserves and in their traditional territories. They indicated, however, that the time period given for review of the discussion papers was too short to thoroughly review or critically analyze the proposed legislation. As per the fiduciary relationship between the Crown and Aboriginal peoples,[9] many respondents requested that the Province provide additional resources for such a review and thus enable greater First Nations participation in the process.

Some respondents also felt that an additional barrier involved the primary use of Internet technologies to communicate information and gather input. First Nations communities in BC can be isolated, with limited Internet access. Relying primarily on an online forum could exclude some communities, particularly community members with specialized local knowledge of water resources (Union of British Columbia Indian Chiefs, 2011: 7). Despite clear direction from both First Nations and Canada's judiciary to engage and consult BC First Nations, the Province did not distinguish First Nations rights and interests from those of other public "stakeholders" as evidenced in their *Report on Engagement* (issued in the first phase of consultation). Moreover, the Province did not follow up on additional feedback received by First Nations during the second and third discussion phases. As a result, there was a distinct change in the tone of First Nations' submissions as the consultation process unfolded. In the early phases, some of the submissions welcomed the opportunity to work together with the Province to shape the legislation to reflect First Nations' interests and values and the new legal realities regarding resource management. One organization called it a "precursor" to consultation, engaging First Nations, while the BC Assembly of First Nations extended a direct offer to help the Province develop a consultative process to engage meaningfully with First Nations across the Province (First Nations Women Advocating Responsible Mining, 2010). However, as the constrained approach to consultation became clear, the tenor became more confrontational. For example, several submissions made in the third phase (the legislative proposal) referred to the legal risk of excluding First Nations from meaningful consultation.

From publicly available documents, it is unclear how the Province chose to consider First Nations submissions into their changing legislative framework – if at all. Indeed, a review of the submissions relative to outcomes confirmed that there were many issues identified where the majority's view of submissions (including from First Nations and input from other stakeholders) did not influence the outcomes. This was particularly the case for rights- and license-related issues for which FITFIR was maintained: the preferences expressed in submissions by industry were heeded (Jollymore *et al.*, 2017).

It is also clear that the legal duty to consult was not perceived by First Nations as merely a matter of legal obligation or procedural requirement owed by the Crown. Rather, First Nations clearly expressed the need for *meaningful* consultation due to the proposed changes' significance for the outdated Water Act and its significant potential to affect Aboriginal rights and title. Their submissions reiterated that First Nations depend on access to clean, safe water for their lives and livelihoods. Poor water quality, degraded aquatic habitat, or even reduced access to water, affects or has the potential to affect the Aboriginal peoples' ability to fish, hunt and harvest traditional foods and medicines. It becomes clear that a change in legislation impacts Aboriginal rights and/or title, which are in fact protected under the Constitution Act of Canada (1982). Numerous First Nations articulated in their submissions that, it is of utmost importance, to these communities, who decides "what" and "how much" water is used in their traditional territories.

Moreover, because of their relationship with their lands and water, First Nations have considerable knowledge about their water systems. As suggested by many respondents, traditional knowledge can help inform development of appropriate site-specific water objectives including critical thresholds, environmental flow needs, and sustainable water allocation plans – all newly enacted provisions in BC's WSA. As noted in the Simpcw First Nation's Water Declaration:

Our relationship with our lands, territories and water is the fundamental physical, cultural, and spiritual basis for our existence. This relationship to our Mother Earth requires us to conserve our fresh waters and oceans for the survival of present and future generations. We assert our role as caretakers with rights and responsibilities to defend and ensure the protection, availability and purity of water. We stand united to follow and implement our knowledge and traditional laws and exercise our right of self-determination to preserve water, and to preserve life.

(Shuswap Nation Tribal Council, 2013: 1)

As stated by the First Nations Summit in a 2013 submission:

First Nations have a rightful governance role in setting principles that guide decision-making in any water management regime. They have important traditional knowledge that would help to establish relevant and necessary standards and thresholds for effective water management.

(First Nations Summit, 2013: 19)

Increasingly recognized as a valuable source of ecological information, traditional knowledge has played a central role in developing co-management systems in northern Canada (Binder and Hanbidge, 1993). However, privileging traditional knowledge cannot simply be an exercise in extracting and appropriating Indigenous peoples' knowledge, to then insert it into colonial frameworks for water governance. Instrumentalizing indigenous knowledge, in other words, should not be mistaken for meaningful consultation and effective engagement and respect for that knowledge. Rather, Indigenous communities have argued for acknowledging the importance of indigenous knowledge, and incorporating this knowledge in a respectful, meaningful manner. Justice-as-recognition, in other words, is just as important as justice-as-inclusion (full participation in decision-making processes).

Consultation and co-governance have been the subjects of ongoing debate in Canada. A summary of critical assessments of consultation (including but extending beyond the legal "duty to consult") is beyond the scope of this chapter. As in international debates, Indigenous peoples in Canada have raised the point that, while consultation is important and perhaps even essential to achieve procedurally just outcomes, it is potentially rife with contradictions. For example, the Canadian Crown's approach to consultation does not systematically incorporate the principles of free, prior and informed consent (FPIC). Although there is no internationally agreed definition of FPIC, the meaning is contained within the four elements that constitute "FPIC": it is the right of indigenous peoples to make free, informed choices about their lands and resources prior to any sort of development. According to the UN Commission on Human Rights (2005) (and sub-committee on the Promotion and Protection of Human Rights Working Group on Indigenous Peoples), the basic principles of FPIC are: to ensure that indigenous peoples are not coerced or intimidated; that their consent is sought and freely given prior to authorizing or starting any activities; that they have full information about the scope and impacts of any proposed developments, and; that ultimately their choices to give or withhold consent are respected.

The four elements of FPIC create a framework that establishes consent as the objective of consultations in order to empower communities to protect their rights and to address situations of power and capacity imbalance between communities and organizations. But the potential of government-led consultation processes to serve as a mechanism for conflict mitigation or resolution is rarely matched by the reality of bureaucratic procedures, particularly when the rules and processes for consultation are created by those in power – and notably when these processes do not incorporate recognition or respect for the full scope of Aboriginal rights. Beyond consultation, there are also increasing calls for co-governance. It is not yet clear what precisely this may look like, or what potential the concept holds, but it is an important task to continue to sketch out the potential for such arrangements in the future, for water governance, and beyond (Simms *et al.*, 2016).

10.4 Reflections

Freshwater allocation in British Columbia embodies historical injustices that are strongly contested by Indigenous communities. Lack of water security and exclusion from governance processes are twin mechanisms of injustice. A key element of the solution, we suggest, is that justice-as-recognition should underpin justice-as-inclusion (full participation in decision-making processes). Without recognition of Indigenous rights, current injustices in regulatory decision-making and water access are likely to persist.

While there were significant missed opportunities with the recent modernization of the Water Sustainability Act, including the maintenance of FITFIR and clear failures with respect to First Nations engagement and consultation, there are clear opportunities that can still be addressed. Among such opportunities, it is worth noting that the FITFIR allocation system does not inherently negate Indigenous water rights per se (excluding the deeper epistemological issue that defining water as an inanimate "resource" to be granted by licenses stands in contradiction to many of the beliefs that Indigenous peoples hold about water, in addition to the upholding of the Crown as the arbiter of allocation and "rights"). These points aside, it is the way in which priority of water access was determined in BC that is particularly controversial, where First Nations were not recognized as the original occupants of the land holding distinct First in Right entitlements to water. We can look to the example of the *Winters* doctrine in the United States, which recognized that surface and groundwater rights were implied by establishing American Indian reservations. In other words, the *Winters* doctrine affirmed that water rights were assigned when reserves were created and that reserve rights were based on historical occupancy, intention, and agreement – not on diversion and use. The doctrine also includes an understanding that reserve water rights were flexible and unquantified, and further, that water rights assigned to reserves were understood to encompass not only existing but also *future* tribal reserve water needs (Shurts, 2000). At least in principle, *Winters* more closely adheres to the notion of First in Time, First in Right in its truest sense: the first peoples of the land generally hold the earliest recorded water licenses, which has "real effects on water allocation decisions, including effects at least partially favorable to the long-term interests of western American Indians" (Shurts, 2000: 8).

Although FITFIR is now written into the WSA and the future of water allocation in BC, other amendments and interpretations are also possible that could begin to address some of the concerns described in this chapter. We suggest that First Nations' rights to water – of sufficient quality and quantity for existing and future needs – must be explicitly recognized and protected within the licensing system, whether or not a particular license holder's ranked priority is currently lower than that of other water users. For instance, the First Nations Summit states in its WSA submission:

Fundamentally, planning for and responding to times of water scarcity must engage First Nations from the outset, on a government-to-government basis. Some First Nations communities experience drought situations regularly. Others may begin to experience scarcity where their water sources are overburdened by user demands. These plans and response mechanisms need to be jointly identified and developed, and a mutual plan for implementation agreed on.

(First Nations Summit, 2013: 1)

Overall, it is critical that water allocation respond to the concerns that First Nations have raised, as such issues will likely only continue to be amplified as water shortages become a tangible reality in many parts of BC. As Brandes and Curran (2008: 4) state: "The vast majority of Aboriginal Rights and Title claims to water in BC have not been finalized and are not factored into the water licensing regime and ecological needs for instream flows. This could have a significant impact on existing allocations in the future."

We also recognize that making progress on meaningful consultation is vital. Meaningful, effective consultation is a resource- and time-intensive task, and it is also particularly difficult given the histories of inequities and lack of trust characterizing the relationship between many First Nations and settler colonial governments. Beginning with this recognition is the only possible way to begin to move forward to build trust, to work on the basis of acknowledging past injustices, and affirming the need for reconciliation and redress. This is a long, difficult road, and often there is no clear "solution." Nonetheless, recognition in the context of justice involves not only acknowledging communities and their unique rights, claims, or histories, but also recognizing that these are sometimes longer-term issues that do not offer a solution. In this way, recognition in the sense of justice (Schlosberg, 2007) is to work from a position of sincerity, respect, and investment – the only place to begin.

Among the many opportunities to move forward on the concerns discussed in this chapter, our final recommendation would be to invigorate and explore options for water co-governance between First Nations and colonial governments. This would involve agreeing upon processes for sharing authority and decision-making related to water. This must necessarily involve including Indigenous conceptualizations of governance and including local, traditional knowledge in decision-making. While co-governance has potential as a route forward toward more meaningful, inclusive watershed governance, there are a number of preconditions for it to truly constitute a meaningful shift away from the status-quo approach. We mention several of these conditions here (explored in more detail in Simms *et al.*, 2016). An initial key condition is to consider not only the resources that First Nations require to address existing capacity constraints, but also the capacities that colonial governments and institutions must build to work respectfully with First Nations in water governance. This could include, for instance, requiring colonial governments to build an understanding of different First Nation's laws and protocols for working together. It is clear that adequate, sustained funding and capacity-building will be essential for both First Nations and colonial governments to engage meaningfully in collaborative processes. It is also important to develop ways and protocols to work together in the interim, notwithstanding fundamental disagreements about jurisdiction and authority. The Kunst'aa Guu–Kunst'aayah Haida reconciliation protocol establishes a possible model: in the protocol, both the Haida Nation and the Province explicitly acknowledge their competing claims to jurisdiction for Haida Gwaii territory. With these competing claims made clear, the protocol (2009: 1) states:

Notwithstanding and without prejudice to the aforesaid divergence of viewpoints, the Parties seek a more productive relationship and hereby choose a more respectful approach to coexistence by way of land and natural resource management on Haida Gwaii through shared decision-making.

Collaborative water governance entails investing time into relationship-building to move beyond the "crisis of confidence" (Goetze, 2005) and establish trust and capacities for collaboration. This must happen at multiple levels, from personal to institutional. Ultimately, reconciliation of the relationship between the colonial governments and Indigenous peoples may not only be a pre-condition, but the *basis* for creating enduring and effective water governance.

10.5 Conclusion

It is clear that the history of water governance in the province of British Columbia is one of exclusion and injustice with respect to Indigenous rights and title. At present, there is a significant legal, ethical and ecological imperative to move in new directions, and to provide more appropriate and meaningful engagement and decision-making authority for Indigenous peoples. Many of the recent policy shifts, including the Water Sustainability Act that was adopted into law in 2016, are calling for a fundamental shift towards co-governance. However, the nature of such shifts, and their tenability, remain uncertain.

That said, there are clear pathways to move forward in addressing, and redressing, some of the key justice challenges laid out in this chapter. First, it is worth noting that while progress has been slow, things are changing very quickly. For example, recent Supreme Court of Canada decisions (such as *Tsilhqot'in Nation* v. *British Columbia*, 2014 SCC 44) may produce fundamental shifts for Indigenous peoples in the Province (and across Canada)

(Mandell Pinder, 2014). Second, there are examples of collaborative water management that have proven to be effective and meaningful. In the Northwest Territories (NWT), the Canadian and territorial governments began working directly with First Nations governments to co-develop a water stewardship strategy. Released in 2010, this strategy privileges Indigenous knowledge and explicitly recognizes Aboriginal rights. The NWT approach acknowledges that future uncertainties and challenges related to water would benefit from sincere learning that acknowledges the tremendous wisdom of Indigenous knowledge. An independent evaluation of the NWT water strategy found the new approach has helped to promote greater collaboration among water partners (Harry Cummings and Associates and Shared Value Solutions, 2015: ii). Other jurisdictions in Canada would do well to learn from, and then adapt and adopt elements of this approach.

Finally, however difficult it may be, we have stressed the need for a politics of recognition that begins by frankly acknowledging past and ongoing injustices in multiple forms. It is only with such acknowledgement that trust and relationship-building can proceed. As such, recognition of injustice, and a shared quest for justice, is as important to our shared water future as any scientific understanding of future water flows or climate change. As this volume stresses, water justice is about understanding and building relations of mutuality, shared norms, and bringing ethics to the fore in our decisions of how to use and manage water. An Indigenous concept of water justice moves beyond technocratic definitions of governance (and, indeed, co-governance) to embody a sense of the sacred trust at the heart of "sustainability." Water justice from this perspective is at once inclusive, non-anthropocentric, and relational – an embodiment of reciprocal relationships among land, water, and animals – redefining human relationships with water and one another.

Acknowledgements

The authors gratefully acknowledge funding support from the Social Sciences and Humanities Research Council of Canada, and the Water, Economics, Policy and Governance Network. We are also deeply grateful to the Chief and Council and community members of the Lower Similkameen Indian Band for collaborating with us as part of this project, and also to Gwen Bridge, Tracy Lawlor, Tessa Terbasket, Tom Perreault, and Sara Amadi for insight and support. The broader project was also supported by several partners: the Center for Indigenous Environmental Resources, Environment Canada, the Federation of Canadian Municipalities and the National Network of Environment and Women's Health. UBC Behavioral Research Ethics approval was granted under certificate number #H12-03691. Any omissions or errors are the sole responsibility of the authors.

Notes

1 In the Canadian context, the term "Indigenous" refers to First Nations, Inuit and Métis peoples, and also serves as an inclusive reference to communities that claim a historical continuity with their original territories (Corntassel, 2003). Inuit refers to Indigenous peoples primarily residing in

Canada's far north. Métis refers to communities of mixed European and Indigenous heritage. The term First Nations has supplanted the previously widely used terms "Indian" and "Native" in policy and public discourse, and is the operational term used to refer to governments of "Indian bands" (the basic unit of governance for reserves under Canada's Indian Act). The term "Aboriginal" is a formal legal term used in the justice system and also by the Canadian government. It is a term criticized by many Indigenous scholars, who tend to prefer the term "Indigenous," as distinct from the formal legal term "Aboriginal" (for example, Indigenous law refers to laws emerging from communities, distinct from Aboriginal law, imposed by the settler colonial state).

2 Aboriginal rights to water have never been explicitly established or disproven through a court ruling in Canada. This situation contrasts to the United States, where the *Winters* doctrine and subsequent *Cappaert* decision (*Winters* v. *United States*, 1908; *Cappaert* v. *United States*, 1976) state that surface and ground water rights are implied by the federal establishment of American Indian reservations, and (further) set standards to which the US Government must adhere in ensuring sufficient water for reservations, thereby recognizing the centrality of water-land interactions to American Indian communities (Shurts, 2000).

3 Canada's Constitution allocates responsibility for issues between the federal Government on the one hand, and provincial governments on the other. Provincial governments have responsibility for most aspects of freshwater management (separate arrangements apply to Canada's three northern territories). Provincial legislative powers include, but are not restricted to: water supply, pollution control, hydroelectric development, flow regulation, and authorization of water use. The federal Government has jurisdiction over fisheries, navigation, federal lands and Indian reserves (including drinking water supply on reserves), and international relations (largely responsibilities related to management of boundary waters shared with the US). Section 91(24) of Canada's Constitution Act (1867) specifically assigns Indigenous issues to federal (rather than provincial) governments. Canada's Indian Act (first passed in 1876 and still in place, with amendments) sets out provisions governing a broad range of issues (including registration for Indian "status," governance, land use, health care, and education) relating to Indian reserves and Indian "bands" (governments and communities). Imposed unilaterally (in contrast to treaties), the Act has been subject to numerous critiques.

4 British Columbia is home to the second-largest Indigenous population in Canada: approximately 200,000 people (5 percent of the Province's population at present).

5 Canadian Confederation refers to the process by which formerly separate British colonies were federally united – initially into the Dominion of Canada in 1867, which was subsequently enlarged. British Columbia joined the Confederation in 1871.

6 Environmental flow needs (EFNs) refer to the quantity and timing of flows in a stream that are required to sustain freshwater ecosystems, including fish and other aquatic life (i.e. to maintain stream health).

7 In 2008 (*Hupacasath*), the Supreme Court distinguished "deep consultation" from lower levels of consultation. The extent of consultation depends on the First Nations' strength of claim and the extent of the potential infringement on Aboriginal rights.

8 From the *Haida* and *Taku River* decisions in 2004, and the *Mikisew Cree* decision in 2005, the Supreme Court of Canada determined that the Crown has a duty to consult and, where appropriate, accommodate First Nations when the Crown is contemplating an action or activity that might adversely impact potential or established Aboriginal or Treaty rights. The degree of consultation with First Nations depends on the strength of the case supporting the existence of Aboriginal rights and title and seriousness of the adverse effect, or potential effects, on the rights and/or title claimed (*Hupacasath First Nation* v. *British Columbia (Minister of Forests) et al.*, 2005 BCSC 1712, para. 138).

9 The fiduciary obligations of the Canadian Crown [state] to Indigenous communities include the duty to consult about resource development on lands subject to Aboriginal rights. The Crown's fiduciary obligations have their origins in negotiations that took place at the time of British colonization of Canada, as enshrined in the Royal Proclamation of 1763. The debate over these fiduciary duties is complicated by divergent views over the nature of the relationship between the Indigenous peoples of Canada and the Canadian state. At the risk of over-simplification, Canadian law frames

this as a nation-to-nation relationship established between sovereign peoples and the Canadian Crown (Canada is a constitutional monarchy; Her Majesty Queen Elizabeth II is the Queen of Canada and Canada's Head of State). Related but distinct, section 35 of Canada's Constitution Act recognizes and affirms Aboriginal rights. Some aspects of this relationship are structured in agreements (e.g. treaties) between Indigenous peoples and the Crown, which are administered via Canadian Aboriginal law, and overseen by the federal Minister of Indigenous and Northern Affairs. Several Supreme Court of Canada decisions (including *Guerin, Sparrow, Wewaykum* and *Ermineskin*) have directly addressed the nature and scope of the Crown's fiduciary duties, which include not only Crown management of reserve lands and resources, but also Crown decision-making and legislative authority over resources and lands subject to Aboriginal rights. The extent of the Crown's fiduciary duties is subject to debate; for example, there is ongoing debate over whether and to what extent Crown public interest obligations may counter, alleviate, or eliminate the Crown's fiduciary obligations to Aboriginal communities. Nonetheless, Canadian law clearly affirms the principle of fiduciary obligations as a constitutional duty, the fulfilment of which is consistent with the Crown's honor (Asch 1984; Borrows 2003; Coulthard, 2014a, 2014b; Daigle 2016; Luk 2013; Macklem and Sanderson 2016; Turpel 1991).

References

Armstrong, J. and Sam, M. (2013). "Indigenous water governance and resistance: A Syilx perspective." In J. Wagner (ed.), *The Social Life of Water*. New York: Berghahn Books, pp. 239–54.

Asch, M. (1984). *Home and Native Land: Aboriginal Rights and the Canadian Constitution*. Agincourt, ON: Methuen.

Bartlett, R. (1998). *Aboriginal Water Rights in Canada: A Study of Aboriginal Title to Water and Indian Water Rights*. Calgary, AB: Canadian Institute of Resources Law.

BCAFN. (2010). "Section 3.31: Water." In J. Wilson-Raybould and T. Raybould (ed.), *British Columbia Assembly of First Nations Govenance Tool Kit: A Guide to Nation Building*. West Vancouver, BC: BC Assembly of First Nations, pp. 443–62.

BCAFN. (2013). Water Sustainability Act Legislative Proposal. Submission Letter to the BC Ministry of Environment, Water Stewardship Division. British Columbia Assembly of First Nations, December 2, 2013, https://engage.gov.bc.ca/watersustainabilityact/files/2013/12/BC-Assembly-of First-Nations.pdf.

Binder, L. N. and Hanbidge, B. (1993). Aboriginal people and resource co-management: The Inuvialuit of the Western Arctic and resource co-management under a land claims settlement. In J. T. Inglis (ed.), *Traditional Ecological Knowledge: Concepts and Cases*. Ottawa, ON: Canadian Museum of Nature, pp. 121–32.

Borrows, J. (2003). Measuring a work in progress: Canada, constitutionalism, citizenship and aboriginal peoples. In A. Walkem and H. Bruce (eds.), *Box of Treasures or Empty Box? Twenty Years of Section 35*, Vancouver, BC: Theytus Books, pp. 223–62.

Brandes, O. M. and Curran, D. (2008). *Water Licences and Conservation: Future Directions for Land Trusts in British Columbia*. Report prepared for The Land Trust Alliance of BC.

Brandes, O. M., Nowlan, L. and Paris, K. (2008). *Going with the Flow? Evolving Water Allocations and the Potential and Limits of Water Markets in Canada*. Canadian Water Network and Conference Board of Canada, www.blue-economy.ca/sites/default/files/reports/resource/09_going_w_flow_1.pdf.

Cappaert v. *United States*, 1976. 426 U.S. 128.

Carstens, P. (1991). *The Queen's People: A Study of Hegemony, Coercion, and Accommodation among the Okanagan of Canada*. Toronto, ON: University of Toronto Press.

Corntassel, J. (2003). Who is Indigenous? "Peoplehood" and ethnonationalist approaches to rearticulating Indigenous identity. *Nationalism & Ethnic Politics*, 9(1), 75–100.

Coulthard, G. (2014a). From wards of the state to subjects of recognition? Marx, indigenous peoples, and the politics of dispossession in Denendeh. In A. Simpson and A. Smith (eds.), *Theorizing Native Studies*. Durham, NC: Duke University Press, pp. 56–98.

Coulthard, G. (2014b). *Red Skin, White Masks*. Minneapolis, MN: University of Minnesota Press.

Daigle, M. (2016). Awawanenitakik: The spatial politics of recognition and relational geographies of Indigenous self-determination. *The Canadian Geographer/Le Géographe canadien*, 60(2), 259–69.

First Nations Summit. (2013). A water sustainability act for BC: Legislative proposal. Submission letter to the BC Ministry of Environment, Water Stewardship Division. December 2013. Unpublished, https://engage.gov.bc.ca/watersustainabilityact/files/2013/12/First-Nations-Summit.pdf.

First Nations Women Advocating Responsible Mining. (2010). Water Act Modernization Initiative. Submission Letter to the BC Ministry of Environment, Water Stewardship Division. April 30, 2010. Unpublished, https://engage.gov.bc.ca/watersustainabilityact/files/2013/10/First-Nations-Women-Advocating-Responsible-Mining.pdf.

Goetze, T. (2005). Empowered co-management: Towards power-sharing and Indigenous rights in Clayoquot Sound, BC. *Anthropologica*, 47(2), 247–65.

Halfway River First Nation v. British Columbia (Minister of Forests) [1999] 4 CNLR (BCCA 470), para.160.

Harris, C. (1997). *The Resettlement of British Columbia: Essays on Colonialism and Geographical Change*. Vancouver, BC: UBC Press.

Harris, C. (2001). *Making Native Space*. Vancouver, BC: UBC Press.

Harris, C. (2004). How did colonialism dispossess? Comments from an edge of empire. *Annals of the Association of American Geographers*, 94(1), 165–82.

Harry Cummings and Associates and Shared Value Solutions Ltd. (2015). *Independent Evaluation of the NWT Water Stewardship Strategy Implementation: Evaluation Report*, www.nwtwaterstewardship.ca/sites/default/files/FINAL%20NWT%20Water%20Strategy%20Implementation%20Evaluation%20Report.%20Sept%20%2022%202015.pdf.

Hupacasath First Nation v. British Columbia (Minister of Forests) et al. [2005] B2005 BCSC 1712 (CanLII), www.canlii.org/en/bc/bcsc/doc/2005/2005bcsc1712/2005bcsc1712.html.

Joe, N., Bakker, K. and Harris, L. (2017). *Perspectives on the BC Water Sustainability Act: First Nations Respond to Water Governance Reform in British Columbia*. Vancouver, BC: Program on Water Governance, www.watergovernance.ca.

Jollymore, A., McFarlane, K. and Harris, L. (2016). *Whose Input Counts? Quantitative Analysis of Public Consultation and Policy Outcomes*. Vancouver, BC: Program on Water Governance, https://watergovernance.ca/wp-content/uploads/2016/02/Policy-brief_Whose-input-counts.pdf.

Kelm, M.-E. (1997). *Colonizing Bodies: Aboriginal Health and Healing in British Columbia 1900–1950*. Vancouver, BC: UBC Press.

Kunst'aa Guu-Kunst'aayah Reconciliation Protocol. (2009). www.haidanation.ca/wp-content/uploads/2017/03/Kunstaa-guu_Kunstaayah_Agreement.pdf.

Laidlaw, D. and Passelac-Ross, M. (2010). Water rights and water stewardship: What about Aboriginal peoples? *Law Now*, 35(1), 1–12.

Luk, S. (2013). Not so many hats: The crown's fiduciary obligations to Aboriginal communities since Guerin. *Saskatchewan Law Review*, 76, 1–49.

Macklem, P. and Sanderson, D. (2016). *From Recognition to Reconciliation: Essays on the Constitutional Entrenchment of Aboriginal and Treaty Rights*. Toronto, ON: University of Toronto Press.

Mandell Pinder. (2014). *Tsilhqot'in Nation v. British Columbia* 2014 SCC 44 Case Summary, www.mandellpinder.com/tsilhqotin-nation-v-british-columbia-2014-scc-44-case-summary/.

Mascarenhas, M. (2007). Where the waters divide: First Nations, tainted water and environmental justice in Canada. *Local Environment*, 12(6), 565–77.

Matsui, K. (2005). "White man has no right to take any of it": Secwepemc water-rights struggles in British Columbia. *Wicazo Sa Review*, 20(2), 75–101.

Matsui, K. (2009). *Native Peoples and Water Rights: Irrigation, Dams, and the Law in Western Canada. Vol. 55*. Montreal, QB and Kingston, ON: McGill-Queen's University Press.

McLaren, M. (2016). "Water systems at risk." *Globe and Mail*. Tuesday, October 30, www.theglobeandmail.com/news/national/indigenous-water/article31589755/.

Miller, J. (2000). *Skyscrapers Hide the Heavens: A History of Indian-White Relations in Canada*. Toronto, ON: University of Toronto Press.

Ministry of Environment (MOE). (1997a). *First Nations Water Rights in British Columbia: A Historical Summary of the Rights of the Lower Similkameen First Nation*. Victoria, BC: Water Management Branch.

MOE. (1997b). *First Nations Water Rights in British Columbia: A Historical Summary of the Rights of the Okanagan Nation*. Victoria, BC: Water Management Branch.

MOE. (2013). *A Water Sustainability Act for BC: Legislative Proposal*. Victoria, BC: Water Management Branch.

Murdocca, C. (2010). "There is something in that water": Race, nationalism, and legal violence. *Law & Social Inquiry*, 35(2), 369–402.

Neegan Burnside Ltd. (2011). *National Assessment of First Nations Water and Wastewater Systems – Ontario Regional Roll-Up Report*, www.aadnc-aandc.gc.ca/DAM/DAM-INTER-HQ/STAGING/texte-text/enr_wtr_nawws_ruront_ruront_1314635179042_eng.pdf.

Phare, M. A. (2009). *Denying the Source: The Crisis of First Nations Water Rights*. Victoria, BC: Rocky Mountain Books.

Ratcliff & Company, LLP on behalf of Coldwater Indian Band. (2010). Water act modernization. Submission letter to the BC Ministry of Environment, Water Stewardship Division. May 4, 2010. Unpublished, https://engage.gov.bc.ca/watersustainabilityact/files/2013/10/Coldwater-Indian-Band2.pdf.

Richard, G. (1999). When the ditch runs dry: Okanagan natives, water rights, and the tragedy of no commons. *British Columbia Historical News*, 32(2), 10.

Sam, M. (2013a). Indigenous water rights: From the local to the global reality. In E. Simmons (ed.), *Indigenous Earth: Praxis and Transformation*. Penticton, BC: Theytus Books.

Sam, M. (2013b). "Oral narratives, customary laws and Indigenous water rights in Canada," PhD diss., University of British Columbia Okanagan.

Schlosberg, D. (2007). *Defining Environmental Justice: Theories, Movements, and Nature*. Oxford: Oxford University Press.

Shurts, J. (2000). *Indian Reserved Water Rights: The Winters Doctrine in Its Social and Legal Context, 1880s–1930s*. Norman, OK: University of Oklahoma Press.

Shuswap Nation Tribal Council. (2013). Legislative proposal for a new water sustainability act. Submission letter to the BC Ministry of Environment, Water Protection and Sustainability Branch. November 15, 2013. Unpublished, https://engage.gov.bc.ca/watersustainabilityact/files/2013/11/Shuswap-Nation-Tribal Council.pdf.

Simms, B.R. (2014). "'All of the water that is in our reserves and that is in our territory is ours': Colonial and Indigenous water governance in unceded Indigenous territories in British Columbia," Master's diss., University of British Columbia, http://circle.ubc.ca/handle/2429/51475.

Simms, R., Harris, L., Joe, N. and Bakker, K. (2016). Navigating the tensions in collaborative watershed governance: Water governance and indigenous communities in British Columbia, Canada. *Geoforum*, 73, 6–16.

Thom, B. (2009). The paradox of boundaries in Coast Salish territories. *Cultural Geographies*, 16(2), 179–205.

Tsilhqot'in Nation v. British Columbia, 2014 SCC 44, [2014] 2 S.C.R. 256, https://scc-csc.lexum.com/scc-csc/scc-csc/en/item/14246/index.do.

Turpel, M. E. (1991). Aboriginal peoples and the Canadian Charter of Rights and Freedoms: Contradictions and challenges. *Canadian Women's Studies*, 10(2) and (3), 149–57.

UN Commission on Human Rights. (2005). Standard-setting legal commentary on the concept of free, prior and informed consent. July 14, www.ohchr.org/Documents/Issues/IPeoples/FreePriorandInformedConsent.pdf.

Union of British Columbia Indian Chiefs. (2011). Water act modernization. Submission letter to the BC Ministry of Environment, Water Stewardship Division. March 9. Unpublished, https://engage.gov.bc.ca/watersustainabilityact/files/2013/10/Union-of-BC-Indian Chiefs1.pdf.

Von der Porten, S. (2012). Canadian indigenous governance literature: A review. *Nga Pae o te Maramatanga*, 8(1), 1.

Von der Porten, S. and de Loë, R. C. (2013a). Collaborative approaches to governance for water and Indigenous peoples: A case study from British Columbia. *Geoforum*, 50, 149–60.

Von der Porten, S. and de Loë, R. C. (2013b). Water governance and Indigenous governance: Towards a synthesis. *Indigenous Policy Journal*, 23(4), 1–12.

Walkem, A. (2004). *Indigenous Peoples Water Rights: Challenges and Opportunities in an Era of Increased North American Integration*. Victoria, BC: University of Victoria Centre for Global Studies.

Winters v. United States, 1908. 207 U.S. 564.

11

Sanitation Justice? The Multiple Dimensions of Urban Sanitation Inequalities

Maria Rusca, Cecilia Alda-Vidal, and Michelle Kooy

11.1 Introduction

The launch of the Water and Sanitation Decade (1980–90) marked the first attempt to place urban sanitation within national governments and international organizations' development agendas. Since then, inclusion of sanitation within the Millennium Development Goals (MDGs), the Sustainable Development Goals (SDGs), and global campaigns such as the UN Sanitation Year (2008), the End of Open Defecation Campaign, and World Toilet Day have institutionalized sanitation as one of the core development goals until 2030 and beyond. However, the results of many of these sanitation development initiatives have been disappointing. Regional statistics show alarming results for Sub-Saharan Africa, where urban population growth has outpaced gains in sanitation coverage since the 1990s; 14 out of 46 countries declined in sanitation coverage (UNICEF/WHO, 2015: 17). The MDGs were unable to eliminate inequalities in access to sanitation between rich and poor urban dwellers in most countries (UNICEF/WHO 2015).

Depressing as this is, the MDGs have focused only on distributive outcomes (access to infrastructure), overlooking other dimensions of sanitation inequality. Failing to address these dimensions hampers development interventions which aim to reduce these inequalities: sub-surface flows of untreated wastewaters contaminate urban poor settlements' shallow groundwater sources, displacing health risks onto the poorest, and reducing developmental opportunities for children and adults who themselves may already be using "improved" sanitation services (Graham and Polizzotto, 2013); building onsite sanitation infrastructure and improved access to latrines without provision for emptying or sludge removal services compromises long-term health benefits from "improved" sanitation (Jenkins *et al.*, 2015; Tsinda *et al.*, 2013). Finally, dimensions of access must include consideration of safety and comfort plus any particular needs of urban poor residents who are marginalized by age, disabilities, gender, or other social relations, so they can use the infrastructure provided (Hulland *et al.*, 2015; Wilbur and Jones, 2014). Ignoring these dimensions of environmental and social inequalities can reverse any positive gains in terms of increased distribution of infrastructure.

In this chapter, we develop the concept of sanitation justice to capture these dimensions of inequalities and their relations, which are often overlooked in debates on development targets for sanitation. In Section 2, we briefly review analyses of inequalities in relation to sanitation already present in the ecological justice literature, specifically urban political ecology and environmental justice. We draw from this literature to define three dimensions of sanitation justice: distributive, procedural, and recognitional justice. In Section 3 of the chapter, we apply the *distributive-justice* dimension to go beyond the usual analysis of inequalities in infrastructure coverage. Examining the dominant infrastructure model used to increase access for low income urban residents, we identify additional dimensions of distributional inequalities related to the sanitation services required for onsite sanitation infrastructure systems. In Section 4, we apply the *procedural-justice* dimension to examine institutional fairness and decision-making inclusiveness: by whom sanitation development targets are set, and with whom sanitation outcomes are to be achieved. Tracing the decisions made through a global sanitation policy reform, we look at who is included in defining what is "adequate" or "improved" sanitation. In Section 5, we apply the *recognitional-justice* dimension to identify the social and relational factors that influence individuals' disproportionate exposure to environmental harm. Seeing sanitation as dignity requires recognition of social categories of gender, age, disability and sexuality as they shape particular sanitary needs, and as they reflect the stigmatization undervaluing some people and places.

Across these three dimensions of justice, we apply the concept to examine how sanitation inequalities are addressed – or not – through current development approaches and policy shaping sanitation interventions in Sub-Saharan African cities. We look specifically at sanitation interventions in low-income settlements, Kawempe District, Kampala, Uganda. This material is drawn from semi-structured interviews held with local government, residents, donors, and international and local NGOs in 2011. Interview material is complemented by documentary analysis of project reports and policy documents from the World Bank and other development agencies, including NGOs and CBOs, involved in in the global and local water and sanitation sector.

11.2 Ecological Justice Frameworks and Sanitation

Urban political ecology (UPE) is a loosely defined body of scholarship concerned with social processes and relations shaping production of environmental problems, which are unevenly experienced. As such, UPE is concerned with distribution of both environmental resources and environmental risks, foregrounding the analysis of social relations shaping environmental inequalities (Cook and Swyngendouw, 2012; Heynen *et al.*, 2006; Swyngedouw, 2004). In UPE, nature and society are recognized as co-produced, meaning environmental problems cannot be understood independently of social processes, and vice versa. This is useful for thinking through urban wastewaters, highlighting the role of social relations and processes producing the contaminated flows. UPE also provides us with a more complex understanding of distributional justice – production of environmental risks may not be spatially contiguous with its environmental impacts, as these are unevenly distributed within urban/peri-urban/rural zones.

The use of environmental justice (EJ) as a framework to understand processes of uneven urbanization in Sub-Saharan African cities in less common than in UPE or water justice literature, which analyze inequalities and injustice in relation to water distribution much more. Most EJ scholarship involving sanitation has focused on US-based cases of social exclusion from wastewater treatment and solid waste management, these facilities' siting and the racialized resulting uneven cost and benefit distributions (Bullard, 1983; Bullard *et al.*, 2008; Perreault *et al.*, 2012). However, the environmental justice literature does more than UPE to explore the patterns that produce socio-spatial environmental inequality (Cook and Swyngendouw, 2012). It broadens analysis of (in)justice to address the role of gender, race, age and sexuality, alongside class in determining unequal distribution of costs and benefits from the environment and its degradation (Buckingham and Kulcur, 2009; Schlosberg, 2013; Schweitzer and Stephenson, 2007; Urkidi and Walter, 2011). Further, EJ scholars have convincingly argued against reductionist definitions of justice exclusively engaging with distribution concerns. Their multidimensional conception of environmental justice entails, first, the *processes* underlying inequitable distribution (Schlosberg, 2007; Urkidi and Walter, 2011). This justice dimension referred to as procedural justice, places meaningful participation, encompassing fairness, inclusion in decision-making and access to information at the core of environmental justice (Holifield, 2012; Holifield *et al.*, 2009; Walker, 2009). Additionally, EJ has made an attempt to include emotional aspects into its concept of justice, by arguing the need to recognize individual rights and collective identities, and their particular needs (Urkidi and Walter, 2011; Walker 2009). In this perspective, rather than focusing on eliminating inequalities, the quest for justice should take the "avoidance of 'humiliation' or 'disrespect'" as a starting point (Honneth, 2004: 351). Last, exploring the relationship between place and stigma, in which misrecognition of people and locality are entwined (Walker 2009), offers important insights into the sanitary and unsanitary city.

Collectively, EJ and UPE frameworks call us to pay attention to wastewater production and distribution and to look beyond distribution inequalities. However, current analytical frameworks are largely based on US contexts (EJ), or focus inequality questions on urban infrastructure (UPE). Most research on dimensions of sanitation justice concerns the presence/absence of infrastructure. In particular, UPE scholars show how sanitation infrastructure reinforces the connections and fragmentations of social groups and the spaces where they live (Iossifova 2015; Morales 2015; Morales *et al.*, 2014). While EJ scholarship has broadened the range of social justice concerns to include climate and food justice, transportation, access to countryside and green space, land use, water quality and availability (Schlosberg, 2013), sanitation justice has not been explored in a context where water systems without sewerage are the norm and most low-income dwellers have access to latrines, or must use other informal disposal methods, such as "flying toilets" (WSP, 2013). Further, unlike sewerage systems, onsite infrastructure (pit latrines and septic tanks) require additional services to ensure safe disposal of fecal sludge. How then to understand sanitation inequalities in the context of rapidly urbanizing Sub-Saharan African cities, characterized by heterogeneous service modalities and extreme socio-economic inequalities? To address this gap within ecological justice scholarship, we apply both UPE and

EJ analytical tools to develop a concept of sanitation justice. We explore the dimensions of justice specifically regarding access to sanitation in Sub-Saharan African cities.

11.3 Distributive Justice: Beyond Unequal Access to Sanitation Infrastructure and Services

11.3.1 Shifting Paradigms from Centralized Public Service to Decentralized Private Services

Up until the 1990s, development interventions in urban sanitation in the global South focused on building and operating large-scale, supply-driven, centralized networked systems (Allen *et al.*, 2006). This approach was reflected in the Drinking Water and Sanitation Decade (1981–90) which delivered hardware interventions through planned expansion of centralized, capital-intensive systems. However, at the end of this Decade, the UN Economic and Social Council (ECOSOC) report suggested that large-scale systems had benefitted only mid- and high-income areas. Later, in 1997, the World Bank's strategy for sanitation identified two other drawbacks of the centralized supply-driven approach based on financial rationality: low cost-efficiency (high investment costs versus low number of people served) and low cost-recovery rates preventing sewerage utilities' service expansion (Wright, 1997). The ECOSOC report had already laid the premises for a new strategy based on decentralized low-cost technologies for sanitation services to lower-income areas (ECOSOC, 1990). The Delhi Declaration, adopted by 115 countries in 1990, formalized the shift in sanitation development approaches, identifying low-cost decentralized sanitation technologies as a more realistic solution for low-income areas.

The sanitation strategy promoted by the World Bank and other segments of the development establishment post-1990 centered on user participation in planning and implementing services, promoting multiple providers, and developing a small-scale private sector. This strategy entailed a shift from macro to micro projects (centralized to decentralized sanitation systems) and required NGOs to play a significant role. The rationality of "unbundling" sanitation projects into smaller-size components was premised on the potential to bring in more private-sector contractors and, thus, more competition and lower prices. In the World Bank's words, matching service level to willingness to pay would achieve "optimum coverage with economic efficiency" (Wright, 1997: 7). However, achieving willingness to pay on the part of urban poor residents who needed access to sanitation also required a user focus: "[a] demand-based approach requires implementing agencies to find out what potential users want and what resources they have, to finance and manage installed systems, and to design systems financing mechanisms, and support structures that are best suited to their needs" (ibid.).

11.3.2 Inequalities in Distributive Justice

As various scholars have noted, for most cities in the global South, unequal access to sanitation is not a new phenomenon, having existed since colonial periods (Kooy and Bakker, 2008; Letema *et al.*, 2014; McFarlane, 2008; Nilsson, 2006). Decentralized onsite sanitation was positioned as an achievable remedy for this inequality, ensuring "adequate" services to all. The MDGs developed a set of criteria for what constituted adequate onsite sanitation and these were pursued through development initiatives.

The sanitation landscape that characterizes most cities in Sub-Saharan Africa, however, is not only disappointing in terms of centralized/decentralized infrastructure distribution but also in distribution of services associated to those infrastructures. Unlike centralized water-borne systems, in which evacuation, transport and treatment of sludge are under the service provider's responsibility, these responsibilities are passed on to the end user of onsite sanitation systems. The user thus becomes responsible for construction and maintenance, including arranging to empty onsite systems. This is problematic because it means that only those end-users who can afford to assume these responsibilities have access to these systems. It also encourages low-income urban residents to opt for cheaper, less environmentally safe, infrastructural solutions (soak pits versus septic tanks), and maintenance (septic tank sludge removal, emptying unlined pit latrines). As reviewed by Jenkins *et al.* (2015) in Dar es Salaam, most residents are unable to access safe pit-emptying services because of either unaffordable prices or unavailable service. Consequently, pit latrines are very rarely emptied and, when emptied, health safety equipment or procedures are improperly followed (Tsinda *et al.*, 2013).

Inequalities in sanitation infrastructure and the services required mean highly uneven exposure to environmental risks. Poor operation and maintenance of onsite sanitation facilities increases exposure to fecal contamination of shallow sub-surface and/or piped water supply and, thus, increases risks of contracting diarrheal diseases (Ashbolt, 2004; Bain *et al.*, 2014; Sarpong *et al.*, 2016). A study of sanitation in Kigali identified the most common sanitation service (traditional pit latrines) as one of the highest public-health risks (Tsinda *et al.*, 2013). These negative health impacts are exacerbated by the fact that inadequate access to sanitation is often coupled with poor access to safe drinking water.

11.3.3 Sanitation Justice in Kampala: Uneven Coverage and Services

The global shift to decentralized onsite sanitation had a direct impact on the Ugandan sanitation sector. Although sanitation infrastructure in Ugandan cities has always been fragmented, the government's declared intention was to provide publicly managed centralized services in the capital city (Letema *et al.*, 2014). When established in 1972, Uganda's national water utility was named the National Water and Sewerage Corporation. The name reflected the ambition that, one day, this parastatal organization would provide wastewater services to the entire country. However, in 1997, responsibility for provision of sanitation was transferred from the state to its citizens. Sanitation became a household responsibility, while the government was responsible for facilitating the private sector. This shift toward "unbundling" sanitation infrastructure in Kampala was institutionalized with the 1997 Strategic Framework for Reform (SFR) for the capital. The strategic framework required downsizing Kampala City Council staff and liberalizing basic service delivery (World Bank, 2007). Kampala's subsequent Declaration on Sanitation (1997, Preamble) that same year envisioned a future where "service delivery will be enhanced through the increased participation of the private and social intermediary sectors (NGOs)."

The district of Kawempe, one of the fastest growing informal settlements in Kampala, provides a clear example of the sanitation reality resulting from the SFR and its Declaration on Sanitation. Surveys in the district reveal very low levels of sanitation coverage: between 64 percent and 75 percent of the approximately 260,000 inhabitants have no access to adequate sanitation (Isunju *et al.*, 2013; Katukizaa *et al.*, 2010; Kiyimba, 2006; SSWARS, 2013; UBOS, 2005), and most residents rely on shared facilities (CIDI and WaterAid, 2008). These low levels of coverage persist despite the ongoing programs of many NGOs working in this district to increase sanitation infrastructure coverage and improve environmental conditions. The approach used in these programs by Plan International, WaterAid, CIDI, SSWARS and others, entails matching infrastructures with willingness to pay by focusing on more affordable or low-cost onsite solutions, such as VIP latrines or EcoSan toilets. The core of this approach is creating a supply chain to fill the (presumed) demand once it is created. This involves training community members to construct "different available sanitation options that are suitable to their pockets" (CIDI and WaterAid, 2009: 5). The distribution of sanitary hardware is, thus, delegated to a community-based private sector: "while doing sanitation marketing we came across people who are willing to construct a toilet, manage it perfectly, make sure it is clean, for those people who have no access."[1]

This strategy, however, raises concerns of affordability for the end-user, who may be unable to sustain costs of construction and maintenance and may opt for very low-cost solutions such as unimproved pit latrines. Further, coverage does not necessarily entail access. The high number of shared facilities users, and the elevated cost and limited availability of pit emptying services poses a significant challenge to safely operate this "adequate" sanitation service. For the customer, the final price of the pit-emptying service depends on the distance and the volume that needs to be discharged. As a result, inhabitants of peripheral areas are being charged a higher price for the service. The public pit-emptier provides the service at lower cost, but "many people are competing for them. Since they are five and the population is quite big, you find that you put in your request and it takes a month."[2] Besides the limited availability, municipal service providers are also considered to be less responsive: "each time you go there they will tell you 'there is no fuel, we have not budgeted for it … [or] the fuel is there but the truck is down' – you find non-responsiveness."[3] As a result, many of the sanitation facilities fill up quickly or are too unsanitary to use, and residents resort to open defecation or the use of plastic bags – the so-called "flying toilets" – that are dumped in open drainage … These last-resort practices are reported as common in all Kawempe District parishes (CIDI and Water Aid, 2010; Katukiza *et al.*, 2010; SSWARS, 2013).

We thus conclude that onsite sanitation approaches that have limited infrastructure, with services provided to customers on a demand-based, commercially-oriented model, are in many contexts distributionally unjust. First, these approaches limit distribution of onsite infrastructure to those who can afford it. Secondly, because they do not adequately account for the distribution of affordable access to necessary emptying and disposal services, the negative environmental effects are concentrated in the poorest settlements, creating uneven distribution of broader development outcomes in relation to environmental health impacts, and subsequently poverty reduction, education, and access to employment.

11.4 Procedural Justice: The Use of Participatory Approaches

11.4.1 The Use of Participation

As the above section has reviewed, the 1990s shift in the sanitation sector has involved transferring responsibility for sanitation service decisions and financing from the state to the individual household. This shift has aligned comfortably with that of grassroots participatory development movements and participatory demand-driven approaches. As has been observed for the water sector, individualizing and commercializing sanitation interventions has worked alongside – rather than against – the use of a wide range of options positioned as participatory development approaches (Bakker, 2007; Harris *et al.*, 2013). That these two distinctly different ideological approaches have become conflated in contemporary development interventions is hardly surprising, according to Mosse (2006: 696), who points out that the "twenty-first-century neoliberal reverse 'rolling-back' of the state" relies on "the 'revival' of community ... management" designed primarily to enhance economic efficiency.

Policy literature thus presents participatory approaches and onsite sanitation infrastructure as an effective combination to ensure pro-poor and demand-driven (i.e. individualized) services (WSP, 2013). Similarly, participatory approaches in sanitation service design and delivery are said to ensure informed infrastructure selection and develop improved, sustainable, demand-driven urban sanitation services (Lüthi *et al.*, 2011). Alongside this use of participation for individualization and commercialization, the move towards participatory approaches is also framed as democratizing sanitation services and shifting away from top-down, technocratic solutions. "Participation," in this definition, is seen as community-driven bottom-up self-management, which has potential to ensure local control over basic services and recognize and pursue collective will (Mitlin, 2008).

In response to this sectoral shift, participatory demand-driven approaches within the sanitation sector have proliferated, although many already existed in the 1980s. Such approaches include household-centered environmental sanitation (HCES), community-led urban environmental sanitation (CLES), community-led total sanitation (CLTS), community health clubs (CHCs), and participatory hygiene and sanitation assessment (PHAST). Although these approaches have their differences, they share a focus on stimulating demand – where it (presumably) does not yet exist – by changing household sanitation behavior. They often rely on peer pressure, and "name and shame" strategies to coerce desired behavioral change from residents. Once the demand for sanitation has been generated, participatory demand-driven interventions aim at ensuring the availability of a diverse range of sanitation products to match different sanitation needs of different urban dwellers (WSP, 2013). For this purpose, these approaches are combined with the promotion of onsite "pro-poor" sanitation infrastructures (Paterson *et al.*, 2007).

However, as has been noted for the water sector, the move towards participatory approaches can also be interpreted as dwindling state commitment to provide services to low-income urban areas. Devolution of sanitation provision responsibilities from the state to individual households is then framed around the more appealing concept of participatory approaches (Mitlin and Thomson, 1995). This more critical perspective raises questions as to the roles and responsibilities of state, communities, households and the private sector in providing sanitation infrastructures, the extent to which communities and households are required to "participate," financial sustainability and, more broadly, residents' perspectives on sanitation, their needs and priorities beyond onsite facilities. It also calls on us to recognize the limits that are placed on a definition of participation used to achieve such a shift. As noted by Joshi *et al.* (2016), residents are not asked whether they want to participate in such programs, nor are they asked to participate in defining what they need in terms of "adequate" sanitation as the range of options are pre-selected, and the possibilities for their participation as consumers is already pre-designed.

11.4.2 Participating as Consumers

The World Bank's strategy for sanitation (World Bank, 1992) describes the traditional supply-driven approach as "one in which planners and engineers assess the needs of the poor, and then decide what type of service will be provided" (Wright, 1997: 1). Here, the traditional supply-driven approach is framed as decision-making that excluded poor citizens. This lack of procedural justice was supposed to be solved by the shift to decentralized sanitation infrastructure, where the poor can decide what option they want. Of course, in this approach, residents are included in the process as clients and customers – rather than citizens – since their decisions about what they want are based on the affordability of different options. Participation in decision-making is thus directly linked to sanitation service commercialization. At the core of the new approach marked by the Delhi Declaration is the idea that

> progress and continuing success depend most on responding to consumer demand. A program's designers and managers must understand that they are selling a product, not provisioning a service. Where sufficient demand exists, the facilities and services offered must be tailored to that demand; where demand is not strong, it must be stimulated.
>
> (World Bank, 1992: 7)

This is made explicit in the Word Bank's *a posteriori* reflection on the Water and Sanitation Decade (1980–90) in which it proposes a "consumer-oriented" approach which views users as customers and sanitation services as products: "a latrine is a product which one seeks to persuade people to acquire" (World Bank, 1992: 43).

11.4.3 Participation and Justice in Kampala

In the strategies and policies adopted in the 1990s in Uganda, participation in sanitation delivery was equated with self-help or individual service provision, rather than being mobilized as an instrument for empowerment. The market-based principles, which are central to social marketing techniques, and support to the private sector to provide facilities and sanitation services completed this consumer-oriented definition of participation.

Following this approach dictated by the Bank and the Ugandan government, NGOs in Kawempe have focused on creating demand through sanitation marketing. As stated by a representative of one local NGO, "we do sanitation marketing to promote ownership and responsibilities and market it as a social good."[4] Participation by individuals in building and maintaining their own sanitation infrastructure is presumed to ensure the project intervention's sustainability. Again, in the words of a local NGO staff, facilities break down because "people don't own them and there is no sense of sustainability." On the other hand, "when you own your own facility you cannot let it die ... If you own the toilet it means that you can take care of it: since you have invested money you cannot let it be destroyed." This opinion echoes the rationalities of global sanitation policy, where financial subsidies are detrimental to sanitation interventions' sustainability and must be eliminated: "at household [level] I do not think subsidies work; it should be left to individual responsibility."[5] These statements conveniently overlook the fact that the minority of upper-class residents, connected to what is often a very small centralized sewerage system in the urban core, have had their infrastructure entirely subsidized.

Demand generation in Kawempe is aggressively pursued by involving landlords: they own the land and generate income from rental, and are considered responsible for providing a toilet facility. Demand is stimulated by highlighting the monetary value of providing a facility to their tenants: "it must be in relation to money. As long as there is money, this is how we are triggering them. Go for a toilet and increase the status quo of your property. This means more income, because then you can increase your tenants' rent."[6] However, while investments by landlords in toilet facilities on their property improves their income and addresses the needs of those residents who can afford these higher rents, this strategy for improving access to sanitation does so to the detriment of the more vulnerable. Higher rental prices are likely to squeeze the poorest out from these units and onto more undesirable land, under more marginal conditions. Further, although the landlord's provision of a toilet does reduce open defecation and health risk, it addresses only a particular step (collection) in the sanitation chain. As noted in Section 11.3, building latrines does not ensure proper wastewater transport treatment and, as such, may not lead to improved environmental conditions. Black and grey waste may still be discharged into open river systems and shallow groundwater used for drinking purposes. Appeals to landlords based on financial benefits of installing toilet infrastructure do not necessarily motivate them to pay for the required sanitation services (transport and treatment), as there is no direct financial reward.

We conclude that procedural justice is often not achieved, despite the rhetoric of participatory approaches. We find that current sanitation development, promoting decentralized and onsite solutions as the only realistic options, are in fact unjust. Participation in decisions on what is adequate sanitation – to what degree it can also encompass aspects of comfort or local needs – or participation in decisions regarding what infrastructure options should be within the selection process, are not on the table. In addition, framing decentralized sanitation options as participatory has shifted sanitation as individualized and commercialized, which has negative impacts on distributional justice. On this basis, we argue that the simultaneously public and private nature of sanitation requires rescaling arenas for participation.

11.5 Recognitional Justice: A Matter of Dignity

11.5.1 Understanding Sanitation Needs

As the momentum for sanitation grows, the development community is doing more to accept and address diverse sanitation issues and needs, including the emotional dimensions of security, comfort, and dignity. This includes the relationship between sanitation and gender-based violence, recognition of menstruation as one aspect of sanitation, and access to sanitation by residents with special needs (Hulland et al., 2015; Wilbur and Jones, 2014). This broadening of the definition of sanitation partly addresses criticism of development priorities that reduce sanitation to defecation, and reduce hygiene to hand washing (Hulland et al., 2015; Rusca et al., 2017). This broadened definition is also accompanied by increased attention to "for whom" services are provided. Disabled, older or female residents, and other categories of social classification determine what sanitation "is" and what residents need it to be.

Other scholarship has also paid attention to the "who" of sanitation, documenting how inequalities in access to sanitation are felt not only in health outcomes but also affect interpersonal relationships and political visibility with subsequent impact on access to income opportunities and land ownership (Jewitt, 2011; Joshi et al., 2011; McFarlane et al., 2014; Morales et al., 2014). The slogan "sanitation is dignity" reflects that most vulnerable citizens' sanitary concerns are related to emotional dimensions, but issues of self-esteem, stigma, and social acceptance are rarely explicitly addressed when approaching sanitation as only a health issue. For example, in many countries in Sub-Saharan Africa, young girls associate communal school toilets with meanings that go far beyond urinating and defecating; accessing the toilet often entails dangers of sexual harassment and abuse (Leach et al., 2003). Further, social norms that situate defecation and other sanitation-related activities (anal cleansing and menstrual hygiene) as taboos create emotional stress, especially for women (Hulland et al., 2015). Although sanitation needs and concerns vary among women from different life stages and locations, menstrual hygiene management is often ranked by many poor urban women as one of the most stressful experiences related to sanitation, together with the lack of private space for conducting hygiene and sanitation routines (Hulland et al., 2015; Joshi et al., 2011; Lahiri-Dutt, 2014). Avoiding smell and dirtiness is also especially important for vulnerable groups such as disabled, elderly or chronically ill people due to the risk of further stigmatization and marginalization (Wilbur and Jones, 2014).

11.5.2 Recognitional Justice in Kampala

In Kampala, sanitation interventions are firmly based on a public-health approach: sanitation means defecation and hand-washing. Interventions begin with the premise that low-income urban residents are unaware of appropriate hygiene practices, and therefore have a low demand for sanitation. Interventions by key actors in sanitation programming in informal settlements focus primarily on increasing demand for sanitation facilities. However, the staff of development organizations working in Kawempe acknowledge that inhabitants already have clear ideas on good hygiene. For local residents, proper hygiene is related to the cleanliness of their private space:

hygiene is more a priority to them than latrines. You can find people living in one room, but well neat, well clean, even the shoes they leave out. They try: hygiene is important. They keep kitchen utensils clean. The children they try to bathe them.[7]

Local residents' existing sanitary practices are therefore observed, but overlooked when it comes to designing interventions. Nor do NGO staff take into account the social or psychological dimensions of their sanitation interventions, and which are likely tied to existing hygiene practices. The influence of social considerations in a decision to use "improper" sanitation instead of a communal latrine is illustrated by a female resident: "when it starts raining, you can only use that latrine when you are sure that you must take a shower; if you do not shower, no one can stand you ... Every time you leave the latrine, the people you meet can tell where you have been. That is why I prefer using the polythene bag at home." (Female resident of Kampala slums, cited in Kwiringira *et al.*, 2014: 5).

Sanitation interventions in Kawempe also use the strategy of positive and negative peer pressure by residents to change their friends and neighbors' hygiene behavior. Peer pressure can come in the form of "shaming" those who continue to practice open defecation, or who use "flying toilets" as a sanitation solution. Other sanitation interventions go beyond the individual household to target schools or churches as (potential) sites for inducing behavioral change. According to development organizations involved in these interventions, these institutions are critical to triggering change: "if we change behavior of institutions (schools, churches), people will change behavior."[8] However, as has been recently cautioned in critical assessments of these programs, using shame to trigger change risks further marginalizing the poorest in communities who may want to invest in proper sanitation solutions, but cannot afford to do so (Engels and Susilo, 2014), and jeopardizes individual human rights (Bartram *et al.*, 2012).

In summary, we highlight the potential injustices associated with programs that socially penalize those who may want to have clean and comfortable facilities to meet their hygiene needs, but lack the means. Our review of the risks for increased social stigmas and exclusionary practices indicates a nascent acceptance of sanitation as dignity, and the dimension of recognitional justice. However, we hold little hope for the current model of sanitation development to go much beyond the existing definition of sanitation as defecation and hand washing, given the already limited financing for the sector.

11.6 Conclusions: From *Improved* to *Just* Sanitation

The development initiatives towards increased access to sanitation under the MDGs have focused on ensuring "improved" access. The Sustainable Development Goals (SDGs) now call for improved access to be affordable, equitable, universal, and sustainable (UN, 2015).

However, in the MDGs, and now under the SDGs, improved access is pursued by promoting onsite sanitation infrastructure within a development approach that individualizes and commercializes the sanitation sector. This approach is rationalized as a realistic, participatory, sustainable alternative to centralized sewerage networks.

We have reviewed this approach through three dimensions of sanitation justice: distributional, procedural, and recognitional justice, developed from EJ frameworks. In our review of the global policy shift towards private decentralized sanitation infrastructure, and its impact within an informal urban settlement in Kampala, we question the justice of both the process, and its impact on developmental outcomes. Our review has raised concerns about both the effectiveness of this sanitation development model in increasing access, but also identifies other dimensions of inequalities which are built into this developmental approach. These inequalities, which lay beyond numerical targets or indicators of sanitation coverage, can be identified by looked at distribution, procedure, and recognition.

Development initiatives under the SDGs should therefore aspire to "just" sanitation, rather than "improved" or adequate sanitation. This calls for development practitioners and policy makers to attend to justice within the development process, as well as the justice of development outcomes, and the relations between them. However, moving toward "just" sanitation first requires an improved understanding of what sanitation is. In the context of Sub-Saharan African cities, sanitation is infrastructure, it is services, and it is dignity. Our analyses of inequalities in sanitation – or what is sanitation justice – must acknowledge these multiple dimensions. This means any assessment of access (affordability, universality, equity, sustainability) must include access to infrastructure, the services required for the infrastructure, and this infrastructure and service's impact on individual dignity. Sanitation is each of these, and all of these together. The inequalities related to these dimensions (distribution, procedural, recognitional) are also therefore related and if they are all addressed under the SDGs, then sanitation access will be affordable, equitable, universal, and sustainable.

Acknowledgments

This project has received funding from INHAbIT Cities, sponsored by the European Union's Horizon 2020 research and innovation programme under the Marie Skłodowska-Curie grant agreement No. 656738.

Notes

1 Deutsche Gesellschaft für Internationale Zusammenarbeit (GIZ), Urban WASH Coordinator, Kampala, May 2011.
2 Community Integrated Development Initiatives, Citizens Actions for Water and Sanitation Coordinator, Kampala, April 2011.
3 Ibid.
4 Community Integrated Development Initiatives, Citizens Actions for Water and Sanitation Coordinator, April 2011.
5 Ibid.

6 Ibid.
7 Urban Poor Liaison Unit – National Water and Sewerage Corporation representative – Kawempe Division, Kampala, May 2011.
8 Community Integrated Development Initiatives, Citizens Actions for Water and Sanitation, Community Mobilizer, April 2011.

References

Allen, A., Davila, J. D. and Hofmann, P. (2006). The peri-urban water poor: Citizens or consumers? *Environment & Urbanization*, 18(2), 333–51.

Bain R., Cronk, R., Wright, J., Yang, H., Slaymaker, T. and Bartram, J. (2014). Fecal contamination of drinking-water in low- and middle-income countries: A systematic review and meta-analysis. *PLoS Medicine*, 11(5), 1–23.

Bakker, K. (2007). The "commons" versus the "commodity": Alter-globalization, anti-privatization and the human right to water in the Global South. *Antipode*, 39(3), 430–55.

Bartram, J., Charles, K., Evans, B., O'Hanlon, L. and Pedley, S. (2012). Commentary on community-led total sanitation and human rights: Should the right to community-wide health be won at the cost of individual rights? *Journal of Water and Health*, 10(4), 499–503.

Buckingham, S. and Kulcur, R. (2009). Gendered geographies of environmental injustice. *Antipode* 41(4), 659–83.

Bullard, R. D. (1983) Solid waste sites and the Black Houston community. *Sociological Inquiry*, 53, 273–88.

Bullard, R. D., Mohai, P., Saha, R. and Wright, B. (2008). Toxic wastes and race at Twenty: Why race still matters after all these years. *Environmental Law*, 38(2), 371–411.

Citizens Action for Water and Sanitation (CIDI) and WaterAid, Community Voices for Accountability in the Water and Sanitation Sector, Newsletter, October 2008–February 2009.

Citizens Action for Water and Sanitation (CIDI) and WaterAid. (2010). Citizens' Report Card on Water Sanitation and Solid Waste Services in Uganda's Informal Settlements, Kampala, December 2010.

Cook, I. R. and Swygendouw, E. (2012). Cities, social cohesion and the environment: Towards a future research agenda. *Urban Studies*, 49(9), 1959–79.

Desai, R., McFarlane, C. and Graham, S. (2014), The politics of open defecation: Informality, body, and infrastructure in Mumbai. *Antipode*, 47(1), 98–120.

Economic and Social Council (ECOSOC). (1990). Report of the economic and social council: Achievements of the international water and sanitation decade, A/45/327. New York, July 13.

Engel, S. and Susilo, A. (2014). Shaming and sanitation in Indonesia: A return to colonial public health practices? *Development and Change*, 45(1), 157–78.

Graham, J. P. and Polizzotto, M. L. (2013). Pit latrines and their impacts on groundwater quality: A systematic review. *Environmental Health Perspectives*, 121, 521–30.

Harris, L., Lucy Rodina, M. and Morinville, C. (2015). Revisiting the human right to water from an environmental justice lens. *Politics, Groups, and Identities*, 3(4), 660–65.

Harris, L. M., Goldin, J. A. and Sneddon, C. (eds.) (2013). *Contemporary Water Governance in the Global South: Scarcity, Marketization and Participation*. London: Routledge.

Heynen, N., Kaika, M. and Swyngendouw, E. (2006). Urban political ecology: Politicizing the production of urban natures. In N. Heynen, M. Kaika and E. Swyngendouw (eds.),

In the Nature of Cities: Urban Political Ecology and the Politics of Urban Metabolism. London: Routledge.

Holifield R. (2012). Environmental justice as recognition and participation in risk assessment: Negotiating and translating health risk at a superfund site in Indian Country. *Annals of the Association of American Geographers*, 102(3), 591–613.

Holifield, R., Porter, M. and Walker, G. (2009). Introduction spaces of environmental justice: Frameworks for critical engagement. *Antipode*, 41(4), 637–58.

Honneth, A. (2004). Recognition, redistribution, and justice. *Acta Sociologica*, 47(4), 351–64.

Hulland, K., Chase, R., Caruso, B., Swain, R., Biswal, B., Sahoo, K., Panigrahi, P. and Dreibelbis, R. (2015). Sanitation, stress, and life stage: A systematic data collection study among women in Odisha, India. *PLoS ONE*, 10(11), 1–17.

Iossifova, D. (2015). Everyday practices of sanitation under uneven urban development in contemporary Shanghai. *Environment & Urbanization*, 27(2), 541–54.

Isunju, J. B., Etajak, S., Mwalwega, B., Kimwaga, R., Atekyereza, P., Bazeyo, W. and Ssempebwa, J. C. (2013). Financing of sanitation services in the slums of Kampala and Dar es Salaam. *Health*, 5(4), 783–91.

Jenkins, M., Cumming, O. and Cairncross, S. (2015). Pit latrine emptying behavior and demand for sanitation services in Dar es Salaam, Tanzania. *International Journal of Environmental Resources and Public Health*, 12, 2588–611.

Jewitt, S. (2011), Geographies of shit: Spatial and temporal variations in attitudes towards human waste. *Progress in Human Geography*, 35(5), 608–26.

Joshi, D., Fawcett, B. and Mannan, F. (2011). Health, hygiene and appropriate sanitation: Experiences and perceptions of the urban poor. *Environment & Urbanization* 23(1), 91–111.

Joshi, D., Kooy, M. and van den Ouden, V. (2016). Development for children, or children for development? Examining children's participation in school-led total sanitation programs. *Development and Change*, 47 (5), 1125–45.

Kampala Declaration on Sanitation, Republic of Uganda National Sanitation Forum, 824 LJG97, October 17, 1997, www.ircwash.org/sites/default/files/824-UG-14784.pdf.

Katukizaa, A. Y., Ronteltap, M., Oleja, A., Niwagaba, C. B., Kansiime, F. and Lensa, P. N. L. (2010). Selection of sustainable sanitation technologies for urban slums: A case of Bwaise III in Kampala, Uganda. *Science of The Total Environment*, 409(1), 52–62.

Kiyimba, J. (2006). *Identifying the Levels of Satisfaction, Need and Representation–Citizens Action Survey Report in Kawempe Division for Improvement of People's Lives. Community Integrated Development Initiatives (CIDI) and WaterAid*. Kampala.

Kooy, M. and Bakker, K. (2008). Splintered networks: The colonial and contemporary waters of Jakarta. *Geoforum*, 39(6), 1843–58.

Kwiringira, J., Atekyereza, P., Niwagaba, C. and Günther, I. (2014). Descending the sanitation ladder in urban Uganda: Evidence from Kampala Slums. *BMC Public Health*, 14(1), 624.

Lahiri-Dutt, K. (2014). Medicalising menstruation: A feminist critique of the political economy of menstrual hygiene management in South Asia. *Gender, Place & Culture*, 22(8), 1158–76.

Leach, D., Fiscian, V., Kadzamira, E. and Machakanja P. (2003). *An Investigative Study of the Abuse of Girls in African Schools*, Educational Papers, Department for International Development, London.

Letema, S., van Vliet, B. and van Lier J. (2014). Sanitation policy and spatial planning in urban east Africa: Diverging sanitation spaces and actor arrangements in Kampala and Kisumu. *Cities*, 36, 1–9.

Lüthi, C., McConville, J. and Kvarnström, E. (2010). Community-based approaches for addressing the urban sanitation challenges. *International Journal of Urban Sustainable Development*, 1(1–2), 49–63.

Lüthi, C., Morel, A., Tilley, E. and Ulrich, L. (2011) Community-led urban environmental sanitation planning: CLUES complete guidelines for decision-makers. Eawag-Sandec/WSSCC/UN-HABITAT.

Matagi, S. V. (2002). Some issues of environmental concern in Kampala, the capital city of Uganda. *Environmental Monitoring and Assessment*, 77(2), 121–38.

McFarlane, C. (2008). Governing the contaminated city: Infrastructure and sanitation in colonial and post-colonial Bombay. *International Journal of Urban and Regional Research*, 32(2), 415–35.

McFarlane, C., Desai, R. and Graham, S. (2014). Informal urban sanitation: Everyday life, poverty, and comparison. *Annals of the Association of American Geographers*, 104(5), 989–1011.

Mitlin, D. (2008). With and beyond the state: Co-production as a route to political influence, power and transformation for grassroots organizations. *Environment & Urbanization*, 20(2), 339–60.

Mitlin, D. and Thompson, J. (1995). Participatory approaches in urban areas: Strengthening civil society or reinforcing the status quo? *Environment & Urbanization*, 7(1), 231–50.

Morales, M. (2015). My pipes say I am powerful: Belonging and class as constructed through our sewers. *WIREs Water*, 3(1), 63–73.

Morales, M., Harris, L. and Aberg, G. (2014). Citizenshit: The right to flush and the urban sanitation imaginary. *Environment and Planning A*, 46(12), 2816–33.

Mosse, D. (2006). Collective action, common property and social capital in south india: An anthropological commentary. *Economic Development and Cultural Change*, 54(3), 695–724.

Nilsson, D. (2006). A heritage of unsustainability? Reviewing the origin of the large-scale water and sanitation system in Kampala, Uganda. *Environment & Urbanization*, 18(2), 369–85.

Paterson, C., Mara, D. and Curtis, T. (2007). Pro-poor sanitation technologies. *Geoforum*, 38(5), 901–07.

Perreault, T., Wraight, S. and Perreault, M. (2012). Environmental injustice in the Onondaga lake waterscape, New York State, USA. *Water Alternatives*, 5(2), 485–506.

Rusca M., Alda-Vidal C., Hordijk M. and Kral, N. (2017). Bathing without water: Stories of everyday hygiene practices and risk perception in urban low-income areas. *Environment & Urbanization*, 29(2).

Sarpong Boakye-Ansah, A., Ferrero, G., Rusca, M. and van der Zaag, P. (2016). Inequalities in microbial contamination of drinking water supplies in urban areas: The case of Lilongwe, Malawi. *Journal of Water and Health*, 14(5), 851–63.

Schlosberg, D. (2007). *Defining Environmental Justice: Theories, Movements, and Nature*. New York: Oxford University Press.

Schlosberg, D. (2013). Theorising environmental justice: The expanding sphere of a discourse. *Environmental Politics*, 22(1), 37–55.

Sustainable Sanitation and Water Renewal Systems (SSWARS). (2013). Baseline study to determine access to basic sanitation in Bwaise I, Bwaise II and Natete parishes, GIZ and German Cooperation, www.sswarsuganda.org/phocadownloadpap/sanitation_baseline_bwaise_and%20_natete_april2013.pdf.

Tsinda, A., Pedley, P. S., Charles, K., Adogo, J., Okurut, K. and Chenoweth, J. (2013). Challenges to achieving sustainable sanitation in informal settlements of Kigali, Rwanda. *International Journal of Environmental Research and Public Health*, 10(12), 6939–54.

UBOS. (2005). *2002 Uganda Population and Housing Census Main Report*. Uganda Bureau of Statistics, Kampala-Uganda, www.ubos.org/onlinefiles/uploads/ubos/pdf%20documents/2002%20Census%20Final%20Reportdoc.pdf.

UN. (2015). Transforming our world: The 2030 Agenda for Sustainable Development. General Assembly, A/RES/70/1, New York, September 25.

Swyngedouw, E. (2004). *Social Power and the Urbanization of Water: Flows of Power*. Oxford: Oxford University Press.

UN-HABITAT. (2003). *Water and Sanitation in the World's Cities: Local Action for Global Goals*. London: Earthscan Publications Ltd.

UNICEF/WHO. (2015). Progress on sanitation and drinking water: 2015 update and MDG assessment. Joint Monitoring Programme (JMP), p.17, www.wssinfo.org/fileadmin/user_upload/resources/JMP-Update-report-2015_English.pdf.

Urkidi, L. and Walter, M. (2011). Dimensions of environmental justice in anti-gold mining movements in Latin America. *Geoforum*. 42, 683–95.

Walker, G. (2009). Beyond distribution and proximity: Exploring the multiple spatialities of environmental justice. *Antipode*, 41(4), 614–36.

Wilbur, J. and Jones, H. (2014). Disability: Making CLTS fully inclusive. *Frontiers of CLTS: Innovations and Insights*, 3, 1–20, Brighton: IDS.

World Bank. (1992). *Sanitation and Water Supply: Practical Lessons from the Decade*. Washington, DC: World Bank.

World Bank. (2007). Kampala institutional and infrastructure development adaptable program loan (APL) project in support of the Strategic Framework for Reform for Kampala urban development program, WB Report No: 35847-UG.

World Sanitation for the Urban Poor. (2013). Getting communities engaged in water and sanitation projects: Participatory design and consumer feedback. Topic Brief, TB#007: 1–12.

World Sanitation Program. (2013). *Targeting the Urban Poor and Improving Services in Small Towns. Poor Inclusive Urban Sanitation: An Overview*. Washington, DC, www.wsp.org/sites/wsp.org/files/publications/WSP-Poor-Inclusive-Urban-Sanitation-Overview.pdf.

Wright, Albert M. (1997). *Toward a Strategic Sanitation Approach: Improving the Sustainability of Urban Sanitation in Developing Countries*. Water and sanitation program. Washington, DC: World Bank Group.

12

Uniting Diversity to Build Europe's Right2Water Movement

Jerry van den Berge, Rutgerd Boelens, and Jeroen Vos

12.1 Introduction

In 2013, the first successful European Citizens' Initiative (ECI), called "Right2Water," collected 1.9 million signatures. This broad European civil-society initiative demanded that the European Commission implement the human right to water and sanitation in European legislation. The campaign was organized by the European Federation of Public Service Unions (EPSU) and supported by a large number of NGOs and water activists throughout Europe.

The Right2Water movement campaigned against the European Commission's intention to further privatize drinking water utilities in Europe, following the UK example. The European campaign faced a large diversity of contexts in each Member State of the European Union. The state of affairs regarding public or private governance of water utilities and those utilities' performance differed per country. Moreover, awareness about the theme of "the human right to water" among the general public, grassroots federations, NGOs, and trade unions was also quite diverse.

Water in Europe is subject to both European and national law and responsibility is shared between the European Union and Member States, making it a suitable issue for an ECI. The initiative aimed to shift the European Commission's focus from their market orientation to a rights-based, people-oriented water-policy approach (Right2Water, 2016). The Right2Water campaign joined in the ongoing struggle for water justice that, in divergent ways, was framed and organized by many civil-society groups, and took a stance against profit-driven water companies with the slogan "Water is a public good; not a commodity!" It urged the European Commission that water services in Europe should not be liberalized.

From April 2012 to September 2013, the ECI "Right2Water" collected signatures across Europe. In 13 countries, the number of signatures passed the EU threshold, which was set by the European Commission for making such peoples' initiatives mandatory to be discussed by the European Commission (see Table 12.1); the first successful ECI ever. The ECI is a tool established by the Lisbon Treaty (EC, 2009) as a means to bring the European Union closer to its citizens. It gives people an opportunity to bring an issue to the European

Table 12.1: *Results of the ECI Right2Water and threshold per country*

Country	Total signatures	Minimum signatories required EU
Austria	64,836	14,250
Belgium	40,912	16,500
Bulgaria	1,602	13,500
Cyprus	3,561	4,500
Czech Republic	7,986	16,500
Denmark	3,547	9,750
Estonia	1,245	4,500
Finland	15,200	9,750
France	22,969	55,500
Germany	1,341,061	74,250
Greece	35,720	16,500
Hungary	20,107	16,500
Ireland	2,959	9,000
Italy	67,484	54,750
Latvia	450	6,750
Lithuania	14,048	9,000
Luxembourg	5,698	4,500
Malta	1,703	4,500
Netherlands	22,065	19,500
Poland	4,807	38,250
Portugal	15,588	16,500
Romania	3,211	24,750
Slovakia	35,075	9,750
Slovenia	21,330	6,000
Spain	65,484	40,500
Sweden	12,258	15,000
United Kingdom	8,578	54,750
Total EU	**1,839,484**	**Total from all EU Member States must be more than 1,000,000**

Source: Right2water, 2016.

political agenda if they manage to collect over one million signatures in one year's time, from at least seven countries, with a minimum for each country (EC, 2011).

In this chapter, we describe how the campaign was organized and analyze how and why Right2Water became successful in some countries, while not in others. Furthermore, we will discuss the political achievements and implications for the global water justice movement.

12.2 The Context of the Right2Water Movement

12.2.1 Neoliberal Threats to Drinking Water Provision in the European Context

One of the European Commission's main objectives is to create a single European market, landmarked in the Maastricht Treaty (1991). This was to be fulfilled by the four movement freedoms in Europe: movement of people, of goods, of capital and of services. This started liberalizing public services such as energy supply, public transport and water services. Before the Treaty, these services in the EU were mostly delivered by governments or local authorities. The European neoliberal point of departure was an economic "laissez faire" perspective, expecting that markets can function best if governments withdraw as much as possible. Accordingly, the European Union has proposed policies to liberalize the economy, including the water sector. Privatization experiments in the UK in the 1980s were seen as an example of how to shift to a more open European market.

In the late 1980s, the idea of private companies managing water emerged on a large scale. Under Margaret Thatcher, the UK government privatized all water companies in England and Wales in 1989 – making it the first European country to do so (Hall and Lobina, 2008; Achterhuis *et al.,* 2010). Coupled with the global emphasis on free-market capitalism after communism's fall, there was a wave of water-utility privatizations. Privatizing water utilities was encouraged by the International Monetary Fund and the World Bank, which made public-to-private takeovers a condition of lending. As a result, the early 1990s saw a rush of cities and countries around the world signing over water resources to private companies (Lawson, 2015). Many European governments followed arguments by the industry and investors that putting water in private hands would improve efficiency and service quality, and that services would be better managed. Selling off water services would also provide an opportunity for governments to gain revenue, and for companies to generate profit. But with profit the main objective, the idea of water as a human right arguably became a secondary concern. Problems with water privatization often began to occur soon after the initial wave of enthusiasm – ranging from lack of infrastructure investment to environmental neglect and socio-economic discrimination and injustices (Bakker, 2010; Lawson, 2015).

The most important document spearheading water policies and politics in the European Union is the Water Framework Directive, referred to by the European Commission as "EU's flagship Directive on water policy" (EC, 2007: 4). In 2012, the European Commission established a "blueprint for water" as its policy to manage Europe's water resources (EC, 2012). This blueprint was founded on an economic and market-based vision that water should be used in sectors where it provides higher economic value (i.e. in agriculture, industries and for electricity production).

In the UK, as mentioned above, in a relative short period, local public companies and other public services were privatized. Multinational corporations quickly absorbed these public entities and local service provision changed drastically, becoming part of transnational big businesses. Trade unions viewed these developments with fear and anger because of losing jobs and public control, as well as increasing multinational corporations' power (Hall and Lobina, 2012a). Based on bad past experiences with water-utility privatization, and with relatively new forms of privatization such as public-private partnerships (PPPs), the Right2Water organizers saw the ECI as a means to counter European governments' neoliberal plans and the threat of pro-privatization policies.

12.2.2 The Movement's Arguments against Privatizing Water Services

Water activists have contested the EU market ideology from social, environmental and human-rights perspectives. Trade unions collaborated with water activists to counter the negative consequences of privatization, which have often resulted in job losses. They saw in the ECI a new tool and an opportunity to revive their struggle (EPSU, 2011). Water services are seen as a public responsibility and government obligation. According to water activists, the vision behind European water policy is based on the urge to commodify water resources and services, expressed in its water-pricing ideology and instruments (EEA, 2010). They heavily criticized this "market approach" to water management, which motivated the ECI's "Water is a public good, not a commodity!" slogan.

Privatizing water-service utilities has been strongly and actively promoted by development agencies and international institutions since the 1990s, and has faced many problems and criticism (Hall and Lobina, 2012b). Neoliberal water policies believe deeply in formal-legal water rights and market incentive structures to determine water users' and managers' behavior, to promote both physical water-use efficiency and economic efficiency in water control and resource use (Boelens and Vos, 2012). New institutional economics (new institutionalism) sees water actors as individualistic, rational decision-makers who aim to generate net economic benefits through resource use and allocative decisions ("individual profit maximizers"). In that way, organizational and political water-management products are viewed as the sum of all rational decisions made by these individuals based on material self-interest, which can be defined objectively and universally (Achterhuis *et al.*, 2010; Bakker, 2007). To "install the right economic incentives," neoliberal water-reform recipes concentrate on three fundamental ingredients: decentralized decision-making (privatized control and management), private, price-valued, transferable property rights (commodified water resources and rights) and market rules (water flows and exchanges governed by market forces) (Boelens and Zwarteveen, 2006). Whenever the incentives are right, private motives of profit maximization and accumulation (that is, positively valued selfishness and greed) will automatically foster the use of water and public funds as efficiently as possible (Achterhuis *et al.*, 2010); private owners and operators will take care of their infrastructure (operation and maintenance) and water will flow to those who (in market terms) most need it: to the water actors who are "most productive" and save water quantity and quality.

Fundamental to this is the belief that "money and water need to follow universal scientific laws and that human beings follow the same rational utility-maximizing aspirations everywhere" (Achterhuis *et al.*, 2010: 45–46; Boelens and Zwarteveen, 2005). But unlike their liberal forerunners, founding fathers Hayek (1944) and Friedman (1962) had already noted that neoliberal civilization, involving an integrated economic and political system in which everyone shares the same modernistic capitalist values and institutions, would not emerge automatically. To emerge and function, the "invisible hand" requires governors' visible power. Rather than "laissez-faire" policies, water neoliberalism calls for a very active, important (indeed, aggressive) role for the state: a "social contract" that guarantees private ownership of water and infrastructure and installs and sustains water-market operation.

Hendriks (1998) and Bakker (2003) have pointed to how water's materiality deeply affects the opportunities for water to operate as a commodified resource under market forces. Its fluidity impedes privatized control and makes it easy to displace negative externalities, and its density makes transport very costly, complex, and arduous. As Hendriks states: "Water is relatively immobile, and expensive and difficult to reallocate" (1998: 300). It is conducive to monopolies, rather than to market competition, which, among other factors, makes it difficult to have production costs reflected in "real-price values." This, coupled with social needs related to public health and environmental issues regarding pollution – important socio-environmental "externalities" that do not respond to market valuation – make it an "uncooperative commodity" (Bakker, 2003). Debunking the myth of the invisible-hand inevitably fostering economically efficient and socio-environmentally prosperous water allocation, water's privatization, commodification and marketization tends to privilege capitalist enterprises and the most affluent population sectors, while neglecting those who cannot afford to pay (e.g. Berge, 2014; Lawson, 2015; Mirosa and Harris, 2012); and elicit a forceful, deep involvement of national and European government structures to accomplish this.

Consequently, creating a market undermines the objective of universal service provision. Market principles bear the risk of excluding those poorer people and less-powerful sectors that cannot afford the newly established water prices (Lawson, 2015). Moreover, in many instances, people lose control over their local sources. If, in these cases, governments subsidized water supply to the poor, they would be subsidizing corporations' profits. The "Cochabamba Water War" may be the most prominent example of a water conflict following from privatizing water services (Schultz, 2001); other cases around the world are documented. Balanyá et al. concluded that "strengthening the democratic, public character of water services is fundamentally at odds with the currently dominant neoliberal model, which subordinates ever-more areas of life to the harsh logic of global markets" (Balanyá et al., 2005: 248). Moreover, privatization's claimed benefits are questioned. A study by the World Bank found that overall evidence suggests "there is no statistically significant difference between the efficiency performance of public and private operators in this sector" (Hall and Lobina, 2005: 4).

Recently, in response, cities and local governments worldwide are increasingly choosing to "remunicipalize" services by taking back public control over water and sanitation services in reaction to private operators' false promises and failure to put the needs of communities before profit (Lobina et al., 2014). Nearly all private water companies in Europe (apart from in the UK) are subsidiaries or partially owned by the two French water multinationals, Veolia and Suez (see Figure 12.1 for the degree of privatization in European

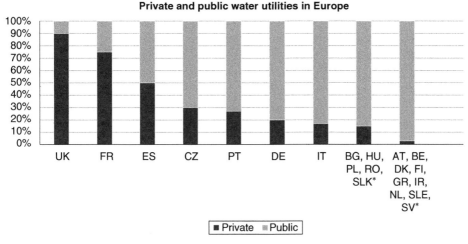

Figure 12.1: Percentage of connections that receive water service from a private water-supply utility per country
Source: by the authors, based on national statistics.

countries). This has major implications for tariffs and particularly for water control and decision-making, the danger of monopolies, technocratic and universalistic governance, setting the rules of the game – even when European countries argue that water remains under democratic control. Nevertheless, the European Commission continued to pursue its path in privatizing water supply and sanitation, especially in the framework of post-crisis austerity measures, as its answer to the economic crisis (Zacune, 2013).

Right2Water came into action at a moment when the European Commission was dealing with further extension of the single EU market and after the financial crisis that hit Europe especially between 2008 and 2012. Many countries were still in recession and the EU announced tough measures to countries that did not comply with financial discipline as agreed in the Eurozone. Austerity policies were enforced across the EU member states, including pressures towards further privatization, especially on the countries in the EU's "periphery" such as Greece and Portugal (Bieler, 2015).

Indebted "peripheral countries" of the EU (especially Greece, Ireland and Portugal) were bailed out by the Troika of the European Commission, European Central Bank (ECB) and IMF, in exchange for imposed restructuring that included labor-market deregulation, cutting public-sector employment, and privatizing public companies (Lapavitsas *et al.*, 2012). Water privatization has been pushed especially vis-à-vis Greece and Portugal: see Box 12.1 (Hall and Lobina, 2012b). Italy too came under pressure in the second half of 2011, when Jean-Claude Trichet, then President of the ECB, and Mario Draghi, who succeeded him in November 2011, urged "the full liberalization of local public services … through large-scale privatizations" (Bieler, 2015: 9).

In 2011, the European Commission again attempted to further liberalize service sectors in Europe by proposing a Concessions Directive stipulating stronger laissez-faire and withdrawing public control and management (EC, 2011). This Directive coincided with the Right2Water campaign and played an important role in public debate on water privatization and signature collection, after a German television documentary by ARD Monitor revealed the conflict between the proposals by European Commissioner Barnier and by Right2Water. As a result, the Commissioner had to give in and excluded water services from the scope of the Concessions Directive (EC, 2013).

Box 12.1 Example of Forced Privatization: The Thessaloniki Drinking Water Company

In 2010 the European Commission, together with the IMF and ECB, ordered the Greek Government to sell public assets as part of a bailout, including the public water companies of Athens and Thessaloniki. Privatizing and selling these water companies met with huge public resistance. People and local organizations formed alliances to keep the water companies under Greek public control. "K136" (Initiative 136) was one of them, as were SOSteTONERO ("SOS for WATER") and "Save Greek Water." All these organizations joined the Right2water campaign to show how EU policy threatened local public control over water. The Commission insisted that it had not forced privatization upon Greece, but the evidence was clear. In a letter to Food and Water Europe, the Commission admitted its support for privatization (CEO, 2012). Although the Greek people were facing many difficulties in their daily lives (huge, increasing unemployment, bank closure, salary non-payment, etc.), they supported Right2Water from the moment they saw the link between EU policy and (national) privatization plans. As Yiorgos Archantopoulos from SOSteTONERO stated: "the forced privatization is providing an opportunity for foreign investors to take over our profitable, well-functioning public utilities for a real bargain" (Y. Archontopoulos, personal communication, 2014). In a short time-span, sufficient signatories were achieved, despite high bureaucratic barriers (ID requirements) for signing an ECI in Greece.

In turn, Right2Water helped Greek organizations hold a referendum against privatizing the Thessaloniki water company (EYATH). The referendum, inspired by the Italian experience, succeeded in achieving the quorum, and 98 percent of participants voted against privatization. The Government tried to declare the referendum illegal, but the Greek High Court ruled the referendum result legally binding on the Government.

Amid the emerging crisis after 2007, and during ongoing governmental plans to extend the European Union, people's attitudes in Europe turned increasingly negative toward the European Union, while nationalist sentiments increased. It is in this context that the ECIs instrument finally came into being: formally, with signing of the Lisbon Treaty in 2009, and in practice, with the adoption of rules for their administration, in 2012 (Parks, 2014).

12.2.3 Building Unity with Diversity in Social Movements: Some Conceptual Notes

Scholars studying social movements addressing water justice issues often ask three main questions: why do social actors start and support a social movement for water justice? How do social movements develop and campaign? And, what are the tangible political effects of water justice campaigning (e.g. Schlosberg, 2004; Spronk, 2007). A parallel theme, related to these three questions, is the alleged need for uniform ideas and strategies in the social movement.

The question of why people start and join a social movement is at the core of discussion on the distinction between "social movements" and "new social movements." The former would in theory be related to a felt sense of deprivation (of access to the means of production in Marxism, or access to basic services in development theories); the latter would in theory be related to ideology, values, morality, and identity. Basically, the social movements would mobilize people who no longer agree to be exploited and deprived of basic services. The labor movement would protest for better wages, and the landless for access to agricultural land. The new social movements (e.g. civil rights; women's rights; indigenous rights, gay, lesbian and transgender rights; peace movement; and environmental movements) fight for recognition and the right to participate in political decision-making.

In reality, this separation is not strict and tends to blur, especially in water-based movements, where the issues at stake are simultaneously material-economic (e.g. water resources, infrastructure, irrigated land, water funds and subsidies, tariffs, etc.), cultural and normative (e.g. water-rights notions, water values, norms, rules and regulations, water management forms, etc.) and political (e.g. water authority and decision-making), always with a strong relation to (politics of) identity and bonds of belonging (Boelens, 2015). Water justice-struggles entwine skirmishes over water resources, water rights and rules contents, legitimate water authority, and water-based worldviews and discourses, and therefore inherently combine struggles for cultural, political, socio-ecological, and socio-economic justice. Moreover, an "unjust" socio-economic situation does not automatically lead to protests and develop a social movement. Several factors have been suggested that influence a protest movement's development. According to Seters and James (2014) four minimal conditions should be met: (1) formation of some kind of collective identity; (2) a minimal shared ideology or normative orientation; (3) a shared drive to change an existing situation; and (4) a connection between like-minded groups over a large time and geographical scale.

For an actual practice of protest to grow into a social movement two issues are key: framing and networking. Framing the problem and its solution is crucial to develop a social movement. Framing problems and solutions is based on world views, as well as on strategic representations of reality. In this sense, social movements are cultural and political, with a strong political role of discourse, values and ideas. A social movement, to develop, needs to mobilize people and resources around shared imaginaries. Therefore, social movements are networks in which cultural and psychological aspects (micro-politics and leadership) play an important role. However, the network's institutional design, in formal tiers or loosely formed groups with their own rules, while composing a multi-actor (e.g. Bebbington *et al.*, 2010) and multi-scalar pattern (e.g. Hoogesteger and Verzijl, 2015; Swyngedouw, 2004), is also an important factor shaping the social movement (Boelens *et al.*, 2016; Hoogesteger *et al.*, 2016). Communication practice links the framing with the networking. Clearly, the digital age has changed communication means and strategies (see e.g. Della Porta and Mosca, 2005), but the notion of mobilizing people by framing ideas that resonate with their convictions remains fundamentally important.

In questions of both why and how social movements appear and develop, the plurality-versus-unity issue is central. Where Schlosberg (2004) (drawing on James, 1977[1909]) and Wenz (1988)) argued that uniformity is not needed for unity inside a social movement, Harvey (1996) would see uniform ideology as a requirement for a social movement's unity and success for environmental justice. We argue that framing the problem and solution can be pluriform: different groups and individuals have different ideas on what exactly the problem is and what might be desired solutions. People might also foreground and background different aspects of the problem definitions and proposed solutions. The movement's organization can also be pluriform over time and in different places. Acknowledging diversity, difference and plurality is an important cornerstone of water justice movements, while their ability to bridge the diverse ways of viewing and struggling for water justice is the inherent challenge as well as the indicator of strength, effectiveness and impact. Strong water movements show engagement and alliances across contexts, continents, scales and differences (Zwarteveen and Boelens, 2014).

In this chapter, we will analyze the Right2Water campaign, looking at the pluriform reasons to join the movement and the pluriform ways to organize the ECI in each European country. We will relate the outcomes of the campaign (in number of signatures per country and political influence) to the unity-with-pluriformity concept.

12.3 Building Europe's Water Movement: The Right2Water Campaign

12.3.1 Uniting Different Cases, Scales and Movements

Public-sector trade unions divided on campaigning about "water." Many of them had little capacity, some knew little about EU policies and regulations, and others had no or little experience in campaigning for an issue that is not about jobs or working conditions. They needed support from non-governmental organizations (NGOs) with campaigning experience. The first objective of Right2Water was to form a broad coalition (Berge, 2014). In November 2011, the first allies were water activists in the existing water justice network: people and organizations that had campaigned against water commodification, commercialization and privatization since the 1990s.

One of the first European groups to embrace the campaign was the European Anti-Poverty Network (EAPN). They committed fully to the campaign, saying that water cannot be a charity for the poor, but is a human right that must be fulfilled for all. For EAPN, lifting people out of poverty started with realizing their rights and basic needs to live a life in dignity. Other European NGO federations followed and together they prepared for an ECI to start on the first possible date: April 1, 2012. These NGOs were: the European Public Health Alliance (EPHA), the European Environmental Bureau (EEB), Women in Europe for a Common Future (WECF), Food and Water Europe (FWE), and the Federation of Young European Greens (FYEG). This broad base enabled a Europe-wide campaign at local and national levels. The very divergent results in different countries (see Table 12.1) show the large diversity of contexts and organizational strengths in each country.

During the campaign, broader support was sought and found among churches or religious groups, in development organizations, consumer organizations, and other civil-society organizations. A total of 149 organizations officially supported Right2Water. Besides this, around one hundred other organizations endorsed the campaign in one way or another: from organizing events to collect signatures to promoting Right2Water in their media. Supporting organizations can be divided into 11 categories (plus one "other"; see Table 12.2). The high percentage of trade unions (47 percent) mirrors labor unions' strong lead in the Right2Water campaign (Lesske, 2015).

The unconventionally broad alliance sometimes raised eyebrows among its members. Trade unions had sometimes worked with environmental groups and sometimes with an anti-poverty group, but never with both at the same time. The same went for trade unions and women's groups or public health campaigners. In civil society, there are many walls between organizations from different backgrounds. The ECI was an unprecedented experiment for the main supporting organizations. The alliance attracted attention of politicians from the entire political spectrum and kept it out of a corner for "trade unionists," "environmental activists" or any other single-issue activist group. The weakness of being hugely diverse turned into a strength when Right2Water was seen as neither left- nor right-wing, and as having support from a wide range of organizations.

It took a lot of discussion and patience to shape the broad coalition into a trusting, cooperating, enduring alliance. Only a few countries actually waged the campaign as a broad alliance. In some countries, it was just a few of them, sometimes cooperating, sometimes each carrying out their own campaign to promote Right2Water. In Scandinavia, water

Table 12.2: *Organizations of Right2Water by type*

	Organizations by category (%)
Union	47
Water Movement	16
Global Justice (including development cooperation)	12
Environment	7
Social/societal	5
Water Supply (public)	3
Political	2
Women	2
Consumer	1
Democracy	1
Other	3

Source: Based on Lesske, 2015.

privatization was not an issue, whereas it was a big issue in Southern Europe. In Western and Eastern Europe, it was an issue only for "water politics" insiders and the context was totally different because Western European countries had a history of market economy and social movements' protests against market economy and privatization excesses, whereas Eastern European countries were new European-market accession countries, with young social movements. In the UK, the debate was about re-nationalization and in France about re-municipalizing water services, since in these countries, privatization had already taken place long ago. On the contrary, in Romania and Bulgaria the issue was about access to and quality of water services, and how that could be best provided.

12.3.2 Building the Campaign and Strategy

The campaign was initiated by EPSU and its board, providing staff and finance to set up the ECI (EPSU, 2009). Decisions on how to campaign were made by the supporting organizations based in Brussels. However, the actual campaigning work needed to be carried out at national and local levels. The EPSU coordinators tried to steer a European campaign from Brussels, but could do no more than support and motivate local campaigners in different countries and facilitate their activities by providing campaign materials and information. An important role in the campaign was for social media.

To reinforce the broad range of supporting organizations, the Right2Water campaigners recruited "ambassadors" in all countries, seeking well-known people's support to give an extra stimulus to people to sign the ECI. Sixty-seven ambassadors made statements in support of Right2Water. Most of them had a link with "water," but some were simply well-known in their country as, for example, a singer, TV-personality or sportsman/woman.

The European coordination team, consisting of people from the European federations, decided to focus on a few countries where mobilizing power was thought to be strong, triggering support and campaigning for three to four months. After that, the focus could shift to other countries, assuming that some basic preconditions had to be present or arranged in every country. Most importantly: a national campaign team should be in place; several organizations would need to be capable of mobilizing people; there had to be a public-policy interest in water and sanitation; and basic public awareness about the European Citizens Initiative's objective and urgency. This last point, in particular, proved to be a major obstacle in mobilizing people and motivating them to sign. High hopes were on Italy, where the Italian Water Movement had organized a referendum against privatization of water in 2011 in which 26 million Italians had voted against privatization. In Italy, the privatization of common goods was considered to be a flight from democracy. Water served as an exponent: "write water but read democracy," was the slogan of the Italian Water Movement (Fattori, 2011).

The strategy did not fully work out as expected. The impression was that it would be easy to achieve one million signatures, but in October 2012, after two months of campaigning, it became clear that all "low-hanging fruit" had been picked. Few people in Europe were aware of the water-privatization problem and the ECI was unknown to most of them.

The first weeks' signatories were the supporting organizations' core activists and the movement did not reach outsiders. Campaign coordinators had to find new ways to reach out to people outside their inner circles. Even in countries where a high turnout was expected, such as Italy, the actual turnout disappointed the campaigners.

The language problem was the first obstacle to be overcome: the website and newsletters made in Brussels needed to be translated in as many languages as possible, in accordance with what was manageable and affordable. Another problem was that the European Commission set up the European Citizens' Initiative mechanism in such a way that it was not possible to campaign through online tools (such as "Avaaz"), to ensure that potential signatories really had to make a personal effort, and prevent bots from automatically generating signatures (EC, 2011). The Right2Water campaigners were the first to experience that they had to explain first "what an ECI was"; second, "how it worked"; and thirdly why they were asking people to sign. This made some campaigners cry out that "it is taking me half an hour to get a signature from one person!" It was clear that, at such a pace, even with a thousand volunteers the campaign would not achieve one million signatures within one year. Coordinators realized they needed to increase social media usage and make the campaign more appealing to citizens.

12.3.3 Different National Realities: Diversity in National Water Movements

European countries were facing different realities in water services and diversity in water movements and activists. In Scandinavia, activism was nearly absent, whereas it was strongly present in Italy, and to a lesser extent in Spain, France, Belgium and the UK. The existing water movement was too small to carry a European campaign. Trade unions were confident about their power, but overestimated their members' support. The challenge was to bring different groups and social movements together and to cooperate across countries and different backgrounds.

The Italian Water Movement (IWM) was expected to spearhead Right2Water. All the conditions were favorable: awareness on water, campaigning capacities, and support for participatory democracy. However, the IWM faced "campaign fatigue" among its members, and could not achieve the mass mobilization for Right2Water that they had for the 2011 referendum. When the referendum failed to yield the expected policy change in Italy, people lost confidence in referendum mechanisms, such as the ECI. Nevertheless, Tomasso Fattori, an IWM leader, stated: "Aside from water, there is another fundamental element which connects the Italian, Berlin and Madrid referenda with the ECI, and that is democracy itself. All of this is about putting democracy before corporate interests and financial markets; the right to water and democracy are closely connected" (Fattori, 2013: 119).

In November 2012, a large social movement gathering in Florence ("Firenze 10+10") should have boosted the ECI Right2Water. Many Firenze 10+10 organizers had been deeply involved in the Italian movement to defend and develop water as a common good, but attention to Right2Water was low. This did not change until summer 2013. Italian trade union FP-CGIL managed to gain the support of three well-known Italians (Stefano Rodota,

Lella Costa, and Ascanio Celestini) for a video clip supporting Right2Water (L'acqua, 2013). This finally gave the Italian campaign a jump to surpass the quorum, following 11 countries that had already passed earlier.

In Germany, the Ver.di trade union took its role seriously as national campaign coordinator and was the first union to build an alliance with national and local groups. It had a campaign plan, made campaign materials and developed a well-functioning website with text in German. It was the first to realize that the water movement needed to be extended to reach people outside activists' circles in order to become successful. Ver.di connected trade unionists to water, social, and environmental activists and got them to join forces (Conrad, 2014).

When a German TV journalist questioned EU Commissioner Barnier about his proposal for a Concessions Directive and the potential conflict with Right2Water, the campaign gained momentum (ARD Monitor: "Geheimoperation Wasser" on December 18, 2012). The Commissioner was not aware of Right2Water and denied that his proposal would favor private companies, but the investigative journalists attracted important attention in Germany. A conflict was born. A TV show by comedian Erwin Pelzig gave the campaign a boost (see Box 12.2).

Box 12.2 The Role of a Comedian in the German Right2Water Campaign

The big success in Germany was related to media attention, accumulating in an explosion on social media after comedian Erwin Pelzig mentioned Right2Water on his TV show "Neues aus der Anstalt" on January 22, 2013.

Pelzig said that French EU Commissioner Barnier had proposed to privatize water services in Europe and ridiculed his proposal: "water is life … it's worrying if Europe wants to leave water to the market or to profit-oriented enterprises. Fortunately," he said, "there is a European Citizens' Initiative trying to prevent water privatisation, which needs one million signatures." At the end of his sketch he showed the Right2Water website in a very clever way by saying that he was not allowed to advertise for a website, but that he was allowed to show a (French) car on which he put a card with the name of the website. The sketch was very popular and went viral in the days after his show. Many people in Germany volunteered to support Right2water and campaigned after the TV show.

From that moment it went like a forest fire across Germany. The alliance managed to leverage their work and reached out to people in all parts and sectors of the country and across the entire political spectrum.

The German water sector was largely managed by municipal companies ("Stadtwerke") and Berlin had just endured a bad experience in their water concession with Veolia. The multinational corporation had signed a secret agreement with the city that guaranteed their profit. Only after a referendum in which inhabitants demanded the agreement's disclosure, did it become clear that the company had boosted its profits from increased prices without investing in improving service. The Stadtwerke and the German population considered the Concessions Directive a dangerous threat, because it could lead to take-overs by the same kind of profit-driven multinationals (Deinlein, 2014). German and Austrian media approached Right2Water and put the campaign in the spotlights.

In Spain, the Platform against privatizing Canal Isabel II and the 15-M (anti-austerity) Movement organized a popular consultation in Madrid region (March 2012) to vote on privatizing Canal Isabel II, with over 160,000 people voting against privatization. Despite this, Spanish organizations supporting Right2Water did not gain momentum until spring 2013. Different groups found each other and started a new movement, *La Marea Azul,* to defend public water-service management. This movement included the *indignados*, a protest movement that had arisen since the May 2012 elections, linked local struggles in Spain to European policies, and managed to reach out to a broader group of people.

In Greece, in spring 2011, Thessaloniki's protest movement organized a mass mobilization: well over half a million people took to the streets to defy austerity and demand democracy, opposing the push for privatization (see also Box 12.1). The water workers of EYATh (Thessaloniki Public Water Company) together with a citizens group, set up "Initiative 136" (K136) to get water into the people's hands. The idea of K136 was that every water user would buy a non-transferable share for 136 euros. This was the result of dividing the 60 million euros for which the company was to be put on the stock market by the number of users (Wainwright, 2013). In spring 2013, this was followed by setting up SOSteTONERO (SOS for WATER) by the same water workers to campaign against privatizing the water company. They made the link between Right2Water and their fight in Greece. They showed it was necessary to combat privatization both locally and at the EU level and gave momentum to the ECI. In 2014, over 50 citizen groups – thousands of people – supported the struggle to make water a public good (Steinfort, 2014).

Organizations supporting Right2Water in France failed to reach the minimum signatures needed, even with help from best-selling writer Marc Levy and prominent politicians. People outside the French water movement did not see the Concessions Directive as a problem, nor did they have a problem with the water service as it was run.

In the UK, many people had frustrations about privatized water services and there was debate about renationalizing water companies. However, trade unions did not manage to mobilize this frustration and or take the initiative to campaign, because they doubted it would turn back UK privatizations. Other civil society organizations were willing to support Right2Water, but waited for trade unions to take the lead.

Attention for Right2Water in the media and from the general public varied widely from country to country. Remarkably, no relation was found between problems in water services and support for the ECI. Whereas problems in water supply were serious in Romania and Bulgaria (also related to privatization), there was very little support. It remains unclear whether this had to do with a lack of interest, a lack of involvement in EU affairs, or a lack of capacity to campaign. In Romania, the country with the poorest water-service quality, nobody seemed interested in the ECI, according to a trade-union representative. As a result, no campaign was launched.

12.3.4 The Forces Countering the Campaign

In France, the two big multinationals in water services viewed the ECI as being against their interests. They pressured French politicians and civil society organizations to not support Right2Water. These multinationals warned their own personnel that a signature for Right2Water would be an act against their company, making workers and unionists afraid of supporting Right2Water openly. They tried to influence EU Commissioner Barnier to maintain the original proposal for the Concessions Directive, to open the door for competition in water services. The multinationals' counter-campaign (after Right2Water achieved over one million signatures) claimed that they were the actual promoters of the human right to water and sanitation, and that Right2water was fueled by a German public lobby (Aquafed, 2013). It proved to be too late to keep the Directive in its original form; water services were excluded from its contents.

In the European Parliament, Right2Water campaigners managed to ensure and maintain broad support during the campaigns' entire period. Although most support was from the left wing, there was significant support among right-wing groups as well. This was confirmed in the own-initiative report by the European Parliament that followed the European Commission's response. The EP accepted a resolution (by 363 to 96 votes) that demanded the European Commission reconsider its response and take more measures to implement the human right to water and sanitation in European law (EP, 2014).

12.3.5 Outcomes of the Campaign

In March 2014, the European Commission responded to the ECI organizers, as Right2Water met with all the conditions for an ECI (see number of signatures per country in Table 12.1). In its response, the Commission recognized "the importance of water as a public good of fundamental value to all Union citizens" (EC, 2014: 5). This was a recognition of what Right2Water had campaigned for and a victory for the movement. However, the response was not what the campaigners had hoped for. The only concrete result was to exclude water from the Concessions Directive. This was decided in June 2013, when Right2Water had already collected over one million signatures and the European Commission felt the urgency to respond immediately to the first ECI that reached this milestone. The official response came in 2014, six months after the end of the one-year campaign. The Commission stated that many of the suggestions made by Right2Water were already part of their policy and that it would not change, amend or propose any legislation (EC, 2014).

Reflecting on the effects that Right2Water had on EU and national policies, there are several achievements. In many countries, it reinforced public water companies' position. Water laws were adapted in Lithuania, Slovakia, Spain, and Greece to guarantee public control and access, affordability, and good-quality water to all inhabitants. Civil participation in water-policy debates increased as many people responded to the EU consultation of the Drinking Water Directive (6,000 compared to 500 in previous and similar consultations), that the Commission held in response to Right2Water (Ecorys, 2015).

Besides the political effect and policy change in EU water policy, the most important achievement was in raising awareness among the European public and members of the European Parliament about water and water policy. Before the ECI, many people did not realize how water services were organized in their country and were not aware of the human right to water and sanitation; the concept of public or common goods; the EU role in water policies; and the EU role in privatizing water services. New alliances were formed that gave new energy and momentum to local struggles against water-service privatization. Right2Water gave the European water movements a political voice at diverse, interlinked scales.

12.4 Discussion and Conclusions

The Right2Water campaign materialized as a pluriform social movement that resonated with the idea that water services are essential services for all people, which must be provided without discrimination, to all inhabitants. The movement sustained that market mechanisms increase the gap between rich and poor in Europe and lead to better services for those who can pay for them and poorer-quality services for those who cannot: markets inherently consolidate inequalities. Thus, public control over water services is essential to ensure availability, affordability, quality, and access for all; i.e. to ensure the human right to water and sanitation. When privatized, control would be handed over to profit-oriented companies that must ensure financial gains, not human rights.

Although the Commission stated several times that it had no intention to privatize water services, its Concessions Proposal was certainly a step towards privatization and opened the door for corporations to demand access to a local water market. Movements and grassroots organizations that had campaigned for Right2Water established new networks and were successful in many European countries in raising awareness among many people about the risks of privatization of water utilities.

When mobilizing for the human right to water and non-privatization of water services, it proved difficult to campaign Europe-wide, so it was necessary to adapt to national and local situations. The ECI Right2Water was the sum of 27 national campaigns, not a single European campaign. Each country had to analyze water policies, people's concerns and active social movements. Available capacity to do this largely determined campaign success. This capacity was lacking in countries such as the UK or Romania, where circumstances were originally seen as favorable for the campaign. The ECI contributed greatly to awareness-raising on water and sanitation as human rights and helped advance a social movement for water justice in Europe and in several member states, such as Spain with *#IniciativAgua2015*.

Interestingly, the campaign raised the largest support in Germany, which had no problem with its water sector, but where people mobilized when they found their water service to be under threat. This was also the case in Austria, Greece, Slovenia, and Slovakia, where people signed to affirm that they wanted to keep local, public control over their water services. The movement was strongest in countries where it linked European policies and local consequences, by explaining the problem and how it affects people in their daily life. In Germany, a comedian sparked a mass mobilization in a few months, whereas in other countries it took lengthy campaigning to raise awareness on the conflict of interests between market policies and human rights. It is likely that the impact would have been less strong if the movement were less diverse. Previous struggles against water privatization remained unknown to the general European public and limited to water justice activists. This time, by engaging trade unions, environmentalists, anti-poverty campaigners and many other civil-society groups, it broke out of this inner circle of water activists and reach an unprecedented audience, focusing on the issue of water-service provision and the threat of privatization.

References

Achterhuis, H., Boelens, R. and Zwarteveen, M. (2010). Water property relations and modern policy regimes: Neoliberal utopia and the disempowerment of collective action. In R. Boelens, D. Getches and A. Guevara-Gil (eds.), *Out of the Mainstream: Water Rights, Politics and Identity*, London and Washington, DC: Earthscan, pp. 27–55.

Aquafed. (2013). "Concessions directive: European Commissioner renounces transparency and equity in public water services to please German public lobbies." Press release June 26, 2013, http://pr.euractiv.com/pr/concessions-directive-european-commissioner-renounces-transparency-and-equity-public-water.

Bakker, K. (2003). *An Uncooperative Commodity: Privatizing Water in England and Wales*. Oxford: Oxford University Press.

Bakker, K. (2007). The "commons" versus the "commodity": Alter-globalization, anti-privatization and the human right to water in the Global South. *Antipode*, 39(3), 430–55.

Bakker, K. (2010). *Privatizing Water: Governance Failure and the World's Urban Water Crisis*. Ithaca, NY: Cornell University Press.

Balanya, B. (2005). Empowering public water: Ways forward. In B. Balanya, B. Brennan, O. Hoedeman, S. Kishimoto, and P. Terhorst (eds.), *Reclaiming Public Water: Achievements, Struggles and Visions from Around the World*. Amsterdam: Transnational Institute, pp. 247–75.

Bebbington, A., Humphreys-Bebbington, D. and Bury, J. (2010). Federating and defending: Water, territory and extraction in the Andes. In R. Boelens, D. H. Getches and J. A. Guevara-Gil (eds.), *Out of the Mainstream: Water Rights, Politics and Identity*. London: Earthscan, pp. 307–28.

Berge, J. van den. (2014). Water and sanitation are a human right! Water is a public good, not a commodity! In C. Berg and J. Thompson (eds.), *An ECI that Works: Learning from the First Two Years of the European Citizens' Initiative*. Alfter, Germany: The ECI Campaign, pp. 19–24.

Bieler, A. (2015). *Mobilising for Change: The First Successful European Citizens' Initiative "Water is a Human Right!"* Paper presented at the ETUI Monthly Forum, Brussels/Belgium, January 22, 2015.

Boelens, R. (2015). *Water, Power and Identity: The Cultural Politics of Water in the Andes*. London: Earthscan, Routledge.

Boelens, R. and Vos, J. (2012). The danger of naturalizing water policy concepts: Water productivity and efficiency discourses from field irrigation to virtual water trade. *Journal of Agricultural Water Management*, 108, 16–26.

Boelens, R. and Zwarteveen, M. (2005). Prices and politics in Andean water reforms. *Development and Change*, 36(4), 735–58.

Boelens, R., Hoogesteger, J., Swyngedouw, E., Vos, J. and Wester, P. (2016). Hydrosocial territories: A political ecology perspective. *Water International*, 41(1), 1–14.

CEO. (2012). "EU Commission forces crisis-hit countries to privatize water." Press release October 17, http://corporateeurope.org/pressreleases/2012/eu-commission-forces-crisis-hit-countries-privatise-water

Conrad, C. (2014). Demokratischer Einfluss auf EU-Ebene? Da geht was! *Forum Wissenschaft*, 31(1), 34–37.

Deinlein, W. (2014). *Kompendium Trinkwasser. Zur Europäischer Bürgerinitiative Right2Water und der EU-Konzessionsrichtlinie*. Karlsruhe: Stadtwerke Karlsruhe.

Della Porta, D. and Mosca, L. (2005). Global-net for global movements? A network of networks for a movement of movements. *Journal of Public Policy*, 25(01), 165–90.

Ecorys SEE. (2015). *Analysis of the Public Consultation on the Quality of Drinking Water*. Draft report ENV.F.1/FRA/2010/0044, http://ec.europa.eu/environment/consultations/pdf/analysis_drinking_water.pdf.

EEA (European Environmental Agency). (2013). *Assessment of Cost Recovery through Water Pricing*. EEA Technical Report No. 16/2013.

EPSU. (2009). "Water campaign action plan," www.epsu.org/sites/default/files/article/files/Water_Campaign_Action_Plan_EN.pdf.

EPSU. (2011). "Right to water and sanitation must be put into action," www.epsu.org/article/right-water-and-sanitation-must-be-put-action-universal-access-government-responsibility.

European Commission. (2007). *Addressing the Challenge of Water Scarcity and Droughts in the European Union*. Communication from the Commission to the European Parliament and the Council. COM(2007) 414 final.

European Commission. (2009). *Lisbon Treaty*, article 11(4), www.lisbon-treaty.org/wcm/the-lisbon-treaty/treaty-on-european-union-and-comments/title-2-provisions-on-democratic-principles/75-article-11.html.

European Commission. (2011). Regulation (EU) no. 211/2011 of the European Parliament and of the Council of 16 February 2011 on the citizens' initiative (OJ L 65, 11.3.2011).

European Commission. (2012). Communication from the Commission to the European Parliament, the Council, the European Economic and Social Committee and the Committee of the Regions. *A Blueprint to Safeguard Europe's Water Resources*. COM(2012) 673 final.

European Commission. (2013). *Exclusion of Water from the Concessions Directive*, Statement by Commissioner Barnier in response to Right2water, http://ec.europa.eu/archives/commission_2010-2014/barnier/headlines/speeches/2013/06/20130621_en.html.

European Commission. (2014). Communication from the Commission on the European Citizens' Initiative "Water and sanitation are a human right! Water is a public good, not a commodity!" COM(2014) 177 final.

European Parliament. (2014). *Follow Up to the European Citizens' Initiative Right2Water*. European Parliament resolution of 8 September 2015 on the follow-up to the European Citizens' Initiative Right2Water (2014/2239(INI)). P8_TA(2015)0294.

Fattori, T. (2011). Fluid democracy: The Italian water revolution. *Transform*, 09/2011, 99–111, www.transform-network.net/en/yearbook/journal-092011.html.

Fattori, T. (2013). The European citizens' initiative on water and 'austeritarian' post-democracy. *Transform* 13/2013, pp. 116–22, www.transform-network.net/journal/issue-132013/news/detail/Journal/the-european-citizens-initiative-on-water-and-austeritarian-post-democracy.html.

Friedman, M. (1962). *Capitalism and Freedom*. Chicago: University of Chicago Press.
Hall, D. and Lobina, E. (2005). *The Relative Efficiency of Public and Private Sector Water*. London: Public Services International Research Unit (PSIRU) and University of Greenwich.
Hall, D. and Lobina, E. (2008). *Water Privatisation*. London: Public Services International Research Unit (PSIRU).
Hall, D. and Lobina, E. (2012a). *Water Companies and Trends in Europe*. Brussels and London: Public Services International Research Unit (PSIRU) and EPSU.
Hall, D. and Lobina, E. (2012b). *Conflicts, Companies, Human Rights and Water: A Critical Review of Local Corporate Practices and Global Corporate Initiatives*. London: Public Services International Research Unit.
Harvey, D. (1996). *Justice, Nature, and the Geography of Difference*. Oxford: Blackwell.
Hayek, F. A. (1944). *The Road to Serfdom*. London: George Routledge.
Hendriks, J. (1998). Water as private property: Notes on the case of Chile. In R. Boelens and G. Dávila (eds.), *Searching for Equity*. Assen, The Netherlands: Van Gorcum, pp. 297–310.
Hoogesteger, J. and Verzijl, A. (2015). Grassroots scalar politics: Insights from peasant water struggles in the Ecuadorian and Peruvian Andes. *Geoforum*, 62, 13–23.
Hoogesteger, J., Baud, M. and Boelens, R. (2016). Territorial pluralism: Water users' multi-scalar struggles against state ordering in Ecuador's highlands. *Water International*, 41(1), 91–106.
James, W. (1977[1909]). *A Pluralistic Universe*. Cambridge, MA: Harvard University Press.
L'acqua. (2013). "ICE per l'acqua diritto umano in Europa – Spot con Stefano Rodotà, Lella Costa, Ascanio Celestini," www.youtube.com/watch?v=2b4NmB7IWaI&feature=youtu.be.
Lapavitsas, C. (ed.) (2012). *Crisis in the Eurozone*. London: Verso.
Lawson, T. (2015). "Reversing the tide: Cities and countries are rebelling against water privatization, and winning," www.truth-out.org/news/item/32963-reversing-the-tide-cities-and-countries-are-rebelling-against-water-privatization-and-winning.
Lesske, K. (2015). "The role of civil society organizations in promoting participation and accountability in transnational water politics: A case study in Europe." Freiburg, MA, Thesis Global Studies Programme, Albert-Ludwigs-University Freiburg.
Limon, R. (2013). La presión social saca el agua de la directiva europea de privatizaciones. *El País*, July 1, 2013, http://sociedad.elpais.com/sociedad/2013/07/01/actualidad/1372702820_342386.html.
Limon, R. (2014). Organizaciones ciudadanas y partidos lanzan un pacto por el agua pública. *El País*, August 23, 2014, http://ccaa.elpais.com/ccaa/2014/08/23/andalucia/1408786978_205688.html.
Lobina, E., Kishimoto, S. and Petitjean, O. (2014). *Here to Stay: Water Remunicipalisation as a Global Trend*. Public Services International Research Unit (PSIRU), Multinational Observatory and Transnational Institute (TNI), London, Paris, Amsterdam, www.tni.org/files/download/heretostay-en.pdf.
Mirosa, O. and Harris, L. (2012). Human right to water: Contemporary challenges and contours of a global debate. *Antipode*, 44(3), 932–49.
Parks, L. (2014). *Framing in the Right2Water European Citizens' Initiative*. Paper prepared for presentation at the ECPR general conference, September 3–6, 2014, Glasgow.
Right2Water. (2016). "About Right2Water," www.right2water.eu/about-our-campaign.
Right2Water. (2016). Results per country, www.right2water.eu//results?lang.

Schultz, J. (2001). "Bolivia's war over water: The dispatches in full." The Democracy Center On-Line, http://democracyctr.org/bolivia/investigations/bolivia-investigations-the-water-revolt/bolivias-war-over-water/.

Schlosberg, D. (2004). Reconceiving environmental justice: Global movements and political theories. *Environmental Politics*, 13(3), 517–40.

Seters, P. van, and James, P. (2014). Global social movements and global civil society: A critical overview. In P. James and P. van Seters (eds.), *Globalization and Politics*, Vol II: *Global Social Movements and Global Civil Society*. Los Angeles: SAGE, pp. vii–xxx.

Spronk, S. (2007). Roots of resistance to urban water privatization in Bolivia: The "new working class," the crisis of neoliberalism, and public services. *International Labor and Working-Class History*, 71(1), 8–28.

Steinfort, L. (2014). "Thessaloniki, Greece: Struggling against water privatization in times of crisis," www.tni.org/en/article/thessaloniki-greece-struggling-against-water-privatisation-times-crisis.

Swyngedouw, E. (2004). Globalisation or "glocalisation"? Networks, territories and rescaling. *Cambridge Review of International Affairs*, 17(1), 25–48.

Tinson, A. and Kenway, P. (2013). *The Water Industry: A Case to Answer*. London: New Policy Institute and UNISON.

Wainwright, H. (2013). "Resist and transform: The struggle for water in Greece," www.redpepper.org.uk/tapping-the-resistance-in-greece/.

Wenz, P. (1988). *Environmental Justice*. Albany, NY: State University of New York Press.

Zacune, J. (2013). *Privatising Europe: Using the Crisis to Entrench Neoliberalism*. Amsterdam: Transnational Institute.

Zwarteveen, M. and Boelens, R. (2014). Defining, researching and struggling for water justice: Some conceptual building blocks for research and action. *Water International*, 39(2), 143–58.

13

Everyday Water Injustice and the Politics of Accommodation

Frances Cleaver

13.1 Domino-Centric Perspectives on Water Justice

This chapter takes as its starting point the call from activists and academics for water justice and for better understanding of how this can be achieved. I argue that, though important and often inspiring, such critical studies often over-focus on struggle and resistance at the expense of understanding the more common ways in which water injustices are routinized, accepted and reproduced by those who suffer from them.

Water justice literature aims to understand the dimensions of injustice and inequality that underpin a perceived world water crisis, and to identify the possibility of alternative water orders. The contributions are varied, examples including: manifestos (Petrella, 2001; Santa Cruz, 2014); moves to theorize water justice and governance (Perreault, 2014); studies of struggles for the right to water (Sultana and Loftus, 2012) and of water "wars" (Graham et al., 2013; Perera, 2012); explorations of anti-privatization movements (Spronk, 2007); water relations viewed through an environmental-justice lens (Mehta et al., 2014; McDonald, 2004); analyses of gendered contestations over water access (Delgado and Zwarteveen, 2007); and examinations of embedding unjust water distributions in culture and politics (Boelens, 2014).

This work is important in our understanding of the politics of water allocations and in scrutinizing policy ideas that equity and water rights can be furthered through markets or through technical and managerial processes (Joy et al., 2014). Often located in political ecology or critical geography approaches, water justice studies advocate context-specific understandings of justice, manifestations of injustice and the "always-contested" nature of water allocations (Zwarteveen and Boelens, 2014). Much of this literature focuses on explicit instances of struggle, and references to protest, opposition, resistance, mobilizations, contestation and conflict abound. Taxonomies of water conflict are offered (Rodríguez-Labajos and Martínez-Alier, 2015) and dimensions of contestation usefully anatomized (Zwarteveen and Boelens, 2014).

There is a strong theme in the literature linking the defense of local water values to achieving water justice, and linking social mobilizations around water to the advance of democracy (Hoogesteger and Verzijl, 2015; Loftus, 2014; Bond and Dugard, 2008). The Santa Cruz Declaration (2014: 249) summarizes this well: "Actions and research for more

---

This chapter was written whilst working on the UPGRO research project "Hidden crisis: Unravelling current failures for future success in rural groundwater supply," funded by the UK's Department for International Development, the Natural Environment Research Council and the Economic and Social Research Council. That support is gratefully acknowledged.

justice need to be explicitly connected to and grounded in people's experiences of injustice and their strategies and struggles to contest and remedy it."

There are two interrelated problems with the tendency to highlight instances of struggle, resistance and conflict. First, the water justice literature can be read as rather dominocentric (Ferguson, 2015) – unintentionally perpetuating a domination/injustice versus resistance/justice binary. The emphasis is often on presenting evidence of powerful actors working to create inequitable allocations and then on looking for the ways in which people oppose or contest these from below. In critical studies, water governance micro-politics usually encompasses the possibility to resist and transform (Sultana and Loftus, 2012). This resistance focus is not confined to water research and seemingly has an enduring academic appeal. Ortner (1995) made similar observations about peasant resistance studies in anthropology, as did Hall *et al.* (2015) about literature on land grabbing and political reactions from below. Hall and colleagues point out that many of the first land-grabbing studies (in the tradition of agrarian-change scholarship) sought out evidence of resistance from local communities disadvantaged by land deals imposed upon them. The authors suggest that responses to land grabbing have in fact been far more varied than this, ranging from organized and everyday resistance to acceptance or even demands to be incorporated into (apparently unjust) land deals. Similarly, I argue here that we need to pay more attention to why many people do not resist unjust water relations, and indeed may accept or even perpetuate them in their relations with powerful people. Understanding such dynamics are important as they contribute to the intractability of everyday water injustices.

Secondly, drawing on an argument made by Brekhus (1998) about sociological research, I suggest that the water justice literature over-focuses on the politically salient, ontologically uncommon features of social life (overt resistance, water wars, even everyday resistance), whilst overlooking, or downplaying, the unmarked mass of non-contested water interactions in social life. Surely many interactions over water are not instances of struggle or resistance but rather of small-scale cooperation, accommodation, and attempts to be incorporated into unequal distribution systems. These everyday practices of compliance in everyday water politics require more exploration and explanation. Brekhus remarks that one problem with the unmarked territory of social life is that it is not sufficiently differentiated in research studies. Explaining similar tendencies, Ortner suggests that this lack of differentiation of everyday interactions in peasant studies is attributable to a sort of "ethnographic thinness." She argues: "resistance studies are thin because they are ethnographically thin: thin on the internal politics of dominated groups, thin on the cultural richness of those groups, thin on the subjectivity – the intentions, desires, fears, projects – of the actors engaged in these dramas" (Ortner, 1995: 190). Ortner's argument is that we should be equally concerned with a range of responses, including the ambiguous aspects of resistance and everyday acts of collaboration and accommodation. For water-justice studies, this would involve explicitly foregrounding the politically unnoticed and taken-for-granted elements of everyday water politics, encompassing the routinized as well as the exceptional. Generating richer understandings of the whole spectrum of everyday water practices, ranging from cooperation to conflict, and all the grey areas of practice in between, would help us to examine whether they produce positive change or perpetuate inequalities.

Various reasons can be suggested for this bias towards resistance. Hall *et al.* identify three different conceptual models which influence such a focus: classic collective action approaches which assume that individuals will not cooperate unless induced to do so; the legacy of Marxist thinking which presumes that shared grievances and experiences of oppression will lead to common political projects of opposition; and heterodox social movement scholarship which, among other explanations, suggests that affective ties produce collective mobilization and struggle. To these explanations, we could add the allure of the extraordinary, as suggested by Brekhus, and the aim of many critical water researchers to actively contribute to water justice by challenging dominant discourses and identifying counter-hegemonic alternatives (Gramsci, 1971; Joll, 1977; Sneddon, 2013; Rodríguez-Labajos and Martínez-Alier, 2015). In this way, large, explicitly political struggles (e.g. against dam-building) receive more attention than do ongoing ordinary efforts to gain water access in social environments riven with intractable inequalities (Joshi, 2015).

13.2 The Tenacity of Everyday Injustices

In this chapter, I concentrate on the everyday politics of water access and why so much of this unfolds as business as usual – with many injustices accepted and accommodated. The sort of injustices I focus on here manifest in local access and allocation and we see numerous examples of them in the watery literature. Examples (elaborated in Cleaver, 2012) might include:

- Only a proportion of community members actually contribute to maintenance fees for the waterpoint – these being the poorer people. Wealthier villagers are rarely sanctioned for non-payment.
- Women waiting hours for men to water cattle before they can fetch water for domestic purposes, even after an NGO project mobilizes women and promotes their role in village public life and water management.
- The exclusion of pastoralists from irrigation management structures (or their inclusion on unequal terms), even though they are also irrigation farmers. The apparent acceptance, even preference, by pastoralists of this form of exclusion.
- Community preference for flat-rate water fees, even where there are large socio-economic variations among households and the quantities of water they use. The poorest, who benefit least, nonetheless argue in favor of a flat-rate tariff.
- Preferential preservation of wealthier people's water rights in crisis situations (such as drought).

The acceptance of such inequalities and injustices in water has been explained by the workings of hegemonic power relations and discourses (Sneddon, 2013). This explanation is compelling, but hegemony is rather a broad-brush category which tends to obscure the importance of quite different, if interrelated factors. For the purposes of this chapter, I have identified these separately as hegemony, pragmatism, connectedness and moral ecological rationality, though they overlap and are difficult to disentangle. Hegemony refers to the ways in which power and the interests of dominant groups become embedded in norms, institutions and discourses and so are taken for granted. Pragmatism concerns the explicit analysis that disadvantaged people make of the workings of power, the high costs of challenging it, and their livelihood dependence on the powerful. Connectedness refers to the imperative for social connection, which is not simply about securing material benefit but about wellbeing, connection to community, and meeting emotional needs. Finally, moral ecological rationalities concern cultural understanding of the links among people, water, and the environment; the worldviews which link proper forms of behavior and social order to the flourishing of human and non-human entities. It could be argued that the first point (hegemony) explains the other three (pragmatism, connectedness and moral rationalities). However, I find it useful to distinguish these points, in an attempt to anatomize the everyday politics of water relations, and to better understand how the parts articulate.

13.3 Hegemony

Accommodation of injustices can be explained by the invisible workings of power and the ways in which it is constituted through ideas and particular forms of knowledge. In his essay exploring such issues, Sneddon (2013) claims that many studies of water injustices adopt a Gramscian idea of hegemony as the wide variety of practices and ideologies associated with the power to influence decisions and thought patterns. In this view, pursuing dominant groups' interests becomes normalized, so they are seen as common sense, even by those they disadvantage (Gramsci, 1971; Joll, 1977). Applied to water studies, hegemonic interests privilege certain ideas about water distribution and water values, shaping the ways people consent to be governed. Sultana and Loftus (2012) deploy such concepts to draw attention to the everyday processes where individuals and groups internalize certain concepts and act in ways inimical to their interests, subconsciously adapting their preferences and values to mirror those of the powerful. Such powerful people have the ability to set agendas, to define the very parameters of social acceptability and debate. Writing of water justice, Zwarteveen and Boelens (2014: 48) cite Foucault's ideas that discourses subtly govern the ways that it is possible to talk about something and so make it difficult to think and act differently. The invisible nature of power helps explain why disadvantaged people do not necessarily see unequal environmental relations as unfair or unjust (Walker, 2014).

A rich empirical case of power's invisible workings is provided by a study of village contributions to rainwater-harvesting infrastructure in India (Cochran and Ray, 2008). The rainwater harvesting structure, which supplies wells with water for livestock and irrigation, requires labor and maintenance payments from villagers. Cochran and Ray attempt to explain why poor people, who neither own livestock nor land requiring water, accept and even promote the idea of equal (flat-rate) payments to the water fund. One explanation the authors offer is that poor people perceive the interests of the village (and of the wealthier villagers) as their own. They do not recognize the particular self-interests embedded in communal arrangements, and it is this misrecognition that allows the unequal social order and associated water practices to continue. This example strongly supports Bourdieu's suggestion that "maintaining power through common ideologies depends on the use of power not being recognized as such, but as legitimate claims on the services of others" (Bourdieu, 1977: 164).

13.4 Pragmatism

However, not all acceptance of unjust relationships can be explained by the unconscious internalization of hegemonic discourses. People *also* practice accommodation because they are painfully aware of the structurally unequal dimensions of their situation. This awareness includes the understanding that such relations are hard and costly to change, and that it is in their overall livelihood interests to maintain rather than challenge them. Such experiences are illustrated in studies of poor households faced with water competition and conflict. For example, in their studies of water competition in Vietnam, Bolivia and Zambia, Funder *et al.* (2012), find a pattern of poor households deploying pragmatic risk avoidance strategies in relation to wealthier households. In particular, the poor shift to less desirable water sources, because the costs of confronting powerful actors over access to better sources are too great. In doing this, the authors suggest, they are prioritizing one aspect of livelihood strategy (good relations with wealthy neighbors) over another (equal access to water).

According to James Ferguson (2015), people may also pragmatically seek out unequal relationships as their best hope of some kind of material redistribution. A great deal of poor people's efforts may go into striking up, asserting or reasserting social relationships, even highly dependent ones, with those better-off than themselves. This deliberately sought social inequality may be preferred as a way of securing access to resources over asocial access routes through resistance, markets, or institutions. Pragmatic strategies will be shaped by the particular social position of the disadvantaged water user; perceived commonality of interests among the marginalized cannot be assumed. In a case of Zambian borehole management (Funder *et al.*, 2012), the community's weaker households pursued different strategies of pragmatic risk-avoidance. In the village, wealthy cattle-owners dominated borehole use and management. Poor men who worked as herders for the cattle-owners could collect water while they watered their employers' cattle, and so did not protest at the unequal arrangements. By acquiescing, they reinforced dominant ideas that cattle should take priority over domestic uses, even to the detriment of their own families. Other households adopted a strategy of discrete lobbying with third parties (such as school teachers, clinic staff) to try to secure more equitable access arrangements, while some – perhaps the more vulnerable – disengaged from the issue by staying away from community meetings where water was discussed and relying on inferior water sources.

Pragmatism may mean that people deliberately choose to further the interests of the powerful. So, for example, they might elect the wealthy or powerful man to chair the water committee, even in the knowledge that he may divert funds and water to his own benefit. Partly this is seen as the rightful order of things, and thus is evidence of the working of a hegemonic social order. However, the wealthier person can also bear the costs of office-holding – he has the material and authoritative resources to make it work, and others can rightfully make claims on him for a share (an unequal share) of the benefits. Such pragmatism may also lead to *not* selecting poor people as representatives, due to a realistic awareness of their material and social limitations. In my work on village water management in Zimbabwe, Tanzania and Malawi, I have heard people saying they don't elect poor people as treasurers of water committees, as they may "eat the cash" and that "it is difficult for a poor person to keep cash in the village." Here they are recognizing that a poor treasurer, holding the cash, may well be compromised by the combination of their own constrained livelihood and their social obligations to others.

Avoiding or bypassing unequal power relations may also be a pragmatic response to unjust water allocations. Anand's (2015) study of how poor people access water by illegally tapping water pipes in Mumbai, India, illustrates this point. He documents how poor Muslim settlers are perceived as unworthy by the engineers, and consequently receive poorer water supplies. Despite their real grievances, they do not forward claims to the engineers; we can presume that they judge that the costs of doing so are too high. These include the opportunity costs of giving up labor time to make the complaint, transport costs, the unlikelihood of having their voices heard, the danger of being branded a troublemaker and losing access to other services and benefits. Instead, the settlers deploy their limited resources to secure water illegally. This could be seen as a pragmatic, imperfect solution to the unjust water situation – one which allows people to access some water *without* having to bear the costs of exercising resistance.

13.5 Connectedness

A focus on the material benefits of obtaining access to water can blind us to the non-material meanings of water relationships. People may seek out patrons, or tie themselves into unequal neighborly relations for *more than* pragmatic reasons of accessing water. Such connectedness also links people into the social fabric (Chabal, 2009; Sayer, 2007); helps to overcome individualism, isolation and insecurity; and helps to sustain the social order and associated world views. Being social really matters beyond immediately instrumental concerns.

Pragmatism and social connectedness are often difficult to disentangle. Zug and Graefe's (2014) study of water-gift relations in an urban neighborhood of Khartoum illustrates both the pragmatic nature of water relationships and their embeddedness in wider social connections and meanings. The concept of water gifting can be used to think through justice issues because gifting emerges as a pragmatic way of securing access, because underlying meanings of fairness and rightful claims and personhood are asserted through the social connections, and because the gift relationship may further unjust allocations. Water sharing is necessary in Khartoum because of the broader framework of water justice – the city water utility does not provide sufficient water to all households. Households that are not connected to the city water supply may purchase water from vendors, or seek gifts of water from their neighbors' taps. Some wealthier households gift water to poorer ones – mostly basic water supplies donated to specific households. The water gifts are often linked to strong social relationships, though the authors claim that the existence of water gifts to strangers and less-liked neighbors shows this is about something more than direct reciprocity. Connections are therefore made and sustained through the water relationship, but this remains unequal in several ways. The authors argue that the gift relationship is predicated on socio-economic heterogeneity, that social rules for water gifts between neighbors are shaped by (unequal) social costs and conditions, and that water gifts reinforce the concept of a rightful share rather than an equal share. Poorer households may seek out these unequal arrangements both for pragmatic reasons, and for the desirability of being more closely connected to their neighbors.

However, there is also contrary evidence in the case that some people try to *avoid* the social costs involved, by shunning the water-gift relationship. This means that potential donors withhold water, despite losing social reputation, and poor people buy more expensive water from commercial vendors, to avoid the transactional costs of the gift relationship. It is clear then that unjust water relationships are embedded in multiple, overlapping social dynamics, and are subject to varying degrees of active facilitation, acceptance, and avoidance. Contestation and resistance are not inevitable, and may take place alongside cooperation and claims for incorporation into unjust relationships. The broader point is that resource use is not separable from the larger set of social transactions in which communities engage and the meaning attributed to them (Cochran and Ray, 2008; Schnegg *et al.*, 2016). It is to the moral dimensions of these social interactions around water that I now turn.

13.6 Moral Ecological Rationality

Understandings of the ways in which moral ideas shape political action from below have long been used to explain resistance to unjust resource allocations. In a classic work on eighteenth-century food riots in England, E.P. Thompson used the moral economy concept to explain how the English crowd's legitimizing notions of entitlement to subsistence, and the establishment's paternalistic obligations to maintain the social order, were deployed to negotiate a social bargain over rights to consumption (Thompson, 1991). In James Scott's usage, the moral economy refers to the ways in which peasants perceive economic justice, exploitation and their definition of tolerable claims made on them by others. For Scott, peasants exercise agency according to a combination of rational reasoning, moral assessment, and ideas of justice shared with others. Adaptations of moral economy have been used to explain water relations and to highlight norms for the right to subsistence and reciprocity in both urban and rural settings (Trawick, 2001; Wutich, 2011).

However, the form of moral economy suggested by Scott has been criticized for overprivileging individual agency at the expense of structural constraint, and for a certain ethnographic thinness (Hyden, 2006; Ortner, 1995). Hyden suggests that evocations of the moral economy are a defensive response to social forces that threaten livelihoods and life worlds, hence the focus on resistance. He suggests, rather, the economy of affection as more real and present in people's everyday lives. This is the network of interactions, between groups of people identified in terms of their relationship to social structure, united by kinship, community, religion, or other affinities. The economy of affection is not simply defensive, but about *maintaining* social relations and social advancement – often effected by the poor seeking out and cultivating those better-placed than themselves. These relationships and the moral norms associated with them constitute the mental framework within which individual action is shaped.

In previous work, I have developed the concept of moral-ecological rationality, in which moral, social, and environmental understandings are linked (Cleaver, 2000). Here I refer to the broader world views or cosmologies within which people's perceptions and actions are framed. Worldviews link ideas about individual and collective behavior, the social order, and connections between the human, natural and spiritual worlds. Concepts of fairness and rightful shares are embedded in such cosmologies, not simply linked to economic or political action. I have illustrated the ways in which moral-ecological rationalities shape people's participation in water governance arrangements, and the tendency for these world views to emphasize cooperation, conflict avoidance, and compromise over public contestation and conflict (Cleaver, 2000). Within the environmental justice movement, such world views, often labeled as indigenous knowledge, are seen as ways of contesting other (often market-based) approaches to water governance. I suggest that they just as often provide the framework within which accommodation of injustice is seen as the proper behavior, to enable the social and natural world to flourish. Such moral rationalities have very positive meanings for people but they also help to embed inequalities, often invisibly sustaining the status quo. This of course relates to the earlier point about hegemony: moralities are related to prevailing patterns of domination, which are legitimized in frameworks of understanding which make them seem natural or desirable (Sayer, 2007; Boelens, 2014).

Let us return to the case of contributions for rainwater harvesting infrastructure in India, discussed earlier. Cochran and Ray (2008) offer three explanations for people supporting a flat-rate payment, despite quite varying ability to pay, and the fact that those with little land and livestock benefit less than those with plenty. They refer to religious duty (the duty to supply water for animals), the *idea* of village unity (as the proper social order) and the ability (in principle, in the future) to claim equal status with other villagers. Cochran and Ray suggest that rather than focusing on the fairness of the benefits achieved from resource management arrangements, it is the *inputs* which are significant to people in this case. By making equal payments and thus notionally sharing in the water, all community members, including the poorer ones, acquired valuable symbolic capital.

We should be wary of seeing such behaviors in individual, instrumentally strategic terms. A justice model based on individual rationality is challenged by Mary Douglas (1987: 124) who claims that "Justice has nothing to do with individual cases – individuals usually offload decisions about justice to institutions." She argues that moral opinions are prepared by institutions and ideas about justice contribute to legitimizing and reinforcing those institutions. She draws on the example of Hindu peasants' adjustments to famine to illustrate the institutionalized ways in which justice works out and define winners and losers from sets of distributive norms.

The application of moral framings which disadvantage the poorest in times of crisis is illustrated in a study of Namibian pastoralists' response to organizing water access in a drought (Schnegg, *et al.*, 2016). In these communities, intentionally just water allocation rules had been introduced by a development project. These benefitted smaller-herd owners by differentially pricing water tariffs between outsiders (often large wealthy cattle owners) and community members. In drought years, however, this rule-based regime was suspended by the pastoralists and replaced by an older model in which cultural assertions of kinship and reciprocity took priority over formal arrangements. Restrictions on access to non-residents were abandoned and a flat rate adopted – benefiting mainly the larger, wealthier herd owners. Here, the authors argue, reciprocity and social relatedness across large scales and a general moral logic of water access prevailed over the introduced rule-based regime that benefitted poorer people. The older, unequal, moral framework for access was perceived as a legitimate, trustworthy arrangement, despite its obviously unequal outcomes.

13.7 Concluding Points

13.7.1 The Tenacity of Unjust Arrangements

The main argument presented here has been that much of everyday water politics involves accepting or actively maintaining inequitable access and skewed benefit distribution. This is due to the invisible workings of power that shape perceptions of interests and fairness; pragmatic approaches to unequal social relationships; a need to be connected; and the moral rationalities that shape ideas about fairness, rightful shares, proper behavior, and social order.

The tenacity of unjust relationships involving water may be reinforced by interconnections among these differing factors and logics. In the case of the case of Namibian pastoralists, the authors suggest that the unfair flat-rate payments come about for a number of reasons: the wealthier community members have greater bargaining power to shift arrangements in their favor; a flat rate is easier to implement and incurs fewer transaction costs than a sliding scale; the likelihood of effective government support for an alternative is slim; and the multiplicity of ties and norms of sharing embed the arrangement in relationships of moral value (Schnegg et al., 2016).

The tendency for governance arrangements (and constructs of what is good and bad) to form around stable/enduring forms of order and authority (Meagher et al., 2014) is strong, which ensures endurance of unjust everyday water politics. People translate such ideas, relationships and governance arrangements through ontological views of the world that tend to emphasize "natural" orders. There is both regularity and variability of social life – but people consciously and unconsciously work hard to maintain regularity – to regularize the irregular. Such ordering is done by calls on tradition, by analogy (to social relations, natural orders, supernatural forces) and helps to legitimize unequal relationships. It is also done by reference to legitimizing authorities – often in the form of existing governance arrangements, powerful people and institutions.

It may seem regressive to focus on these examples of the tenacity of inequalities rather than on the stirring examples of contestation and resistance. But paying attention to everyday manifestations of water injustice does not necessarily mean valorizing or celebrating them. We need this focus in order to understand the 'stickiness' or path-dependence of unjust arrangements – and the ways in which positive change may easily become embedded in existing inequalities.

13.7.2 Scope for Change?

What then are the prospects for change in the direction of water justice? This chapter has established that people might adopt any of a whole spectrum of practices to secure water. How can we better understand when the practices of accommodating injustices shift into more explicit challenge or collective action? What are the changes which disrupt views of the rightful order, which open up space to challenge hegemonic power structures, and through which alternative ideas about fair arrangements are able to gain traction and authoritative legitimacy? There are several possible hints and directions to pursue here.

Under ordinary conditions, the moral ecological framework (the symbolic universe, in Berger and Luckmann's terms, 1966) is self-maintaining and self-legitimizing. However, when problems arise – droughts, economic shocks, generational shifts – then specific articulation, reinforcement, or adjustment of authoritative arrangements is required to maintain a perceived order. The question that arises here is the extent to which changing conditions open up the field for alternative visions, more explicit articulations – changes and challenges to the status quo. And at what point does maintaining a specific aspect of the moral framework become impossible to sustain?

One possible entry point for subjecting unequal water relationships to scrutiny comes from De Herdt and Olivier De Sardhan's (2015) ideas about practical norms. They identify these as existing somewhere between explicit norms (social or official, legitimately accepted, enforceable) and effective practices (numerous and diverse latent regulations – the logic behind actual practices). Key to their conceptualization is that explicit norms may be asserted as the right way of doing things but actual practices reflect other ways. A watery example could be that official legislation for minimum amounts of water *and* common African cultural assumptions of universal access are both explicit norms which are often asserted but patchily implemented. In this case, practical norms may be evident – the various forms of uneven, selective, and distorted application of universal water access principles in which some (wealthier or important people) are allowed preferential access over others (outsiders, or the 'lazy' poor, for example). The question then arises as to what extent both practical and explicit norms work to limit or to enlarge the room for social action.

This line of thinking seems to be in keeping with suggestions that the very plurality of water governance arrangements (of actors, rights, norms and institutions) may provide room to maneuver in terms of shifting the taken-for-granted order of things. "Outsider" policies and normative principles (the right to water, development for all) as well as "insider" principles (water belongs to the people/the Chief/the ancestors, all have a right to make a living) may provide the resources for assembling water governance arrangements, underlying which are different concepts of justness and fairness. Potentially the plurality of these governance resources provides material to transform unjust relationships – through processes of adjustment, re-articulation and blending, in addition to explicit negotiation and contestation.

Responding to the Santa Cruz Declaration (2014), K.J. Roy suggests that certain larger normative principles (such as the human right to water) can help in working with people to develop a socially just agenda. Zwarteveen and Boelens (2014) argue for the need to connect transcendent views of what justice should be to local understandings and articulations. These different perspectives may well provide the material for reinterpreting, shifting or recasting local notions of fairness, and they might provide the authoritative space for questioning unfair local arrangements. We need richer understandings of deeply held ideas about rightfulness, proper orders, obligations as well as rights, and the ways these are put into practice. These would help us to better understand both the tenacity of everyday water injustices and what changes might enable a shift towards reshaping them in progressive directions.

References

Anand, N. (2015). Leaky states: Water audits, ignorance and the politics of infrastructure. *Public Culture*, 27(2), 305–30.
Berger, P. L. and Luckman, T. (1966). *The Social Construction of Reality: A Treatise on the Sociology of Knowledge*. Garden City, NY: Anchor Books.
Boelens, R. (2014). Cultural politics and the hydrosocial cycle: Water, power and identity in the Andean highlands. *Geoforum*, 57, 234–47.
Bond, P. and Dugard, J. (2008). Water, human rights and social conflict: South African experiences. *Law Social Justice and Global Development Journal*, www.go.warwick.ac.uk/ elj/lgd/2008_1/bond_dugard.
Bourdieu, P. (1977). *Outline of a Theory of Practice*. Cambridge: Cambridge University Press.
Brekhus, W. (1998). A sociology of the unmarked: Redirecting our focus. *Sociological Theory*, 16(1), 34–51.
Chabal, P. (2009). *Africa: The Politics of Suffering and Smiling*. London: Zed Books.
Cleaver, F. (2000). Moral Ecological rationality, institutions and the management of common property resources. *Development and Change*, 31(2), 361–83.
Cleaver, F. (2012). *Development through Bricolage: Rethinking Institutions for Natural Resource Management*. London: Routledge.
Cochran, J. and Ray, I. (2008). Equity re-examined: A study of community-based rainwater harvesting in Rajastahan, India. *World Development*, 37(2), 435–44.
De Herdt, T. and Olivier de Sardhan, J. P. (2015). *Real Governance and Practical Norms insub Saharan Africa: The Game of the Rules*. London: Routledge.
Delgado, J. V. and Zwarteveen, M. (2007). The public and private domain of the everyday politics of water. *International Feminist Journal of Politics*, 9(4), 1–9.
Douglas, M. (1987). *How Institutions Think*. London: Routledge and Keagan Paul.
Ferguson, J. (2015). *Give a Man a Fish: Reflections on the New Politics of Distribution*. Durham, NC: Duke University Press.
Funder, M., Bustamante, R., Cossio, V., Huong, P. T. M., van Koppen, B., Mweemba, C., Nyambe, I., Phuong, L. T. T. and Skielboe, T. (2012). Strategies of the poorest in local water conflict and cooperation: Evidence from Vietnam, Bolivia and Zambia. *Water Alternatives*, 5(1), 20–36.
Graham, S., Desai, R. and McFarlane, C. (2013). Water wars in Mumbai. *Public Culture*, 25(1), 115–41.
Gramsci, A. (1971). *Selections from the Prison Notebooks of Antonio Gramsci*. New York: International Publishers.

Hall, R., Edelman, M., Borras, S. M., Scoones I, White, B. and Wolford, W. (2015). Resistance, acquiescence or incorporation? An introduction to land-grabbing and political reactions "from below." *Journal of Peasant Studies*, 42(3–4), 467–88.

Harris, L., Goldin, J. and Sneddon, C. (eds.) (2013). *Contemporary Water Governance in the Global South: Security, Marketisation and Participation*. Abingdon: Earthscan/Routledge.

Hoogesteger, J. and Verzijl, A. (2015). Grassroots scalar politics: Insights from peasant water struggles in the Ecuadorian and Peruvian Andes. *Geoforum*, 62, 13–23.

Hyden, G. (2006). The moral and affective economy. *African Studies Quarterly*, 9(1+2), 1–8.

Joll, J. (1977). *Gramsci*. London: Fontana.

Joshi, D. (2015). Like water for justice. *Geoforum*, 61, 111–21.

Joy, K. J., Kulkarni, S. and Roth, D. M. (2014). Repoliticising water governance: Exploring water re-allocations in terms of justice. *Local Environment*, 19(9), 954–73.

Joy, K. S. (2014). The Santa Cruz Declaration on the Global Water Crisis: A good beginning to repoliticize the water discourse, Comments on the Santa Cruz Declaration. *Water International*, 39(2), 256–58.

Loftus, A. (2009). Intervening in the environment of the everyday. *Geoforum*, 40(3), 326–34.

Loftus, A. (2014). Water (in)security: Securing the right to water. *The Geographical Journal*, doi: 10.1111/geoj.12079.

Lu, F., Ocampo-Raeder, C. and Crow, B. (2014). Equitable water governance: Future directions in the understanding and analysis of water inequities in the global South. *Water International*, 39(2), 129–42.

McDonald, D. A. (ed.) (2002) *Environmental Justice in South Africa*. Cape Town: University of Cape Town Press.

Meagher, K., De Herdt, T. and Titeca, K. (2014). Hybrid governance in Africa: Buzzword or paradigm shift? *African arguments*, April, http://africanarguments.org/2014/04/25/hybrid-governance-in-africa-buzzword-or-paradigm-shift-by-kristof-titeca-kate-meagher-and-tom-de-herdt/.

Mehta, L., Allouche, J., Nicol, A. and Walnycki A. (2014). Global environmental justice and the right to water: The case of peri-urban Cochabamba and Delhi. *Geoforum*, 54, 158–66.

Ortner, S. B. (1995). Resistance and the problem of ethnographic refusal. *Comparative Studies in Society and History*, 37(1), 173–93.

Perera, V. (2012). From Cochabamba to Colombia: Travelling repertoires in Latin American water struggles. In F. Sultana and A. Loftus (eds.), *The Right to Water: Governance, Politics and Social Struggles*, Earthscan Water Text Series, London: Routledge.

Perreault, T. (2014). What kind of governance for what kind of equity? Towards a theorization of justice in water governance. *Water International*, 39(2), 233–45.

Petrella R. (2001). *The Water Manifesto: Arguments for a World Water Contract*. London: Zed Books.

Rodríguez-Labajos, B. and Martínez-Alier, J. (2015). Political ecology of water conflicts. *WIREs Water*, 2, 537–58.

Santa Cruz Declaration. (2014). Santa Cruz Declaration on the Global Water Crisis. *Water International*, 39(2), 246–61.

Sayer, A. (2007). Moral economy as critique. *New Political Economy*, 12(2), 261–70.

Schnegg, M. and Bollig, M. (2016). Institutions put to the test: Community-based water management in Namibia during a drought. *Journal of Arid Environments*, 124, 62–71.

Schnegg, M., Bollig, M. and Linke, T. (2016). Moral equality and success of common pool water governance in Namibia. *Ambio*, 45(5), 581–90.

Scott, J. C. (1985). *Weapons of the Weak: Everyday Forms of Peasant Resistance*. New Haven, CT: Yale University Press.

Sneddon, C. (2013). Water, governance and hegemony. In L. Harris, J. Goldin and C. Sneddon (eds.), *Contemporary Water Governance in the Global South: Security, Marketisation and Participation*. Abingdon: Earthscan/Routledge.

Spronk, S. (2007) Roots of resistance to urban water privatisation in Bolivia: The new working class, the crisis of neoliberalism and public services. *International Labour And Working Class History*, 17(1), 8–28.

Sultana, F. and Loftus, A. (2012). *The Right to Water: Politics, Governance and Social Struggles*. London: Earthscan/Routledge.

Sultana, F. and Loftus, A. (2015). The human right to water: Critiques and conditions of possibility. Wiley Interdisciplinary Reviews – *Water*. 2.2, 97–10.

Thompson, E. P. (1991). *Customs in Common*. London: Penguin.

Trawick, P. (2001). The moral economy of water: Equity and antiquity in the Andean Commons. *American Anthropologist*, 103(2), 361–79.

Walker, G. (2014). Editorial: Environmental justice as empirical and normative. *Analyse and Kritik*, 36(2), 221–28.

Wutich, A. (2011). The moral economy of water reeaxamined: Reciprocity, water insecurity and urban survival in Cochabamba, Bolivia. *Journal of Anthropological Research*, 67(1), 5–26.

Zug, S. and Graefe, O. (2014). The gift of water: Social redistribution of water among neighbours in Khartoum. *Water Alternatives*, 7(1), 140–59.

Zwarteveen, M. and Boelens, R. (2014). Defining, researching and struggling for water justice: Some conceptual building blocks for research and action. *Water International*, 39(2), 143–58.

14

Sharing Our Water: Inclusive Development and Glocal Water Justice in the Anthropocene

Joyeeta Gupta

14.1 Introduction

Evolving freshwater justice ideas over thousands of years' governance have mostly focused on who has access to a minimum quantity of water and how the remaining usable water should be allocated among uses/users, including through usufructuary rights, "priority of use" of water and, over time, sovereignty principles (Dellapenna and Gupta, 2009). While these elements remain key water justice themes even today, I argue that, in the Anthropocene, water justice approaches must proactively anticipate and pre-empt future water injustice (cf. Gupta, 2016a) by adopting an inclusive development (ID) approach (Gupta et al., 2015). An ID approach requires taking the ecological aspects as a starting point (i.e. ecological inclusiveness), assessing relational inclusiveness in terms of examining challenges to governance, and ensuring social inclusiveness by guaranteeing universal access to water resources and ensuring the just distribution of rights, responsibilities, and risks.

The Anthropocene, a term proposed by natural scientists, describes the current age, in which there is an overwhelming human footprint on the Earth's environment (Bogardi et al., 2012; Crutzen, 2006; Steffen et al., 2004). Its anticipated future impacts include exacerbated water stress (EC, 2012). This implies that local water injustices need to be studied in the context of major global trends and their underlying past, current and potential future inequities (cf. Johnson, 2014; Nicholson and Jinnah, 2016; Winter, 2006). Hence, this chapter addresses the question: how can humans anticipate the evolving nature of water-justice challenges in the Anthropocene and deal with them? It first discusses ecological and relational inclusiveness, and identifies water justice questions for the Anthropocene, before drawing conclusions.

14.2 Ecological and Relational Inclusiveness: Water Ecospace and Governance

14.2.1 Ecological Inclusiveness and Water Ecospace

An ID perspective on water justice requires ecological inclusiveness through an understanding of water ecospace. The Anthropocene requires humans to recognize the limits of "water ecospace" or water availability, water sinks and water-flow manipulation, given its role in global ecosystems. Like ecospace or environmental utilization of space, which conveys resource limits for human use and sinks for emitting human wastes (cf. Gupta, 2016a; Opschoor, 2009), water ecospace conveys the notion of limits on the use/abuse (i.e. pollution) of water, its flows and related ecosystem services and is derived from the hydrosphere's carrying capacity.

Fresh water is a tiny subset of the total water available, and much of it is locked in as ice. It is fairly fixed in quantity, while global demand grows because of the exponential rise in spatially differentiated production, consumption and demographic patterns. This implies an ever-decreasing amount of fresh water availability per capita, which already raises justice issues, to say nothing of the related distributional challenges and human-health impacts when water is polluted.

Furthermore, water circulates through solid, liquid and gaseous states, and from one place to another. Each physical phase makes its own unique contributions to biodiversity and ecosystem services (supporting services through e.g. nutrient circulation; regulating services through e.g. flood regulation, disease control, and climate regulation; provisioning services through e.g. providing water and fish; and cultural services) which, in turn, support human wellbeing (Millennium Assessment, 2005).

Human–nature interactions lead to fresh water (blue surface and groundwater, green water in soils and plants) being increasingly extracted in one area and being returned in the form of gray polluted water, black sewage water or cleaned water to other areas. Blue groundwater extraction beyond recharge levels, by pumps and pipes, might lead to land subsidence where it is extracted, and contribute to marginal sea-level rise elsewhere. Water circulation is affected by anthropogenic climate change and deforestation. White (ice, snow) water with a high albedo effect is melting into blue water with a low albedo effect, thus absorbing more heat from the sun as well as causing rising flows in rivers and rising sea levels (Hayat and Gupta, 2016; IPCC, 2014). Where the climate and flood-regulation tasks of water are affected, this can increasingly expose humans and other life forms to extreme weather events.

Hence, limiting this use and abuse of water and its flows, and the ecosystems it supports, is a key element of water justice and is captured through the notion of not overusing our water ecospace, as this will have impacts on others today and in the future (cf. Rockstrom et al., 2009).

The notion of limits is often criticized as neo-Malthusian. Technological optimists argue that technologies may address shortage aspects by converting salt water to fresh water, seeding clouds so that they will rain, developing drought-proof seeds, conveying ice glaciers to supply fresh water needs, etc. However, desalination, cleaning polluted water, and replacing natural flow systems by artificial ones (e.g. river linking, irrigation, cloud seeding) is expensive, energy-intensive, and socially and ecologically problematic. While technologies can help shape water flows, address quality and quantity issues, and reduce water injustices, thereby ostensibly erasing the idea of limits, they also convert nature's free ecosystem services into commodities, directly or indirectly, available only against payment, and requiring risk insurance. The problem is that it is primarily the rich and powerful, with their water-intensive production and consumption lifestyles that are directly extracting and polluting large quantities of water, manipulating water flows and consuming water-intensive traded products, while their large-scale fossil-fuel use is also causing climate change. Those affected are often the poor and marginalized, whose traditional access to the resource is hampered and who do not have the resources to pay for the technological solutions or insure against water-related risks. This raises a key justice challenge because such technologies and related commodification affects access and allocation.

14.2.2 Relational Inclusiveness: Discursive and Glocal Decision-Making to Share Water

In terms of justly governing water ecospace, four issues are critical. First, defining the global-to-local limits of water ecospace needs to be science-based and democratically and discursively defined, to be legitimate and effective. Second, water and water flows are affected by direct and indirect drivers at multiple governance levels (Gupta and Pahl Wostl, 2013), making fresh water not just a local issue but also global (Gleick, 2003; Pahl-Wostl et al., 2008; Vorosmarty et al., 2015). Teleconnections connect phenomena by cause-and-effect chains over vast distances. In other words, the causes of water injustices are often beyond the water basin, in which case they cannot be dealt with at basin level. Such causes include direct human behavior elsewhere but also indirect influences in terms of production, distribution, and consumption patterns, and underlying prominent global trade and investment discourses and laws. Moreover, local land, environmental, and agricultural issues that affect water use and flows may also themselves be embedded in global dynamics. At the same time, local action can have global repercussions. Local water-related issues (e.g. irrigation networks and dams; lack of access to water) may cumulatively add up to problematic global trends requiring also global-level action. As water quality and quantity reliability deteriorates and flows change, this may affect food production, human migration, or even the flights of migratory birds (UNEP, 2012; Vorosmarty et al., 2010). Framing water as an exclusively local-justice issue externalizes the impacts of local water use and abuse and how this is embedded in global discourses. Framing water exclusively as a global-justice issue tends to hide past, present, and potential future everyday and structural injustices in water access and allocation (Gupta et al., 2013). Hence, water justice requires a glocal[1] approach that takes the local to global aspects into account and challenges the "politics of scale" whereby actors choose a level to suit their own interests (Gupta, 2014a).

Third, water can be used directly and indirectly to enhance economic growth and GDP. Within the neoliberal capitalist framework, states may argue that they can achieve "growth" by selling the land, water, air,[2] minerals, trees, plantations, and food crops; by taming water through large-scale dams and inter-basin transfers; and by commodifying climate-adaptation services. After all, in colonial times, that is in line with what the colonial exploiters did to become rich. But they did so at their colonies' expense. Do current states wish to promote capital accumulation at the cost of the well-being of their own people and nature (cf. Moore, 2015), at the cost of losing complete control over these water resources, flows and related ecosystem services, especially given that trade and investment regimes will promote and protect capitalizing and privatizing this ecospace? This means that water justice also needs to engage with the wider debate on what development is, who defines it and for whom; and what this means for how to define and share water ecospace. Hence, this chapter has used the inclusive-development approach, which focuses more on socio-ecological wellbeing rather than on economic growth.

Fourth, the idea of limits embodied in water ecospace and its potential implication for "growth" may lead actors and countries to monopolize access, through water liberalization, which allows access to other countries' waters; water privatization, which allows water ownership; or water technologies and datafication, where water-use infrastructure and data (big data) are combined with large-scale modeling techniques to control water and obscure it from others; water sovereignty, which allows nation-states to use power to determine water-sharing and water-flow manipulation; and water securitization, which empowers countries to control water even through military means (Fisschendler, 2015; Gupta et al., 2015). There are rising global trends toward monopolizing water through spatial, technological, and infrastructural activities nested within governance responses (neoliberal, hegemonic, securitized) and hence water ecospace is increasingly subject to jurisdictional, ownership and control patterns that may make it difficult to limit and share (Gupta, 2016a). These background water governance rules may make it impossible to share water within the context of sustainable, inclusive development and thus exacerbate glocal injustice and prevent social inclusiveness.

14.3 Social Inclusiveness: Access and Allocation of Rights, Responsibilities and Risks

14.3.1 The Water Justice Questions Framework

The social-inclusiveness component of the water justice framework contests water monopolization, water flows, and their related ecosystem services by seeing them as public goods (cf. Kaul et al., 2003) from which no one should be excluded, and that use/enjoyment by one should not, in theory, reduce others' chances for use/enjoyment. The concept of water as a public good itself requires us to think in terms of water ecospace, to reduce the risks of compromising on water's ecosystem services.

The social-inclusiveness component can be explored using the access (i.e. meeting human needs/rights) and allocation (i.e. distributing property/tenure rights, responsibilities and risks) framework (Gupta and Lebel, 2010) of the Earth System Governance project of Future Earth[3] (Biermann et al., 2010; see Table 14.1).

14.3.2 Access: Guaranteeing Minimum Water

I will now try to address four water justice challenges raised in Table 14.1 on access (to basic needs/human rights) and allocation of rights, responsibilities and risks.

Table 14.1: *Water justice (WJ) questions in the Anthropocene*

	Elaborating on water justice (WJ) for inclusive development				
	Inclusive				Development
	Access	Allocation	Responsibilities	Risks	
	How to meet the basic water rights of all?	How to define water ecospace and share water ecospace in terms of (property) rights, responsibilities and risks?			How to define implications of water ecospace for development and human wellbeing?
Global	Rights/needs What do human rights mean for WJ of individuals, communities and indigenous peoples? How are the rights to participate	Rights Are commodification, monopolization, privatization, property rights compatible with water justice?	Responsibilities How is priority of use organized? Who is responsible for damaging the ecosystem services of water and water flows and how can they be held accountable?	Risks How are global water related risks shared and how does this affect WJ?	
Transboundary		How do debates on sovereignty and equitable use affect WJ?	How do debates on not causing harm to others affect WJ?	How are transboundary risks shared?	
National	organized within and across levels?	How do rights arranged through land rights, permits, infrastructure, and purchase affect WJ?	How do responsibilities with respect to protected areas and maintaining ecosystem service affect WJ?	How do droughts/floods and other water related risks affect WJ?	
Local		How are customary access rules affected by and affect WJ?	How are customary pollution rules affected by and affect WJ?	How are customary risk management practices affected by and affect WJ?	

What is the role of science, social actors/movements, and power in promoting discourses?

The first challenge is about guaranteeing minimum water access. Throughout history, humans have settled near water bodies and their access was regulated at community level and institutionalized in community law, such as the right to thirst in Islamic law (Caponera, 1992; Naff, 1999 on the Islamic tradition; see Kornfeld, 1999 on Mesopotamia; Laster et al., 1999 on the Jewish tradition). Sharia law guarantees access to pure drinking water (Naff, 1999).

In most industrialized countries, majority access to water and sanitation services was institutionalized in the twentieth century, although there are still pockets of people (gypsies, refugees, outdoor workers, homeless, and the elderly) without access to improved sanitation services. In the developing world, this is a significantly larger problem, as 2 billion people still lack access to sanitation services and about half that number lack access to drinking water. Population growth may exacerbate these numbers. Access has been dealt with in several political documents by calling for state and international responsibility, the need to develop pro-poor approaches in the 1990s, the Millennium Development Goals, which called for halving the population without sustainable access to safe drinking water and sanitation between 1990 and 2015, and now the Sustainable Development Goals (UNGA, 2015) which call for universal access to water and sanitation by 2030. The role of participation was emphasized in 1992 in the Dublin Water Declaration and has since become an integral part of the integrated water resource management discourse.

The needs and pro-poor discourse framed water access in terms of paternalistic state responsibility. Many felt that this does not go far enough, as access to water and sanitation should be treated as a human right, which the state *must* meet, directly or indirectly. This led to legally binding international treaties that spelled this out for women (CEDAW, 1979) and children (CRC, 1989). In 2010, the UN General Assembly and the UN Human Rights Commission declared that access to water and sanitation services is a human right (UNGA, 2010; UNHRC, 2010). This right is gradually becoming legally binding customary law (Obani and Gupta, 2016). However, this access to a bare minimum of services is increasingly compromised by political discourses and institutionalized rules regarding ownership, control and allocation, and climate change impacts.

Historically, the Islamic law concept of "priority of use" discussed how water should be allocated among different priorities, including for animals within a community, and animals in neighboring communities, to promote water justice in times of water shortage. This concept has been used in different national and international policies. However, the 1997 UN International Watercourses Convention explicitly discarded the concept, as countries were not willing to prioritize access to water in other riparian countries. Simultaneously, the global water, trade, and investment arena adopted neoliberal discourses which treated water as an economic good that should be subject to pricing, commodification, and explicit and implicit privatization. Thus, instead of regulating water use through socially determined priorities, these priorities could be determined by the market.

14.3.3 Allocation: Distribution of Rights

The second challenge is how rights to water ecospace are allocated among multiple uses and users. Such rights to larger quantities of water can be through (land) ownership, prior appropriation (the principle of first-come, first-served or first-in-time, first-in-right), permits and payments. Ancient Hindu law viewed water as indivisible; water could not be owned. However, people could use water, and the King was responsible for protecting public works (the Laws of Manu, c.200–100 BCE; Kautilya, c.350 BCE–150 CE). Islamic law also viewed water (and land) as indivisible, while limited ownership resulted from labor to dig a well (Naff, 1999). However, in common law prevalent in former British colonies, water could be owned through riparian ownership, which entitles the land owner to water access under and along the land (Caponera, 1992). In Roman Law, which prevails in many former colonies of continental Europe, navigable rivers were public waters, but small water bodies could be privately owned (Caponera, 1992). The concept of prior appropriation, developed in the US, allowed the use of technology to acquire water rights, even when there was no land ownership involved, by simply staking a claim until there was no water left to claim. In Communist countries, water is owned by the state. The current situation globally is that water is either owned privately when linked to land, by the state when it concerns large navigable rivers, by communities when it concerns small local water bodies, or through technological and financial investments in water. However, there are many spatial and contextual variations in ownership with consequent implications for water justice (Dellapenna and Gupta, 2009).

In interstate relations, countries have claimed absolute sovereign rights over water in their jurisdiction, sometimes based on how much they used historically. Downstream countries have countered this with the principle of absolute territorial integrity where water flows are part of the downstream country's territory (Dellapenna, 2001). Since 1966, the principle of equitable and reasonable water-sharing between states has been articulated in academic documents (ILA, 1966, 2004) and the 1997 UN Watercourses Convention (UNWC, 1997; Kaya, 2003). However, although the latter has entered into force, only 36 water-rich, mostly downstream countries have ratified the Convention. Thus, most countries have not accepted equitable sharing and are probably leaning towards sovereignty, securitization, and hegemonic approaches (Gupta, 2016b). So, despite the discourse of transboundary water-sharing, in practice, sovereign control still remains the dominant approach. If there is limited water ecospace, this could also exacerbate the tendencies of states to adopt hegemonic water strategies (Cascão, 2008; Daoudy, 2008; Warner, 2008).

Furthermore, with globalization and liberalization, foreigners and foreign companies can control water in other countries through land-ownership contracts, public-private partnerships, or permits to extract water. For example, Coca Cola's investment in India enabled it to extract water at neighboring farmers' expense. Their protests have now reached the Supreme Court of India. Nestlé's purchase of land and associated water rights in Maine, USA, has also led to local discontent. These examples are part of a growing trend of water-grabbing at local people's expense (Rulli *et al.*, 2013).

If water ecospace is globally scarce, in a business-as-usual approach, land and water ownership, hegemonic behavior, securitization, and grabbing will become a globally unstoppable phenomenon and will compromise water justice issues locally and at transboundary level. It is difficult to reverse this trend, as it requires expropriating water rights and compensating those who lose ownership rights. For example, in 1985, Spain declared all groundwater to be public but, as there was no wish to pay expropriation costs, it was declared that those who voluntarily gave up their ownership rights would have state protection to use the water for the next 50 years, as compensation (de la Hera *et al.*, 2015).

A corollary to ownership is the issue of pricing water. Under ancient Islamic law, water is held to be God's gift, and this does not allow water pricing (Caponera, 1992); Hindu law, however, allows for water service pricing (Cullet and Gupta, 2009). However, economic conceptualization of "water scarcity" led economists to promote water pricing and the 1992 Dublin Declaration on Water (Dublin Declaration, 1992) stated that water is an economic good. This might imply simply being conscious that water is a limited resource requiring careful management or, and which is more often the case, that water is a commodity that should be subject to market-based pricing as promoted by integrated water-resource management. This can lead to greater use-efficiency but also higher profits for those who own or control it. However, even low-level water-trading practices, as in Chile, can lead to hoarding practices during droughts (Hurlbert, 2016).

Governments that engage in public-private partnerships for water infrastructure often find that related investments are shrouded in contractual secrecy, are subject to investment treaties and contract law (Klijn *et al.*, 2009; Merme *et al.*, 2014), reduce their ability to independently govern the private party, and may de facto transfer control over a glocal commons to private ownership and freedom to price through growing water-sector financialization (Ahlers and Merme, 2016). The Indian government's build, own, operate, transfer (BOOT) 22-year concession agreement with Kailash Soni of Radium Water Ltd gives Soni de facto control over 23.6 km of the Sheonath river, while Soni expects to make huge profits through water-supply infrastructure (Outlook, 2002). The United States experience leads Dellapenna (2009: 382) to conclude that: "The attempt to commodify water thus far has generated the inequities that follow from markets without bestowing the benefits that markets at their best can provide – rational management and efficient use." Once water can be owned/controlled de jure or de facto, it is a short step to maximizing returns on water investments. For example, the growing global bottled water and drinks industry is rapidly depleting local groundwater.

In a business-as-usual scenario, the combination of de facto and de jure ownership combined with the right to price water may make water a monopoly business, out of reach for the poor and lower-middle classes: it might lead to water moving from low financial-return situations such as flowing in a canal or water body to high-end returns such as golf estates (Bernasconi-Osterwalder and Weiss, 2005).

14.3.4 Allocation: Distributing Responsibilities

The third water justice question addresses responsibilities for polluting water. Ancient cultures protected and still protect holy water/rivers for their own sake. Early community water law called on water controllers to maintain their water bodies and had detailed rules on water protection (cf. Kautilya, c.350 BCE–150 CE; Laws of Manu, c.200–100 BCE). Subsequent common and civil law also punished those whose activities led to pollution.

Over time, rules regarding environmental impact assessments and pollution reduction have been developed, but these generally only focus on the provisioning ecosystem services of water, and not on supporting, regulating or cultural services. While the UNECE Convention (UNECE WC, 1992) tries to address some of these environmental impacts and the Ramsar Convention (UN Ramsar Convention, 1972) protects wetlands of global significance, the UN Watercourses Convention (UNWC, 1997) does not give significant attention to water's ecological aspects except to prohibit substantial harm to water and call for states to compensate for the harm caused. Many countries have marine protected areas (MPAs) and protected wetlands, but the way in which they are designed and demarcated often disproportionately affects the poor in these regions. Water justice approaches thus also call for designing approaches that ensure that water's supporting, regulating and cultural services and water flows are also maintained.

14.3.5 Allocation: Distributing Risks

Finally, who takes responsibility for the risks associated with water? Water hoarding has always taken place, and when dams burst downstream, riparian peoples were flooded. Early community law in India called on water managers to prevent flooding (Kautilya, c.350 BCE–150 CE; Laws of Manu, c.200–100 BCE) and twentieth-century case law in the UK compensated downstream riparians for man-made floods. Floods and droughts are now much more common, partly because of climate variability and change, but also because of changing spatial and physical planning and infrastructure, entailing high risks. Climate change, in fact, manifests itself through water: enhanced water evaporation, reduced groundwater recharge, changing rainfall patterns, rapid snow melting, rising sea levels and salt water intrusion, and droughts and floods. Even if the long-term objective of the Paris Agreement (2015) to the Climate Change Convention is met, climate change impacts will continue into the future and be felt unevenly worldwide, with the poorest most likely to suffer the most, as they are the ones who live in precarious housing in times of storms and have limited access to water (IPCC, 2014). All this calls for adaptation and adaptive governance in the face of uncertainty (Palmer, 2008) and climate-proofing water agreements (Heather and Gleick, 2011).

However, limited formal resources for adaption have only been available since 2008. These resources have fallen short of past expectations and are more than likely to fall significantly short of future expectations. This is because the promised funds (US$100 billion) for 2020 are neither in the Paris Agreement's legally binding component, nor is it clear who should pay (Gupta, 2016c). As a result, current official development assistance is being relabeled as climate assistance and research shows that the private sector is not willing to pay for adaptation (Timmons Roberts and Wiekmans, 2017). Financing adaptation within a domestic context also raises justice issues. The costs of adapting to floods is increasingly being transferred to home owners, as in the US (Bergsma, 2017), or is scarcely being discussed, as in Canada, Chile and Argentina (Hurlbert and Gupta, 2016). This shifts the risk burden from those who gain from fossil fuel use, to those who are affected.

Adaptive justice would also require drastically reducing greenhouse-gas emissions, and that those who have contributed most to the problem of climate change should pay their fair share (Gupta, 2014b; UNFCCC, 1992: Art. 3). However, mitigation proceeds slowly and so there remain serious water risks.

14.4 Conclusion

Through human history, water justice issues have concerned access to clean water and water rights allocation, responsibilities and risks. However, this chapter argues that, in the Anthropocene, we need to revisit water justice understanding. It argues in favor of an inclusive-development perspective, focused on ecological, relational and social inclusiveness. By focusing on social wellbeing as opposed to economic growth, it aims to counter the dominance of the neoliberal capitalist paradigm.

This chapter submits that ecological inclusiveness requires recognizing that fresh water is a scarce, priceless eco-systemic medium that should be allowed to flow as a socio-natural entity, and that there is an overall limit to water use and abuse, its flows and related ecosystem services (water ecospace).

Relational inclusiveness argues that water ecospace needs to be (a) discursively defined at multiple levels, and (b) governed simultaneously at local to global levels as the direct/indirect causes and impacts will be felt at all levels. (c) Furthermore, since water is closely linked to "growth" prospects, there will be pressure to commodify water and (d) monopolize control over water through water liberalization, privatization, technocratization, or securitization, which will adversely affect prospects for water justice and social inclusion.

Instead, a socially inclusive approach calls for recognizing water's role in development, and understands that water is a glocal public good that must be shared. Principles for sharing include rules regarding just access and rules for just allocation of property/tenure rights, responsibilities and risks.

Such water ecospace should be managed in the public interest by the community and/or state as the people's institutionalized representative who takes responsibility for just access and allocation, that these institutionalized representatives need to share the water ecospace with others; and that states need to individually and collectively take responsibility for climate change. If states are not careful and continue in a business-as-usual manner, they will have outsourced their natural resources to private companies and will have very little control over the very resources that they were charged, as society's institutionalized representatives, to hold in trust.

Such outsourcing may make it impossible to actually implement the human right to water and sanitation, let alone proposed water rights (Boelens, 2015), such as the right to cultural difference and normative plurality (as the need to encompass rights in context), the right to governance and political co-decision (as encompassed in participatory rights), the right to socio-ecological integrity and intergenerational stewardship (or responsibilities) and the right to distributive equality (as encompassed in access and allocation). Uneven water access and allocation will become even more uneven in the future, as water prices reach monopoly levels, as water moves from low-return to high-return uses; and as investment, trade, and contract laws and rules entrench the private sector with rights and power, and the ability to shift money across the globe in a fraction of a second. Reversing ownership through expropriation will become increasingly more expensive for states. Furthermore, de facto access and allocation were, are, and will increasingly be also affected by environmental pollution and climate change.

Water justice approaches must demand limits on water ecospace and call for a halt to increasing water privatization and commodification and expropriate now, rather than later, water resources where it has been de facto or de jure privatized. Water justice also implies investing in a state that is accountable to its people and to other riparians as well as to global citizens, and not just a state that is so poor that it depends on privatizing its resources and outsourcing its responsibilities for public and merit goods. Furthermore, while glocal water justice requires plural approaches to water governance, research shows that such plural approaches may often embody contradictions that are de facto resolved in favor of the rich. While plural approaches are necessary and possibly inevitable, contradictions should be resolved through principles embodied in a glocal constitutional approach that prioritizes access and allocation over monopolization and commodification, that revisits the importance of accountable, just government over governance, and that re-imagines a definition of development that goes beyond simply capitalizing all natural resources and their ecosystem services, while also exacerbating climate change in an ever-continuing race for increasing economic growth. Unless they are hijacked by neoliberal capitalists, the Sustainable Development Goals could be a first step in the direction toward a new global imagination.

Notes

1 The "glocal" and "glocalization" concepts refer to the intertwining of local and global economic, cultural, political and institutional scales. Erik Swyngedouw introduced the terminology in 1992 referring to re-scaling the nation-state. According to a more recent paper by the same author (Swyngedouw, 2004: 25) the concept of glocalization "refers to the twin process whereby, firstly, institutional/regulatory arrangements shift from the national scale both upward to supra-national or global scales and downward to the scale of the individual body or to local, urban or regional configurations and, secondly, economic activities and inter-firm networks are becoming simultaneously more localised/regionalised and transnational."
2 A Canadian company has started to sell clean Canadian air in spray cans to China (CNN, 2017).
3 Future Earth is a global research platform to advance Global Sustainability Science funded by national science foundations and other agencies; their only research project on governance is the Earth System Governance project.

References

Ahlers, R. and Merme, V. (2016). Financialisation, water governance and uneven development. *WIREs Water*, 3, 766–774. doi: 10.1002/wat2.1166.

Bergsma, E. (2017). "From flood safety to risk management: The rise and demise of engineers in the Netherlands and the United States." PhD thesis manuscript, University of Amsterdam.

Bernasconi-Osterwalder, N. and Brown Weiss, E. B. (2005). International investment rules and water: Learning from the NAFTA experience. In E. B. Brown Weiss, L. Boisson de Chazournes and N. Bernasconi-Osterwalder (eds.), *Fresh Water and International Economic Law*. Oxford: Oxford University Press, pp. 263–88.

Biermann, F., Betsill, M. M., Vieiera, S. C. *et al.* (2010). Navigating the Anthropocene: The Earth System Governance Project Strategy Paper. *Current Opinion in Environmental Sustainability*, 2(3), 202–08.

Boelens, R. (2015). *Water Justice in Latin America: The Politics of Difference, Equality and Indifference*. Amsterdam: University of Amsterdam.

Bogardi, J. J., Dudgeon, D., Lawford, R., and Vörösmarty, C. (2012). Water security for a planet under pressure: Interconnected challenges of a changing world call for sustainable solutions. *Current Opinion in Environmental Sustainability*, 4(1), 35–43.

Bondre, N. and Wilke, S. (2014). Beyond the Anthropocene's common humanity. *Geocritique*, www.geocritique.org/beyond-anthropocenes-common-humanity-politicizing-anthropocene/.

Caponera, D. (1992). *Principles of Water Law and Administration*. Rotterdam: Balkema Publishers.

Cascão, A. E. (2008). Ethiopia: Challenges to Egyptian hegemony in the Nile Basin. *Water Policy*, 10(S2), 13, http://doi.org/10.2166/wp.2008.206.

Castree, N. (2015). Unfree radicals: Geoscientists, the Anthropocene, and left politics. *Antipode*, 49(S1), 52–74. pp 1–23, doi: 10.1111/anti.12187.

CEDAW. (1979). *Convention on the Elimination of All Forms of Discrimination Against Women*. New York, December 18.

CRC. (1989). *Convention on the Rights of the Child*. New York, November 20.

Crutzen, P. J. (2006). The Anthropocene. In E. Ehlers and T. Kraft (eds.), *Earth System Science in the Anthropocene*. Berlin/Heidelberg: Springer, pp. 13–18.

Cullet, P. and Gupta, J. (2009). India: Evolution of water law and policy. In J. Dellapenna and J. Gupta (eds.), *The Evolution of the Law and Politics of Water*. Dordrecht: Springer Verlag, pp. 157–74.

Daoudy, M. (2008). Hydro-hegemony and international water law: Laying claims to water rights. *Water Policy*, 10(S2), 89–102.

Dellapenna, J. W. (2001). The Customary International Law of Transboundary Fresh Waters. *International Journal of Global Environmental Issues*, 1(3–4), 264–305.

Dellapenna, J. W. (2009). The market alternative. In J. W. Dellapenna and J. Gupta (eds.), *The Evolution of the Law and Politics of Water*. Dordrecht: Springer Verlag, pp. 373–90.

Dellapenna, J. and Gupta, J. (eds.) (2009). *The Evolution of the Law and Politics of Water*. Dordrecht: Springer Verlag.

Dublin Declaration. (1992). The Dublin Statement and Report of the International Conference on Water and the Environment Conference. In *International Conference on Water and the Environment: Development Issues for the 21st century*. Dublin, Ireland.

European Commission. (2012). *Global Europe 2050 Full Report*. Brussels: European Commission.

Fiscchendler, I. (2015). The securitization of water discourse: Theoretical foundations, research gaps and objectives of the special issue. *INEA*, 15(3), 245–56.

Gleick, P. H. (2003). *Water in Crisis: A Guide to the World's Fresh Water Resources*. Oxford: Oxford University Press.

Gupta, J. (2014a). Glocal politics of scale on environmental issues: Climate, water and forests. In F. J. G. Padt, P. F. M. Opdam, N. B. P. Polman and C. J. A. M. Termeer (eds.), *Scale-sensitive Governance of the Environment*. Oxford: John Wiley & Sons, pp. 140–56.

Gupta, J. (2014b). *The History of Global Climate Governance*. Cambridge: Cambridge University Press.

Gupta, J. (2016a). Geopolitics of the new earth: Towards sharing our ecospace. In S. Nicholson and S. Jinnah (eds.), *New Earth Politics*. Cambridge, MA: MIT Press, pp. 271–92.

Gupta, J. (2016b). The watercourses convention, hydro-hegemony and transboundary water issues. *The International Spectator*, 51(3), 118–31, http://dx.doi.org/10.1080/03932729.2016.1198558.

Gupta, J. (2016c). The Paris Climate Change Agreement, China and India. *Climate Law*, 6, 171–81.

Gupta, J. and Bavinck, M. (2014). Towards an elaborated theory of legal pluralism and aquatic resources. *COSUST*, 11, 85–93.

Gupta, J. and Lebel, L. (2010). Access and allocation in global earth system governance: Water and climate change compared. *INEA*, 10(4), 377–95.

Gupta, J. and Pahl-Wostl, C. (2013). Global water governance in the context of global and multi-level governance: Its need, form, and challenges. *Ecology and Society*, 18(4), 53, http://dx.doi.org/10.5751/ES-05952-180453.

Gupta, J., Pahl-Wostl, C. and Zondervan, R. (2013). "Glocal" water governance: A multi-level challenge in the Anthropocene. *Current Opinion in Environmental Sustainability*, 5, 573–80, http://dx.doi.org/10.1016/j.cosust.2013.09.003.

Gupta, J., Pouw, N. and Ros-Tonen, M. (2015). Towards an elaborated theory of inclusive development. *European Journal of Development Research*, 27, 541–59.

Hayat, S. and Gupta, J. (2016). Kinds of freshwater and their relation to ecosystem services and human wellbeing. *Water Policy*, 18, 1229–46.

Heather, C. and Gleick, P. H. (2011). Climate-proofing transboundary water agreements. *Hydrological Sciences Journal*, 56, 711–18. DOI: 10.1080/02626667.2011.576651.

Hera, A. de la, Fornés, J. M., Custodio, E., Llamas, M. R. and Sahuquillo A. (2015). Groundwater use changes in Spain since 1950: Past and present challenges and opportunities, paper for the Congress on Water Tensions in Europe and in the Mediterranean: Water crisis by 2050? October 8–9, 2015, Paris-Marne la Vallée, www.researchgate.net/publication/283345775.

Hurlbert, M. and Gupta, J. (2016). Adaptive governance, uncertainty, and risk: Policy framing and responses to climate change, drought, and flood. *Risk Analysis*, 36(2), 339–56.

International Law Association. (1966). Helsinki Rules on the Uses of the Water of International Rivers (52nd Conf.). Helsinki, Finland: Reprinted in S. Bogdanovic (2001). *International Law of Water Resources – Contribution of the International Law Association (1954–2000)*. The Hague: Kluwer Law International.

International Law Association. (2004). Berlin Rules on Water Resources. In *Fourth Report of the Berlin Conference on Water Resources*. Berlin: Berlin Conference on Water Resources Law.

IPCC. (2014). *Climate Change 2014: Impacts, Adaptation, and Vulnerability*. Cambridge and New York: Cambridge University Press.

Johnson, E. (2014). "Beyond the Anthropocene's common humanity," www.geocritique.org/beyond-anthropocenes-common-humanity-politicizing-anthropocene/.

Kaul, I., Conceicao, P., Le Goulven, K. and Mendoza, R. (2003). Why do global public goods matter today? In I. Kaul, P. Conceicao, K. Le Goulven and R. Mendoza (eds.), *Providing Global Public Goods*. New York and Oxford: Oxford University Press.

Kautilya. (*c.* 300 BCE). *The Arthashastra* (L.N. Rangarajan, trans. 1992). Delhi: Penguin Books.

Kaya, I. (2003). *Equitable Utilization: The Law of Non-Navigational Uses of International Watercourses.* Aldershot: Ashgate Publishing.

Klijn, A. M., Gupta, J. and Nijboer, A. (2009). Privatising environmental resources: The need for supervision. *Review of European Community and International Environmental Law*, 18(2), 172–84.

Kornfeld, I. E. (1999). Mesopotamia: A history of water and law. In J. W. Dellapenna and J. Gupta (eds.), *The Evolution of the Law and Politics of Water.* Dordrecht: Springer Verlag, pp. 21–36.

Laster, R., Aronovsky, D. and Liveny, D. (1999). Water in the Jewish legal tradition. In J. W. Dellapenna and J. Gupta (eds.), *The Evolution of the Law and Politics of Water.* Dordrecht: Springer Verlag, pp. 53–67.

Laws of Manu. (1991 [*c.*200 BCE]), eds. W. Doniger and B.K. Smith. New Delhi: Penguin Books.

Meinzen-Dick, R. (2007). Beyond panaceas in water institutions. *Proceedings of the National Academy of Sciences of the United States of America*, 104(39), 15200–05.

Merme, V., Ahlers, R. and Gupta, J. (2014). Private equity, public affair: Hydropower financing in the Mekong Basin. *Global Environmental Change*, 24(1), 20–29.

Millennium Ecosystem Assessment. (2005). *Ecosystem Services and Human Wellbeing: Policy-responses, Millennium Ecosystem Assessment*, Vol. 3. Washington, DC: Island Press, pp. 489–523.

Moore, J. W. (2015). *Capitalism in the Web of Life: Ecology and the Accumulation of Capital.* New York: Verso.

Naff, T. (1999). Islamic law and the politics of water. In J. W. Dellapenna and J. Gupta (eds.), *The Evolution of the Law and Politics of Water.* Dordrecht: Springer Verlag, pp. 37–52.

Nicholson, S. and Jinnah, S. (eds.) (2016). *New Earth Politics: Essays from the Anthropocene.* Cambridge, MA: MIT Press.

Obani, P. and Gupta, J. (2016). Human right to sanitation in the legal and non-legal literature: The need for greater synergy. *Wires Water.* doi: 10.1002/wat2.1162.

Opschoor, H. (2009). *Sustainable Development and a Dwindling Carbon Space.* Public Lecture Series 2009/1, The Hague.

Outlook India. (2002). This man owns the river, B. Kang, September 23, www.outlookindia.com/magazine/story/this-man-owns-the-river/217326.

Pahl-Wostl, C., Gupta, J. and Petry, D. (2008). Governance and the global water system: A theoretical exploration. *Global Governance*, 14(4), 419–35.

Palmer, M. A., Reidy-Liermann, C., Nilsson, C., Flörke, M., Alcamo, J., Lake, P., and Bond, N. (2008). Climate change and the world's river basins: Anticipating management options. *Frontiers in Ecology and the Environment*, 6, 81–89.

Paris Agreement. (2015). UNFCCC. *Paris Agreement*, http://unfccc.int/files/home/application/pdf/paris_agreement.pdf.

Rulli, M. C., Saviori, A. and D'Odorico, P. (2013). Global land and water-grabbing. *Proceedings of the National Academy of Sciences*, 110(3), 892–97.

Serageldin, I. (2009). Water: Conflicts set to arise within as well as between state. *Nature*, 459, 163.

Steffen, W., Sanderson, A., Tyson, P. D. *et al.* (2004). *Global Change and the Earth System: A Planet under Pressure.* New York: Springer.

Swyngedouw, E. (2004). Globalisation or "glocalisation"? Networks, territories and rescaling. *Cambridge Review of International Affairs*, 17(1), 25–48.

Timmons Roberts, J. and Weikman, R. (2017). Postface: Fragmentation, failing trust and enduring tensions over what counts as climate finance. *International Environmental Agreements: Politics, Law and Economics*, 17(1), 129–37.

Turton, A. and Ohlsson, L. (1999). *Water Scarcity and Social Stability: Towards a Deeper Understanding of the Key Concepts Needed to Manage Water Scarcity in Developing Countries*. London: SOAS.

UN Ramsar Convention on Wetlands of International Importance especially as Waterfowl Habitat, Ramsar 2.2.1971, as amended by the Paris Protocol of 3.12.1982 and the Regina amendments of 28.5.1987.

UNECE (United Nations Economic Commission for Europe). (1992). *United Nations Economic Commission for Europe Convention on the Protection and Use of Transboundary Watercourses and International Lakes*i, 1936 UNTS 269 (March 17, 1992, entered into force October 6, 1996).

UNEP. (2012). Global Responses, *Global Environmental Outlook – 5*. Nairobi: UNEP.

UNFCCC. (1992). *United Nations Framework Convention on Climate Change* (UNFCCC), 31 I.L.M. 849, May 9, 1992.

UNGA. (1986). *Declaration on the Right to Development*, Resolution A/RES/41/128, December 4, 1986.

UNGA. (2010). *Resolution on Human Right to Water and Sanitation*. A/64/292, July 28, 2010, www.un.org/News/Press/docs/2010/ga10967.doc.htm\.

UNGA. (2015). *Transforming Our World: The 2030 Agenda for Sustainable Development*, Draft resolution referred to the United Nations summit for the adoption of the post-2015 development agenda by the General Assembly at its sixty-ninth session. UN Doc. A/70/L.1 of September 18.

UNHRC. (2010). *UN Human Rights Council Resolution*, 15/9, UN Doc. A/HRC/RES/15/9, October 6, 2010.

UNWC. (1997). *United Nations Convention on the Law of the Non-Navigational Uses of International Watercourses*, Reprinted in 36 ILM 700 (May 21, 1997, entered into force August 17, 2014).

Vörösmarty, C. J., Hoekstra, A. Y., Bunn, S. E., Conway, D. and Gupta, J. (2015). What scale for water governance: Fresh water goes global? *Science*, 349(6247), 478–79.

Vörösmarty, C. J., McIntyre, P. B., Gessner, M. O. *et al.* (2010). Global threats to human water security and river biodiversity. *Nature*, 467, 555–61.

Warner, J. (2008). Contested hydro-hegemony: Hydraulic control and security in Turkey. *Water Alternatives*, 1(2), 271–88.

Winter, G. (ed.) (2006). *Multilevel Governance of Global Environmental Change: Perspectives from Science, Sociology and the Law*. Cambridge and New York: Cambridge University Press.

Part IV

Governmentality, Discourses and Struggles over Imaginaries and Water Knowledge

Ecuador.
Photo by Julio García.

Introduction: Governmentality, Discourses and Struggles over Imaginaries and Water Knowledge

Tom Perreault, Rutgerd Boelens, and Jeroen Vos

Standing Rock

The protesters called themselves "water protectors." They came from all corners of the US and Canada, and included members of over 100 Native American tribes and their non-indigenous allies. At the peak of the protest, in early autumn 2016, they numbered in the thousands, but by late February, when the Governor sent some 200 police officers to clear the encampment, only a few dozen remained. Those who left burned their temporary shelters in an act of defiance, and the embers were still smoldering when the police moved in and arrested the few who stayed behind (Smith and Blinder, 2017). For a brief period, it seemed the protesters might have won: the Obama administration halted construction of the Dakota Access pipeline and asked the US Army Corps of Engineers, which was overseeing the project, to find an alternate route. But then the political winds shifted. A right-wing populist was elected President and the winter cold brought with it a form of revanchist politics that had little patience for Native Americans and their atavistic attachments to land and water (Healy and Fandos, 2016; Smith, 2017).

The protesters had gathered on the banks of the Cannonball River, near its confluence with the Missouri, where the Texas-based Energy Transfer Partners planned to build the last section of the pipeline. Construction was nearly complete, and all that remained was a short stretch to be built under the Missouri River. The pipeline route consisted entirely of private land, but because it had to cross a navigable waterway, it required the approval of the US Army Corps of Engineers (New York Times, 2016a). Despite opposition from the US Environmental Protection Agency, the US State Department and the Sioux Nation's Tribal Historic Preservation Office, the Army Corps seemed determined to fast-track the pipeline's approval. Contrary to federal requirements, it approved construction without meaningful consultation with the Standing Rock Sioux, whose lands and waters the pipeline would impact (Archambault, 2016; Sammon, 2016). Although the pipeline would not cross the Standing Rock reservation, instead passing half a mile north of the reservation boundary, tribal members argued that it crossed land they considered sacred (and which had formerly been part of Sioux territory, of which they were dispossessed in the 1870s when they were confined to the reservation). Crucially, the tribe was also concerned about oil spills from the pipeline contaminating their source of drinking water, a concern that was amplified by the fact that the pipeline was re-routed from its original path, just upstream from the city of Bismarck (North Dakota's capital), because federal regulators thought it posed too great a risk to the city's water supply (New York Times, 2016b). Downstream from the Standing Rock reservation, the Missouri River provides drinking and irrigation water for millions of people, across a vast swath of the Great Plains and Midwest, before joining the Mississippi River at St. Louis.

The pipeline is only the latest episode in a long history of white encroachment onto Sioux land. The Sioux signed treaties with the US Government in 1868 and 1951, only to see them ignored. The sacred Black Hills, in what is now South Dakota, were plundered by gold miners and then turned into the tourist attraction of Mount Rushmore (New York Times, 2016b). Years later, the Army Corps dammed the Missouri River in 1958 to create Lake Oahe, flooding forests, orchards and farmland on Sioux territory (Archambault, 2016). Now Sioux water and sacred lands are put at risk in the interest of oil extraction.

The nearly 1200-mile Dakota Access pipeline will transport crude oil from the Bakken shale field in remote western North Dakota, where it is extracted by means of hydraulic fracturing (or "fracking"), to distribution terminals in Illinois (from where it will be sent to refineries on the Gulf coast). With the advent of hydraulic fracturing as a means of accessing previously inaccessible oil and natural gas, North Dakota's Bakken shale – like the Marcellus shale formation in Pennsylvania – has become a site of frenetic activity, and a centerpiece of the ongoing boom in hydrocarbons production the United States. The pace of oil extraction has far exceeded the construction of adequate infrastructure to transport crude oil to refineries, which are located thousands of miles away. Natural gas is flared off for lack of infrastructure needed to capture and convey it to distant markets, turning this sparsely populated corner of the northern plains into a major emitter or greenhouse gases. Oil extraction began before a pipeline was in place. As a consequence, for years, producers transported crude oil by rail, using miles-long chains of tanker cars, a method of oil transport that is both expensive and dangerous (in 2013 a train loaded with Bakken crude crashed in the town of Lac-Mégantic, Quebec, resulting in a catastrophic explosion and fire that killed 47 people; McNish and Robertson, 2014). The solution to this dilemma, as proposed by the Texas-based Energy Transfer Partners, was the Dakota Access Pipeline. Built at a cost of US$3.7 billion, the pipeline will transport more than a half-million barrels of oil per day (Healy and Fandos, 2016). The state of North Dakota strongly backed the pipeline and state officials have done little to hide their contempt for the Sioux and their concerns. Protesters found some sympathy in the Obama administration, but it proved to be too little, too late and when the new President assumed office, he ordered the Army Corps to stop looking for alternate routes and approve construction of the final stretch of pipeline. Trump himself owns stock in Energy Transfer Partners (Healy and Fandos, 2016), and completion of the pipeline aligns with the Republican Party's objectives of maximizing fossil fuel production and weakening environmental regulations. It is also entirely consistent with the Party's callous disregard for the rights and interests of Native Americans and other minority groups. During the protest, state police and private security forces were heavily militarized, and at various times deployed attack dogs, rubber bullets, pepper-spray and water cannons (in sub-freezing temperatures) on the protesters (Sammon, 2016; Smith and Blinder, 2017). The state government's sympathies were on full display just an hour after the protest camp was cleared by police, when

North Dakota's governor signed four bills into law that together expanded the scope of trespassing laws, made it illegal to wear masks or otherwise cover the face during the commission of a crime, and increased the penalties for protest and riot offenses (Smith, 2017).

Resource Extraction and the State

The events at Standing Rock are reminiscent of other such struggles around the world. Native peoples protesting resource extraction and the threats it poses to water and land, coupled with criminalizing protest and militarizing policing, have their counterparts in Ecuador, Nigeria, Peru and elsewhere (Sawyer, 2004; Watts, 2004). In Ecuador, indigenous peoples have long protested the damaging effects of oil development in the country's Amazon region, and more recently indigenous and mestizo residents have objected to plans for large-scale mine projects in the cloud-forests region in the country's northern Andes (Billo and Zukowski, 2015). In both cases, protesters have been met with police and military repression, and while their struggles have gained international attention, it has done little to slow the Government's intensifying extractionist agenda. Similarly, in Nigeria, decades of oil extraction in the Niger Delta have left the region environmentally and socially devastated (Kashi and Watts, 2010). These protests, and others like them around the world, point to broader political economic structures that underlie resource extraction and the threat it poses to indigenous peoples and their lands and waters: the state's dependence on extractivist regimes of accumulation. In the United States, petroleum is foundational not only to economic growth, but to the "American way of life," rooted in individuated freedoms, mass consumption and auto-mobility (Huber, 2013). Throughout Latin America, mining, oil and gas undergird national governments on both the political right and the political left. Resource extraction is as vital to the neoliberal governments of Chile, Peru and Colombia as it is to left-wing governments of Venezuela, Bolivia and Ecuador. The relationship between resource extraction and state power draws attention to what Watts (2004: 199) has termed the "oil complex": "a configuration of firm, state and community through which petro-capitalism operates." Watts was specifically concerned with Nigerian oil extraction (largely performed by Royal Dutch Shell) and the political and social pathologies it facilitates, but the analogy can be extended far beyond the Niger Delta. In his analysis of the "resource-state nexus," Bridge (2013) argues that, far from being settled questions or static entities, both resources and states are best understood as emergent properties. As such, the most interesting questions are not what states and resources are, in a realist sense, but rather "how they come to be – i.e. the formative processes through which resources and states are generated as 'effects,' and these effects' consequences to organize socio-natural relations" (Bridge, 2013: 119). A crucial aspect of these socio-natural relations is producing national imaginaries, often rooted in particular territorial forms that can extend well beyond the state's own boundaries.

What do these national imaginaries mean for water justice? To the extent that national development programs are rooted in understanding progress and, ultimately, national identity, they necessarily reflect powerful actors' vested interests, most often excluding marginalized groups. National imaginaries, in this sense, are always partial and exclusionary in promoting specific development visions. These are reflected in producing more-or-less stable "resource regimes," understood as "territorialized complexes of political economic institutions, socio-natural relations, social norms and knowledges, which support particular forms of resource extraction and the interests of those involved, over an extended period" (Marston and Perreault, 2017: 255–56).

Part II has already illustrated how different actors and entities deploy and generate multiple complementary or alternative territorial imaginaries, all of which interact, encounter, overlap or question the others. As a result, in one and the same political-geographical place we find multiple "territorial layers" that mix but also compete with and oppose each other, for instance, based on indigenous histories and imaginaries, state-administrative organization, territorialized resource extraction, or river-basin governance structures. As Ferguson (2005) and Vos and Hinojosa (2016) show, transnational networks and companies add yet another "territorial complex" profoundly challenging or dynamizing locally or nationally existing imaginaries regarding territorial boundaries and resource regimes. In this sense, resource regimes' stability and reproduction rely on constructing hegemony, rooted in popular consent and sustained by certain forms of "governmentality" (see Chapter 1 of this volume) but always underwritten by the threat of force. Subaltern, insurgent or otherwise unauthorized forms of water knowledge and management are tolerated only to the extent that they do not pose a threat to hegemonic resource governance. What emerges from this scenario is the understanding that water governance is always already political and cannot be otherwise. Hegemonic water governance regimes and forms of governmentality serve both to reflect and reproduce uneven relations of social power; environmental and water justice alliances that seek to materialize alternative hydro-territorial projects crucially depend on building counter-images and discourses, and mobilizing political power at different interconnected scales. Such is the case with the water protectors at Standing Rock, who have vowed to continue their struggle, this time on a national stage and with a dense network of alliances that links them to activists across the continent and throughout the world.

Structure of Part IV

The chapters in Part IV examine the relationship among state, elite and expert-based power and forms of alternative/subaltern water knowledge and practice. They explore the complex power relations involved in contesting and accommodating hegemonic forms of water knowledge and associated practices by state and capital. Ideas, imaginaries, values, norms, and perceived benefits and drawbacks of water governance are steered by discourses about what constitutes good or bad water management. They tend to present "efficient water management," "effective rules and institutions," "environmentally just practices," or "gender-sensitive water development" as naturally given concepts and goals. But to what extent are these commonly defined and accepted conventions? Is there a shared public water interest and who is entitled to frame it? How much can one pollute a river in the name of progress? Can one compensate or pay for an environmental service? In much thinking and acting on water, such societal notions about justice remain implicit. In debates on water policies and projects, different actors frame their arguments according to strategic deliberations and a pragmatic assessment of how well their ideas resonate in relevant policy-making arenas or with stakeholder groups. This section presents chapters exploring such mechanisms of water governmentality, the politics behind the art of government. Particularly, the chapters explore how existing water communities' specific cultural histories, knowledge, and collective memories may be erased by national water plans and global policy ideas. The chapters illustrate that water conflicts are indeed about allocating water and conserving natural resources, but are also about the collective values and meanings attached to hydro-territorial imaginaries, water ontologies and epistemologies shaping water's substance and relationships, the divergent ways in which actors conceptualize, think, dream and reminisce about water's complex socio-nature.

Chapter 15 ("Neoliberal Water Governmentalities, Virtual Water Trade, and Contestations") by Jeroen Vos and Rutgerd Boelens, examines the enormous worldwide increases in virtual water trade during recent decades. Increased water extraction, consumption and pollution by agribusiness and extractive industries – in most instances with global financial institutions' financial backing and local and national states' political support – directly affect local communities by depleting fragile water balances, re-patterning local water flows and livelihoods, and altering the cost-benefit structure associated with water use. Often, water governance restructuring is accompanied by emerging global water stewardship discourses that have profound effects on local water-user communities' ideas about water governance and justice. This chapter examines how different governmentality mechanisms foster general acceptance of these water stewardship discourses, paying particular attention to new phenomena such as green accounting, water stewardship standard-setting, and certification schemes.

In Chapter 16 ("Critical ecosystem infrastructure: governing the forests-water nexus in the Kenyan Highlands") Connor Joseph Cavanagh examines state water and forest governance in Kenya, and the emerging tensions and contradictions associated with state attempts to implement the "green economy" amid widespread corruption and political division. In Kenya, a country that is increasingly drought-prone, five upland watersheds provide an estimated 75 percent of renewable surface water, together supporting the livelihoods of millions of small-scale farmers and pastoralists, as well as a growing number of commercial agribusinesses, and hydroelectricity-generating schemes. As a result, a growing number of national and international actors have called for the conservation of these areas, as vital to Kenya's emerging "green economy." On this basis, numerous state agencies have sought to reconsolidate their control over the country's upland forest estate, deploying military and paramilitary forces to violently evict several traditionally forest-dwelling indigenous groups. Such efforts suggest a shifting conceptualization of these areas not simply as commercially valuable ecosystems, but increasingly as a kind of critical ecosystem infrastructure, which must function effectively for profitably "green" economies and potentially also for political stability as such. As the chapter demonstrates, however, certain elements within the state are simultaneously colluding with both commercial and artisanal loggers to illegally deforest portions of these same protected areas, in some cases allocating the newly converted land to political supporters. Thus, the chapter interrogates the "Janus-faced" nature of state environmental governance, and considers the tensions within Kenya's emerging "green economy."

In Chapter 17 ("The meaning of mining, the memory of water: collective experience as environmental justice"), Tom Perreault uses collective memory as a lens to examine power differentials in water governance and resource extraction in Bolivia. Memory, expressed verbally as spoken and written narratives, or visually through public art and monuments, plays a fundamental role in how people come to understand environmental degradation, its causes and potential remedies. In Bolivia, mining is memorialized as central to collective national experience, constructing a national identity as a *país minero* (mining country). Memory is similarly important, though less public, for populations impacted by mining contamination and their claims for reparations. This chapter considers the case of indigenous peasant communities and their exposure to mine-related water pollution on the Bolivian Altiplano. Drawing on ethnographic research, the author argues that memory – as stories told about past experience – necessarily requires selective remembering and forgetting, and is best viewed as a political and ideological resource in its own right. As representations of the past, memory is always also a representation of the present, and a reflection of contemporary realities, which in turn informs political demands. The author argues that memory is a vital conceptual tool for theorizing environmentally just futures.

Finally, in Chapter 18 ("New spaces for water justice? Groundwater extraction and changing gendered subjectivities in Morocco's Saïss region"), Lisa Bossenbroek and Margreet Zwarteveen shed a feminist light on the mechanisms and implications of water reallocation from low- to high-value uses in Morocco's fertile Saïss plain. The authors argue that gendered social relations, forms of labor and ways of being, both shape and are shaped by new agricultural and groundwater dynamics. They argue that groundwater abstraction dynamics and gender relations are mutually constitutive, and cannot be analytically separated. The chapter demonstrates that intensifying groundwater use, and the broader changes it entails, are enabled by the gendered organization of labor and space. Complex configurations of class, generation, and gender intersect with both existing agricultural practices, and new policy directions and market opportunities. Together, they produce a variety of arrangements for farming and water use. Drawing on ethnographic documentation of people's everyday experiences with changing groundwater dynamics, the authors suggest that there is merit in theorizing water dynamics and gender dynamics as intimately linked materially (through labor and property relations), and symbolically (through norms, meanings and symbols).

References

Archambault II, D. (2016). Taking a stand at Standing Rock. *The New York Times*, www.nytimes.com/2016/08/25/opinion/taking-a-stand-at-standing-rock.html.

Billo, E. and Zukowski, I. (2015). Criminals or citizens? Mining and citizen protest in Correa's Ecuador. *NACLA Report on the Americas*, https://nacla.org/news/2015/11/02/criminals-or-citizens-mining-and-citizen-protest-correa's-ecuador.

Bridge, G. (2013). Resource geographies II: The resource-state nexus. *Progress in Human Geography*, 38(1), 118–30.

Ferguson, J. (2005). Seeing like an oil company. *American Anthropologist*, 107(3), 377–82.

Healy, J. and Fandos, N. (2016). Protesters gain victory in fight over Dakota Access oil pipeline. *The New York Times*, December 4, www.nytimes.com/2016/12/04/us/federal-officials-to-explore-different-route-for-dakota-pipeline.html.

Huber, M. (2013). *Lifeblood: Oil, Freedom and the Forces of Capital*. Minneapolis: University of Minnesota Press.

Kashi, E. and Watts, M. (2010). *Curse of the Black Gold: 50 Years of Oil in the Niger Delta*. New York: Powerhouse Books.

Marston, A. and Perreault, T. (2017). Consent, coerción and *cooperativismo*: Mining cooperatives and resource regimes in Bolivia. *Environment and Planning A*, 49(2), 252–72.

McNish, J. and Robertson, G. (2014). "The deadly secret behind the Lac-Mégantic inferno." *Globe and Mail*, www.theglobeandmail.com/report-on-business/industry-news/energy-and-resources/the-hazardous-history-of-the-oil-that-levelled-lac-megantic/article15733700/?page=all.

New York Times. (2016a). Power imbalance at the pipeline protest. *The New York Times*, November 23, www.nytimes.com/2016/11/23/opinion/power-imbalance-at-the-pipeline-protest.html.

New York Times. (2016b). Time to move the Standing Rock pipeline. *The New York Times*, November 3, www.nytimes.com/2016/11/04/opinion/time-to-move-the-standing-rock-pipeline.html.

Sammon, A. (2016). A history of Native Americans protesting the Dakota Access pipeline. *Mother Jones*, September 9, www.motherjones.com/print/313506.

Sawyer, S. (2004). *Crude Chronicles: Indigenous Politics, Multinational Oil and Neoliberalism in Ecuador*. Durham, NC: Duke University Press.

Smith, M. (2017). Standing rock protest camp, once home to thousands, is razed. *New York Times*, February 23, www.nytimes.com/2017/02/23/us/standing-rock-protest-dakota-access-pipeline.html.

Smith, M. and Blinder, A. (2017). North Dakota arrests 10 as pipeline protest camp empties. *The New York Times*, February 22, www.nytimes.com/2017/02/22/us/a-deadline-looms-for-dakota-protesters-to-leave-campsite.html.

Vos, J. and Hinojosa, L. (2016). Virtual water trade and the contestation of hydrosocial territories. *Water International*, 41(1), 37–53.

Watts, M. (2004). Antinomies of community: Some thoughts on geography, resources and empire. *Transactions of the Institute of British Geographers*, 29, 195–216.

15

Neoliberal Water Governmentalities, Virtual Water Trade, and Contestations

Jeroen Vos and Rutgerd Boelens

15.1 Introduction

Global virtual water trade has increased enormously during recent decades. Agricultural, mining and hydrocarbon exports are promoted and increasingly governed by international financing, free-trade agreements, and water stewardship policies, upstaging local and national decision-making institutions. Increased water extraction, consumption and pollution by agribusiness and extractive industries affect many local communities directly, depleting fragile water balances, re-patterning local water flows and livelihoods, and altering the local-national-global structure of costs and benefits associated with water use (GRAIN, 2012; Mena-Vásconez et al., 2016; Smaller and Mann, 2009, Vos and Hinojosa, 2016).

To better understand how growing transnational trade in agricultural and extractive industry products affects society and the environment, the concept of "virtual water" can be useful. Virtual water is the water consumed in the production process (Allan, 1998, 2003). With steeply increasing trade in agricultural products, virtual water exports are also booming, with manifold effects for exporting regions: export agriculture generates jobs, income for land owners and investors, but also depletes and contaminates water resources. Furthermore, it changes global and local water governance structures, affecting water justice.

Looking at water justice issues from the virtual water trade perspective provides an opportunity to link inequalities both in and among different hydrosocial territories (Boelens et al., 2016), relating the producing geographies with the consuming places. Which groups and regions benefit from intensifying virtual water trade, and which ones stand to lose? From a political-ecology point of view, it is important to look at cost-benefit distribution, degradation and marginalization, environmental conflict, conservation and control (Perreault et al., 2015; Vos and Hinojosa, 2016). Beyond virtual water export impacts on economic and welfare distribution issues, there are aspects of socio-ecological integrity, representation in decision-making on water affairs, and recognizing plural water values, norms and governance forms – including those of subaltern groups (Zwarteveen and Boelens, 2014; cf. Fraser, 2005).

Although mainstream economics would predict virtual water to flow from relatively wet to relatively dry places, according to the laws of comparative advantage – the free or open world market would "automatically" contribute to solving water scarcity – various studies have shown that virtual water often flows in the opposite direction (Vos and Boelens, 2016). Indeed, water scarcity is not a good predictor for virtual water trade and flow patterns. In many cases, agricultural products are exported from relatively water-scarce places, such as Peru's desert coast and western India, and imported by relatively wet regions, such as northern Europe (GRAIN, 2012; Progressio, 2010; Roth and Warner, 2008; Sojamo et al., 2012).

Virtual water export is often actively stimulated by international financing agencies and governments, through trade policies or subsidies for large water and transport infrastructure. The resulting economic returns and negative effects are distributed unequally over groups in society (Smaller and Mann, 2009; Solanes and Jouravlev, 2007; Swyngedouw, 2005). Although the virtual water concept certainly has its limitations, it enables three important insights regarding water justice. First, virtual water trade generates negative ecological and social effects in relatively dry export regions, and in geographies already characterized by large water-distribution inequities (quantity and quality). Over-exploitation and pollution of surface and groundwater tends to affect poor local communities disproportionally (GRAIN, 2012; Mehta et al., 2012; Smaller and Mann, 2009). Second, virtual water flows, around the world, are increasingly governed by private multinational companies, such as agro-export companies and multinational retailers. Private investment decisions, water use and contamination, and water stewardship certification schemes set by those companies become increasingly influential, thus modifying international water governance at the expense of local and national democratic control (Van der Ploeg, 2008, 2010; Sojamo et al., 2012; Vos and Hinojosa, 2016). Third, the rationale behind virtual water export is promoted actively through what this chapter will call "neoliberal water governmentalities": socio-material, techno-political governance rationalities that deploy water-market forces, institutions and techniques, and proliferate neoliberal "securitization languages," in order to push for national policies that stimulate agro-export, reallocate water to "highest added-value" producers, and make all water players want to join the water market and virtual water export regime. With arguments of "modernization," "water-use efficiency" and "international competitiveness," water is reallocated to private companies at the expense of local water-user communities (Sojamo et al., 2012; Vos and Boelens, 2014; Vos and Hinojosa, 2016).

This chapter examines how different governmentality (government-rationality) mechanisms (Foucault, 1991) foster general acceptance of these water stewardship discourses, paying particular attention to new phenomena such as green accounting, water stewardship standard-setting and certification schemes. New water governance ontology, epistemology and materiality are composed, installed and made commensurable through myriad governance techniques. We show how global extractive industry and agrifood chains endeavor to re-pattern hydrosocial territories' material and hydrological foundations; further, through neoliberal water governmentalities, they also produce particular subjectivities vis-à-vis water consumption. Next, by normalizing and naturalizing corporate business's water governance supremacy, they strategize to also spread and overlay a new water justice imaginary and discourse. These new corporate water justice discourses, however, do not go uncontested. The chapter concludes with several examples of "movements against the current," challenging the ever-subtler corporate water strategies and discourses.

To scrutinize the mechanisms and technologies of virtual water trade and neoliberal governmentalities, the chapter is structured as follows. We first set the stage with a brief overview of increased virtual water trade's consequences. Then we explain the notion of governmentality in relation to virtual water trade. Next, we present three main forms of neoliberal water governmentalities related to agricultural (virtual) water export. Thereafter, we present an overview of various forms of contestation against neoliberal water governmentalities. We end the chapter with some concluding remarks.

15.2 Effects of Virtual Water Export

International trade in agricultural commodities has grown tremendously over recent decades. Figure 15.1 shows the export volume index of selected commodities since 1961, accelerating especially since the turn of the century. In particular, soy exports from South America to China and India have increased, but also, for instance, fresh fruits and vegetables from the Global South to Europe and the USA have boomed. Free-trade agreements and other neoliberal policies, together with changing consumption patterns and consumers' increased purchasing power, have quadrupled agricultural product exports worldwide since 1990 (Vos and Hinojosa, 2016).

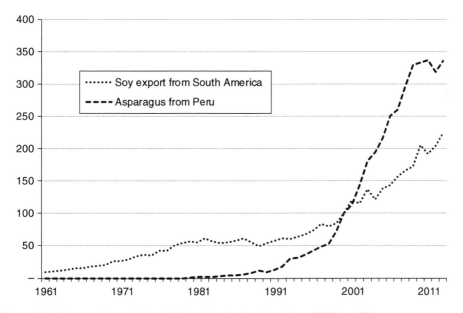

Figure 15.1: Export volume index of selected agricultural products (2000 = 100)
Source: FAO STAT, 2017.

Consequently, exports of "virtual water" embedded in the agricultural exports traded have grown tremendously (Dalin *et al.*, 2012). Currently, some 30 percent of worldwide water withdrawal is for export agriculture (Chen and Chen, 2013). Allan (1998) and others (e.g. Zhao *et al.*, 2009) have argued that importing virtual water is a good alternative to domestic production for water-deficit nations such as Egypt. However, in reality, net virtual water flows often run from relatively dry to relatively wet places (De Fraiture *et al.*, 2004; Ramírez-Vallejo and Rogers, 2010). Negative impacts for export regions include surface and groundwater depletion and water pollution (Dabrowski *et al.*, 2009; Dauvergne and Lister, 2012; Hoekstra and Chapagain, 2008; Khan and Hanjra, 2009; Hoogesteger and Wester, 2015). Moreover, in agro-export regions, water resources tend to be concentrated in the hands of a few, excluding local communities from access to water (GRAIN, 2012; Mehta *et al.*, 2012; Vos and Boelens, 2014).

Water removal from rivers has increased so much that many major rivers no longer reach the ocean for part of the year, a phenomenon known as "basin closure" (Molle *et al.*, 2010). Groundwater is also depleted by irrigation. In many agro-export regions, groundwater levels drop more than a meter per year. Worldwide, the use of groundwater grew from approx. 100 km^3 in 1950 to some 1,000 km^3 in the year 2000 (Shah *et al.*, 2007). Countries with rapidly increasing groundwater extraction include India, China and Bangladesh (Wada *et al.*, 2010). Groundwater depletion is also clearly linked to agro-export in places such as Northern Mexico, the Coast of Peru and northern Chile, North Africa, western India, Pakistan, Southern Thailand and Occupied Palestine Territory. Another major problem with agricultural production and especially export agriculture is water contamination by agro-chemicals such as fertilizers, herbicides and insecticides (Devine and Furlong, 2007).

The negative effects of virtual water export for local communities can be witnessed in many of the above places. Both water depletion and contamination deprive those communities of access to clean water for domestic and productive uses. Notwithstanding creating jobs and income for marginalized groups (including many women) in agri-business, social inequality and exploitation are common (Bee, 2000; Pearson, 2007).

15.3 Governmentalities and Neoliberalizing Virtual Water Flows

15.3.1 Governmentality

Many water policies, taken-for-granted academic conventions and generally held ideas on "good" water governance can be related to neoliberal water governmentalities, alluding to efficiency and productivity ideas from the industrial era, Adam Smith's "invisible hand," and the approach of stakeholder dialogues to forge compromise and consensus based on universalistic notions of modern progress and harmonizing (inter)national water interests. Often the "modernization/efficiency/productivity" argument draws upon the Malthusian threat: population growth and scarce natural resources. Therefore, the "natural solution" – commodified water resources and services; privatized, tradable water property rights; and installing water resource and service markets to make water flow to the most efficient users (who are assumed to be most interested in investing in and maintaining water facilities) – is commonly presented. Indeed, those ideas are "naturalized" in national water-policy documents and university water-studies curricula, and in international organizations and private companies' water stewardship programs (Achterhuis *et al.*, 2010; Roth *et al.*, 2015; Smaller and Mann, 2009). Before we discuss three neoliberal water governmentality schemes, we first briefly overview the governmentality concept.

Governmentality was defined by Michel Foucault as "the conduct of conduct" (Foucault, 1991: 48, 87–104; Dean, 1999), "the myriad ways in which human conduct is directed by calculated means" (Ferguson and Gupta, 2002: 989). Rather than just by legal force or state-endorsed violence, state and non-state entities have increasingly developed a range of multiform government tactics to productively and economically *manage and direct* society: "it is no longer the right to kill, to employ force, that forms the essence of the figure of the governor ... but rather the knowledge of things, of the objectives that can and should be attained, and the disposition of things required to reach them" (Foucault, 1991: 96). In this Foucauldian line of thought, Burchell *et al.* (1991: 102) describe governmentalities as "the ensemble formed by institutions, procedures, analyses and reflections, calculations and tactics that enable the exercise of this very specific albeit complex form of power, which has as its target *population*, as its principal form of knowledge *political economics*, and as its essential technical means *apparatuses of security*."

Governmentality, beyond a mode of only state power, involves governance technologies that may be deployed by a multitude of powerful entities (de facto "governors"), e.g. multinational companies, intergovernmental organizations, NGOs, or terrorist organizations (Ferguson and Gupta, 2002); in other words, it contains "methodological devices through which the publicly observable rationalities, procedures and techniques of state and non-state actors can be read as proxies for processes of subject-formation" (Barnett *et al.*, 2008: 625).

This government practice rationality becomes manifest in dominant interests and strategies to investigate, document, interpret, classify, and (re)shape society – thereby giving order and substance to, for instance, convenient territory and territoriality (as in "hydrosocial territories," see Boelens *et al.*, 2016; Hommes *et al.*, 2016). As Foucault observed, "government is the right disposition of things ... the things with which in this sense government is to be concerned are in fact men, but men in their relations, their links, their imbrication with those other things which are wealth, resources, means of substance, the territory with its specific qualities, climate, irrigation, fertility, etc.; men in their relation to that other kind of things, customs, habits, ways of acting and thinking, etc." (Foucault, 1991: 93). Therefore, Rogers *et al.* (2016: 429) argue that governmentality "attempts to render space governable" and Li (2011: 57) states that it concentrates, crucially, on "rendering society technical."

Foucault (2008: 82) differentiates among diverse modes or "arts of government," looking at the techniques and strategies by which a society is rendered governable. While he first reserved the governmentality concept for recent, liberal-modernist modes of governance, in his Birth of Biopolitics lectures (2008[1979]) he distinguishes among four historical forms (see also Chapter 1, in this volume). The first form of governmentality relates to the art of government according to Truth. Truth governmentality derives from transcendental, "unquestionabilized" (or rather, "super-naturalized") forms of powers, such as in religious belief systems (see, e.g. Boelens, 2014). The second concerns the art of government according to sovereign rule; associated with setting rules top-down and enforcing obedience and compliance with those rules. Third, there is ruling through discipline: the normalizing power that deploys moral and behavioral rules, as subtly shifting norms in relational webs that invoke guilt, desires to become "normal," and self-surveillance by "subjectified" subjects. The fourth mode and rationale of governmental power is neoliberal governmentality, which functions by setting incentives to make a "free market" function in different realms of society (Ferguson and Gupta, 2002; Fletcher, 2010). Therefore, neoliberal governmentality simultaneously builds on and conflates individual responsibility to make choices and deploys calculated economic rationality to steer society. In that respect, Lemke (2007: 12) asserts that "one key feature of the neoliberal rationality is the congruence it endeavors to achieve between a responsible, moral individual and an economic-rational individual."

Neoliberal governmentality, as Foucault asserted, "should not ... be identified with laissez-faire, but rather with unceasing vigilance, activity, and intervention" (2008: 132). Indeed, rather than the state's "retreat" or "rolling-back," key government functions are transferred to non-state entities (who join the effort of governmentalizing subjects) while the state itself takes the task of both facilitating market operation and transforming its core institutions (hospitals, schools, social security, national security, police, etc., including natural-resource management and water governance) into market enterprises.

15.3.2 Neoliberal Water Governmentality

In this chapter, we use the term "neoliberal water governmentalities" for that particular art of government comprising socio-material, discursive techniques and practices to promote and legitimize water use for agribusiness and extractive industry exports, building and steering a hydro-political society based on transnational marketing of material and virtual water. In water governance, this variegated set of governmentalities consists of truth regimes aiming to (re)produce socio-natural order and acceptance via transnational market-based positioning of and control over water, natural resources, infrastructure, investments, knowledge, and ultimately, populations. Neoliberal water governmentalities aim to organize and direct water users' behavior by approaching users as rational, enterprising agents who economically benefit from water development – more precisely: as individual utility-maximizers who strategically calculate costs and benefits to materialize personal interests (see Achterhuis et al., 2010; Duarte-Abadía and Boelens, 2016; Li, 2011). Accordingly, national and transnational governments and corporations set out to support and install the right economic incentive structures to incorporate local water users into the market's hydro-political order.

Consequently, neoliberal water policies set out to create hydrosocial territories that redesign existing, vernacular territories to facilitate market-driven water allocation, in particular to export companies (Boelens *et al.*, 2015, 2016). Building large and medium-sized dams, large inter-basin transfers, deep tube wells, and distribution networks all reorder existing territories (Swyngedouw, 2015; see also Chapter 6 of this volume). Neoliberal national legislation, subsidies for export agriculture, free-trade agreements, and slack enforcement of conservation/environmental legislation all foster virtual water export. The imaginaries of a "modern" watershed and "enlightened" techno-economic managerial governance norms and practices are powerful means to promote this "modernist, leading-edge" hydrosocial grid, often connected to developing hydropower, mining, transport and other capitalist-enterprise infrastructure (Hommes *et al.*, 2016; Hommes and Boelens, 2017).

As mentioned above, not only national governments promote virtual water trade; international organizations and transnationalizing private companies equally advocate these ideas of "good water governance." Permissive environmental legislation and corporate social responsibility standards for "water stewardship" certification also encourage the hydro-political (but de-politicized) interests of large agro-export companies (Fulponi, 2007). This promotion goes hand-in-hand with stigmatization of "inefficient," "unproductive" smallholders' "non-competitive" farming practices. Training programs offered to smallholders or, for example, subsidies for low-cost drip and sprinkler technology, suggest that smallholders use irrigation water inefficiently and unproductively. This also legitimizes reallocating water from the inefficient smallholder sector to industrial, domestic, tourism and export agriculture sectors (e.g. Bossenbroek, 2016; Venot *et al.,* 2017; Vos and Boelens, 2016). Therefore, neoliberal water governmentalities promote a particular idea of "efficient," "productive," "modern" water use and governance. Subaltern groups have no say in these matters. Rather, as we have shown (Vos and Boelens, 2014, 2016), neoliberal water governmentalities lead to a process of "subjectification" (Foucault, 1980), forcing people into normative categories of "unproductive" farmers or "wasteful" consumers – categories, however, that – together with the discursive legitimization of virtual water export – are increasingly being contested.

15.3.3 Three Neoliberal Water Governmentality Discourses

Neoliberal water governmentality schemes set out to induce water-management regimes and water-use behavior whereby water users, consumers, water experts and policy makers internalize discourses and practices related to water's universally calculable economic value, technified water-use efficiency criteria, transferring water to the highest economic value and (therefore) the most efficient users, following scientific water-sector authorities' prescriptions, and feeling guilty if not adhering to these standards (Boelens and Vos, 2012).

In this next section, we illustrate three types of (often interrelated) neoliberal water government-rationalities:

- the rationality of "highest economic water-use value and technical water-use efficiency;"
- the rationality of "corporate water stewardship;"
- the rationality of "water footprint" or "water accounting."

Highest Economic Water-Use Value and Technical Water-Use Efficiency

The main argument of the "highest economic water-use value" is that water should be allocated to those users who produce the highest economic value with that particular water resource. High-value use is associated not only with export agriculture, but also with "modern" agriculture, high water-application efficiency, high water-use productivity, and high investments in major water infrastructure.

Trade in virtual water is supposed to optimize use of the planet's scarce water resources. "Virtual water" should flow from relatively "water-rich" countries to relatively "water-poor" countries. Hoekstra and Chapagain (2008: 138) phrased the neoliberal assumption about virtual water trade in this way: "Liberalization of trade seems to offer new opportunities to contribute to a further increase of *efficiency* in the use of the world's water resources" (emphasis added). Trade liberalization would only diminish pressure on water resources if water were priced according to relative scarcity. Full water costs should include production costs and externalities. In that case, the market's "invisible hand" would result in producing goods with high water productivity in relatively water-scarce areas and producing goods with low water productivity in relative wet areas. According to Hoekstra and Chapagain, "such a protocol would also contribute to *fairness*, by making producers and consumers pay for their contribution to water depletion and pollution." (ibid.: 142, emphasis added).

Many neoliberal governments pursue policies that actively promote virtual water export, but they are not alone. Socialist governments, private agribusinesses, development-consultancy firms and mining companies equally tend to be fervent promoters. Private business networks often organize nationally or globally to lobby for policies favoring their export facilities, implicitly involving virtual water. An example is AGAP, the organization of agribusiness companies in Peru. They actively promote the image of agribusiness being efficient and productive, and lobby against legislation that would make them pay a fee to use groundwater. Furthermore, transnational organizations such as the WTO, the FAO (e.g. Farsund *et al.*, 2015) and the World Bank (e.g. Chaherli and Nash, 2013) promote agricultural-export development.

Neoliberal governments can foster reallocation to "high-value water uses" by national legislation and subsidies. In Chile, tradable water rights were established in 1981 (Achterhuis *et al.*, 2010; Bauer, 2005) and, in Peru and Colombia, legislation rewards the installer of water-saving technology by higher priority in water rights allocation (Roa-García, 2014). However, virtual water export policies become most manifest in large-scale water infrastructure projects such as large dams, large inter-basin water transfers and related large roads, ports and electricity grids, whereby watersheds are "re-territorialized"; i.e. hydrosocial territories are fundamentally "governmentalized" (Boelens *et al.*, 2016). Examples are the Ica, Olmos and Chavimochic projects in Peru (Lynch, 2013) and the Daule-Peripa project in Ecuador (Hidalgo *et al.*, forthcoming), in which water and land are exclusively allocated to big agribusiness companies. In many semi-arid and arid places, governments actively promote groundwater use by subsidizing energy for pumping water (Venot *et al.*, 2017), leading to groundwater over-extraction.

Many governments seek a solution for groundwater depletion from more efficient water use by inefficient farmers. For example, in Pakistan where much irrigation water is used to grow cotton for export, the Executive Director of the Sustainable Development Policy Institute, Abid Qaiyum Suleri, considers that: "improving water storage will be important. But [he] believes another way to combat the growing water shortage is to teach farmers how to effectively irrigate crops with less water. 'The Government should introduce water-efficient seed varieties in water-stressed areas and train farmers on using drip irrigation and sprinkle irrigation to save the scarce resource,' he said." (Reuters, 2015: 1).

Another example of neoliberal government planning is the Moroccan national agricultural development program: the *Plan Maroc Vert* (Green Morocco Plan), developed by the McKinsey private consultancy company (2011) and financed partly by the World Bank. McKinsey recommends for the Moroccan Government to develop export agriculture by extracting groundwater and using drip irrigation. The Moroccan Government has now started a program to subsidize private deep-tube wells, which has led to over-extraction from the aquifer, concentrating land in the hands of the (urban) elite, destroying communal irrigation traditions, and empowering young men to the disadvantage of women, who were not associated with "modern" agriculture (Bossenbroek, 2016).

The "efficiency discourse" is powerful, not only in agricultural water governance but also in other virtual water export domains. Governmental land-use planning in Peru and Mexico, for instance, assures that rules will open up fragile landscapes and food-based community livelihoods to introduce and favor open pit mining to export gold (Preciado-Jeronimo *et al.*, 2015; Stoltenborg and Boelens, 2016).

The efficiency-argument logic resonates with beliefs held by many people in society who care about the environment. Who can be against efficiency? However, the impacts for water user communities in local watersheds are devastating: they lose control over their water resources and territories. Meanwhile, public subsidies and governmental discourses are mobilized to finance large infrastructure, fostering groundwater exploitation, conversion to drip irrigation and "modernizing" irrigation systems. Subtly, in many places, spreading, de-politicizing, and naturalizing the "efficiency ideology" assures that small and indigenous farmers – hardest-hit by the neoliberal efficiency myth and resulting water dispossession – start to blame themselves and feel guilty about their "inefficiency" (Boelens and Vos, 2012). The remedy is enhancing water market rules: the self-fulfilling prophecy of neoliberal water governmentalities.

Corporate Water Stewardship

"Water stewardship" rationality refers to private companies' claim that they do not negatively affect water resources during their production process (or they compensate for those negative effects elsewhere). This is part of the broader corporate social responsibility (CSR) discourse deployed by agribusiness, retailers and mining corporations (e.g. Dalton and Newborne, 2016). Companies claim they comply with countries' environmental and social legislation where they operate, and claim they establish voluntary measures to conserve water resources. Most of those measures are technological innovations such as installing drip irrigation, recycling process waters, desalinizing seawater, or using drought-resistant crops. Those measures might indeed "save" water in the particular processes; nevertheless, social and environmental impacts might be negative, as those "technofixes" might use more energy, encourage more overall use of water, and serve to "greenwash" the company's image (Blowfield and Dolan, 2008; Boelens and Vos, 2012; Mutersbaugh and Lyon, 2010; Vos and Boelens, 2014).

Four types of "water stewardship governmentalities" may be distinguished: (1) private water stewardship standards, (2) water offsetting, (3) claiming increased water-use efficiency, and (4) techniques to make water use more visible.

The first "water stewardship governmentality" mode concerns private water stewardship certification schemes. In many cases, they are part of wider environmental and social responsibility standards, such as GlobalGAP (Global Good Agricultural Practices), the Rainforest Alliance, BCI (Better Cotton Initiative), the RTRS (Roundtable on Responsible Soy) and the RSB (Roundtable on Sustainable Biomaterials). An example is how, in 2002, Nestlé – together with two other major food companies (Unilever and Danone) – created the Sustainable Agriculture Initiative (SAI) Platform: "a non-profit organization to facilitate sharing, at a precompetitive level, knowledge and best practices to support development and implementation of sustainable agriculture practices involving stakeholders throughout the food value chain" (SAI, 2017: 1). The SAI standard on water use reads: "Farm enterprise activities shall not knowingly deplete available water resources, beyond the recharge capacity of the watershed/catchment, by direct abstraction," and: "The most efficient, commercially practical water-delivery system is used. In addition, water-saving practices should be adopted and water should be reused or recycled where possible" (SAI, 2017). Other examples of international organizations and platforms that foster 'water stewardship' standards are: the CEO Water Mandate (UN Global Compact), World Business Council for Sustainable Development (WBCSD) Water, Alliance for Water Stewardship (AWS), and multinational NGOs such as The Nature Conservancy (TNC), International Union for Conservation of Nature (IUCN), World Wildlife Fund (WWF) (e.g. AWS, 2011).

As we have detailed elsewhere (Vos and Boelens, 2014), they all have some major water-injustice problems in common, regarding water stewardship certification: technocratically defined standards; low stakeholder participation (and no small-holder or indigenous federations involved) in setting the standards; excluding smallholders from producing for supermarket chains; insignificant environmental benefits; vulnerability to poor monitoring and fraud; and therefore the risk of simply greenwashing corporate images. Moreover, the discourse on "inefficient smallholder" farmers tends to induce water legislation that legitimizes and formally legalizes taking away their water rights. Export farmers who become privileged in obtaining water rights are portrayed and present themselves as "efficient" (using drip irrigation) and "productive" (obtaining high yields). However, this does imply using large water volumes, which can be witnessed in many places around the world, where these "efficient" water stewardship champions devastatingly deplete aquifers. In Peru, for instance, high-tech agribusiness companies earn "water efficiency certificates" giving them water-rights prioritization, which enables them to extend their irrigation areas with high-consumption high-value crops, draining local water resources and drying out smallholder wells (Roa-García, 2014). Thus, in many cases, water-saving technology supported by water stewardship discourses does not save water; it merely shifts water use from smallholders to big export agribusiness (Boelens and Vos, 2012).

The second type of water stewardship governmentalities are schemes to offset negative effects on water resources. The idea is to compensate for negative effects elsewhere. Compensation in another place presumes the ability to quantify ecological and social damage and benefits. This presumes commensurable valuation of both ecological and social damage and benefits. Oftentimes economic valuation or ecosystem services approaches are used. Habitually, the stakeholders affected are not invited to evaluate the effects and benefits, and no democratic procedures are established to facilitate negotiation over the value and distribution of costs and benefits to social groups.

Coca Cola watersheds are an example of offsetting (Raman, 2007). In some regions, local Coca Cola companies install drinking water supply systems in villages as a "compensation" for the water they extract from the watershed. Another example is wetland-credit certification regulated by the US Clean Water Act. Private parties can sell wetland credits to another private company, to offset a wetland area it affected. This does not account for the value of biodiversity, complex and interlinked ecological functions or people's territorial values (Robertson and Hayden, 2008).

The third type of water stewardship governmentalities can be found in the domain of private companies that claim they have improved (or will improve) their water use efficiency and therefore use less water. Many major manufacturing and mining companies (and golf courses) have recently claimed that their operations are becoming more water efficient. Nestlé, for example, has reported that they achieved a 41 percent reduction in 2015 in direct water withdrawal per ton of product versus 2005 (Nestlé, 2016). This refers, however, only to the manufacturing factories, not to the agricultural inputs used to manufacture products, which is most of the water being used to produce food. Under this "water use efficiency" label, water extraction is booming.

Many other multinational companies have their own water stewardship programs, e.g. Unilever, Heineken, SABMiller, Coca Cola and Levi's. They mostly focus on factory production and not on large water use and contamination while producing their agricultural inputs.

The fourth water stewardship governmentalities are those that make water use visible. One way is satellite (remote sensing) observation of evapotranspiration to determine water use by different users. Several international organizations and national governments use these data in decision-making regarding water-resource policies. As the World Bank writes:

Advances in the use of remote sensing technologies are now making it possible to cost-effectively estimate crop evapotranspiration (the sum of evaporation and plant transpiration to the atmosphere) from farmers' fields and to improve water accounting and management at the regional and basin-wide levels. Since 2010, China has adopted this approach in the Xinjiang Turpan Water Conservation Project in the arid northwest region of the country.

(World Bank, 2017: 1).

Notwithstanding the potential usefulness of obtaining data on crop water use, the satellite technology has serious pitfalls, as it produces data without any context regarding the populations involved, water-use costs and benefits for different social groups, nor the environmental effects of water use. Data generated by these sorts of high-tech methods are seldom discussed or validated together with the water user groups who are directly involved. Therefore, data are presented to policy makers, who draw conclusions bearing little connection to actual problems and opportunities "on the ground."

Water Footprint

Water footprint rationalities are a governmentality technology for identity-crafting ("subjectifying") consumers. The water footprint is water accounting that calculates water use and contamination caused by generating a certain product or service. Water footprints can be calculated for entire sectors, specific food products, nations and geographical regions, or individual consumers (Hoekstra et al., 2013).

The water footprint accounting system is promoted by the Water Footprint Network and other organizations such as ISO, World Economic Forum, Water Futures Partnership, the UNEP WaFNE program, WWF and GIZ.[1] Its rationality hints at individual consumers' responsibility to reduce virtual water use. The water footprint concept aims to shape individuals' water consumption behavior. This starts with the idea of consumer choice: water footprint calculations make consumers responsible for the environmental and social impact of their buying behavior.

Other concepts that link individual buying choices to distant social and environmental impacts are Life Cycle Analysis and True Pricing. This "responsibilization" is increasingly done by private companies promoting "water stewardship" slogans (Barnett et al., 2008). Multinational Procter and Gamble (P&G) clearly demonstrates this on its webpage: "At P&G, water is of crucial importance for both production and use of our products. Thus, our approach to water aims at *responsible use* by both our Company and *the consumers* who use our products." And: "We will also seek innovative ways *to reduce water use by our consumers* as well as *educate them* about the opportunity to save water. This will positively impact the cost of water to the Company, our consumers and the communities in which we operate" (P&G, 2017: 1, emphasis added).

In a Foucauldian way, reducing water footprints rationality is presented as a productive, positive objective; it certainly may become a powerful tool to induce the "water-saving mentality."[2] In fact, following normalization logic, companies set norms and targets, while consumers "feel" that they have been "involved." Companies aim to make consumers "accomplices" in their extractive production and product marketing game, and address morals, guilt and shame – subtly entwining disciplinary and neoliberal governmentalities. As Li (2011: 58) states, this neoliberal governmentality focuses on "setting conditions under which people are encouraged to take responsibility for their own improvement by engaging with markets, learning how to conduct themselves in a competitive arena and making appropriate choices." These water footprint discourses influence consumer and company decisions, as part of the same Foucauldian, relational web. They are pushed to perceive shared problems and solutions, and they supposedly make "common" decisions on establishing water-saving norms. So, the products that "can be bought safely" tend to have a large impact on producers (as smallholder groups) who have no say in these discussions. As we have demonstrated (Vos and Boelens, 2014), marginalized producer groups are the first to be excluded, since they cannot live up to the new norms regarding technology use, efficiency image, and uniform production output.

15.4 Contestation

Growth in virtual water exports and increased water-intensive cropping by agro-businesses have major implications for how and where water is governed, and how water-user populations and territories are being reframed and re-patterned. This extractivism increasingly meets with resistance. For instance, many local demonstrations and conflicts occur around the direct negative effects of water depletion and contamination. Sometimes such local protests jump scales, and national or international alliances are forged. Local protests by water users' communities against water extraction by agribusiness have been documented in many places, e.g. Ethiopia, Senegal, Tunisia, Mexico, Chile, Peru and Ecuador (cf. Boelens, 2015; Harris and Roa-García, 2013; Vos and Hinojosa, 2016). Nevertheless, compared to mobilizations against mining-sector water extraction, so far grassroots actions against water-extractive agribusiness exports have been limited at best, largely because of strong, deeply asymmetrical dependence ties within that sector (Vos and Boelens, 2014).

In some cases, local protests against water-extractionist export policies or activities of agro-export companies grow into a wider societal movement (e.g. Bebbington *et al.*, 2010; Boelens *et al.*, 2015; Hoogesteger and Verzijl, 2015; Hoogesteger *et al.*, 2016). The "new water culture" movement in Spain is an illustrative example. In 2001, the central Government of Spain launched the National Hydrological Plan (PHN). The main idea was to bring water from the relatively wet north of Spain to the relatively dry south, transferring some 850 million cubic meters of water per year from the Ebro river to the agro-export and tourism regions in southern Spain (see Swyngedouw and Boelens, in this volume). Some years earlier, a broad coalition of Spanish civil-society organizations and universities had been formed in opposition to large-scale inter-basin water transfers and attempts to commensurate economic rationalities in all spheres of water governance. This organization become known as the *Fundación Nueva Cultura del Agua* (Foundation for a New Water Culture, FNCA). That same year, several large protest marches were organized in the streets of Zaragoza, Barcelona and Brussels. These protests forced the Government to change the PHN and abandon the inter-basin transfer plan (instead, the new 2003 PHN would opt for other neoliberal water governmentalities, constructed around investing in improving irrigation efficiency and desalinizing seawater (Swyngedouw, 2015). The FNCA further developed into a national social movement, currently addressing water governance from an ecological and social point of view and pushing for a new, trans-disciplinary, multicultural, cross-collaborative water perspective (see www.fnca.eu). Issues raised by the New Water Culture movement in Spain address, for instance, the fact that rivers and other water bodies have many values far beyond only economic cost-benefit analysis. This is expressed in the idea of *fluviofelicidad* ("enjoying the living river"). They also argue that river basins should be the basis for decision-making and that water management and governance should gather all stakeholders, not just official government authorities and so-called "water experts." Environmental governance needs to assure that rivers are restored and allowed to flow through cities and landscapes in clean, healthy ways.

"New water culture" movements and initiatives also develop in other parts of the world. Examples of local or national organizations include CENSAT Agua Viva in Colombia, the Forum for Policy Dialogue on Water Conflicts in India, the National Water Forum in Ecuador, the Patagonia Without Dams movement in Chile, and many others. They share their objectives with environmental justice and indigenous movements that defend water quality in their territories – such as in the US and Canada (for example, the alliance in Standing Rock, protesting against an oil pipeline planned to go under the Missouri River adjacent to (and upstream from) Sioux territory; see Introduction to Part IV). International NGOs or alliances that also promote similar ideas are, for instance, International Rivers, Food and Water Watch, Earth Rights International, Friends of the Earth International, and the Justicia Hídrica/Water Justice alliance. In different ways and to different degrees, they challenge the discourses and forces that promote and induce neoliberal water governmentalities.

15.5 Concluding Remarks

Hugely increasing international trade in agricultural products also intensifies negative effects for local environments, communities and marginalized population groups. Booming agro-export is legitimized by rationalities alluding to notions of "efficiency," "productivity," and "modernity." Those notions are promoted via neoliberal water governmentalities fostering virtual water trade. Modernist national water sector policies, international financing programs, and private water stewardship standards, among others, provide important substance and force to these governmentalities, also driven by generally held notions on the desirability, universality and simplicity of "efficient" production, and "water-saving" technology.

Though fully impregnated by discourses and strategies of "participation," "inclusion" and "integrated" water resource management, these water governmentality rationalities, norms, and practices are not developed democratically. Water stewardship standards are set by representatives of large agribusiness companies and retailers. Water experts and government advisors define national and international water policies. Many NGOs enthusiastically and uncritically support de-politicization of neoliberal water governmentalities by training water-user collectives how to "rationally and efficiently" use their water resources, subtly inducing market-based notions of water control and rendering hydrosocial territories "understandable, containable and governable." Indeed, although in some cases these governmentality technologies are designed and applied consciously and strategically, most actors deploy and thereby unconsciously reinforce neoliberal water rationalities. These governmentalities gain strength through capillary wicking, participatory capacity-building, inclusive development, and "resonance" within society.

In water governance, control, and use practice, these governmentalities mostly serve the elites, and have very real effects on water flows: water is reallocated to "high-value" export crops and privileged users who could afford to install "water-saving" technologies. This commonly worsens rates of depleting resources and concentrating access to water and other productive resources in the hands of a few capitalist producers. In a rapidly growing number of countries and cases, these "virtual water rationalities" legitimize and expand agro-export companies' use of large volumes of water because they have water stewardship certification. Last but not least, these government-rationalities also reinforce the power of hydrocracies and technocratic water experts, who serve those water users, impoverished and impacted by neoliberal water rules, by explaining to them that the only way to escape from material and mental poverty is to further deploy neoliberal water rationality norms and rules.

Nevertheless, neoliberal virtual water trade rationalities are contested by local water-user households, communities and international alliances. Some local protests articulate at broader governance levels as multi-scale contestations. Many, however, go unnoticed by national and international media and the academic community. They express their dissatisfaction with neoliberal water governmentality silently, through everyday crafting, use and proliferation of non-commodified, non-marketized undertow water rules, norms and practices. They may potentially question the neoliberal water dream's hegemonic rationalities and virtual water export realities.

Notes

1 ISO (International Organization for Standardization), UNEP WaFNE (United Nations Environment Programme Water Footprint, Neutrality and Efficiency), GIZ (German Federal Enterprise for International Cooperation).
2 What water footprints largely neglect is that water use by particular actors, products or services is not proportional to environmental damage (which is specific to modes of extraction) or water depletion (which is influenced by many aspects of water availability). Universal notions obscure particular local circumstances and interests of specific social groups.

References

Achterhuis, H., Boelens, R. and Zwarteveen, M. (2010). Water property relations and modern policy regimes: Neoliberal utopia and the disempowerment of collective action. In R. Boelens, D. Getches and A. Guevara-Gil (eds.), *Out of the Mainstream: Water Rights, Politics and Identity*. London, Washington, DC: Earthscan, pp. 27–55.

Allan, J. A. (1998). Virtual water: A strategic resource: Global solutions to regional deficits. *Ground Water*, 26(4), 545–46.

Allan, J. A. (2003). Virtual water: The water, food, and trade nexus: Useful concept or misleading metaphor? *Water International*, 28(1), 106–13.

AWS. (2011). *The Alliance for Water Stewardship Water Roundtable Process*. Final Draft, April 20. London: Alliance for Water Stewardship.

Bauer, C. (2005). In the image of the market: The Chilean model of water resources management. *International Journal of Water*, 3(2), 146–65.

Barnett, C., Clarke, N., Cloke, P. and Malpass, A. (2008). The elusive subjects of neoliberalism: Beyond the analytics of governmentality. *Cultural Studies*, 22(5), 624–63.

Bebbington, A., Humphreys, D. and Bury, J. (2010). Federating and defending: Water, territory and extraction in the Andes. In R. Boelens, D. Getches and A. Guevara (eds.), *Out of the Mainstream: Water Rights, Politics and Identity*. London, Washington, DC: Earthscan, pp. 307–27.

Bee, A. (2000). Globalization, grapes and gender: Women's work in traditional and agroexport production in Northern Chile. *The Geographical Journal*, 166(3), 255–65.

Blowfield, M. and Dolan, C. (2008). Stewards of virtue? The ethical dilemma of CSR in African agriculture. *Development and Change*, 39(1), 1–23.

Boelens, R. (2014). Cultural politics and the hydrosocial cycle: Water, power and identity in the Andean highlands. *Geoforum*, 57, 234–47.

Boelens, R. (2015). *Water, Power and Identity: The Cultural Politics of Water in the Andes*. London: Earthscan, Routledge.

Boelens, R. and Vos, J. (2012). The danger of naturalizing water policy concepts: Water productivity and efficiency discourses from field irrigation to virtual water trade. *Agricultural Water Management*, 108, 16–26.

Boelens, R., Hoogesteger, J. and Baud, M. (2015). Water reform governmentality in Ecuador: Neoliberalism, centralization and the restraining of polycentric authority and community rule-making. *Geoforum*, 64, 281–91.

Boelens, R., Hoogesteger, J., Swyngedouw, E., Vos, J. and Wester, P. (2016). Hydrosocial territories: A political ecology perspective. *Water International*, 41(1), 1–14.

Bossenbroek, L. (2016). "Behind the veil of agricultural modernization: Gendered dynamics of rural change in the Saïss, Morocco." PhD thesis, Wageningen University, Wageningen.

Burchell, G., Gordon, C. and Miller P. (1991). *The Foucault Effect: Studies in Governmentality*. Chicago: University of Chicago Press.

Chaherli, N. and Nash, J. (2013). *Agricultural Exports from Latin America and the Caribbean: Harnessing Trade to Feed the World and Promote Development*. Washington, DC: World Bank, https://works.bepress.com/aparajita_goyal/29/.

Chen, Z. M. and Chen G. Q. (2013). Virtual water accounting for the globalized world economy: National water footprint and international virtual water trade. *Ecological Indicators*, 28, 142–49.

Dabrowski, J., Murray, K., Ashton, P. and Leaner, J. (2009). Agricultural impacts on water quality and implications for virtual water trading decisions. *Ecological Economics*, 68(4), 1074–82.

Dalin, C., Konar, M., Hanasaki, N., Rinaldo, A. and Rodriguez-Iturbe, I. (2012). Evolution of the global virtual water trade network. *Proceedings of the National Academy of Sciences*, 109(16), 5989–94.

Dalton, J, and Newborne, P. (2016). *Water Management and Stewardship: Taking Stock of Corporate Water Behaviour*. Gland: IUCN.

Dauvergne, P. and Lister, J. (2012). Big brand sustainability: Governance prospects and environmental limits. *Global Environmental Change*, 22(1), 36–45.

De Fraiture, C., Cai, X., Amarasinghe, U., Rosegrant, M. and Molden, D. (2004). Does international cereal trade save water? The impact of virtual water trade on global water use. *IWMI Research Report 4*. International Water Management Institute, Colombo, Sri Lanka.

Dean, M. (1999). *Governmentality: Power and Rule in Modern Society*. Thousand Oaks, CA: Sage Publications.

Devine, G. and Furlong, M. (2007). Insecticide use: Contexts and ecological consequences, *Agriculture and Human Values*, 24, 281–306.

Duarte-Abadía, B. and Boelens, R. (2016). Disputes over territorial boundaries and diverging valuation languages: The Santurban hydrosocial highlands territory in Colombia. *Water International*, 41(1), 15–36.

FAO STAT. (2017). www.fao.org/faostat/en/#data/TP.

Farsund, A. A., Daugbjerg, C. and Langhelle, O. (2015). Food security and trade: Reconciling discourses in the Food and Agriculture Organization and the World Trade Organization. *Food Security*, 7(2), 383–92.

Ferguson, J. and Gupta, A. (2002). Spatializing states: Toward an ethnography of neoliberal governmentality. *American Ethnologist*, 29(4), 981–1002.

Fletcher, R. (2010). Neoliberal environmentality: Towards a poststructuralist political ecology of the conservation debate. *Conservation & Society*, 8(3), 171–81.

Foucault, M. (1980). Power/knowledge. In C. Gordon (ed.), *Foucault. Power/knowledge: Selected Interviews and Other Writings 1972–1978*. New York: Pantheon Books.

Foucault, M. (1991[1978]). Governmentality. In G. Burchell, C. Gordon and P. Miller (eds.), *The Foucault Effect: Studies in Governmentality*. Chicago: University of Chicago Press, pp. 87–104.

Foucault, M. (2008). *The Birth of Biopolitics*. New York: Palgrave Macmillan.

Foucault, M. (with Sellenart, M. and Burchell, G.) (2007). *Security, Territory, Population*. New York: Palgrave Macmillan.

Fraser, N. (2005). Mapping the feminist imagination: From redistribution to recognition to representation. *Constellations*, 12(3), 295–307.

Fulponi, L. (2007). The globalization of private standards and the agri-food system. In J. Swinnen (ed.), *Global Supply Chains, Standards and the Poor*. Wallingford: CABI Publications, pp. 5–18.

GRAIN. (2012). Squeezing Africa dry: Behind every land-grab is a water-grab, www.grain.org/e/4516.

Harris, L. and Roa-García, M. C. (2013). Recent waves of water governance: Constitutional reform and resistance to neoliberalization in Latin America (1990–2012). *Geoforum*, 50, 20–30.

Hidalgo, J. P., Boelens, R. and Isch, E. (forthcoming). Hydro-territorial configuration and confrontation: The Daule-Peripa Multipurpose Hydraulic Scheme in Coastal Ecuador.

Hoekstra, A. and Chapagain, A. (2008). *Globalization of Water: Sharing the Planet's Freshwater Resources*. Oxford: Blackwell Publishing.

Hoekstra, A., Chapagain, A., Aldaya, M. and Mekonnen, M. (2013). *The Water Footprint Assessment Manual: Setting the Global Standard*. London: Earthscan.

Hommes, L. and Boelens, R. (2017). Urbanizing rural waters: Rural-urban water transfers and the reconfiguration of hydrosocial territories in Lima. *Political Geography*, 57, 71–80.

Hommes, L., Boelens, R. and Maat, H. (2016). Contested hydro-social territories and disputed water governance: Struggles and competing claims over the Ilisu Dam development in southeastern Turkey. *Geoforum*, 71, 9–20.

Hoogesteger, J. and Verzijl, A. (2015). Grassroots scalar politics: Insights from peasant water struggles in the Ecuadorian and Peruvian Andes. *Geoforum*, 62, 13–23.

Hoogesteger, J. and Wester, P. (2015). Intensive groundwater use and (in)equity: Processes and governance challenges. *Environmental Science & Policy*, 51, 117–24.

Hoogesteger, J., Boelens, R. and Baud, M. (2016). Territorial pluralism: Water users' multiscalar struggles against state ordering in Ecuador's highlands. *Water International*, 41(1), 91–106.

Khan, S. and Hanjra, M. (2009). Footprints of water and energy inputs in food production: Global perspectives. *Food Policy*, 34, 130–40.

Lemke, T. (2007). An indigestible meal? Foucault, governmentality and state theory. *Distinktion*, 15, 43–64.

Li, T. M. (2011). Rendering society technical: Government through community and the ethnographic turn at the World Bank in Indonesia. In D. Mosse (ed.), *Adventures in Aidland: The Anthropology of Professionals in International Development*. Oxford: Berghahn, pp. 57–80.

Lynch, B. (2013). River of contention: Scarcity discourse and water competition in highland Peru. *Georgia Journal of International and Comparative Law*, 42, 69–92.

McKinsey. (2011, April). "Four lessons for transforming African agriculture," www.mckinsey.com/industries/public-sector/our-insights/four-lessons-for-transforming-african-agriculture.

Mehta, L, Veldwisch, G. J. and Franco, J. (2012). Introduction to the special issue: Water-grabbing? Focus on the (re)appropriation of finite water resources. *Water Alternatives*, 5(2), 193–207.

Mena-Vásconez, P., Boelens, R. and Vos, J. (2016). Food or flowers? Contested transformations of community food security and water use priorities under new legal and market regimes in Ecuador's highlands. *Journal of Rural Studies*, 44, 227–38.

Molle, F., Wester, P. Hirsch, P. (2010). River basin closure: Processes, implications and responses. *Agricultural Water Management*, 97, 569–77.

Mutersbaugh, T. and Lyon, S. (2010). Transparency and democracy in certified ethical commodity networks. *Geoforum*, 41(1), 27–32.

Nestlé. (2016). *Nestlé in Society: Creating Shared Value and Meeting Our Commitments 2015*, www.nestle.com/asset-library/documents/library/documents/corporate_social_responsibility/nestle-csv-full-report-2015-en.pdf.

P&G, Procter and Gamble. (2017). "Environmental policies and practices," "Water," http://us.pg.com/sustainability/environmental-sustainability/policies-practices/water.

Pearson, R. (2007). Beyond women workers: Gendering CSR. *Third World Quarterly*, 28(4), 731–49.

Perreault, T., Bridge, G. and McCarthy, J. (eds.) (2015). *The Routledge Handbook of Political Ecology Routledge International Handbooks*. Abingdon: Routledge.

Preciado-Jeronimo, R., Rap, E. and Vos, J. (2015). The politics of land use planning: Gold mining in Cajamarca, Peru. *Land Use Policy*, 49, 104–17.

Progressio. (2010). *Drop by Drop: Understanding the Impacts of the UK's Water Footprint through a Case Study of Peruvian Asparagus*. London: Progressio, CEPES, WWI.

Raman, R. (2007). Community–Coca-Cola interface: Political-anthropological concerns on corporate social responsibility. *Social Analysis*, 51(3), 103–20.

Ramírez-Vallejo, J. and Rogers, P. (2010). Failure of the virtual water argument: The case study of Mexico and NAFTA. In C. Ringler, A. K. Biswas and S. A. Cline (eds.), *Global Change: Impacts on Water and Food Security*. Berlin: Springer-Verlag (Chapter 6).

Reuters (2015, June 10). "Rapid groundwater depletion threatens Pakistan food security," Aamir Saeed http://in.reuters.com/article/pakistan-water-food-idINKBN0 OQ16920150610.

Roa-García, M. C. (2014). Equity, efficiency and sustainability in water allocation in the Andes: Trade-offs in a full world. *Water Alternatives*, 7(2), 298–319.

Robertson, M. and Hayden, N. (2008). Evaluation of a market in wetland credits: Entrepreneurial wetland banking in Chicago. *Conservation Biology*, 22(3), 636–46.

Rogers, S., Barnett, J., Webber, M., Finlayson, B. and Wang, M. (2016). Governmentality and the conduct of water: China's South–North water transfer project. *Transactions of the Institute of British Geographers*, 41, 429–41.

Roth, D. and Warner, J. (2008). Virtual water: Virtuous virtual impact? *Agriculture and Human Values*, 25, 257–70.

Roth, D., Boelens R. and Zwarteveen, M. (2015). Property, legal pluralism, and water rights: The critical analysis of water governance and the politics of recognizing "local" rights. *Journal of Legal Pluralism and Unofficial Law*, 47(3), 456–75.

SAI, Sustainable Agriculture Initiative Platform. (2017). "Who we are," www.saiplatform.org/about-us/who-we-are.

Shah, T., Burke, J. and Villholth, K. (2007). Groundwater: A global assessment of scale and significance. In D. Molden (ed.), *Water for Food, Water for Life: Comprehensive Assessment of Water Management in Agriculture*. London: Earthscan, pp. 395–423.

Smaller, C. and Mann, H. (2009). *A Thirst for Distant Lands: Foreign Investment in Agricultural Land and Water*. Winnipeg: IISD.

Sojamo, S., Keulertz, M., Warner J. and Allan, J. A. (2012). Virtual water hegemony: The role of agribusuness in global water governance. *Water International*, 37(2), 169–82.

Solanes, M. and Jouravlev, A. (2007). *Revisiting Privatization, Foreign Investment, International Arbitration, and Water*. Santiago: UN/ECLAC.

Stoltenborg, D. and Boelens, R. (2016). Disputes over land and water rights in gold mining: The case of Cerro de San Pedro, Mexico. *Water International*, 41(3), 447–67.

Swyngedouw, E. (2005). Dispossessing H2O: The contested terrain of water privatization. *Capitalism, Nature, Socialism*,16(1), 81–98.

Swyngedouw, E. (2015). *Liquid Power: Contested Hydro-Modernities in 20th Century Spain*. Cambridge, MA: MIT Press.

Van der Ploeg, J. D. (2008). *The New Peasantries: Struggles for Autonomy and Sustainability in an Era of Empire and Globalization*. London: Earthscan.

Van der Ploeg, J. D. (2010). The food crisis, industrialized farming and the imperial regime. *Journal of Agrarian Change*, 10(1), 98–106.

Venot J. P., Kuper, M. and Zwarteveen, M. (eds.) (2017). *Drip Irrigation: Untold Stories of Efficiency, Innovation and Development*. London: Earthscan/Routledge.

Vos, J. and Boelens, R. (2014). Sustainability standards and the water question. *Development and Change*, 45(2), 1–26.

Vos J. and Hinojosa, L. (2016). Virtual water trade and the contestation of hydro-social territories. *Water International*, 41(1), 37–53.

Vos, J. and Boelens, R. (2016). The politics and consequences of virtual water export. In P. Jackson, W. Spiess and F. Sultana (eds.), *Eating, Drinking: Surviving*. New York: Springer, pp. 31–41.

Wada, Y., van Beek, L., van Kempen, C., Reckman, J., Vasak, S. and Bierkens, M. (2010). Global depletion of groundwater resources. *Geophysical Research Letters*, 37(L20402), 1–5.

World Bank. (2017). "Irrigation and drainage, overview." "Results," www.worldbank.org/en/topic/irrigationdrainage/overview#3.

Zhao, J. Z., Liu, W. H. and Deng, H. (2009). The potential role of virtual water in solving water scarcity and food security problems in China. *International Journal of Sustainable Development & World Ecology*, 12(4), 419–28.

Zwarteveen, M. and Boelens, R. (2014). Defining, researching and struggling for water justice: Some conceptual building blocks for research and action. *Water International*, 39(2), 143–58.

16

Critical Ecosystem Infrastructure? Governing the Forests–Water Nexus in the Kenyan Highlands

Connor Joseph Cavanagh

16.1 Introduction

In Kenya, approximately 80 percent of land area is classified as either arid or semi-arid, and an estimated 70 percent of the population resides on the relatively more fertile 20 percent (World Bank, 2005; Government of Kenya [GoK], 2012). Within the latter, an estimated 75 percent of renewable surface water resources originate from five highland watersheds: Mount Elgon, the Cherangani Hills, the Mau Forest complex, Mount Kenya, and the Aberdares range (UNEP, 2012). In such a context, these areas take on considerable significance not only to conserve biodiversity, but also more broadly for sustainable development and public administration. Known in Kenyan popular discourse as the country's "five water towers," these highland forests support the livelihoods of millions of small-scale farmers, pastoralists, and freshwater fishing communities. As the most recent "status report" from the Kenya Water Towers Agency – the state organization responsible for forested water catchment areas – puts it, rather strikingly:

> The term "Water Towers of Kenya" refers to montane forests – the mountainous regions that are sources of water. Through their watershed function, Kenya's Water Towers supply Kenya's drinking water, its energy, water for irrigation, industry, water for food and other purposes. But all this is threatened by a combination of factors, including human encroachment, climate change, environmental degradation, and unplanned dams and irrigation projects that have a devastating effect on the *ecological infrastructure*. The outcome is a real threat to Kenya's water security, stability and surfeit.
> (KWTA, 2015: 1, emphasis added).

And yet, the critical importance of such "ecosystem infrastructure" is not limited to their contribution to water security alone. For example, the UN Environment Program recently estimated economic losses from deforestation in these areas – resulting from the degradation or diminishment of related ecosystem services to downstream industries – to be approximately US$35 million per year, or more than 2.8 times the direct value of profits accruing from sales of timber and other forest products in these areas (UNEP, 2012: 38). Such calculations are characteristic of emerging "natural capital accounting" approaches to the green economy, which are increasingly embraced by both multilateral organizations, donors, and Kenyan state agencies (e.g. GoK, 2013, 2015; KWTA, 2015; UNEP, 2012, 2014). Indeed, conservation and environmental management agencies in the country now perceive such a green economy transition as crucial not only for sustaining economic growth, but also for realizing Kenya's pursuit of a related "low-carbon, climate-resilient development pathway" in the face of increasingly ominous processes of both global and regional environmental change (GoK, 2013, 2015).

Such rhetoric suggests a shifting conceptualization of these areas not simply as intrinsically valuable "wildernesses," nor merely as stocks of commercially valuable resources, but also as what I will term – following the KWTA (2015) itself – a kind of *critical ecosystem infrastructure*. "Critical," because these ecosystems' contribution to both subsistence and commercial production is such that they are perhaps of *biopolitical* rather than merely economic importance; that is, their functioning is essential for the economy's prosperity, and more broadly for the population's physical health and wellbeing (see also Bakker, 2013). Further, these forests are a form of "infrastructure," as well, because the factors contributing to their survival in many ways now arise equally from social, political, and ecological processes. Though, of course, East Africa's highland-forest ecosystems vastly pre-date the modern, colonial, and pre-colonial polities that have emerged in the region, their current form and extent owe much to an intertwined legal, institutional, and economic apparatus intended to prevent conversion to alternative land uses. This apparatus – consisting of laws, policies, gazetted and demarcated forest blocks, funding modalities, and of course the day-to-day operations of state conservationists and other civil servants – must function effectively to sustain these ecosystems' critical benefits.

Accordingly, while Kenya's "water towers" certainly constitute public goods in this regard, the stakes of (re)asserting state control and property rights to ostensibly protect these areas are increasingly high. Indeed, in some cases, this has begun to resemble the protection and maintenance of more conventional forms of infrastructure: transportation, communication, electricity generation, and so forth. Not least, such efforts are illuminated by recent deployments of military and paramilitary forces to violently evict both traditionally forest-dwelling groups and communities at the forest margins, who are frequently now construed as threats to protected forest stocks and water security (e.g. Kenya Forest Service [KFS], 2007). Though no official statistics have been kept about the scale of these evictions in recent decades, the Kenyan Centre on Housing Rights and Evictions (COHRE) notably estimates that – between 2004 and 2006 alone – evictions from a sample of only eight of Kenya's approximately 300 gazetted forests affected nearly 100,000 people (COHRE, 2006, see also Masinde and Karanja, 2011).[1] Conversely, however, given the undeniable significance of the water and other ecosystem services provided by these forests to both commercial and subsistence economies, the question is not so much *whether* but rather *how* these ecosystems might be conserved in ways that are both socially and environmentally just.

Examining both challenges and opportunities to achieve the latter, this chapter proceeds in three sections. First, I situate these contemporary issues in relation to colonial histories of forest governance in Kenya, which decidedly foreground the dynamics of an intricate, shifting choreography of interests that has ultimately shaped and reshaped prevailing forms of conservation policy and practice in the country. Importantly, careful attention to these dynamics illuminates certain historically recurring tensions and fault lines between different branches within the state, between local and international capital, and among various segments of the Kenyan population – tensions that also resonate into the present. Secondly, however, I discuss the ways in which these histories are further complicated by what we might call the informalization of Kenyan forest governance in recent decades, and the widespread integration of forest resources into systems of patronage-based political support (see also Klopp, 2012). As we will see, these processes' empirical contours undoubtedly challenge our understanding of a coherent "state" that stalwartly defends certain ecological infrastructures, even if they do underscore the recurring deployment of violence against certain population segments on these grounds. Finally, in light of these complex histories, I argue that contemporary attempts to govern Kenyan forests for current and future generations' true benefit will require substantial land and resource tenure reform, as well as alliances between progressive elements within the state and locally rooted movements for environmental justice. Without such alliances, local mobilizations will inevitably be construed as an idiosyncratic form of community-driven land grabbing: community enclosures *for* rather than public or private enclosures *of* a specifically local and otherwise exclusionary commons (e.g. Astuti and McGregor, 2016; Castree, 2004). Yet without the latter alliances, conservation agencies will likely replicate past failures with state "fortresses" of ecological control and their well-demonstrated misadventures in the country (e.g. Klopp, 2012; Southall, 2005), missing rich opportunities to find common ground with grassroots movements supporting rather than opposing conservation in the region (e.g. Forest Dwelling Communities, 2014; Kimaiyo, 2004; Nixon, 2006).

16.2 Watershed Conservation and/as State Formation

In Kenya, government concern with the security of upland watersheds is nearly as old as the (colonial) state itself. In the first instance, this involved rising white settler agriculture in the British East Africa Protectorate (Kenya Colony and Protectorate only after 1920). Land alienations for these settlers were most concentrated in the central and western highlands, eventually amounting to an area of at least 31,000 km^2 (e.g. Morgan, 1963). Though comprising a comparatively small proportion of Kenya's terrestrial surface area – especially relative to the extent of alienations for white settlement in Southern Rhodesia (now Zimbabwe) and South Africa (see Lützelschwab, 2013) – the Government of Kenya's Njonjo Commission (2002) would later estimate this area to include approximately 75 percent of the highest potential agricultural lands in the country. In short, however, as such alienations proceeded apace, concerns about both settler and native conversion of forests to agriculture and other land uses also multiplied (Ofcansky, 1984). By 1960, the result was that a further 22,000 km^2 had been alienated and reserved as protected areas for both forest and wildlife conservation (East African Statistical Department, 1964). In part, this was due to the demands by a growing "preservationist" movement throughout the British Empire (e.g. Neumann, 1996), but also to the local administration's much more pragmatic concerns about timber supplies and the sustainability of water resources for settler enterprises in the highlands and elsewhere (e.g. Hutchins, 1907, 1909; Nicholson, 1931).

One should note that alienating these lands and imposing British rule more broadly was, to put it lightly, far from uncontested. In Lonsdale's (1992: 28–29) analysis of the 1893–1911 period, for instance, he observes that a sample of only 18 documented "punitive expeditions" resulted in some 5,858 African casualties, and "less than half a dozen" British ones. A considerable number of other raids appear to have gone unrecorded, failed to report the extent of casualties, or resulted in reports that simply alluded to "many" casualties (ibid.). In short, the actual extent of this violence may have been even more severe. For instance, in Ogot's (2006) reconstruction of the British campaign against the highland Nandi population during this period in western Kenya – drawing upon the diaries of one of the campaign's officers, Richard Meinertzhagen (1983), rather than official records – he cites Meinertzhagen's estimate that more than 100,000 Nandi were killed and 1,250 square miles of their land appropriated. Moreover, Ogot (2006) notes that a "shoot-on-sight" order was subsequently put in place with immediate application to any Nandi caught outside the confines of their newly demarcated native reserve. Recalling an incident in which he discovered an attempt to settle outside these boundaries, Meinertzhagen (1983: 296) himself writes that:

I attacked them at once, killed 14 and captured 28 goats. They were so taken by surprise that they offered but little resistance and I had only one casualty. They had apparently come to stay, as they had even built new huts for themselves. These I burned.

Admittedly, Meinertzhagen's recollections were possibly fabricated or at least exaggerated – even if so, they are indicative of the attitudes of certain members of the administration and colonial military. Conversely, it is likewise possible that the casualties reported in the official records that Lonsdale examines were underestimated, under-reported, or otherwise strategically downplayed. Nonetheless, these records and recollections collectively remind us of the severity of both the discursive and physical violence deployed throughout the process of colonial state formation to depopulate territories intended for either white settlement or conservation.

Though extensive historical literature examines the nature and consequences of European colonization and settlement in Kenya, as well as early efforts to initiate conservation measures (e.g. Anderson, 1987; Ofcansky, 1984), development-induced *deforestation* remains one of colonization's less-interrogated effects. For instance, Klopp and Sang (2011: 133) estimate that Kenya's forest cover declined from roughly 30 percent in the late 1890s, to approximately 3 percent in 2010, noting that "[l]arge amounts of forest loss occurred during the colonial period." The drivers of such deforestation were a source of substantial debate between Kenya Colony's fledgling Forest Department, the broader colonial administration, and the European settler community, frequently pitting representatives of each against each other. Particularly in the highlands, the economic opportunity costs of forest conservation for the settler community were perceived to be prohibitively high – lucrative tea and coffee estates could be carved from these forests, and demand for timber during both world wars was generally seen to trump preservationist concerns (e.g. Mwangi, 1998). In this sense, conservation for conservation's sake was often a "hard sell," and precipitated pragmatic or anthropocentric justifications for preservation within the Forest Department, usually regarding the sustainability of timber and water supplies necessary to fuel the colony's economic development.

Particularly before 1920, settler interests would have enjoyed significantly more influence over the protectorate's administration relative to the Forest Department's preservationist concerns (Fanstone, 2015). Kenya's European settlers were diverse, in both class and origin. They included Boers of modest means from southern Africa, as well as wealthy English families with connections to (or membership of) the peerage. Encouraged by the chauvinism of Commissioners such as Sir Charles Eliot (1905), who famously declared Kenya to be a "white man's country," the first wave of settlers frequently perceived themselves as the incipient state's true economic, political, and perhaps even civilizational vanguard. Indeed, in the early twentieth century, it would have still seemed possible that Kenya would eventually become a white settler colony on the scale of Southern Rhodesia (now Zimbabwe), French Algeria, or even South Africa, perhaps also destined for white minority rule (Ogot, 1968). Any attempt by the colonial administration to oppose settler economic interests, therefore – whether ostensibly unfavorably delineating native reserves, prohibiting arbitrary use of severe violence against African labor, or restricting expanding settler farms into the highland forests – was often met with significant frustration, protest, and complaint (see, *inter alia*, Berman and Lonsdale, 1980; Shadle, 2010, 2012).

Here, the initially blurry distinction between settler and state, as well as potential conflicts of interest between settler and state forester, are illustrated by the case of one Ewart Grogan (e.g. Anderson, 1987). Grogan, well-connected to the protectorate's "Happy Valley" set of aristocrats – yet without a truly equivalent pedigree himself – was perhaps the colony's foremost settler-turned-logging magnate. With the help of Sir Charles Eliot, Grogan and his partners managed to acquire concessionary rights to some 186,000 acres of forest in the western highlands on terms that entailed only a relative pittance in taxes for state coffers, much to the chagrin of the early Forest Department (Ofcansky, 1984: 138). In many ways, the deal's brokering illuminates the Forest Department's decidedly insecure position versus such economic interests. Indeed, the Department had only opened its first offices in Nairobi in 1902, with a staff of precisely four: the Conservator of Forests himself, one C.F. Elliot, and three newly hired rangers (Colony and Protectorate of Kenya, 1962). In short, if the Department was to justify its provisioning in such a context, or its interference in the affairs of private capital, it would have to position itself carefully amid the interests of settlers, "natives," and the administration. Moreover, it would have to demonstrate its potential for economic self-sufficiency, to subvert the early twentieth-century stereotype of Forest Departments as merely "a drain on the limited financial resources of the colonial state in Africa" (Anderson, 1987: 253), preoccupied with the apparently quixotic ideals of preservation for preservation's sake, and aloof from the political-economic grit of colonial state formation proper. As we will see, such concerns about the "fiscality" of conservation – as well as the potential for its costs and benefits to accrue asymmetrically for different segments of the population – are dimensions of Kenya's fraught conservation politics that endure to date.

16.3 Forests, Patronage, and the Shifting Choreographies of Conservation

In practice, however, both settler and conservationist attempts to stake a claim to these highland forests were complicated by the presence of their longstanding inhabitants, hunters, pastoralists, and foragers known to the Maasai as *il-torobo,* or those too poor to own cattle (Chang, 1982). For the most part, the administration engaged these forest-*dwelling*, as opposed to forest-*adjacent*, communities decidedly as though they resided beyond the civilizational pale: as ostensibly in pressing need of forcible modernization – construed as an expression of paternalistic colonial benevolence – but effectively amounting to their coercive separation from land and resources. As then-Chief Native Commissioner O.F. Watkins (1934: 213) put it, such communities apparently could not "exist in the modern world as forest dwellers without danger to forest and so to water, already a scarcening commodity in Eastern Africa." In accordance with recommendations by a National Land Commission led by one Sir Morris Carter (see Colonial Office, 1934), these communities were slated for assimilation into nearby communities of agriculturalists or agro-pastoralists, an approach intended to conclusively depopulate forest reserves and other protected areas, but also to "civilize" some of the colonial state's most allegedly backward subjects. Underlying the immediate issue of forest access and use, then, was also the deeper question of precisely which communities – and which modes of livelihood or socio-ecological relations – could be accepted as candidates for full citizenship within the framework of an eventually both independent and "modern" Kenyan state (Cavanagh, 2017).

As one can imagine, however, these initiatives to both dispossess and assimilate forest dwellers did not go uncontested. The latter were loath to acquiesce to a resettlement program that would undoubtedly marginalize them within larger communities' existing native reserves. Similarly, forest-adjacent farmer and agro-pastoralist communities, who were equally dependent on forests and upland glades for both fuelwood and seasonal grazing, likewise adamantly opposed their de jure exclusion (Anderson, 2002). Yet in some cases, the contingent interaction of these different interests – that of different African communities in retaining access to forests, the Forest Department in improving its position relative to private capital, and the central administration in both maximizing its revenue and demonstrating its "trusteeship" over Africans – occasionally resulted in favourable outcomes for local people, even if temporarily so. For instance, Anderson (e.g. 1987) nicely shows how the Forest Department was somewhat counter-intuitively instrumental in obtaining then-Governor Coryndon's intervention on behalf of Tungen families in the vicinity of Grogan's concession at Lembus forest. Indeed, Coryndon pressed for these communities' forest access and use rights, even though this largely contradicted the state's own forest ordinances, which favored their wholesale exclusion (Mwangi, 1998). In other words, Coryndon's intervention effectively preserved subsistence access for several hundred families at a time when such rights were being sharply curtailed throughout the colony. Similarly, archival records in Kenya suggest that, when it came down to the work of actually rounding up forest dwellers, settling them elsewhere, and subsequently policing their access to forests, local officials sometimes informally opted to allow certain access to forest resources for subsistence purposes.[2]

Rather than an isolated chapter in Kenya's colonial forest history, such circumstantial beneficence would become an enduring feature of forest management after independence in 1963 as well. Differently put, Coryndon's tolerance of legally tenuous forest "squatters" for reasons of political expediency in Lembus foreshadows the postcolonial state's use of forests as patronage on a much larger scale. Particularly during the regime of President Daniel arap Moi (1978–2002), informal networks within diverse government agencies opened a new agricultural frontier within the nation's forest reserves, redistributing both profits from deforestation and the newly converted agricultural land to their political clients. The Ndung'u Commission's (2004) *Report of the commission of inquiry into the illegal/irregular allocation of public land* would later estimate that such dubious concessions totaled nearly 300,000 ha of forest lost between independence and 2002. These and similar practices would become the focus for much writing on the themes of "institutional decay," "neopatrimonialism," and "the criminalization of the state" in Kenya more broadly (e.g. Klopp 2000, 2012; Southall 2005; Mueller 2011; Manji 2012). Too often, however, such analyses would implicitly construe these phenomena as the postcolonial corruption of a competent – if decidedly authoritarian – bureaucracy, one that supplanted the colonial politics of modernization and development with a cynical and environmentally irresponsible *politique du ventre* (Bayart, 2009).

However, President Moi's overarching strategy had deeper roots. The colonial state's first clients included a constituency of pale-skinned foreigners, hungry for land and eager to prove their worth on a new agricultural frontier. Their fortunes, too, were carved from the highland forests of central Kenya and the Rift Valley. Unsurprisingly, then, these similar *consequences* of past and present were clearer to an historian of forestry than to many political analysts. Indeed, Thomas Ofcansky (1984: 137) laments that Europeans had already carved out an estimated 264,400 acres of Kenya's highland forests for their own agricultural use just between 1895 and 1908. Just "[l]ike their African counterparts," Ofcansky (ibid.) writes, "the white settlers were guilty of clearing away large tracts of forest for farming [and] for industrial purposes." Admittedly, settlers' properties did not always come at the expense of the forest estate *per se*, as it largely did not yet exist in the form of protected areas, and especially so in the years before 1920. Nonetheless, the de jure forest estate was in many cases designed and emerged in subordination to or shaped around their interests – a strategy that the postcolonial state could ultimately only ever hope to wield in reverse.

In brief, the point is that the terrain of forest conservation in Kenya is, and has been since the emergence of the contemporary state, the outcome of an intricate and shifting choreography of interests – one deeply rooted in the politics of statecraft itself. Historically, the extension of state authorization to benefit from forest resources – both formal and informal – reflects the ebb and flow of its conception of precisely which elements of the population are most deserving of such beneficence. Yet the nature and terms of this choreography undoubtedly shift along with wider political, economic, and ecological dynamics, in the emergence of colonial conservation and its alleged "decline" under Moi, just as surely as in its reconsolidation under more recent logics of "green" development and growth, to a discussion of which I now turn.

16.4 Conservation Resurgent? Security, Territory, (De)population

After President Moi's regime finally ended in 2002, President Mwai Kibaki's government took initiatives ostensibly intended to correct Moi's malign legacies from both authoritarian and later nominally democratic periods. These included the above-mentioned Ndung'u Commission (2004), whose decidedly quite damning findings, coupled with growing pressure from both donors and civil society – such as the Green Belt Movement led by Nobel laureate Wangari Maathai – prompted significant forest-sector legislative and policy reform (e.g. Nixon, 2006).

Initially, Kibaki promptly terminated more than 800 Forest Department employees in 2003, explicitly for involvement in corruption and mismanagement of the forest estate (see BBC, 2003). More significant, however, was the adoption of new legislation in the form of the 2005 Forest Act, the content of which again seems to reflect a complex balance of interests between different branches of the state, the influence of key civil society figures such as Maathai, and the imperatives of bilateral donors. Crucially, the Act withdrew ministerial powers to degazette public forests and convert them to other land uses, which had previously only required 28 days' advance notice via publication in the Kenya Gazette. Under this new legislation, degazettement would require parliamentary approval (Klopp 2012) – if not undermining prevailing strategies for the disposal of forests as patronage outright, then certainly vastly complicating their execution. Likewise, Kenya's new Constitution of 2010 and subsequent legislation further solidified parliamentary oversight over the disposal over both public land and forests, giving rise to a perception amongst donors and certain development practitioners that it is now "unlikely that Kenya will again witness the sort of corrupt excisions of public forests that have occurred in the past" (e.g. Standing and Gachanja 2014: 12).

As a consequence, however, the passing of the Act has generally precipitated a reterritorialization of state control over the protected area estate, and the forcible eviction of remaining "encroachers" or "squatters." This has decidedly entailed certain ironies. Firstly, while many of these often-impoverished forest settlers now find themselves dispossessed, the state has made limited efforts to enact the Ndung'u Commission's (2004) recommendations to repossess larger-scale estates illegally carved out from public forests. The technicalities of doing so would be admittedly complex: given the well-connected standing of many of the alleged beneficiaries (several of whom continue to hold public office), there is apparently little "political will" to pursue such repossessions through legal means (see also Southall, 2005). Secondly, many of these ill-gotten properties appear to have been laundered through Kenyan land markets, meaning that several transactions may have transpired between degazetting and the acquisition of current ownership (Boone, 2012).

Conversely, however, the state and its rebranded Forest Department – now known as the Kenya Forest Service (KFS) – have zealously pursued the dispossession of informal beneficiaries of forest patronage, namely communities of farmers and agro-pastoralists at the forest margins. Likewise, caught up in this resurgence of state control are the descendants of Kenya's remaining il-torobo (or 'Dorobo' in Swahili or English parlance) communities. As a result, groups such as the Ogiek of Mount Elgon and the Mau forest complex, as well as the Sengwer of the Cherangani Hills have been doubly marginalized – first, by the state-facilitated plunder of their customary forest territories, and now their wholesale eviction alongside the illegitimate beneficiaries of the former process (e.g. Tiampati 2015).

Forcible evictions and related forms of violence throughout Kenya's forest estate are now once again justified as securing public goods – the critical "ecosystem infrastructure" invoked by state agencies such as the KFS (see KFS and UNEP, 2012) and the Kenya Water Towers Agency (2015). Like the old concerns about sustaining settler productivity, these forests' water and other ecosystem services are now perceived as crucial to support downstream enterprises throughout Kenya's transition to a "green economy," entailing both high economic growth rates and transition to "low-carbon, climate-resilient" production (e.g. UNEP, 2012; Government of Kenya, 2013). Increasingly large volumes of international aid and credit are facilitating these objectives, including financial schemes intended to expedite readiness activities to design a Reducing Emissions from Deforestation and Forest Degradation (REDD+) program.

Although numerous civil-society groups have criticized such financial schemes' economic incentives to dispossess Kenya's indigenous forest communities (Forest Peoples Programme, 2014; Global Justice Ecology Project, 2014; No REDD in Africa Network, 2014), state agencies and development practitioners continue to deny such a link. As Standing and Gachanja (2014: 27) would have it, such a perspective "mistakenly links forced eviction of [communities like] the Ogieks in the Mau with REDD+, although the government's decision to evict the Ogieks was not motivated by REDD+ financing." It is indeed the case that the trials and tribulations of Kenya's forest dwellers precede contemporary debates about the effects of REDD+ by more than a century. Conversely, one cannot underestimate the capture of new forms of 'green rent' provided by bilateral and multilateral actors as a motivating force for such evictions. This is particularly so if their volume will be tied in part to quantities of carbon sequestered in the protected area estate, or the capacity of the state to guarantee its ostensible 'permanence' via the exclusion of human influence (see also Cavanagh et al., 2015). What is certain, however, is that such finances will articulate with the continuously shifting choreography of interests behind the contemporary shape of forest conservation in Kenya, bringing the tensions and contradictions between some of these into increasingly sharp focus.

16.5 Conclusion: Towards Water and Environmental Justice in the Kenyan Highlands

A perspective rooted in historical political ecology decidedly illuminates how contestations about the *form* rather than the *principle* of forest conservation in what is now Kenya precede the modern state's embryonic emergence in 1895. Indeed, virtually all the region's pre-colonial societies retained customary institutions to regulate forest resource use, albeit without drawing schematic distinctions between "forest" (or nonhuman "nature" more generally) and "society" (e.g. Mwangi 1998: 2–4), and continuously adapting to changing political-economic and ecological contexts. In other words, given widespread dependence on forests for water and other ecosystem services that continue to sustain both agrarian and pastoralist livelihoods, few rural communities in Kenya oppose conservation outright. Still today, widespread forest destruction is ultimately antithetical to the livelihoods of most Kenyans. Rather, what a growing swathe of the population increasingly challenges are the specific institutional arrangements underpinning conservation activities, as well as the historically unequal political economy of conservation governance.

Notably, for example, the most prominent grassroots mobilizations in the country post-independence have been *for* rather than *against* conservation, most notably perhaps in the form of Maathai's Green Belt Movement (e.g. Nixon, 2006; Njeru, 2010). Here, the objective was not to impede conservation, but to *democratize* it to serve most of the population, rather than a cabal of Nairobi-based elites and their clients elsewhere. But other movements have and are pressing for such outcomes as well, notwithstanding little international limelight. The descendants of Kenya's forest-dwelling communities, for instance, continue to contest the legitimacy of their dispossession from customary forest territories, perceiving themselves as among the most capable stewards – even under contemporary conditions – of Kenya's forests (see Forest Dwelling Communities, 2014; Kimaiyo, 2004). Again, what is being challenged is the current *form* of conservation, even though these communities view forest conservation as necessary indeed, both for their own livelihoods and for millions of their fellow Kenyans downstream.

Following the promulgation of Kenya's 2010 Constitution, these latter initiatives have been especially emboldened by a new legal category of "Community Land" tenure, constitutionally enshrined to include "land lawfully held, managed, or used by specific communities as community forests, grazing areas or shrines," as well as "ancestral lands and lands traditionally occupied by hunter-gatherer communities" (Government of Kenya, 2010: §63(2)(d)). Likewise, Kenya's newly expanded Bill of Rights now includes specific protections for "marginalized communities," defined to encompass "an indigenous community that has retained and maintained a traditional lifestyle and livelihood based on a hunter or gatherer economy" (Government of Kenya, 2010: §260). In principle, these provisions, as well as recent legislation, such as the Community Land Act and Forest Conservation and Management Act, allow for converting public forests – or portions of public forests – to a type of community-*owned* forest or community protected area. This is quite distinct from previous iterations of "community forest" – which were owned by local government and held "in trust" for communities – as well as "collaborative" conservation arrangements that allowed community *access* to government-*owned* forests under conditions prescribed by state authorities.

Whether or not the promise of such institutional reforms will be realized in practice is still very uncertain. State agencies such as the Kenya Forest Service will likely continue to face substantial incentives to capture "green rents" for centralized protected-area management, whether from a future REDD+ program or more conventional donor support for conservation governance. Likewise, the state's executive branch will encounter substantial pressure from county governments to grant them ownership of formerly "trust" land and forest, likely with a view to formally or informally using such lands as patronage, even if the most blatant strategies for doing so are no longer viable. Further still, communities of current and former forest-dwellers themselves have undergone complex livelihood changes over the last several decades, their systems for customary forest management gradually eroded by de facto tenure insecurity and state-facilitated forest resource plunder. Consequently, for Kenyans to agree to implement such novel forms of community-*led* rather than community-"based" watershed conservation, they must be assured that they will not merely become an idiosyncratic form of community-driven land- and forest-grabbing.

As I have sought to illuminate in this chapter, struggles over Kenya's forests have always found themselves at the shifting locus of conflicting and often mutually exclusive interests. This is not to say that a more benign choreography of such interests cannot emerge, perhaps even one that ameliorates rather than exacerbates prolonged histories of injustice for Kenya's forest-dwelling and dependent communities in particular. Transition to an unprecedented model of community land and forest ownership is undoubtedly riddled with potential for various kinds of challenges and crises. Yet the state-centric alternative effectively proposes to continue a model that has at times presided over a largely unmitigated disaster for both Kenyan forests and the rural communities that depend so heavily upon them, and particularly so in the last several decades. From the long struggle against colonialism, autocracy, and state criminality, Kenyans know all-too-intimately that they must combat injustice to overcome it, and that failure to consolidate today's victories might sow the seeds of tomorrow's defeats. Recent institutional reforms offer us a glimmer of hope for more socially and environmentally just futures in Kenya's forested watersheds. Whether or not that vision is realized, how or how not, to whose benefit and to whose detriment, will inevitably be the stuff of political ecologies to come.

Notes

1 Of course, such estimates are both contested and difficult to verify. However, one should note that these problems with verification directly result from the state's sustained refusal to document the full social impacts of its long-standing program of forest evictions.
2 See, *inter alia*, Kenya National Archives[KNA]/PC/CP/8/2/2 – "General Correspondence: Wandorobo"; KNA/DC/NKU/1/6/2 – "Olenguruone Settlement"; KNA/PC/NZA/3/31/9 – "Tinderet Forest."

References

Anderson, D. (1987). Managing the forest: The conservation history of the Lembus, Kenya, 1904–1963. In D. Anderson and R. Grove (eds.), *Conservation in Africa: People, Policies and Practice*. Cambridge: Cambridge University Press, pp. 249–68.

Anderson, D. (2002). *Eroding the Commons: The Politics of Ecology in Baringo, Kenya, 1890s–1963*. Oxford: James Currey.

Astuti, R. and McGregor, A. (2016). Indigenous land claims or green grabs? Inclusions and exclusions within forest carbon politics in Indonesia. *Journal of Peasant Studies*, 44(2), 445–66.

Bakker, K. (2013). Constructing "public" water: The World Bank, urban water supply, and the biopolitics of development. *Environment and Planning D: Society and Space*, 31(2), 280–300.

Bayart, J. F. (2009). *The State in Africa: The Politics of the Belly*. 2nd edn. Cambridge: Polity.

Berman, B. J. and Lonsdale, J. M. (1980). Crises of accumulation, coercion and the colonial state: The development of the labor control system in Kenya, 1919–1929. *Canadian Journal of African Studies/La Revue canadienne des études africaines*, 14(1), 55–81.

Boone, C. (2012). Land conflict and distributive politics in Kenya. *African Studies Review*, 55(01), 75–103.

British Broadcasting Corporation (BBC). (2003). Kenya suspends forestry staff. *BBC News* (25.10.2003), http://news.bbc.co.uk/2/hi/africa/3214073.stm.

Castree, N. (2004). Differential geographies: Place, indigenous rights and "local" resources. *Political Geography*, 23(2), 133–67.

Cavanagh, C. J. (2017). *Anthropos* into *humanitas*: Civilizing violence, scientific forestry, and the "Dorobo question" in eastern Africa. *Environment and Planning D: Society and Space*, 35(4): 694–713.

Cavanagh, C. J., Vedeld, P. and Traedal, L. T. (2015). Securitizing REDD+? Problematizing the emerging illegal timber trade and forest carbon interface in East Africa. *Geoforum*, 60, 72–82.

Center for Housing Rights and Evictions (COHRE). (2006). "Forest evictions – a way forward?" http://humanrightshouse.org/noop/page.php?p=Articles/7465.html&d=1.

Chang, C. (1982). Nomads without cattle: East African foragers in historical perspective. In E. Leacock (ed.), *Politics and History in Band Societies*. Cambridge: Cambridge University Press.

Colonial Office. (1934). *Report of the Kenya Land Commission (including Evidence and Memoranda)*. London: HMSO.

East African Statistical Department. (1964). *Economic and Statistical Review. No. 16*. Nairobi: East African Statistical Department.

Eliot, C. (1905). *The East Africa Protectorate*. London: Edward Arnold.

Fanstone, B. (2015). Sowing the seeds of modernity: The colonial Kenyan forestry department. *Quarterly Journal of Forestry*, 109(4), 42–46.

Forest Dwelling Communities. (2014). *Forest Dwelling Communities Position Statement: Securing Our Rights, Our Lands, and Our Forests*. Submission to the Taskforce on Historical Land Injustices, National Land Commission (11.09.2014). Unpublished (copy with author).

Forest Peoples Programme (FPP). (2014). "Kenya government's forced evictions threaten cultural survival of the Sengwer," www.forestpeoples.org/topics/rights-land-natural-resources/news/2014/02/kenyan-government-s-forced-evictions-threaten-cult.

Global Justice Ecology Project. (2014). "Why REDD is wrong," http://globaljusticeecology.org/publications.php?ID=472-edn.

Government of Kenya. (2010). *The Constitution of Kenya, 2010*. Nairobi: National Council for Law Reporting.

Government of Kenya. (2012). *Master Plan for the Conservation and Sustainable Management of Water Catchment Areas in Kenya*. Nairobi: Ministry of Environment and Natural Resources.

Government of Kenya. (2013). *National Climate Change Action Plan, 2013–2017*. Nairobi: Ministry of Environment and Natural Resources.

Government of Kenya. (2015). *Green Economy Strategy and Implementation Plan*. Nairobi: Ministry of Environment and Natural Resources.
Hutchins, D. E. (1907). *Report on the Forests of Kenya*. London: HMSO.
Hutchins, D. E. (1909). *Report on the Forests of British East Africa*. London: HMSO.
Kenya Forest Service (KFS). (2007). *Forest Law Enforcement and Governance in Kenya*. Nairobi: KFS.
Kenya Forest Service (KFS) and UNEP. (2012). *Report of the High-Level National Dialogue on Kenya Water Towers, Forests, and Green Economy*. Nairobi: Kenya Forest Service.
Kenya Water Towers Agency (KWTA). (2015). *Kenya Water Towers Status Report 2015*. Nairobi: Kenya Water Towers Agency.
Kimaiyo, J. T. (2004). *Ogiek Land Cases and Historical Injustices*. Nakuru: Ogiek Welfare Council.
Klopp, J. M. (2000). Pilfering the public: The problem of land-grabbing in contemporary Kenya. *Africa Today*, 47(1), 7–26.
Klopp, J. M. (2012). Deforestation and democratization: Patronage, politics and forests in Kenya. *Journal of Eastern African Studies*, 6(2), 351–70.
Klopp, J. M. and Sang, J. K. (2011). Maps, power, and the destruction of the Mau Forest in Kenya. *Georgetown Journal of International Affairs*, 12(1), 125–34.
Lonsdale, J. (1992). The conquest state of Kenya, 1895–1905. In B. Berman and J. Lonsdale (eds.), *Unhappy Valley: Conflict in Kenya and Africa, Part One – State and Class*. Oxford: James Currey, pp. 13–45.
Lützelschwab, C. (2013). Settler colonialism in Africa. In C. Lloyd, J. Metzer and R. Sutch (eds.), *Settler Economies in World History*. Leiden: Brill, pp. 141–68.
Manji, A. (2012). The grabbed state: Lawyers, politics and public land in Kenya. *Journal of Modern African Studies*, 50(3), 467–92.
Masinde, S. and Karanja, L. (2011). The plunder of Kenya's forests: Resettling the settlers and holding the loggers accountable. In Transparency International (ed.), *Global Corruption Report: Climate Change*. London: Earthscan, pp. 280–83.
Meinertzhagen, R. (1983). *Kenya Diary, 1902–1906*. London: Eland Books.
Morgan, W. T. W. (1963). The 'white highlands' of Kenya. *The Geographical Journal*, 129(2), 140–55.
Mueller, S. D. (2011). Dying to win: Elections, political violence, and institutional decay in Kenya. *Journal of Contemporary African Studies*, 29(1), 99–117.
Mwangi, E. (1998). *Colonialism, Self-governance, and Forestry in Kenya*. Research in Public Affairs Working Paper V590. Bloomington: Indiana University.
Ndung'u Commission. (2004). *Report of the Commission of Inquiry into the Illegal/Irregular Allocation of Public Land*. Nairobi: Government printer.
Neumann, R. P. (1996). Dukes, earls, and ersatz Edens: Aristocratic nature preservationists in colonial Africa. *Environment and Planning D: Society and Space*, 14(1), 79–98.
Nicholson, J. W. (1931). *The Future of Forestry in Kenya*. Nairobi: Government printer.
Nixon, R. (2006). Slow violence, gender, and the environmentalism of the poor. *Journal of Commonwealth and Postcolonial Studies*, 13(2), 14–37.
Njeru, J. (2010). "Defying" democratization and environmental protection in Kenya: The case of Karura Forest Reserve in Nairobi. *Political Geography*, 29: 333–42.
Njonjo Commission. (2002). *Report of the Commission of Inquiry into the Land Laws System of Kenya*. Nairobi: Government printer.
No REDD in Africa Network. (2014). "Forced relocation of Sengwer proves urgency of canceling REDD," www.no-redd-africa.org/index.php/2-uncategorised/101-press-release-forced-relocation-of-sengwer-people-proves-urgency-of-canceling-redd.

Ofcansky, T. P. (1984). Kenya forestry under British colonial administration, 1895–1963. *Journal of Forest History*, 28(3), 136–43.

Ogot, B. A. (1968). Kenya under the British, 1895 to 1963. In B. A. Ogot (ed.), *Zamani: A Survey of East African History*. Nairobi: Longman.

Ogot, B. A. (2006). Review article: Britain's gulag. *Journal of African History*, 46, 493–505.

Shadle, B. (2010). White settlers and the law in early colonial Kenya. *Journal of Eastern African Studies*, 4(3), 510–24.

Shadle, B. (2012). Cruelty and empathy, animals and race, in colonial Kenya. *Journal of Social History*, 45(4), 1097–1116.

Southall, R. (2005). The Ndungu report: Land and graft in Kenya. *Review of African Political Economy*, 32(103), 142–51.

Standing, A. and Gachanja, M. (2014). *The Political Economy of REDD+ in Kenya: Identifying and Responding to Corruption Challenges*. U4 Issue 2014(3). Bergen: U4 Anti-Corruption Resource Centre/CMI.

Tiampati, M. (2015). Kenya. In *The Indigenous World 2015*. Copenhagen: International Working Group on Indigenous Affairs (IWIGA).

UN Environment Program (UNEP). (2012). *The Role and Contribution of Montane Forestsand Related Ecosystem Services to the Kenyan Economy*. Nairobi: UNEP.

UNEP. (2014). *Green economy assessment report: Kenya*. Nairobi: UNEP.

Watkins, O. F. (1934). The Report of the Kenya Land Commission, September, 1933. *Journal of the Royal African Society*, 33(132), 207–16.

World Bank. (2005). *Agricultural Productivity and Sustainable Land Management (KAPSLM) Project: Project Brief*. Washington, DC: World Bank.

17

The Meaning of Mining, the Memory of Water: Collective Experience as Environmental Justice

Tom Perreault

17.1 Introduction: Memory, Environment and Justice

This chapter examines the relationship between memory and environmental justice. It does so by exploring the memory narratives of people affected by acute mine-related water contamination in the Huanuni river valley on the Bolivian Altiplano. Although there has been mining activity in the Huanuni valley for centuries, technological changes in ore processing during the 1960s, and more recently the mining boom of the 2000s, have resulted in widespread, acute water contamination. Residents of the valley recall the period before the river was polluted as a time of plenty, with clear water and verdant pastures. As I argue below, however, memory narratives filter the past through collective cognitive frames that tell us much about present conditions. Further, I suggest that memories of the past can serve as conceptual bases for making claims to environmentally just futures. In this sense, the chapter examines water justice by exploring the relationship between the past and the present, and the individual and the collective, as captured in memory narratives.

Individual memories are recalled, interpreted, understood and represented in the context of contemporary social relations, what Maurice Halbwachs referred to as "collective frameworks." Halbwachs (1992: 40, cited in French, 2012: 339) argued that, "The past is not preserved but is reconstructed on the basis of the present … Collective frameworks are … precisely the instruments used by the collective memory to reconstruct an image of the past which is in accord, in each epoch, with society's predominant thoughts." In this view, even individual recollection is a social act, insofar as personal memories can only be understood in the context of collective discourse and representation (Canessa, 2012). As Molden (2016) argues, the past is always represented in a way relevant to, and even in the service of, the present. This understanding of collective memory helps shift analytical focus from individual recollection to constructing collective cognitive frameworks that filter, shape and give meaning to personal memories. Such collective memories are often given material form and literally take place as monuments, memorial sites or landscapes.

In his study of memorialization practices in post-dictatorial Latin America, Andermann (2015: 6) notes that "places of memory" such as monuments and museums must be understood both in terms of the physical, material sites they occupy, and the symbolic forms and practices such sites represent (e.g. Foote and Azaryahu, 2007). In Latin America, considerable attention has been paid to such sites and their associated memorial practices in the context of post-conflict reconciliation processes (e.g. Gómez-Barris, 2009; Meade, 2001). Less attention has been paid, however, to the ways that "natural" landscapes are remembered individually and collectively, and the potential – often unmet – that these memories hold for political action (Legg, 2007). One exception is the work of Auyero and Swistun (2009), who examine environmental suffering in a Buenos Aires barrio called *Inflamable* ("Flammable"), which is surrounded and acutely contaminated by petrochemical plants. Older residents, who experienced the barrio before the chemical industry's arrival, recall swimming and fishing in a pristine river, and cultivating gardens on fertile green fields. Auyero and Swistun note, however, that such hazy memories likely reflect a certain nostalgia for an idealized past. As they put it, 'one way to convey the present uneasiness is to contrast it with a time and place that may have never existed in quite the form it is remembered, but the need to do this strongly indicates the depth of one's present discomfort' (Auyero and Swistun, 2009: 56). Such idealized representations may be thought of as examples of what Blunt (2003: 722) calls "productive nostalgia" which, as opposed to forms of nostalgia commonly disparaged as maudlin and apolitical (Lowenthal, 1989), holds potential for collective political action. As I discuss below, idealized representations of past landscapes uncontaminated by mining activity are similarly common in Bolivia. Far from mere nostalgia for a past long-since-vanished, these representations must be understood in the context of present social relations and possibilities for alternative futures. That is, idealized individual memories can serve as a basis for collective political action and claims making for environmental justice.

In what follows, I examine the ways that mining is remembered and memorialized in Oruro, Bolivia. This chapter is based on ethnographic fieldwork I have conducted in Oruro beginning in 2009. This work has consisted of household surveys in 14 indigenous-*campesino* communities in the Huanuni river valley; participant observation in community meetings and other events; semi-structured interviews with state and mining officials, as well as researchers and activists in the region; and chemical analysis of water and soil samples taken at various points in the valley. The chapter begins with a brief overview of the region's mining history and mining activity's environmental implications. I then examine the ways mining is officially memorialized in Oruro, through murals and monuments. I argue that this memorialization both reflects and produces the region's shared mining experience. I then discuss the ways that mining activity, and particularly the environmental effects of mining waste, is recalled through the memory of residents of the Huanuni Valley. Such recollections represent what Legg (2007: 460) calls "counter-memory," which stand in contrast to the official, collective memory of mining. However, widespread collective action is notably absent so I end the chapter by considering memory's promise and limitations as a political basis for environmental justice.

17.2 Mining and Water Contamination in Oruro

Oruro's mining regions are home to numerous Quechua-speaking *agro-minero* (mixed agro-pastoral and mining) indigenous-*campesino* communities. This region's residents are among Bolivia's most impoverished, with some 46.3 percent of its population living in what the United Nations Development Program characterized as "extreme poverty" (as compared to 32.7 percent for Bolivia as a whole). The city of Oruro itself has relatively low extreme poverty rates (at 33 percent), but outside the departmental capital, every municipality has an extreme poverty rate of over 70 percent (UPADE, 2010). In this high-altitude region (Oruro's elevation is 3800m), agriculture is difficult under any circumstances. The central Altiplano is characterized by a cold, semi-arid climate, and in many areas its soils are highly saline (one of the world's largest salt flats, the Salar de Uyuni, is located south of Oruro). Rural residents have long engaged in semi-subsistence agriculture, typically growing potatoes and other tubers, fava beans, and quinoa, along with a mix of vegetables (including onions, carrots, and turnips) where soil and water conditions permit. Residents raise sheep, cattle and sometimes llamas, and sell milk, yogurt and soft, fresh cheese (*queso fresco*) for local consumption. Discouraged by scanty economic opportunities and the difficulty of rural life, many residents leave their communities for urban centers such as Oruro, La Paz, Cochabamba, Buenos Aires, or São Paulo.

Mining in Oruro dates to the seventeenth century, but for most of the colonial period, Oruro's mining economy was overshadowed by that of Potosí, and its enormous silver deposits. When world silver markets crashed in the late nineteenth century, activity shifted to Oruro and northern Potosí, whose mountains are rich in tin (Madrid *et al.*, 2002; Nash, 1993). In the early twentieth century, the Huanuni, Uncía, Catavi and Siglo XX mines represented the Bolivian economy's center of gravity, controlled by a small cadre of mining elites. Following years of violent repression, miners played a central role in Bolivia's 1952 Social Revolution that swept the National Revolutionary Movement (MNR) and its exiled leader, Víctor Paz-Estenssoro, into power. The MNR government nationalized mines and created the state mining company, COMIBOL (Dunkerley, 1984). Ironically it was not only the MNR, but Paz Estenssoro himself, in his fourth and final term as president, who implemented the first wave of neoliberal measures in the mid-1980s, which closed the state-run mines and largely dismantled COMIBOL. The sudden loss of some 25,000 mining jobs, and thousands more in ancillary employment, devastated the region.

Neoliberal reforms in the 1980s and '90s restructured the conditions for mine ownership, labor and rent distribution. Foreign investment in Bolivia's mining sector slowly increased during this period, as part of a broader international mining boom (Bebbington, 2009; Kaup, 2013). With the election of Evo Morales and his Movement to Socialism (MAS) party in 2005, the sector again underwent an important shift. Following his election, and the May 2006 "nationalization" of hydrocarbons (Kaup, 2010), Morales reconstituted COMIBOL, giving it a much larger role in managing national mining production – even if it is still a shadow of what it was from the 1950s and '60s. With its reconstitution, COMIBOL increased its workforce tenfold to about 5,000 miners. Although this number has since declined to roughly 3,500, the mine's rapid expansion both enabled and necessitated a dramatic increase in production.

With this expansion, the Huanuni tin mine has become the region's single largest source of mine-related water and soil contamination. In its nearly 100 years of operation, it has never had an adequate retention facility for containing mining waste, and instead has dumped untreated waste material directly into the Huanuni river, which flows through (or adjacent to) dozens of indigenous *campesino* communities downstream on its way to Lake Uru Uru. In addition to the direct release of contaminated water and chemicals used in the processing of ore, soils and surface waters are affected by acid mine drainage, which lowers the pH of rivers and lakes. Quintanilla and García (2009) identified heavy metals such as lead, arsenic, cadmium, iron and zinc in rivers throughout the Poopó watershed – in all cases exceeding levels permissible under Bolivian law (see also López, 2011; López et al., 2010). Water and soil testing in several communities has demonstrated low pH and the presence of heavy metals in levels exceeding permissible quantities (Montoya et al., 2010; Perreault, 2013). In 2009 President Evo Morales, responding to calls from environmental activists, declared an environmental emergency in the lower Huanuni river watershed. The declaration (Supreme Decree 0335) set in motion an inter-agency remediation effort. In the years since the decree was passed, however, little has been done and environmental conditions remain largely unchanged.

Mining activity consumes relatively large quantities of water, primarily in the processing of ore. Modern techniques of mineral processing rely on chemical processes to separate mineral from ore, whether in the form of cyanide heap-leach mining (commonly employed in large-scale open pit mining) or the froth flotation process (also known as "sink and float"), which relies on chemical reagents that bind with ores. Both techniques are chemical-intensive and can be highly contaminating unless proper environmental controls are in place (Bridge, 2004). In addition to their impact on water quality, these processing techniques have dramatic implications for water *quantity* as well. The Huanuni mine consumes over 28 million liters of water per day, roughly the same quantity as the 300,000 residents of the city of Oruro. Meanwhile, Bolivia's largest mine, the massive open pit San Cristóbal gold mine, consumes over 45 million liters per day, in the arid and impoverished southern Altiplano in Potosí department (Emilio Madrid, Colectivo CASA, personal communication). While some of the water consumed in processing is reused by the mines, mines continually add to their water supplies by diverting water from surrounding watersheds. The Huanuni mine captures water from the river above the mine, leaving the community of Q'intaya with no water to irrigate crops. Below the mine, water is diverted from a spring in the community of Yaku Pampa and conveyed via open canal to a small *ingenio* (processing plant) near the town of Machacamarca. Deterioration in water quality and quantity are reflected in the memory narratives of the area's residents, as discussed in the following section.

17.3 Mining, Meaning and Memory

Bolivia has long been known as a *país minero*, a mining country. This reputation harkens back to the colonial period, when, during the seventeenth and eighteenth centuries, the mines of Potosí were a source of spectacular wealth for the Spanish empire (Brown, 2012). It also recalls Oruro's importance as a mining center during the twentieth century, when Bolivia was one of the world's leading producers of tin and the miners themselves were a potent political force (Dunkerly, 1984). Nationally, mining's economic and political importance has since declined, and export earnings from natural gas now far exceed those from mining. Locally and regionally, however, mining retains its economic importance. In the Altiplano departments of Oruro and Potosí, mining also has potent symbolic significance, as integrally connected to Bolivia's Social Revolution and thus to the emergence of the modern Bolivian nation. Moreover, since the vast majority of Bolivia's miners (like most of Bolivia's poor and working classes) are of indigenous descent – mostly Quechua and Aymara speakers – the intertwined histories of mining and the Bolivian nation are further imbricated with a generalized and often idealized sense of Andean indigenous cultural history. These histories are represented visually in murals and monuments found in cities throughout the Altiplano, which serve to memorialize miners and place them squarely in the national story. Such public memorials – typically commissioned by local governments and displayed in prominent public spaces – reflect commonly accepted views of the heroic miners and their centrality to Bolivian national history. These officially sanctioned and very public displays of collective memory – examples of what Molden (2016) refers to as "mnemonic hegemony" – are part of an effort to maintain a specific understanding of mining's central role in Bolivian nation building (see also Brockmeier, 2002). The construction of these official national narratives necessarily involves selective remembering and, equally important, selective forgetting, whether "willful, organized or unconscious" (Blight, 2009: 239). What is forgotten in the monuments and heroic murals is the memory of everyday experience, and in particular of those negatively affected by mining's toxic legacy. It is to these memories that the chapter now turns.

17.4 Memories of Water: Everyday Experiences of Mining Contamination

The inherent difficulties of agricultural production that farmers experience on the Bolivian Altiplano – cold temperatures, high altitudes, intense solar radiation, saline soils, dry climate (with highly pronounced seasonal fluctuations in precipitation) – have been exacerbated by acute mine-related water and soil contamination. Environmental degradation is experienced in a myriad of ways by those living downstream and downwind. In Oruro's southern neighborhoods, which abut the tailings and processing facilities of the Baremsa metallurgy plant, situated below the Itos mine, residents breathe the dust that blows from the slag heaps during the dry season (April–October), while in the rainy season (November–March), acid runoff drains downstream, polluting soil and groundwater. In the communities of Mallku Cocha and Huayra, downstream from COMIBOL's Huanuni mine, sediments laden with heavy metals, salts and chemical toxins are deposited at the mouth of the Huanuni river, adjacent to agricultural fields and pasture (Perreault, 2013). Residents complain of respiratory and skin ailments, and have watched their crops wither and their livestock die. Stories of deformed calves and lambs (or deformed stillborn fetuses) are commonplace. These profound environmental changes frame people's everyday conversations, and provide a touchstone by which people come to understand their lives. For residents of the Huanuni valley, narratives of past lives and landscapes represent both a moral order and a way of making sense of their current experiences. Don Braulio,[1] an indigenous authority from one of the communities most affected by mine-related contamination, explained it this way,

Before, when I was still a child, we had wells that were good, good. Now, in comparison, no. Before, we made cheese, we had 20 or 30 cows, but now no. There aren't any. The forage, with the contamination, the *totora* (reeds) that the cows eat, they don't give any nutrition. There's no milk, they die. This is coming from Huanuni. It's contaminating everything ... The water itself is totally contaminated with it. It's ruining us. The grasslands used to be, ahhh, so good, so tall. But now there's not one left. There's nothing. It looks like a salt flat [*salar*]. When it rains, this salty water, it's not drinkable, it's not good for anything ... With the pollution we have to carry water from (the town of) Machacamarca to the countryside to drink and cook. Before, we used to get water from a well, but not anymore. When we drink this water now we get sick.

(Author interview, May 30, 2011)

Similarly, Don Gerardo, an official with the departmental government at the time of our interview, described his community, Mallku Kocha, which is among the most acutely contaminated by mining waste. The deposition of sediments laden with heavy metals has made agriculture in the community all but impossible:

Look, this place used to be an orchard [*vergel*], filled with *totora* [reeds], this whole place. All this area that is now (barren) plain, really used to be an orchard. The rivers that are now clogged with sediment, that you've seen yourself, were deep rivers. I used to see fish, no problem, it was crystal clear. There were duck eggs [a locally important source of food] and eggs of other birds too – I used to go gather them in quantity. There were cattle. I had my boat – there was so much water that I had to travel by boat. Look, these lands that really produced well have converted into a desert. And it's advancing more and more.

(Author interview, May 20, 2011)

The practice of gathering wild duck eggs – once an important source of protein – has all but vanished, along with most of the area's once-abundant waterfowl, and the area Don Gerardo describes is now a barren plain. Don Miguel Quispe Condori, an indigenous authority from the community of Urku Pampa, near where the Huanuni river flows into Lake Uru Uru, had similar memories of his childhood and reminisced about catching fish:

And my father told me that in the Huanuni river – we called it the San Juan river – my father told me there used to be fish in it. We would catch and eat them. The water was very clear because there wasn't much mining. There was mining, but on a small-scale.

(Author interview, May 30, 2016)

As the water became progressively more polluted, people experienced this transformation bodily, through the taste and feel of the water. Mining contamination was not an abstraction, to be measured only in terms of biochemical parameters such as pH, dissolved oxygen, or heavy metal content. Water was experienced as thicker and as tasting bad. As Doña Eugenia Sula Canki put it, "[The pollution] has affected us this way: the water, the land too, now it's not like it was before, it's completely spent, the land is all white; white, ugly, salty [*salado*]. And now the water has also turned salty" (author interview, June 2, 2016). Residents discuss the past as a time of clean water and fertile pastures, and as a time of plenty. Doña Celestina Mamani told me of her childhood:

When I was a little girl, the community was beautiful, no? In the first place, we had lots of livestock: we had sheep, cows, pigs, burros. We had lots of them. The pastures were good. There were all types of pasture, and food for the sheep ... the water was clean [*era dulce*]. So with all this we had large numbers of livestock and crops also. We also had plenty to eat: milk, cheese, lots of meat, also potatoes, *chuño* [dried potato], quinoa, dry fava beans, no? Fresh fava beans. We never lacked for food. But as the years went by, our life changed.

(Author interview, June 2, 2016)

Indeed, life for residents of the valley has changed in a number of ways. Out-migration has increased dramatically, as part of a broader trend toward urbanization in Bolivia (and throughout Latin America), and it is now common for households to be divided between multiple places – with older people remaining full-time in the community, children often away at school in Oruro, and younger adults working in one of Bolivia's larger cities or doing menial labor abroad. While these processes are complex and are driven by a number of factors, the degradation of the region's environment has severely limited the livelihood options for rural residents. A similar story was related by Doña Eugenia:

When I was a little girl, the land was good, it produced potatoes, quinoa, fava beans – everything produced well: barley, grains, everything produced well. My father grew everything well and we helped, all us little kids. Everything produced well, by the sack-full. We had cows, sheep and also burros ... There was so much *totora* [reed] that people would get lost. That's how it was. Cows would enter the water and we would lose the cows [in the *totora*]. It was good before, everything was good. When I was a little girl I saw all this, same when I was a teenager. But now, now it's not like that with the pollution.

(Author interview, June 2, 2016)

As opposed to sudden changes caused by an oil spill or flood, the accumulation of mining waste is a form of "slow violence" (Nixon, 2013) to which people gradually adjust their lives and which seldom merits government attention. Don Miguel told me of how technological changes at the mines irreparably altered the quality of water and the quality of life for communities downstream:

So as the years passed, the problem was when the tin barons, the owners here in the valley, in this watershed, only worked with calcination. They burned the tin. This red water that passed [down the river] still served for irrigation. And the people in the community of Urku Pampa irrigated with this water. There was production of fava beans, everything. With this water, nothing happened. From there, the new technology that arrived in this time they changed to chemical reagents ... From there started the reactivation [of mining] then, the new technology, they started to use xanthate [*xantato*], the famous cyanide, and later sulfuric acid, and later copper sulfate ... So, little by little, little by little, year by year, what's happening? The water began to change its flavor. The water started to get a little thicker, yeah? It changed its flavor.

(Author interview, May 30, 2016)

Again, changes in water quality are related not in terms of chemical parameters (even though Don Miguel is well aware of the chemical processes involved in the water's contamination), but rather in terms of affective, experiential and embodied qualities: taste, texture, and appearance. The pictures painted by these residents draw on images and experiences common in such descriptions, and which I heard repeatedly during in my conversations with local residents: crystalline waters, wetlands full of *totora* reeds, plentiful fish and waterfowl, bountiful agricultural production. I do not doubt that there is truth in these representations, and it is certainly the case that mining contamination has resulted in profound environmental and social deterioration. However, one must interpret such reminiscences with care. Given the long history of mining in the region, environmental conditions are unlikely to have been as idyllic as these residents describe.[2] Moreover, their communities, which are among the most affected by mining contamination, were also subject to the forced labor and racialized subjugation of the *hacienda* [large feudal estates] system, until its abolishment by the agrarian reform of 1953. Most of the people I interviewed were born after the end of the *hacienda* system, but their parents were all subject to it. While the 1950s and '60s was a time of social progress, it was also a time of widespread and entrenched poverty, racism and frequent political turmoil. But as Javier Auyero and Débora Swistun (2009) note, statements about the past are never *only* about the past, and are always also reflections on the present. Such reminiscences, then, serve to measure what people experience as a much-diminished present, by holding it up as a lens through which they view an idealized, vanished past. In this sense, memory plays a central role in the production of meaning, and the way people come to understand the water contamination that has so affected their lives.

17.5 Memory, Landscape and Environmental Justice

There is no direct or necessary connection between memory and justice, and yet sites and practices of collective memory figure prominently throughout Latin America and elsewhere in efforts to promote social reconciliation and healing in the aftermath of dictatorship, civil war and other forms of violence (Andermann, 2015; see also Brockmeier, 2002). By contrast, collective memory of natural landscapes is seldom invoked in conceptualizations of environmental justice. As a field of academic inquiry and social action, environmental justice is rooted in the struggles of social movements and their allies in the church and the academy, especially in the southern United States, against the siting of factories, sewage treatment plants, solid waste incinerators and other facilities that disproportionately affect African-American neighborhoods and other communities of color (Holifield, 2015). Not surprisingly, then, early research into environmental justice emphasized quantitative analysis of the spatial/racial distribution of polluting sites (Bullard, 1983, 1990). While the distributional frame is still prevalent, analyses of environmental justice have moved beyond it to consider such social dynamics as white privilege (Pulido, 2000), varieties of environmentalism (Guha and Martínez-Alier, 1997), social marginalization (Tschakert, 2009), and, drawing on the work of Amartya Sen (1999), capabilities and freedoms (Goff and Crow, 2014; Schlosberg, 2007; Crow, this volume). These examples are all rooted in "ideal types" of justice, however: normative understandings of what *should* be, perhaps best exemplified by Rawls' (1971) "veil of ignorance."

But as Zwarteveen and Boelens (2014: 147, emphasis in original) assert, "understandings of justice cannot be based only on abstract notions of 'what should be,' but also need to be anchored in how injustices are *experienced*. They need to be related both to the diverse 'local' perceptions of equity and to the discourses, constructs and procedures of formal justice." In this view, justice must be understood in dialectical fashion, as a historically constituted relationship between, on the one hand, abstract forms of formal justice (as encoded institutionally in laws and norms), and, on the other hand, historically- and place-specific forms of justice, rooted in embodied experience and situated, subjective knowledge. This view shares much with the notion of "moral economy" as developed in the work of E.P. Thompson (1993), James Scott (1977) and others. Thompson uses the term to describe the English peasantry's sense of fairness during times of food scarcity. In this case, a "fair" price paid for bread was deemed preferable and more just, as compared to a price determined by market principles of supply and demand. Crucially, Thompson's analysis is historically specific and examines conceptions of justice in the context of the transition to capitalism. Thus, while the concept of the moral economy does not pertain equally to all social relations, it is particularly salient at the historical meeting place between subsistence and capitalist relations of production.

Such is the case with communities in the Huanuni river valley. Here, no residents are subsistence farmers, but few can be considered engaged in fully capitalist forms of production. Nearly all residents engage in semi-subsistence agriculture and livestock raising, in combination with small-scale market-oriented agriculture and/or occasional wage labor. Dairy products such as fresh cheese and yogurt are the most common agricultural commodities produced. This petty commodity production supplements the raising of quinoa, potatoes, fava beans and other crops for household consumption. People's livelihoods – their ability to eat and earn a living – depends in large part on access to clean land and water. The loss of water, cropland and pasture due to mine-related contamination has had dramatically negative implications for residents of the valley, and their narrated memories of clean water, green pastures, fertile fields and abundant birdlife speak not just to past experience but to the present and very palpable sense of loss. As Auyero and Swistun (2009) argue, the ways in which environmental suffering is experienced are intimately bound up with relations of domination and social exclusion. Collective memories of water, soil, pasture, fish and birds are directly connected to contemporary experiences and understandings of injustice and, perhaps less directly, to conceptualizations of, and occasional calls for, justice.

Although residents have at times been involved with the efforts of the Oruro-based environmental group CORIDUP, there has been no sustained social mobilization or large-scale collective action in response to mining contamination among community members themselves (see Chapter 13 by Cleaver, this volume). Many residents of the Huanuni valley express skepticism that conditions will ever change, and have grown cynical in the face of seemingly endless meetings and announcements that bring no tangible improvement in their condition. As a result, many communities are largely de-mobilized, and residents have resigned themselves to living with contamination, and to the hopeful waiting that comes diminished expectations (Auyero, 2012). Many residents are also conflicted about mining itself and its role in their lives. This was evident in a discussion I had with Doña Lucila, in the courtyard of her home in the community of Chuspa. Doña Lucila asserted that the Huanuni Mining Company has never done any remediation projects in Chuspa, nor have any of the residents received compensation for the damage done to their lands and water, let alone for the disruption caused to their families and way of life. She told me she believes there was no way to remedy the problem of mining contamination. The problem was too great, she said, and the government was powerless to act. When I asked her if there is a solution to the pollution problem, she replied, "I don't believe so. It's contaminated, contaminated. The wind continues to blow dust (from mine tailings), and the acid drainage [*copajira*] continues. There's no solution." She then said something I had not anticipated. In spite of all the pollution and official neglect, she said that Oruro could not survive without mining. She commented that the modest social welfare payments [*bonos*] funded by mining royalties have improved life for many people. Doña Lucila's statement captures a sense of resignation and contradiction that I heard many times in my conversations with people in the Huanuni valley. Many residents and local activists acknowledge the daunting scale of the pollution problem, and the fact that the local environment, and the livelihoods it once supported, have been irreparably altered. But Doña Lucila also expressed a sense of tension within Bolivia's extractive economy, and the benefits – however meager – they provide to the country's poor.

17.6 Conclusions

Individual memories, such as those recounted above, are also collective memories in the sense of Halbwachs (1992), who argued that once narrated, such memories are filtered through cognitive frames informed by present, collectively shared experience: memories *in* the group, as opposed to memories *of* the group (Wertsch, 2009). Though individual in character, such memories are fundamentally *social* to the degree that our understandings of the past are filtered through a conceptual lens shaped by current experiences. Landscapes and the natural environment provide a powerful medium for collective experience. As French (2012: 342) states, "the landscape comes to index the past for those who inhabit it in the present." This indexing operates in both directions, with present experience shaping our understanding of the past even as past experiences give meaning to the present (Legg, 2007).

In this chapter I have considered the role memory plays in narrating the transformation of landscapes and environments. In particular, the chapter has examined the way residents of Bolivia's Huanuni river valley recount memories of flowing waters and lush, fertile fields uncontaminated by mining waste. Such memories stand in stark contrast to the contemporary reality of acute water pollution and the toxic sediments and barren lands to which it has led. These personal memories also contrast with the official and very public memorialization of mining. Whereas murals and monuments found in cities like Oruro, Potosí and Huanuni represent miners and mining at the center of the national story, the personal narratives recounted here represent counter memories that disrupt the "mnemonic hegemony" of official histories (Molden, 2016). There is no direct or necessary relationship between memory and justice; rather, this relationship is contingent and open-ended, and rooted in personal, situated and embodied experience.

Similarly, our understandings of water justice must be rooted not only in formal notions of "ideal type" justice (such as distributional and procedural justice), but also in the affective, experiential and embodied nature of water. In arid and impoverished regions such as the central Altiplano, people experience water not only through its presence but also, and perhaps especially, through its absence. The drying of wells and rivers or their severe contamination by mining waste, mark a landscape of want and deprivation for residents of the Huanuni valley. In this context, memory narratives of fertile landscapes, lush fields and crystalline waters provide a conceptual framework by which people make sense of their lives, and tell us as much about contemporary lived experiences as they do about past realities. Such memories have the potential to form a conceptual basis for understanding water justice, and for collective action aimed at achieving more just environmental futures. Indeed, many members of the environmental group CORIDUP come from rural communities and urban neighborhoods severely impacted by mine-related contamination, and have told me similar stories of the clean waters and productive fields they experienced as children. At the very least, such memories of the past serve to inform collective action oriented toward an environmentally just future.

Acknowledgements

I owe a debt of gratitude to friends and colleagues at the Centro de Ecología y Pueblos Andinos (CEPA) for their patience and support, and to the residents of the Huanuni Valley for sharing their stories with me.

Notes

1 The honorifics "Don" (for men) and "Doña" (for women) are used for respected elders. All names of interviewees and indigenous-*campesino* communities in this chapter are pseudonyms. All interviews were conducted in Spanish, and transcribed and translated by the author.
2 The central Altiplano is semi-arid, and the Huanuni valley, Lake Uru Uru and the Desaguadero River (which drains Lake Titicaca and is the main tributary of Lakes Uru Uru and Poopó) are among the few reliable year-round sources of fresh water in the region. Waterfowl such as ducks, ibises and flamingos are still common in the area, although their populations have been greatly diminished.

References

Andermann, J. (2015). Placing Latin American memory: Sites and the politics of mourning. *Memory Studies*, 8(1), 3–8.

Auyero, J. (2012). *Patients of the State: The Politics of Waiting in Argentina*. Durham, NC: Duke University Press.

Auyero J. and Swistun D. (2009). *Flammable: Environmental Suffering in an Argentine Shantytown*. New York: Oxford University Press.

Bebbington, A. (2009). The new extraction: Rewriting the political ecology of the Andes? *NACLA Report on the Americas*, 42(5), 12–20.

Blight, D. W. (2009). The memory boom: Why and why now? In P. Boyer and J. V. Wertsch (eds.), *Memory in Mind and Culture*. Cambridge: Cambridge University Press, pp. 238–51.

Blunt, A. (2003). Collective memory and productive nostalgia: Anglo-Indian homemaking at McCluskieganj. *Environment and Planning D: Society and Space*, 21, 717–38.

Bridge G. (2004). Mapping the bonanza: Geographies of mining investment in an era of neoliberal reform. *The Professional Geographer*, 56(3), 406–21.

Brockmeier, J. (2002). Remembering and forgetting: Narrative as cultural memory. *Culture & Psychology*, 8(1), 15–43.

Brown, K. W. (2012). *A History of Mining in Latin America: From the Colonial Era to the Present*. Albuquerque, NM: University of New Mexico Press.

Bullard, R. D. (1983). Solid waste sites and the Black Houston community. *Sociological Inquiry*, 53, 273–88.

Bullard, R. D. (1990). *Dumping in Dixie: Race, Class, and Environmental Quality*. Boulder, CO: Westview Press.

Canessa, A. (2012). *Intimate Indigeneities: Race, Sex and History in the Small Spaces of Andean Life*. Durham, NC: Duke University Press.

Dunkerley, J. (1984) *Rebellion in the Veins: Political Struggle in Bolivia, 1952–1982*. London: Verso.

Foote, K. E. and M. Azaryahu (2007). Toward a geography of memory: Geographical dimensions of public memory and commemoration. *Journal of Political and Military Sociology*, 35(1), 125–44.

French, B. M. (2012). The semiotics of collective memories. *Annual Review of Anthropology*, 41, 337–53.

Goff, M. and Crow, B. (2014). What is water equity? The unfortunate consequences of a global focus on "drinking water." *Water International*, 39(2), 159–71.

Gómez-Barris, M. (2009). Mapuche mnemonics: Beyond modernity's violence. *Memory Studies*, 8(1), 75–85.

Guha, R. and Martínez-Alier, J. (1997). *Varieties of Environmentalism: Essays North and South*. New York: Earthscan.

Halbwachs, M. (1992 [1951]). *On Collective Memory*. Chicago: University of Chicago Press.

Holifield, R. (2015). Environmental justice and political ecology. In T. Perreault, G. Bridge and J. McCarthy (eds.), *The Routledge Handbook of Political Ecology*. London: Routledge, pp. 585–97.

Kaup, B. Z. (2010). A neoliberal nationalization? The constraints of natural gas-led development in Bolivia. *Latin American Perspectives*, 37(3), 123–38.

Kaup, B. Z. (2013). *Market Justice: Political Economic Struggle in Bolivia*. Cambridge: Cambridge University Press.

Legg, S. (2007). Reviewing geographies of memory/forgetting. *Environment and Planning A*, 39, 456–66.

López, E. (2011). Bolivia: Agua y minería en tiempos de cambio. In P. Urteaga (ed.), *Agua e Industrias Extractivas: Cambios y Continuidades en los Andes*. Lima: Instituto de Estudios Peruanos, pp. 61–88.

López, E., Cuenca, A., Lafuente, S., Madrid, E. and Molina P. (2010). *El Costo Ecológico de la Política Minera en Huanuni y Bolíviar*. La Paz: PIEB.

Lowenthal, D. (1989). Nostalgia tells it like it wasn't. In C. Shaw and M. Chase (eds.), *The Imagined Past: History and Nostalgia*. Manchester: University of Manchester Press, pp. 18–32.

Madrid, E., Guzmán, N., Mamani, E., Medrano, D. and Nuñez, R. (2002). *Minería y Comunidades Campesinas ¿Coexistencia o Conflicto?* La Paz: PIEB.

Meade, T. (2001). Holding the junta accountable: Chile's "Sitios de la Memoria" and the history of torture, disappearance and death. *Radical History Review*, 79, 123–39.

Molden, B. (2016). Resistant pasts versus mnemonic hegemony: On the power of collective memory. *Memory Studies*, 9(2), 125–42.

Montoya, J. C., Amusquívar, J., Guzmán, G., Quispe, D., Blanco R. and Moll, N. (2010). *Thuska Uma: Tratamiento de Aguas Ácidas con Fines de Riego*. La Paz: PIEB.

Nash, J. (1993). *We Eat the Mines and the Mines Eat Us: Dependency and Exploitation in the Bolivian Tin Mines*. New York: Columbia University Press.

Nixon, R. (2013). *Slow Violence and the Environmentalism of the Poor*. Cambridge, MA: Harvard University Press.

Perreault, T. (2013). Dispossession by accumulation? Mining, water and the nature of enclosure on the Bolivian Altiplano. *Antipode* 45(5), 1050–69.

Pulido, L. (2000). Rethinking environmental racism: White privilege and urban development in southern California. *Annals of the Association of American Geographers*, 90, 12–40.

Quintanilla, J. and García, M. E. (2009). Manejo de recursos hídricos-hidroquímica de la Cuenca de los lagos Poopó y Uru Uru. In P. Crespo Alvizuri (ed.), *La Química de la Cuenca del Poopó*. La Paz: UMSA/DIPGIS – Instituto de Investigaciones Químicas, pp. 117–43.

Rawls, J. (1971). *A Theory of Justice*. Cambridge, MA: Belknap Press.

Schlosberg, D. (2007). *Defining Environmental Justice: Theories, Movements and Nature*, Oxford: Oxford University Press.

Scott, J. C. (1977). *The Moral Economy of the Peasant: Rebellion and Resistance in Southeast Asia*. New Haven, CT: Yale University Press.

Sen, A. K. (1999). *Development as Freedom*. Oxford: Oxford University Press.

Thompson, E. P. (1993). *Customs in Common: Studies in Traditional Popular Culture.* New York: New Press.

Tschakert, P. (2009). Digging deep for justice: A radical re-imagination of the artisanal gold mining sector in Ghana. *Antipode*, 41, 706–40.

UPADE. (2010) "Human development in the Department of Oruro." Unidad de Análisis de Políticas Sociales y Económicas (UPADE) and United Nations Development Program, www.udape.gob.bo/portales_html/ODM/Documentos/Boletines/Bol_2010_04_Eng.pdf.

Wertsch, J. V. (2009). Collective memory. In P. Boyer and J. V. Wertsch (eds.), *Memory in Mind and Culture*. Cambridge: Cambridge University Press, pp. 117–37.

Zwarteveen, M. Z. and Boelens, R. (2014). Defining, researching and struggling for water justice: Some conceptual building blocks for research and action. *Water International*, 39(2), 143–58.

18

New Spaces for Water Justice? Groundwater Extraction and Changing Gendered Subjectivities in Morocco's Saïss Region

Lisa Bossenbroek and Margreet Zwarteveen

18.1 Introduction

For the last couple of decades, groundwater tables in the Saïss agricultural plain, mid-north Morocco, have been dropping. This is partly due to intensified groundwater use in agriculture and changing cropping patterns. Specifically, since the 1980s, farmers have increasingly dug wells to engage in commercial agriculture. From growing rain-fed crops, they have shifted to higher-value irrigated crops, notably onions and other vegetables. Changing tenure relations (land privatization, most notably) and new agricultural policies that actively promoted modernization have further intensified groundwater use. Today, water-intensive crops produced in the region are sold on regional, national and international markets. They produce new material and cultural linkages between hitherto-unconnected life worlds, interactively reshaping socio-natural relations. They open up new possibilities of living and being, but also close off old ones.

In this chapter, we describe and analyze these changes to interrogate them from a feminist-justice perspective. We use material from extensive ethnographic fieldwork, conducted over a two-year period from 2011 to 2013, complemented by various recurring several-week visits over 2014 and 2015. We have focused on anecdotes regarding the changing groundwater and irrigation dynamics that illustrate how they are intimately linked to changing gender relations. Rather than merely assessing or mapping the gendered impacts of intensifying groundwater use, our purpose is to tease out how prevailing gendered ways of being, working, and relating co-shape and in turn are shaped by new agricultural and groundwater dynamics, in ways that are not always straightforward or easy to predict. Hence, our argument is that groundwater abstraction and gender relations dynamics co-constitute each other, which also means that one cannot be comprehended without the other. We show this by focusing on how intensifying groundwater use, and the larger changes it provokes, is made possible through and shaped by changing the gendered organization of labor and space.

Contemporary agrarian policies in Morocco intended to promote a new kind of green revolution are significantly premised on promoting a particular gendered organization of farming (and consequently of families), attributing specific roles, responsibilities and rights to men and to women. Hence, the two main protagonists of agricultural modernization that the Moroccan government focuses on and wants to support are new agricultural investors, and those existing farmers who have the spirit and willingness to modernize. In public policies, these protagonists are clearly cast as men – with farms figuring as enterprises that operate with a logic of profit maximization. Through agricultural support programs, extension services and subsidies, the farming policy model and farmers co-shape what happens on agricultural fields both materially and discursively. Yet, a closer look at how different people engage with changing farming realities and groundwater dynamics reveals a much larger diversity of practices and experiences, many of them hybrids combining peasant traditions and entrepreneurial logic. Complex intersections of class, generation and gender mingle with existing ways of doing agriculture, on the one hand, and new policy directions and market opportunities, on the other, to contingently produce a range of possible farming and water-use configurations: each comes with different gender arrangements – different options for relating and performing as men or as women. The chapter presents some of these configurations and arrangements, based on ethnographic documentation of people's everyday experiences and engagements with changing groundwater dynamics. We do this to suggest that there is merit in theorizing water dynamics and gender dynamics as intimately linked: materially (through labor and property relations), and symbolically and discursively (through norms, meanings and symbols), with "gender" and "water" continuously defining and redefining each other – moving together, as in a perpetual dance.

We first briefly review our theoretical sources of inspiration for understanding interactions between gender relations, gendered subjectivities and groundwater dynamics. We then provide a brief description of the Saïss agricultural plain and discuss the various processes that have contributed to intensifying groundwater use. In the third section we show how the Moroccan Government's current agricultural modernization plans are founded on a particular farming model and farming identities – a model that both co-shapes contemporary agrarian dynamics, and serves as the mainstream frame of reference that many researchers use to make sense of it. We then continue by using our empirical material to interrogate this model, showing how gendered meanings and identities evolve and are actively (re)negotiated and performed through changing ways of dealing with groundwater. We conclude the chapter with a brief feminist reflection on what these observations mean for the feminist analysis of, or feminist struggles for, water justice.

18.2 The Interlinkages Between Gender and Environmental Change

Linkages between gender and the environment have been a recurrent topic in feminist scholarship and debate. Much of this work starts from the original definition of gender itself, a definition that, by separating someone's biological sex from someone's social gender, is premised on the analytical possibility to distinguish nature from society. Early feminist work on the environment, much of which was done under the broad umbrella of eco-feminism, attempted to demonstrate how, in modern societies, this separation (between nature and society) justifies and produces social-valuational hierarchies, between people and animals, between people of different ethnicities, but importantly also between men and women. This work showed how modern societies organize activities (work), identities and much else through binary distinctions in which (some) men, culture, public life, rationality, productive work (etc.) are considered as separate, opposed and often superior to women, nature, private life, emotion and domestic work (etc.). In a feminist attempt to revalue the feminine side of this binary, much eco-feminist effort was invested in showing the importance of all that was associated with women. This often happened as part of a more encompassing critique of modernity, in which feminists formed productive alliances with a rapidly emerging environmental movement. Eco-feminists thus convincingly argued and showed how modern progress presupposes exploiting both women and nature. Maria Mies' work is seminal here: in her book *Patriarchy and Accumulation on World Scale* (1986) she demonstrates how capital accumulation is premised on subordinating and exploiting women, nature and colonies[1] (Mies, 1986: 2). For Mies, just as for other eco-feminist scholars, the logical implication was that women were the protagonists *par excellence* of social movements for resistance and change (see Mies and Shiva, 1993). This was importantly founded on the idea that women, because of how societies relegate them to the "nature" side of the binary, have a closer relationship to nature than men.

While ecofeminist analyses provide a powerful, profound feminist critique of modernity and science, they are less useful for understanding everyday gendered experiences of living with the environment. Further, the idea that womanhood is a good basis for environmental activism proved questionable in practice: people's everyday relationships and dealings with the environment are only very partly determined by their gender. In this chapter, we therefore propose complementing eco-feminist thinking with two other strands of feminist thought.

The first is a body of work that questions a priori assumptions about nature–society distinctions that inform both some eco-feminist as well as early feminist political-ecology analyses. The argument is that these distinctions do not exist prior to the analysis, history or human experience, but are instead themselves the product of ways of being, relating and conceptualizing that are always context-specific. The boundary between nature and society, in other words, is not dictated by a world that exists irrespective of how human beings make sense of it: this boundary comes into being through the very conceptual categories and definitions used to describe it. For gender, this line of thinking has most famously been articulated by Judith Butler. She took issue with how the definition of gender continues to assume some unchangeable biological core, which gets its context-specific and changeable feminine or masculine shape and meaning through processes of socialization. Butler instead posited that being or behaving as a man or a woman is always a performance, something that is not necessarily linked to biology. Gender is not something one *"is"* but something one *"does."* The performance of particular gender identities is based on "the tacit collective agreement to perform, produce and sustain discrete and polar gender as cultural fictions" (Butler, 1990: 138). Conceptualizing gender as performative implies that gender identities are never complete(d); they are continuously becoming. Identities, therefore, are necessarily fragmented and provisional; they are wrought through the interplay among various fields of power and regulatory frameworks (see also Resurreccion and Elmhirst, 2008: 15). As Butler (1990) argues, the coexistence (clashing or convergence) of various discursive injunctions itself enables reconfiguration or redeployment. It leads to theorizing the everyday social practices that constitute gender relations and environmental change as a spiral in which the two constantly (re)define each other. Rather than theorizing gender subjectivities or relations as pre-existing or separate from the environment, this theorization posits that changes in the environment – in our case, changes in groundwater use, access and control – always happen and are defined through changing gender subjectivities and relations: gender and the environment move together.

A second strand of feminist work that we build on for our analysis entails a plea for taking seriously women's and men's own creativity, experiences, actions and self-perceptions in their dealings with others and with environmental (or more generally capitalist) transformations. Dissatisfied with analyses that reduce these experiences to Marxist false consciousness or interpret them as the result of Foucauldian disciplining and normalization, this work suggests that there may not be one larger structure or totality that holds reality together. Rather than helping uncover or expose this larger structure or totality, therefore – to either further strengthen it or to more effectively contest and resist it – feminist scholars in this tradition use ethnographic methods to trace the multiple ways in which realities or worlds are constituted and enacted (in practices), accepting that there may be clashes, convergences or overlaps between these worlds, making them more or less durable or mobile. Methodologically, this work starts by emphatically acknowledging and learning from people's own ways of navigating and making sense of the multiple force fields that they belong to. Rather than using this as empirical material to either provide evidence for or dismiss larger theories, these scholars propose mobilizing it to *interfere*, *think with* or *destabilize* them in search of alternative ways of articulating and creating realities (Domínguez-Guzmán, forthcoming). The study by Katherine Gibson, Lisa Law and Deirdre McKay (2001), although focusing more generally on global capitalist economic developments, provides an interesting example here. It focuses on the actions and life-trajectories of female migrant workers in the Philippines, illustrating these women's multi-faceted experiences. To make sense of these experiences, the authors move beyond existing neoliberal or Marxist frames of analysis, which tend to respectively frame female migrant worker as either "heroes" of national development, emancipation and modernization, or "victims" of capitalist growth and exploitation. Their attempts to do more justice to migrant workers' own stories reveal the paradoxes and ambiguities in their experiences. Female migrant workers face hardship and exploitation, but also consciously strive to purchase particular assets and capacities that allow them to move upward socially. This careful ethnographic account of capitalist economic expansion experiences illustrates how gendered subjectivities and relations become redefined through new economic possibilities, with female workers assuming new aspirations and roles by identifying economic and cultural possibilities to support their home communities and realizing their own dreams in their host community.

Similarly, the work of Priti Ramamurthy attempts to grasp the ambiguous experiences and subjectivities of female rural wage workers (2003) and smallholders (2011) in South

India within capitalist growth. She mobilizes the perplexity concept to mark tension among overlapping, opposing and asymmetric force-fields, expressing people's puzzlement as they experience both the joys and aches of global everyday life, often simultaneously (2003: 525). She argues that their perplexity captures their experiences of capitalism's contradictions through a structure of feeling, combining on-going struggles against existing socio-cultural norms and existing socio-economic structures (2011: 1054) with pleasures in the benefits and luxuries that come with new opportunities for wage work or more purchasing power. Taking inspiration from these bodies of thought, we place the ambiguous experiences and aspirations of men and women who experience groundwater intensification at the heart of our attempt to make a feminist contribution to thinking about water justice.

Taken together, one clear implication of these theorizations for thinking about feminist water justice is that it becomes more difficult to identify winners and losers from environmental and capitalist change. Questions of (in)justice, examined close by, often lose the neatness they may have when observed from the comfortable teleological distance of more structural frames of analysis. While situated within larger power structures, analytical/methodological insistence on taking seriously people's own sometimes contradictory or indeed perplexing feelings and experiences usefully directs attention to how they themselves subtly maneuver, adapt or navigate situations of injustice, transforming and repositioning themselves in the process. While such everyday transformations may not immediately add up to fuel larger change movements, they may provide pragmatic, more modest inspirations for alternative ways of engaging and intervening in water – ways that are more friendly, caring, convivial and companiable, or indeed more just.

18.3 Intensifying Groundwater Abstraction in the Saïss Agricultural Plain

Over the last century, groundwater abstraction in the Saïss agricultural plain has gradually intensified. In particular since the 1980s, farmers have increasingly accessed and used water through wells and deep-tube wells, partly prompted by liberalization policies, which provoked changes in tenure relations, facilitated access to new irrigation technologies (such as deep tube-wells and drip irrigation), and integrated the Saïss rural economy into national and globalizing markets. Groundwater extraction further intensified during the last two decades, partly as a result of proactive government policies to restructure the agricultural sector. Most notably, in 2006, the Government decided to privatize the lands of 93 state cooperatives (created in the early 1970s as part of a national land-reform program). These reforms had redistributed land that had been owned by foreign settlers during the French protectorate (1912–56). Some peasant families received an individual land plot with land-use rights, organized into state cooperatives, while others became laborers working in kolkhoz-inspired state cooperatives.[2] The 1970s land reforms had aimed to create a class of peasant farmers, who would contribute to agricultural development. These plots could not be sold until 2006, when former members could acquire a formal land title from the state and became landowners.[3] Many families were subsequently forced to sell their land, as they were either in debt or had inheritance problems, attracting urban investors and new "entrepreneurs" to the region. Today, land fetches ever-higher prices (45,000–55,000 euros a hectare) with brokers who actively help connect willing buyers and sellers. In the same vein, the recently established (2004) private-public partnership arrangements for formerly state-owned land, mostly favor those investors who are able to set up "productive," "modern" farm projects.

New possibilities to obtain land combine with availability of water and (female) labor at relatively low prices to attract many urban investors and "entrepreneurs" to the region. As they can afford to pay much more, they make it ever more difficult for those born and raised in the region to access land and water. Newcomers' farm projects tend to be highly water-intensive. As soon as they have purchased the land, they fence it, dig new tube-wells, install drip irrigation and usually plant high-value crops such as grapes or fruit trees. In contrast to the resident peasant families, who irrigate only between 30 and 40 percent of their plots, new investors usually irrigate up to 80 to 90 percent of their newly purchased land. Altogether, these transformations have had drastic consequences for the groundwater tables, which over the last three decades have dropped 10 meters (Kuper *et al.*, 2016). This was even more noticeable in the confined aquifer, where the groundwater table dropped up to 2.6 meters a year (ibid.).

New farm projects not only require water, but also strongly rely on relatively cheap female labor, with women often working without contracts and under poor labor conditions. They are hired because of their nimble fingers and other supposedly feminine skills that makes them particularly suitable for the tasks they need to do. This is one way in which, as we show in more detail below, intensifying groundwater use is premised on changing the gendered organization of labor and space.

18.4 The Intimate Relationship Between Gendered Norms/Practices and Groundwater Dynamics

Intensifying groundwater use, as described above, not only impacts prevailing gendered ways of being, but is also itself shaped by gendered practices and norms. The new agrarian policies, intending to promote a new kind of green revolution, are significantly premised on a particular gendered organization of and distinction between farming and families. The Moroccan Government identifies two main protagonists of the modernization of the agricultural sector: new farm "entrepreneurs" and peasant farmers who are willing and able to modernize. Both are clearly imagined as men, while their ideal performance assumes a particular gendered distinction between farms and families. While co-shaping actual farm realities, for instance, through targeted subsidy programs, our analysis of contemporary rural actors' experiences and farming practices show that groundwater dynamics provide a much wider range of new opportunities for performing farming identities and constructing social relations than imagined or foreseen by policy frameworks. New possibilities for being and relating are forged, while old ones are closed off. This happens along intersecting axes of gender, class and generation.

18.4.1 A New Green Revolution Founded on a Particular Farming Model and Farming Identities

These new farm projects epitomize the Moroccan Government's ambition, articulated in its 2008 Green Morocco Plan (PMV or Plan Maroc Vert): to modernize the agricultural sector, turning it into a profitable part of the national economy. As the plan is based on a particular capital- and resource-intensive form of entrepreneurial farming, it should come as no surprise that it favors the more well-to-do, larger farm enterprises and private investors, to peasant farms' neglect (for a critical analysis, see Akesbi, 2011; Kadiri and El Farah, 2013). Private investors are therefore the PMV subsidy system's main beneficiaries. In contrast, peasant/family farming – which in Morocco is referred to as "the second pillar of agriculture" – receives much less state support. The little that it does receive falls under solidarity, rather than productivity: it is meant to assist small- and medium-scale farmers in marginal regions to survive.

Most government officials that we conversed with align with the ideas and ideals of the PMV: they proudly refer to the unfolding agrarian transformations as modernity and progress. For many of them, the PMV offers much-needed institutional and technological support to speed up Morocco's development. It is clear that the main protagonists of this vision of modernity are the new farming "entrepreneurs" who buy and invest in land in the region. Government officials refer to these new "entrepreneurs" as individuals "who contribute to the real development and who bring investments to the agricultural sector." Yet, existing farmers are also allowed to play a potentially important role in helping bring about modernity, provided they adopt new technologies and increase their productivity. These two main protagonists of the current transformations – farming entrepreneurs and modernizing farmers – are clearly (although largely implicitly) cast as men. We conducted an interview with a government ministry of agriculture official in which we asked whether there were also female farmers, or female new entrepreneurs. His answer is telling: "Yes, I know one, but she is more man than a man. She has about 2,000 hectares and everything is equipped with drip irrigation." Promoting current agrarian developments is based on distinct gendered ideas about the preferred way to organize farming: the agricultural sector is expected to largely modernize through some audacious men's work and initiative, who are pictured as heading modern farm enterprises, enterprises that are separate and distinct from families and homes. This representation foregrounds the work and achievements of some – the male farmers who supposedly direct the farm enterprise – while rendering the work, efforts and strivings of all others – including women – invisible. It is also an account that, by focusing on marketable profits, leaves much social and environmental farming costs – and particularly water – out of the equation.

18.4.2 New Possibilities for Groundwater Access, Use and Control and Performing New Masculinities

New groundwater access and use modes are key to realizing the new "modern" farm projects and becoming new farm "entrepreneurs." As previously mentioned, these new farm projects produce high-value crops, are fully equipped with drip irrigation, and rely on several deep tube-wells for their water access. Their owners usually do not live on the farms, nor do they themselves engage in physical farm work. They only come every once in a while, driving to and through the area in fancy four-wheel-drive cars and dressed in distinctly urban outfits. They visit occasionally just to inspect their projects or to spend a weekend of leisure in the countryside. Hassan (aged 46) typifies this new farm entrepreneur. He is an architect who lives in the city of Fes. Two and half years ago (in 2014), he bought a plot because, according to him, "the agricultural sector offered interesting business opportunities and I like to spend my weekends here." He planted one hectare of peaches, one hectare of nectarines, one hectare of apples and four hectares of prunes. On his newly acquired land, he also built a two-storied house with a terrace overlooking a swimming pool.

While these farm projects are a hobby or investment for people like Hassan, for people who are originally from the region, farming is the backbone of their livelihoods. A number of residential families started their own farms after having benefited from the early 1970s land reforms. They, or their parents, became members of the socialist-inspired state cooperatives as laborers, and started working collectively on the land. In the early 1990s, these large farms were divided into smaller state cooperative farms with members receiving an individual plot with land-use rights. From that moment onwards, families started to work on their own plot and became responsible for their own farms. Some, in particular the wealthier ones, began to dig wells and to diversify their production by planting irrigated crops such as onions, peppers, carrots and potatoes. These were the people who were relatively better-positioned to benefit from land privatization after 2006. There were many others who failed to hold onto their plots. In particular, those families who were already quite heavily indebted were forced to sell their land. Some decided to move to the city, often using the money earned by selling the land to repay their loans. Others remained living in the countryside and became laborers on the land that they used to own. Families who obtained their private land titles usually continue to farm as they used to, irrigating 30–40 percent of their plot. Some have installed drip irrigation and have the aspiration, one day, to dig a second tube-well.

To these farming families' young men, newcomers' agricultural projects are a source of inspiration. The new farmer-entrepreneurs are exciting role models for them, embodying new, more modern ways of doing agriculture and being a farmer. Young men may refer to these newcomers when articulating dreams of starting a fruit-tree farm, renting extra land to realize their own farm project, or engaging in biofarming. They are fully aware of how important groundwater access is to realize these aspirations. Young male farmers, for instance, state: "access to groundwater opens up new horizons"; "now we can engage in real agriculture"; "now we can install a drip system and plant fruit trees." On the face of it, this vision of farming, built on the intensive water use, contrasts with older generations' farming vision. These older farmers believe in mixed-cropping patterns in which water is used with care, and with consideration for longer-term futures. Some elderly men even reject drilling deeper tube wells, along with the fruit trees and grapes, altogether. This is not because they cannot afford them, but because they wish to continue with their existing farming practices; it is part of who they are and what works for them (e.g. Bossenbroek *et al.*, 2015). For younger male farmers, instead (potential) groundwater access opens up attractive new farming horizons, fueling dreams of planting more lucrative crops, and of using new technologies such as drip irrigation. The difference between the generations may nevertheless be less stark than it appears at first sight. Many younger men who aspire to modernize, also wish to respect their parents' farming practices and values (Bossenbroek and Kadiri, 2016). They wish to make changes, but not to the detriment of their key values of sharing and caring for each other and their environments. Where they imitate new entrepreneurial farmers' dress, behavior, or advanced technologies, to appear modern, they also continue to express strong adherence to their territory and family. Farming is part of who they are, and sustains their livelihood. They see themselves as *fellah* (farmer) and *rajal âamal* (businessman) at the same time (see Bossenbroek *et al.*, 2015).

New ways to access and use groundwater play an essential role in performing new masculine farming identities. Groundwater access gives young men the opportunity to work with drip irrigation, which they hope to actively take up to reinvent more modern versions of themselves. In these young men's eyes, drip irrigation makes the job of irrigating cleaner, more technical and less tiring, as the following quote of Mohammed (aged 23) illustrates: "Before, when my father irrigated with the *sequia* [small earthen channels through which water is conducted to crops in the fields], irrigating took a lot of time, sometimes four to five hours. His clothes were muddy and dirty. Today, I only have to open the valves and, in the meantime, I can do something else. I can call my clients, for example, or oversee the laborers working on the field." Mohammed considers drip irrigation not just as a technology that facilitates irrigating, but also as a way he can positively distinguish himself from the older generation of peasant farmers. Khalid (aged 26) likewise considers drip irrigation as offering new possibilities to farm, while creating a new farmer identity. Drip irrigation is "a new way of farming," he explains: "with drip irrigation I can incorporate the fertilizer into the irrigation water. Previously, this was done manually. Drip irrigation also gives the possibility to plant other crops." When we asked if there were other benefits to drip irrigation, he replied: "Yes, I don't get dirty anymore and can start to irrigate from my house without going to the field, while engaging in other activities." One day, we passed by his land when he had just opened the drip-system valves. He was heading home, where he was working on his future farm project: a small tree farm. Like some other young male farmers we interviewed, Khalid was neatly shaven and proudly featured a fashionable hair style to match his sporty jacket and boat shoes, complemented by a ring and a silver chain. This costume is not very different from young people in the city; it marks Khalid as cool and in vogue, thereby also distinguishing him from the older generation of male farmers.

New farming possibilities based on accessing groundwater thus provide exciting new opportunities for becoming a modern rural man. "The farm" plays an important role in performing these new professional masculine identities. As a newly designated, distinct space for engaging in agricultural activities, it enables performing particular gender identities (Pratt, 1998). When compared to the old peasant farm, the new farm is increasingly becoming defined as a masculine space, with farming redefined as a masculine activity. As we explain in more detail in the following section, this redefinition of gendered organization of work and space happens by more strictly separating the private home women's domain and the public work men's domain.

18.4.3 Changing Gendered Meaning of Space, New Labor Practices and Performing New Femininities

Alongside and partly because of how current groundwater dynamics redefine farms and the activity of farming in more strictly masculine terms, definitions of domestic and "feminine" work are also changing. During the period of the state cooperative, women used to engage in agriculture almost matter-of-factly and casually, considering farm activities as a more-or-less logical extension of their domestic duties. Today, the domestic or private sphere seems to become ever more narrowly defined and spatially separated, with farm enterprises that symbolize the productive and public domain, contrasting ever more starkly with the protected, bounded family home space. This forces women into a homebound traditionalism that is newly invented, even when justified by supposedly traditional norms of female virtue and chastity. We use this section to further illustrate how different rural women actively take up and navigate this newly bounded "private" home domain to recreate their femininity, among others by taking up new "bourgeois" female identities. Whereas many belonging to the older generation proudly engage in those farming activities that can be done within the safe homestead space, such as husbandry activities, many younger women prefer to abandon farm work altogether. They instead prefer to engage in what they perceive as "clean" work, such as craft skills. There is also a large group of women, who belong to households with no/little land, for whom it is much more difficult to perform clean, modern forms of rural femininity. Struggling to make ends meet, they are forced to accept low-prestige, low-paying wage-work. For them, it is nearly impossible to find public recognition or personal pride in what they do. They therefore hesitate to talk about it, and literally try to prevent others from seeing them doing it, by covering their faces behind veils that reveal only their eyes.

Women's work used to be embedded in collective family-based farm organization, in which "private" and "public" were deeply entangled (e.g. Bossenbroek and Zwarteveen, 2015). Farmers' wives and their daughters recall how, when they cultivated rainfed crops, they would work in the field alongside their husbands, fathers and other male and female family members. They would be in charge of sowing, weeding, husbandry and help with harvesting, activities that they fluidly combined with housework such as cooking, cleaning and educating children. Although there was some gender division of labor, all collaborated in and were responsible for ensuring family and farm prosperity and wellbeing. Women's identity was significantly co-formed, enacted and reaffirmed through their active contributions to farm work. Through their everyday domestic work, as well as by helping with collective work on the land, farm women continuously reaffirmed their commitment to their marriage, while also asserting their productive value by helping assure family farm reproduction (Bossenbroek and Zwarteveen, 2015). Their identities as a wife, mother and farmer were thus closely interwoven. Likewise, for young unmarried women, work on the land was often regarded as "simply part of our responsibilities."

With intensified farming and irrigation, and with men actively occupying a newly emerging public sphere to perform their masculinity and professionalism, women's involvement in farm work is gradually changing. Ilham (aged 32) for example, explains that when her family was still engaging in rainfed crops; "Everyone helped during the harvest period and women would be in charge of harvesting and weeding. Today we have laborers and I hardly help in the fields and my mother's biggest concern are the stables." In a similar vein, Aziza (aged 42) explained how she used to walk "far" to work in the fields. Today, however, she is relieved of those duties as she is now mainly looking after what have come to be seen as strictly domestic duties: "My husband is in charge of the irrigation. For harvesting or for weeding he hires laborers, usually women." The experience of Ilham and Aziza is typical of women belonging to land-owning families, who have obtained their private land titles during the privatization process. These women appreciate the fact that they do not have to work in the fields anymore as something positive. It is a source of pride to them, something that reflects their new social standing; from being laborers they now have become co-owners of an irrigated land plot of seven to 13 hectares, with husbands who can afford to hire laborers to do the work.

By actively taking up more narrowly defined domestic activities, they not only show off their new social standing but also reinvent what it means to be a good (rural) women, defining it through emphasizing what used to be typical more traditional urban bourgeois virtues of femininity: smoothly running the "private" domain of the household while remaining relatively invisible and unseen by the public eye. Although their physical presence on the farm has become much less, they continuously emphasize the importance of their household activities for the reproduction of the farm. Aziza (aged 43) for instance explains: "I am in charge of cooking for the laborers and for my family, raising the children, washing the clothes. Of course it's work, it's the motor of the farm!" In a similar way, Halima (aged 48) is proud to tell about her full responsibility for all the husbandry activities, consisting of feeding and milking the cows, and cleaning the stable. All this, according to her, is "a key activity of the farm." With a twinkle in her eyes, she tells us how she assisted a cow to give birth: "It was in the early morning at four o'clock that the contractions started. Around six in the morning I helped her. It was very difficult and you have to be careful." Proudly she adds, "I did it all by myself." She aspires to start a milk cooperative, together with her husband. In their emphasis on the continued dependencies and linkages between farm and family, and between their own work and that of their husbands, the stories and experiences of Halima and Aziza challenge the masculinist accounts of farming – with farms being likened to continuously innovating enterprises – that dominate policy documents as well as much research. Their professional pride does not reside in identifying with what increasingly comes to be seen as the masculine farm domain, with mastery of technology and networking skills as the defining characteristics. Instead, they emphasize their hard work, and express the importance of this work for the prosperity of both farm and family. This is similar to what Haugen (1998: 143) observed when studying changing rural femininities in Northern Europe.

Many belonging to the younger generation, while often being full of admiration and respect for the hard work of their mothers, seem less invested in portraying themselves as part of the farm. They instead articulate a clear desire to distance their future feminine selves from physical farm work. They refer to their mother as being "strong," "hardworking" and "caring," but also note that their mothers' work is very exhausting and demanding, while also being "dirty." Many therefore opt for expanding and redefining domestic work, as an avenue for performing new professional rural female identities. Hence, Saïda (aged 28) started a business baking sweets that she sells at her elder brother's café in town. Woman-owned home-based enterprises are a known phenomenon in the city, but so far remain a rarity in the Saïss countryside. Saïda explains that it was her own idea; she had to persuade her brother to support it, by arguing that she could save money so they could buy a car together. Although she had a driving license, she admits that "he will probably use it more often, but at least it helped to persuade him." Also Hind (aged 23) is eager to learn new things and take up activities that may generate some income: "I love to cook and to learn to cook new dishes. I also would like to learn how to use a sewing machine; I am currently taking classes in the city." When we asked her what kind of clothes and models she intended to sew, she replied: "Modern clothes, like pants and modern djellabas." Other young women we interviewed also aspired to sew, embroider, or weave carpets. For them, such activities more closely align with what they have come to consider as appropriately feminine work.

While seemingly traditional, young women also express subtle forms of rebellion when engaging in virtuous feminine crafts. When weaving carpets, for instance, young women do not necessarily reproduce the previous generation's same motifs or patterns. Fatima's last creation (aged 28) consisted of vast bright pink, bright yellow and blue rectangles; quite a contrast to the dark red, white, and dark blue carpets that decorate most houses. For her, the motifs represented "my thoughts" and "dreams"; the carpet mirrors her creativity and emotions. The different patterns and colors can also reflect a dialogue between generations: by reproducing particular patterns taught by their mothers or grandmothers, young women identify with their roots. On the other hand, innovating or including other patterns and colors is a way of self-expression and renegotiating one's femininity (cf. Merini, 2008). Some young women want to sell their carpets through a yet-to-be-formed cooperative, indicating their sense of adventure and entrepreneurialism. Although such cooperatives exist in various other regions in Morocco, in the Saïss such cooperatives are rare.

All this illustrates that women, particularly those belonging to landowning families, consciously attempt to give a positive interpretation to what it means to be a modern rural farm woman. The younger generation's activities seem to further reinforce an emerging division between the private-home-women's domain and the public-work-men's domain, also reinscribing a gendered distinction between different types of work. Their wish to engage in home-based "clean" skills and crafts expresses their desire to distance themselves from dirtier, outdoors, physical labor. They reinvent what it means to be a virtuous woman, also setting standards that are difficult to achieve for those who cannot afford to refrain from physical, "dirty," public labor. It is here where class, gender, and generation intersect to create new hierarchies and exclusions: emerging ideals of womanhood are unattainable for a growing group of female agricultural wageworkers, many of whom belong to landless families. It is telling that women from landowning families would often depreciatingly talk about these women, particularly those who find their jobs through *mouquéfs*,[4] referring to them as illicit women. Malika (aged 42), whose husband's family has eight hectares of land, says: "These women work hard and it is not rewarding. Sewing or weaving is much better. It is clean, you don't have to work under the burning sun. These women even have to hide their faces, as they don't want to be recognized."

Indeed, most women working for wages wear their scarves so they reveal only their eyes. This has earned them the nickname of *ninjas*. This all-covering outfit marks their paradoxical situation: although newly set up farm projects rely heavily on their relatively cheap workforce, their contributions tend to go unnoticed and even they want to be invisible. While explaining that they dress that way to protect themselves from the sun, dust, and pesticides, when prompted many also admitted that they use their scarves to remain invisible and anonymous. Their work and income are a source of shame and embarrassment, of inner struggle and frustration, rather than something they can take professional pride in. Indeed, many prefer to use other sources of identification when presenting themselves (see Bossenbroek *et al.*, forthcoming). Women wage workers would often complain about poor work conditions (low wages, not being paid, no work contract, no insurance, long working days, etc.) and common harassments. Nevertheless, even for these women, the new work opportunities may offer new possibilities for being and becoming, as the following example of Donja (aged 27) illustrates. For her, agricultural work offered a much-sought-after chance to earn an independent income, while also enabling her to "meet other women" while going to work, and exchange experiences and ideas with them. Especially for younger female wageworkers, agricultural wage work often provides one of the few opportunities to leave the confined spaces of their homes. Hence, female wageworkers' experiences combine feelings of happiness and satisfaction, along with feelings of distress and confusion.

18.5 Conclusion

In this chapter, we have illustrated how changing groundwater-use practices, and the larger changes it provokes and forms part of, happen through and are accompanied by renegotiating gendered identities, social relations and spaces. Groundwater dynamics alter farms' gendered labor organization, as well as the gendered meanings of labor and space. These changes open up new possibilities of living and being, but also close off others.

New water-intensive farming futures and identities are actively promoted by the Moroccan Government; clearly favoring particular actors and very specific forms of farming. New farm "entrepreneurs" and peasant farmers who have the willingness and ability to modernize, both clearly cast as men, mostly benefit from the Government's various subsidies. Yet, representing contemporary processes of agrarian change in Moroccan policy terms (also the language preferred by many researchers who favor modernization) reproduces the idea that it is mainly through the work and achievements of the male farmers and entrepreneurs who supervise the farm enterprise that change happens. It makes the work, efforts and strivings of all others – including women – invisible. Additionally, it is an account that, by focusing on marketable profits, leaves much of farming's social and environmental costs – and in particular water – out of the equation. We have reconstructed the story of contemporary groundwater intensification in Morocco from a feminist angle, based on an ethnographic study of people's own engagements with and interventions in groundwater.

Our analysis shows that only a privileged few young men want and can pursue those modern farming identities, mimicking entrepreneurial stereotypes, actively seizing new technological opportunities to invent new ways of combining manhood with rurality. New investors' emerging farms also modify existing gendered organization and meanings of farm work, as they heavily rely on (cheap) wage labor for the labor-intensive activities of planting, weeding and harvesting, much done by women. Female engagement in casual wage work for others ("strangers") provokes and happens through a re-articulation of the divide between "private" and "public" spaces. The overall effect is that farming increasingly becomes or is seen as the exclusive professional domain of men, with "the farm" being reinvented as an exclusive space to perform a new type of farming masculinities. This space becomes increasingly distinct and separated from the private family sphere of the home, which in the process becomes reinvented as women's location to perform a new kind of femininity. Some women, particularly those from landowning families and of the older generation, proudly assume a more "bourgeois" feminine identity by taking up and performing supposedly more domestic roles. Younger women instead creatively reinvent the domestic domain to carve out new, and perhaps more appropriately feminine, professions for themselves. By engaging in what they call "clean" work, such as craft-skills or small businesses, they create new, more "graceful" rural feminine ways of being and behaving. For women belonging to landless families, who rely on agricultural wage work to make ends meet, it is difficult to comply with these newly emerging normative definitions of womanhood. They take little personal pride in what they do, and many even try hiding their wage-work involvement from the public eye.

In line with Mies (1986), we could use this analysis to show that economic progress and modernity are enabled by exploiting nature – water resources – and women's labor. The conclusion from such an analysis could be that female wage-workers are both the victims of new capitalist processes of agrarian change, and protagonists of movements for more social and environmental justice. This is an interesting analysis, but one that is difficult to translate or explain to the men and women we studied and therefore also of little direct use to identify entry-points for change. This is why we instead tried to emphatically acknowledge and learn from people's own ways of navigating the multiple force-fields that they are part of. This yields an account in which winners and losers are less easily identifiable, one in which perplexities and paradoxical experiences abound. Categories and differences – including those based on gender – are remade and reinvented in production and trade, in ways that are often contingent and not always clear-cut or consistent.

Without denying the existence of larger power structures and without denying that some actors are relatively better-off than others, our study shows how all rural actors actively navigate existing spaces or create new ones to fuel, co-shape, and fulfill emerging desires. This happens, for instance, through negotiations about labor practices and the meaning of labor, negotiations that are always deeply entwined with re-articulations of class, gender or generational differences. Paraphrasing an older article by Katz, simultaneous existence of diverse, articulated structures of dominance, exploitation, and oppression means that there are multiple, connected positionings from which to confront and potentially transform various forms of domination (1992). Documenting these diverse grounded positionings and spaces may pave the way for pragmatic and perhaps modest strategies of interrogating challenging injustices; strategies that risk being overlooked by more structural analyses.

Notes

1 Inspired by Rosa Luxemburg's work (1923) Mies uses the term "colonies" to refer to "non-capitalist" milieus and areas to appropriate more labor, more raw materials, and more markets.
2 Only three such state cooperatives were nationally created, and in the 1990s these lands were redistributed: members received an individual plot with land-use rights.
3 On the condition that former state-cooperative members pay off their debts, they can now own the land they used to work on, in return for a fee of approximately 70,000 Dirham (approximately 6,250 euros), depending on land size and soil quality. Former cooperative members could thus become private landowners, with the land becoming private property.
4 Places where wageworkers gather in order to find a job for the day, which are usually situated at the outskirts of the Saïss' various small agricultural centers.

References

Akesbi, N. (2011). Le Plan Maroc Vert: une analyse critique. In Association Marocaine de Sciences Économiques (ed.), *Questions d'économie marocaine*. Rabat: Presses universitaires du Maroc.

Bossenbroek, L. and Kadiri, Z. (2016). Quête identitaire des jeunes et avenir du monde rural. *Economia* 27, 46–50.

Bossenbroek, L. and Zwarteveen, M. (2015). "One doesn't sell one's parents": Gendered experiences of land privatization in the Saïss, Morocco. In C. Archambault and A. Zoomers (eds.), *Global Trends in Land Tenure Reform: Gender Impacts*. London and New York: Routledge, 152–68.

Bossenbroek, L., van der Ploeg, J. D. and Zwarteveen, M. (2015). Broken dreams? Youth experiences of agrarian change in Morocco's Saïss region. *Cahiers Agricultures*, 24(6), 342–48.

Bossenbroek, L., Zwarteveen, M. and Errahj, M. (forthcoming). Agrarian change and gendered wage-work in the Saïss, Morocco. *Gender Place & Culture*.

Butler, J. (1990). *Gender Trouble: Feminism and the Subversion of Identity*. London and New York: Routledge.

Domínguez Guzmán, C. (forthcoming). Grandes narrativas, pequeños agricultores: Explorando desbordes académicos en el caso de los excedentes de agua en el Valle de Motupe, Perú. *Estudios Atacameños*.

Gibson, K., Law, L. and McKay, D. (2001). Beyond heroes and victims: Filipina contract migrants, economic activism and class transformations. *International Feminist Journal of Politics*, 3(3), 365–86.

Haugen, M. (1998). The gendering of farming: The case of Norway. *The European Journal of Women's studies*, 5, 133–53.

Ingold, T. (2000). *The Perception of the Environment: Essays in Livelihood, Dwelling and Skill*. London: Routledge.

Kadiri, Z. and El Farah (2013). *L'agriculture et le rural au Maroc, entre inégalités territoriales et sociales*. Rabat: Blog Scientifique Farzyates/Inégalités du Centre Jacques Berques.

Katz, C. (1992). All the world is staged: Intellectuals and the projects of ethnography. *Environment and Planning D: Society and Space*, 10, 494–510.

Kuper, M., Faysse, N., Hammani, A., Hartani, T., Hamamouche, M. F. and Ameur, F. (2016). Liberation or anarchy? The Janus nature of groundwater on North Africa's new irrigation frontier. In T. Jakeman, O. Barreteau, R. Hunt, J. D. Rinaudo and A. Ross (eds.), *Integrated Groundwater Management*. The Hague: Springer, pp. 583–615.

Merini, A. F. (2008). Rêve et transmission: de la tradition à la créativité. In F. Mernissi (eds.), *A quoi rêvent les jeunes*. Rabat: Marsam, 49–53.

Mies, M. (1986). *Patriarchy and Accumulation on a World Scale*. London: Zed.

Mies, M. and Shiva, V. (1993). *Ecofeminism*. London and New York: Zed Books.

Naji, M. (2009). Gender and materiality in-the-making. *Journal of Material Culture*, 14(1), 47–73.

Pratt, G. (1998). Grids of difference: Place and identity formation. In R. Fincher and J. M. Jacobs (eds.), *Cities of Difference*. New York: Guilford.

Ramamurthy, P. (2003). Material consumers, fabricating subjects: Perplexity, global connectivity discourses, and transnational feminist research. *Cultural Anthropology*, 18(4), 524–50.

Ramamurthy, P. (2011). Rearticulating caste: The global cottonseed commodity chain and the paradox of smallholder capitalism in south India. *Environment and Planning A*, 43(5), 1035–56.

Resurreccion, B. P. and Elmhirst, R. (2008). *Gender and Natural Resource Management Livelihoods, Mobility and Interventions*. London and Sterling, VA: Earthscan.

19

Conclusions: Struggles for Justice in a Changing Water World

Tom Perreault, Rutgerd Boelens, and Jeroen Vos

19.1 Conceptualizing Justice Frameworks

This book has examined the multi-faceted problematic of water justice. As the preceding chapters have demonstrated, struggles over water are as pervasive as they are diverse. Not surprisingly, calls for water justice similarly take a variety of forms, from struggles over access to drinking water and sanitation (Chapters 5, 12); to the dynamics of water-grabbing and virtual water trade (Chapters 3, 15); (re)configuring hydrosocial territories (Chapters 6, 8); dam building and dispossession (Chapter 9); sanitation and water pollution (Chapters 11, 17); contested water knowledges (Chapters 4, 10, 18); and competing visions of conservation and environmental governance (Chapters 7, 16). As this brief (and far from exhaustive) list indicates, water injustices and their attendant struggles abound. While focused primarily on examining the various forms that such struggles can take, the book has also endeavored to advance thinking about the very nature of such struggles, and to formulate a positive vision of water justice itself (see, particularly, Chapters 1, 2, 13, 14). In this final chapter, we return to the themes that began this book, by considering the nature of water justice in the twenty-first century.

Our conceptualizations of water justice are rooted in a plural vision of social justice. There is no single, unitary water justice principle, any more than there is a single, unitary water-*in*justice experience. The environmental justice field and its connection to the transdisciplinary domain of political ecology (e.g. Forsyth, 2003; Neumann, 2005; Perreault *et al.* 2015; Robbins, 2004; Schlosberg, 2004, 2007) provide inspiration as well as a conceptual foundation for our understanding of water justice. An important current has its roots in the environmental struggles of communities of color in the United States, starting in the 1980s. In addition, and independent from the US water justice struggle, environmental protests over water pollution and dam-building have emerged in many other places around the world throughout the past century. Some of the water-related protests that became known worldwide include Japan's protest against the Ashino copper mine in 1907 (Martínez-Alier, 2003); or more recently, India's protests from 1993 onwards in reaction to plans to build the Narmada dam (Dwivedi, 1999); protests against mining in Papua New Guinea, Peru and Ecuador in the 1990s (Martínez-Alier, 2003); and protests against privatizing the drinking water utility in Cochabamba (1999 and 2000) (Assies, 2003). In the US, seeking to address socio-spatial inequities in exposure to environmental "bads" (e.g. garbage incinerators, industrial sites, solid-waste dumps, sewage-treatment plants) and access to environmental "goods" (e.g. clean air and water, green spaces, trees), and taking inspiration (and organizational lessons) from the US civil-rights movement, environmental-justice activists sought to demonstrate the structural bias inherent in land use decisions (UCCCRJ, 1987). Sociologist Robert Bullard's pioneering work on this topic provided empirical evidence to support grassroots activists' claims regarding the spatial relationship between solid-waste facilities and African-American communities, first in Houston, Texas (Bullard, 1983), and then in other cities and rural communities throughout the southern United States (Bullard, 1990). Owing largely to the practical, strategic need to demonstrate the uneven spatial distribution of environmental burdens, and to support legal claims made by community-based activists, this early work emphasized statistical analysis and was rooted almost entirely in a distributive understanding of justice (Holifield, 2015). In the decades since its emergence, environmental justice research has diversified considerably, to embrace a broad range of empirical topics, analytical approaches, and theoretical frameworks. These include a focus on the multi-scalar nature of environmental injustice (Sze *et al.*, 2009), attention to the structural dynamics of racism, including white privilege (Pulido, 2000), and the lasting effects of white supremacy (Pulido, 2015). Reed and George (2011) point out that environmental justice activism and scholarship remain largely focused on the United States, and on uneven distribution of environmental "goods" and "bads" in society. As this book's chapters attest, however, the environmental justice paradigm does not have one origin, as it has been simultaneously developed in different parts of the world, with different expressions, strategies, effects, and theoretical conceptualizations.

To the extent that environmental processes are necessarily spatial in their expression, environmental justice research has traditionally focused on uneven socio-spatial distribution of environmental burdens and benefits. A subsequent current, most closely identified with philosopher John Rawls' work (1971, 2001), focuses a distributional, procedural understanding of justice on how essential goods, including wealth and certain personal freedoms, are allocated (Israel and Frenkel, in press). As such, Rawls' notion of a just society, conceptualized from behind a hypothetical "veil of ignorance" (Rawls, 1971), is rooted in a liberal, ahistorical understanding of individual rights. As central (and indispensable) as it is to environmental justice research and activism, these distributive and procedural paradigms have been roundly criticized as insufficiently attentive to the diversity of experiences and axes of social difference in a plural society, and inadequate for analyzing structural power relations (such as oppression, domination, and marginalization). While distributive and procedural justice paradigms may be necessary for environmental (and water) justice analyses, they are clearly not sufficient.

Research and activism have thus moved to embrace other spheres of justice (Cole and Foster, 2001). One such sphere is representational justice: the modes and quality of participation in which populations are involved or from which they are excluded. As with distributive notions of justice, the emphasis on representation has roots, among others, in liberal theories of participatory democracy (Holifield, 2004). As Edwards *et al.* (2016)

note, if distributive, procedural, and representational views of justice are largely rooted in liberal democracy and individualized rights theories (and as such, belong to the social contract tradition in justice theory), the concept of *recognition* – of social difference and the legitimacy of rights claims – derives from work in feminist and postcolonial justice theories (Fraser, 1997). In recent years, scholars have increasingly embraced a plural conception of justice (Olson and Sayer, 2009; Walker and Bulkeley, 2006). Schlosberg (2004, 2007), for instance, argues for a *trivalent* view of justice, in which environmental justice (and social justice more generally) must be understood as simultaneously involving distribution, participation, and recognition.

Such a three-dimensional view is similarly held by feminist political theorist Nancy Fraser (1997). In her analysis of social exclusion, Fraser elaborates a justice theory that she refers to as "parity of participation," requiring attainment of three conditions: (a) equitable *distribution* of essential resources; (b) equal opportunity for *participation* in political processes; and (c) equal *recognition* for all participants, with equal opportunity to attain social esteem (Fraser, 2010: 365). Fraser's focus is on social exclusion from the distributive-justice, democratic-representation, and social-recognition conditions necessary for a just society. These three dimensions – distribution, participation and recognition – attend to economic, political, and cultural concerns, respectively, none of which is reducible to the others. Fraser argues that social exclusion arises where these three dimensions of justice intersect, operating across multiple spatial scales. To this we would add a fourth dimension, crucial in the struggle for water and environmental justice: "socio-ecological integrity," which adjoins the notion of "socio-ecological justice" to the trivalent conception of "distributive, representational, and cultural justice." The fourth domain integrates a focus on caring for and nurturing socio-natural environments, sustaining livelihood security for current and future generations of humans and non-humans (see Chapter 1; Boelens, 2015; Zwarteveen and Boelens, 2014). Thus, understandings of justice must consider scale production and interactions as they shape economic, political, cultural, and ecological processes.

A complementary framing of justice is the "capabilities" approach, developed by Amartya Sen (2009) and Martha Nussbaum (2006; see also Sen and Nussbaum, 1993) as a critique of social-contract justice theories. Rather than seeing justice as "fairness" in the Rawlsian sense, the capabilities approach is centrally concerned with people's ability to live lives they consider meaningful and worthwhile (Edwards *et al.*, 2016. See also Chapter 2 by Dik Roth *et al.* and Chapter 5 by Ben Crow in this book). For Sen, social justice necessitates understanding *in*justice, which may be ascertained from (historically constituted and culturally rooted) shared meanings. Thus, for Sen, justice is comparative, relational, and may be assessed by reference to existing social relations. In this view, achieving justice does not require a fully formed justice theory based on *a priori* assumptions (Barnett, 2010). For Sen and Nussbaum, justice is a function of *capabilities*, a positive sense of freedom that refers to the opportunities one has regarding the life they lead. Justice, in this view, is achieved when people enjoy the freedom to live in a way that they consider meaningful and valuable (Edwards *et al.*, 2016). As Walker notes, a fundamental strength of the capabilities approach "is that is has an internal pluralism, incorporates a diversity of necessary forms of justice ... and retains flexibility in how functionings and flourishings are to be secured" (Walker, 2009: 205). While Sen's notion of capabilities has been critiqued for being individualistic, and thus aligned politically with liberal concepts of rights (Stewart, 2005), it may be argued that capabilities are emergent properties of the social context in which one lives, and thus may be considered social or collective conditions (Anderson, 2010). Indeed, Edwards *et al.* (2016) argue that one of the core strengths of the capabilities approach is its capacity for achieving communalist, societal wellbeing.

19.2 Conceptualizing Water Struggles

The plural conceptualization of social or socio-environmental justice as rooted in distribution, participation, recognition, and socioecological integrity, may be operationalized through and roughly mapped onto what Boelens (2008, 2015) and Zwarteveen and Boelens (2014) refer to as an Echelons-of-Rights Analysis. According to this model, there are four echelons, or levels of abstraction, involved in struggles over rights. The first of these echelons is *struggles over water itself, as a material resource*. These can entail struggles over access to water for irrigation or consumption where people lack secure rights to water, formal or otherwise. Struggles may also entail contests over water quantity, where people might have formal access to water, but in insufficient amounts. Conversely, people may struggle over exposure to excess water, if they have inadequate protections from flooding, or protests may concern water quality, in cases where people suffer from water contamination (Perreault *et al.*, 2012; Sultana 2011). Moreover, struggles over water access, quality, and quantity may center on the hydraulic infrastructure necessary for water delivery, water diversions, or water containment (e.g. canals, dams, levees, water meters, pipes, etc., see Sneddon, 2015), and may also refer to other water-based assets such as irrigated land or financial resources. Regardless of their precise focus, these are essentially struggles over the distribution of water and water-related resources: who gets access, of what quality and in what quantity. As such, this echelon asks what are essentially economic questions, and maps onto justice in the distributive sense.

The second echelon of rights analysis involves *struggles over operational rules, laws, and norms* governing water management, and the formal and informal mechanisms used to secure rights to water resources. These are contests over the quotidian practices involved in managing water and water systems and have fundamentally to do with allocation: how water is delivered, to whom, when, at what cost. Closely related is the third echelon of water rights analysis, *struggles over legitimacy and authority in water governance*. These are essentially struggles over knowledge and power: who holds legitimate authority to define, enact, and enforce water rights? Which social groups have the acknowledged right to participate in decisions over water allocation, and which do not? Whose interests are privileged and at whose expense? These are, fundamentally, questions of social difference – of gender, race, class, caste, nationality – but also of institutional configuration. Crucially, questions of authority and legitimacy are multi-scalar, and are as significant in local water systems' governance (for instance, where men are often granted greater authority and their interests are commonly privileged over those of women) as they are in governance at broader scales, such as regional irrigation schemes or municipal drinking water and sanitation systems (for example, where neoliberal privatization and marketization ideologies may overwhelm communal interests). Struggles over operational norms and rules (echelon 2) and legitimate authority in governance (echelon 3) concern political questions of democratic participation – who can and who cannot fully participate in decision-making processes – and thus map onto understanding of procedural justice, albeit in different ways.

The fourth echelon, *struggles over discourse,* refers to the symbolic meanings and conceptual framings that shape how we come to understand and value water (as a natural resource, a commodity, a human right, or what-have-you). In addition to shaping our understandings of water, representation regimes also authorize or de-legitimize various forms of water knowledge and practices involved in water management. This fourth water-struggle echelon is the most abstract, and concerns not only material practices (as with the first echelon of analysis), but also and especially cultural and symbolic understandings of water. The struggle is over installing and proliferating (or alternatively, challenging) particular world views; a discourse that convincingly and coherently connects the other three echelons in an overarching, 'naturalizing' framework. As such, struggles over discourse map onto the notion of recognition, insofar as it concerns production of meaning.

Far from existing as discrete, independent characteristics of water struggles, these four echelons reproduce and reinforce one another. Irrespective of their specific details, struggles over water inevitably interact among multiple echelons. As Boelens (2015: 308) notes,

> Water rights, established by societal interactions and power structures, inform the design and use of hydraulic technology and water flows; in turn, norms embedded in such distribution systems/techniques co-structure organizational, legal, cultural and political relations in water-control society. Further, in water-rights disputes, all four rights echelons are involved simultaneously and "chained" together in particular ways ... that either strengthen or challenge the status quo.

It is therefore vital to emphasize water struggles' inherent complexity. Here, we wish to make four points. First, *water struggles' analyses must account for history*. Struggles over water access (or, conversely, exposure to polluted water or flooding) are often rooted in long, complex histories of land settlement, uneven power relations, symbolic meanings, and institutional configurations. These complex, sedimented histories must be weighted as factors that shape existing rules and practices and contemporary understandings and representations of water (see Chapter 17 by Tom Perreault, on the importance of memory and environmental imaginaries in Bolivian water justice).

Second, *water struggles and water justice are, by their very nature, trans-scalar*. Water governance is inherently scalar in focus: the watershed is paradigmatic in this regard. Whatever the faults of the watershed scale for water resource management (Budds and Hinojosa, 2012; Molle, 2009), the nested character of watersheds illustrates well how water issues are trans-scalar. Moreover, as Fraser (2010: 363) has argued, "sensitivity to scale ... is the *sine qua non* for understanding, let alone overcoming, core injustices in the twenty-first century." Water struggles may center on water access in a neighborhood, in multi-community irrigation system management, in municipal water and sanitation service administration, in large dams' regional effects, or in multi-country governance arrangements. These scales interact and influence one another. Similarly, scalar politics are invaluable for social movement actors as they mobilize for water justice (Perreault, 2005).

Third, *water issues are inherently interdisciplinary and transdisciplinary*. As an element of nature, water may be understood in geo-hydrological and ecological terms. However, the natural sciences, while necessary to understand water's physical qualities, are far from sufficient to understand the complexities of struggles over water and water justice. Water is impounded, conveyed, and distributed through highly engineered hydraulic infrastructure systems. Its allocation is adjudicated through complex systems of laws, rules, customs, and norms. Its governance involves multi-scalar forms of social organization and collective labor. Finally, water is imbued with multiple and at times contradictory and often highly contested meanings which animate and inform social struggles. As such, analysis of water use, water governance and water justice must carefully and rigorously integrate natural sciences, engineering and technology, legal studies and social sciences, and therefore requires joint, transdisciplinary efforts by actors with diverse backgrounds. Water and its allocation are too important to be left to the scholars, lawyers, and politicians, and must actively involve social movements, activists, and common people.

Fourth, *analyses of water struggles and water justice questions require rigorous attention to the nature of power*. Like water itself, power relations are multiply scaled, and their analysis must be sensitive to the production and politics of scale. Power geometries may operate across multiple axes of social difference: gender, race, ethnicity, caste, class, and sexuality. As both factor and product of human labor, water reflects and helps reproduce social relations of power (Bakker, 2002, 2004; Swyngedouw, 2015; UNDP, 2006; Vos and Boelens, 2014).

19.3 Water Justice: Themes from the Book

In this book, we have examined varied themes related to water justice and social struggles. Part I ("Re-politicizing water allocation") begins by discussing lead contamination in Flint, Michigan (Introduction to Part I). This section's chapters examine conflicts over water access. These are, at heart, questions of water governance: the institutional frameworks and forms of social organization, operating at various spatial scales, through which water is allocated (Bakker, 2010; Norman *et al.*, 2014). Water governance, like environmental governance more broadly, is fundamentally concerned with managing water as a resource and, crucially, with producing particular social orders in relation to water (cf. Bridge and Perreault, 2009). This insight is a leitmotif that recurs throughout the book, and is one of the central themes of Section I, the chapters of which emphasize the need for re-politicizing water governance analyses. Much mainstream water governance discourse serves to "render technical" (Li, 2007) the processes of water governance, and the uneven power geometries it entails. Through its use of technical, de-politicized language, and an emphasis on "getting the policies (and prices) right," mainstream water policy discourse maintains a veneer of rationality, objectivity, and apolitical neutrality (Boelens et al., 2010). This dispassionate, technocratic discourse, however, is itself fundamentally political and deeply conservative in its propensity to authorize and underwrite dominant power hierarchies in water governance. As Andrea Gerlak and Helen Ingram demonstrate in Chapter 4 of this volume, the interdisciplinary field of public policy – in which much mainstream water governance is rooted – consciously adopts an objectivist stance that aims for political neutrality. Such a position elides fundamental questions of equity and justice, rendering mainstream policy discourse wholly inadequate for addressing questions of social exclusion in water governance (cf. Fraser, 2010). The need to re-politicize water governance analyses is similarly highlighted in Chapter 2, by Dik Roth, Margreet Zwarteveen, K.J. Joy, and Seema Kulkarni, who call for water governance (and those involved in governance decisions) to engage directly with social-justice questions. The authors argue that water governance must specifically address processes of distribution, participation, and recognition, which in turn requires (re)politicizing the terms through which water is allocated. Ben Crow extends this understanding of justice in Chapter 5 by drawing on Sen's notion of capabilities (Sen, 2009). Crow conceptualizes water access as essential to the urban poor's wellbeing, and thus their ability to live lives of dignity and value. This perspective shifts focus from distributional aspects of water allocation to questions of individual and collective wellbeing, as measures of a just society (Edwards et al., 2016).

Discussing water grabbing, Gert Jan Veldwisch, Jennifer Franco, and Lyla Mehta (Chapter 3) argue that the legal voids and institutional complexities of global water and agro-food governance have fostered local ambiguity that have in turn facilitated land and water grabbing. One fundamental element of this ambiguity is water's own complex socionatural character. Water's fluid nature, its widespread (if highly uneven) distribution, and the fact that it exists in multiple material forms (in rivers, lakes, and oceans, in aquifers, as soil moisture, groundwater, and precipitation) means that it is relatively easy to impound but difficult to contain and expensive to move. These factors, combined with the institutional complexity of global water governance, facilitate water grabbing. Similarly, Roth, Zwarteveen, Joy, and Kulkarni (Chapter 2) call for careful attention to water's hydrosocial character. The authors argue that water's materiality should be accounted for in any critical analysis of governance, and must figure centrally in any consideration of water justice.

The chapters in Part II ("Hydrosocial de-patterning and re-composition") examine the making and remaking of hydrosocial territories. The institutional configurations, forms of knowledge, scalar arrangements, and normalized practices involved in water governance are inherently spatial in their enactment and effects. The production of hydrosocial territories, as particular forms of imagined communities (cf. Anderson, 1983), entails the production of subjectivities and the arrangement of the relations between people and things in ways that are convenient to dominant modes of water governance (Foucault, 1991). The (re)production of hydrosocial territories requires place-based forms of knowledge, ecology, power relations, and modes of practice, which often supersede or subsume pre-existing socio-natural relations. This theme is taken up in various ways in each of the chapters in this section. In their conceptual exploration of the concept of hydrosocial territories, Lena Hommes and her co-authors (Chapter 8) argue that these are best understood as spatial-political configurations of people, institutions, water flows, hydraulic technology, and non-human natures, revolving around the contested processes of water control. The other chapters in the section further develop this concept through in-depth case study analyses. Chapter 6, by Erik Swyngedouw and Rutgerd Boelens, examines the changing national imaginaries of the Spanish waterscape over the course of the twentieth century, and how the reconfiguration of hydraulic relations – most notably through the projects of *regeneracionismo* and *franquismo* – entrained differing visions of justice, which both reflected and reproduced the territorial forms of hydraulic relations promoted by the Spanish state.

Similarly, Chapter 7, by Renata Moreno and Theresa Selfa, examines contested knowledge and competing interests in the Cauca river basin, Colombia. Afro-Colombian smallholder farmers and large-scale industrial sugarcane producers have fundamentally divergent visions of wetlands, a fact that has triggered fierce conflicts over water rights and management practices. Water conflict and protest politics, features of the hydrosocial territory of the Cauca River basin, are similarly characteristic of Guatemala, as recounted in Chapter 9, which examines conflicts over the Chixoy dam. In this chapter, Barbara Rose Johnston analyzes the multi-scalar social and ecological impacts of large-scale hydro-development and electricity generation. Here, as in the Cauca valley, the hydrosocial territory is marked by social conflict over divergent visions of water use and development. Notably, Colombia and Guatemala share long, complex histories of armed conflict and extreme violence (which is ongoing in various ways in each country), which shape the political and discursive forms that social mobilization takes.

Part III of the book ("Exclusion and struggles for co-decision") examines the multi-scalar and highly contested processes of water governance. A core aspect of this is the various scales of social mobilization, examined most directly in Chapter 12 by Jerry van den Berge, Rutgerd Boelens, and Jeroen Vos. This chapter examines the European Citizen's Right2Water Initiative, which sought to incorporate the Human Right to water into European water law as a means of fighting drinking water utility privatization in the European Union. The process of mobilizing across multiple countries and collecting nearly two million signatures, was, in essence, an exercise in scalar politics: contesting national and regional water policy through a trans-national alliance, which in turn drew on a global discourse. A similar politics of scale is at work in Joyeeta Gupta's examination of water governance in the Anthropocene (Chapter 8). The author stresses the enormous influence humans have on the Earth's hydrology, a feature of the Anthropocene. Water for human use becomes available in the "ecospace" which is simultaneously local and global. These chapters serve as reminders that water governance is historically specific and highly contested, and requires careful consideration of spatial and temporal scales.

Indeed, the contested nature of water governance is similarly at the heart of Chapter 10 by Karen Bakker, Rosie Simms, Nadia Joe, and Leila Harris. This chapter examines the regulatory injustices to which Canada's indigenous peoples are subjected, including the notion of First in Time, First in Right (FITFIR, or the so-called doctrine of prior appropriations; Wilkinson, 1993) as a principle of water allocation. Although indigenous peoples indisputably have historical precedent over whites and other populations, their water uses are seldom clearly defined in law and are thus afforded lower priority than those of agricultural, industrial, and urban users. As Bakker and her co-authors argue, the vulnerability of indigenous water rights necessitates a new conceptualization of water rights and greater attention to how principles of justice may be incorporated into water governance arrangements. A similar call to reconceptualize water justice is contained in Chapter 11 by Maria Rusca, Cecilia Alda Vidal, and Michelle Kooy. In this chapter, the authors examine urban sanitation in Kampala, Uganda and call for a multi-dimensional understanding of "sanitation justice." Drawing on the above multi-dimensional understanding of social justice (Fraser, 2010; Schlosberg, 2004), the authors argue that sanitation justice for the poor requires adequate access to sanitation infrastructure, meaningful participation in decision-making regarding sanitation, and recognizing the emotional dimensions of sanitation access. The cases of indigenous water rights in Canada and sanitation justice in Uganda – neither of which involve sustained social mobilization nor open conflict – raise important questions regarding the everyday practices of water-injustice accommodation and acceptance. These issues are raised in Chapter 13 by Frances Cleaver. In her critique of the water justice literature, Cleaver argues that scholars tend to over-emphasize open conflict and the antagonistic relations of social protest, while eliding the subtle but arguably more common practices of compliance and acquiescence. Diverting attention from these everyday practices, the author argues, provides a distorted view of the workings of power. Careful attention is warranted, therefore, not only to overt processes of domination and resistance, but also to the quotidian practices through which people come to live with and accommodate power relations in which they are enmeshed (cf. Ortner 1995; Auyero and Swiston, 2009).

The chapters in Part IV ("Governmentality, discourse, and struggles over imaginaries and water knowledge") examine the political economies and cultural politics of water imaginaries. The section began by discussing the protests at Standing Rock, North Dakota (USA), where Sioux residents of the Standing Rock reservation, together with their indigenous and non-indigenous allies, protested the extension of the Dakota Access oil pipeline under the Missouri River in 2016–2017. This protest encapsulates the struggles over water, and over the different meanings and values ascribed to hydrosocial territories, which are at the heart of this section. The authors of these chapters take as axiomatic that social power is inseparable from the forms of knowledge that give it meaning. Social struggles, water governance, and the everyday practices through which water is known and managed, are all rooted in social imaginaries of water: what water is and should be, what qualities makes it valuable, who has a legitimate right to water and who does not. At their core, these are normative questions about the way the world should be and, crucially for our purposes, what constitutes a just society. Such questions are taken up in Chapter 18, by Lisa Bossenbroek and Margreet Zwarteveen, who provide a feminist analysis of water allocation and capitalist agrarian change in Morocco. The authors argue for a more rigorous vision of water politics, grounded in principles of equity and justice. Drawing on ethnographic analysis, they argue that groundwater management and gendered forms of power are inseparable and mutually constitutive. A rather different exploration of environmental imaginaries is offered by Connor Cavanaugh (Chapter 16), who examines conflicts over what he terms the forest-water nexus in the context of ecological modernization in Kenya. Here, the imaginaries of Kenya's highland forests as "water towers for the nation to develop a green economy" bump up against the realities of corruption, violent rural dispossession, the national elite's interests, and international financial flows related to forest conservation for carbon sequestration. This is a clash of environmental visions, between place-based rural livelihoods and global framings of ecological modernization and global integration. Something similar is at work on the Bolivian Altiplano, as discussed in Chapter 17 by Tom Perreault. In this case, the hegemonic imaginary is of Bolivia as a mining country (*país minero*), in which the mining economy figures centrally in understanding national sovereignty and economic development. State-promoted images of the (heroic, indispensable) extractive economy are intimately bound up with understandings of the nation and national belonging. By contrast, residents of communities negatively affected by mine-related water contamination have a different imaginary altogether, rooted in collective memories of clean water and productive agriculture.

Underlying the environmental imaginaries that animate struggles over water justice are political economies of export agriculture, resource extraction and conservation. In Chapter 15, Jeroen Vos and Rutgerd Boelens examine global virtual water trade, arguing that underlying assumptions of modernist terms such as "efficiency," "hydrologic offsets," and "sustainable landscapes" serve to conceptually frame and ideologically legitimize export-oriented industrial agriculture. Similarly, as Cavanaugh notes in Chapter 16, hydro-electricity development, export-oriented agriculture, ecotourism, and other exclusionary ventures are discursively legitimized by recourse to the modernist imaginaries of Kenya's emerging green economy. These chapters demonstrate that material struggles over water cannot be separated from symbolic meanings with which water and other natural resources are imbued, and suggest that analyses of and calls for water justice must similarly take into account the material and symbolic aspects of water use.

19.4 What is to be Done? Setting an Agenda for Water Justice

How then, do we understand "water justice?" Surely it is easier to grasp water *in*justice. Water grabbing, disproportionate exposure to water pollution, non-access to sufficient, clean water, and the inability to participate in water governance, are all undeniably unjust conditions. As with environmental injustice more broadly, inequity in water resource distribution, lack of full, meaningful participation in decision making, and non-recognition of marginalized populations' rights and claims, are intuitively identifiable as cases of injustice. However, while it is relatively easy to identify what justice is *not*, it is rather more difficult to formulate a positive vision of what water justice *is*. How do we understand water justice as a problematic of theory and praxis?

In Chapter 1, we presented a definition or understanding that conceptualizes water justice by elaborating on the four dimensions of justice with which we began this chapter: distribution, participation, recognition, and socio-ecological integrity. These may be entwined with the four-part Echelons of Rights Analysis used to assess water struggle (Boelens, 2008, 2015; Zwarteveen and Boelens, 2014): struggles over water and water-related assets (resources); water-right contents, operational norms, and organizational forms and practices of water systems (rules); the legitimacy of decision-making authority (regulatory control); and discourse (regimes of representation). Consequently, water justice conceptualizations must integrate the economic (distribution), cultural (recognition), political (participation) and ecological (socioecological integrity) dimensions of justice. In complementary fashion, the capabilities model of justice, as developed by Sen and Nussbaum (1993; see also Nussbaum, 2006 and Sen, 2009), holds that the ultimate measure of a just society is wellbeing, which is a function of the freedoms people have to live meaningful lives (Edwards et al., 2016). For Sen and Nussbaum, "capabilities" refer to people's ability to attain basic needs, as mediated by societal norms and social relations, and thus infers access to other bundles of rights. In this view of justice, access to water or the ability to participate in water governance are means to a broader end, rather than ends in themselves. A capabilities approach emphasizes positive freedoms – the right or ability to access or achieve something desired – as opposed to negative freedoms, understood as freedom from an external constraint.

We echo these calls for a view of water justice that is integrative, relational and multiscalar, and which emphasizes water justice as a means of achieving social wellbeing. In addition to the above justice models, we may identify four principles that we believe apply to water justice and to justice struggles more broadly. First, *concerns of water justice are historically rooted and historically contingent*. Rather than viewed as transcendental, universal principles of justice, water justice questions must be understood dialectically, in their historical and geographical context. To be clear, this is not a call for moral relativism or a defense of NIMBYism, or what Raymond Williams has termed "militant particularisms" (see Harvey, 1996). Rather, it is a call for careful attention to the socio-natural context of water injustices, and calls for justice. Second, and following from the above, *concerns of water justice are relational*. Water justice is, in essence, a socio-natural and social-power relationship; both between people and water, *and* between powerful and less-powerful people, as mediated by broader social and institutional contexts that systematically disadvantage certain groups by preventing them from accessing water they need, exposing them to water that is potentially harmful, preventing them from participating fully in decision-making about water management, or marginalizing subaltern forms of water knowledge or management. In this sense, justice and justice claims exist in relation to someone or something, and should be viewed less as matters of abstract principle than as embodied in material practice. As Iris Marion Young (1990: 25) notes, "Rights refer to doing more than having, to social relations that enable or constrain action." Thus, water justice questions are questions of social relations that enable or constrain peoples' ability to attain the water they need and their ability to organize for and decide over water control. Consequently, the empowerment of deprived groups is essential to attain water justice.

Third, *water justice (like water injustice) is inherently multi-scalar*. Like environmental justice (and injustice) more broadly, questions of water justice are fundamentally spatial in character. It follows, then, that they are also inherently scalar. Just as watersheds exist at multiple, nested, and hierarchically arranged spatial scales, from puddles to large river basins, water injustices are produced through multiply-scaled processes and relations, connecting broad patterns of financial flows, knowledge production, and international trade, to more localized processes of policy making, infrastructure development, and land tenure relations. Analyses of and struggles for water justice must account for the multi-scalar character of water injustices. Fourth, *water justice is a means to an end, rather than an end in itself*. Important as water justice is, it must be viewed as an essential component of broader struggles for social and environmental justice. Here, capabilities analysis is useful (Sen, 2009). Water is essential for life and for living, individually and collectively. Edwards *et al.* (2016) point out that the capabilities approach is grounded in the concept of wellbeing, which may be thought of in social, collective, and communal terms (Olson and Sayer, 2009). It is to this sense of capabilities, rooted in social wellbeing and collective flourishing, that we appeal.

As this book goes to press, the world is experiencing the effects of climate change on an unprecedented scale, with shrinking ice caps in the Arctic and average global temperatures setting records during each of the last three years. While the place-specific impacts of climate change are manifold and difficult to predict, the most pronounced effects have to do with water. Droughts have become longer and more severe. Localized flooding events have become more frequent and intense. In some senses, climate change may be viewed as the spatial and temporal reallocation of water on a massive scale, with potentially disastrous implications for vulnerable populations. Accelerating climate change and the growing threat to democratic governance together pose grave challenges for water justice. The coming years will test even the most dedicated and astute activists and scholars. Water justice challenges are greater than ever, and the need for creative, collaborative, dedicated action has never been more urgent.

References

Anderson, B. (1983). *Imagined Communities: Reflections on the Origin and Spread of Nationalism*. London: Verso.

Anderson, E. (2010). Justifying the capabilities approach to justice. In H. Brighouse and I. Robeyns (eds.), *Measuring Justice: Primary Goods and Capabilities*. Cambridge: Cambridge University Press, pp. 81–100.

Assies, W. (2003). David versus Goliath in Cochabamba: Water rights, neoliberalism, and the revival of social protest in Bolivia. *Latin American Perspectives*, 30(3), 14–36.

Auyero, J. and Swistun, D. (2009). *Flammable: Environmental Suffering in an Argentine Shantytown*. New York: Oxford University Press.

Bakker, K. (2002). From state to market? Water *mercantilización* in Spain. *Environment and Planning A*, 34, 767–90.

Bakker, K. (2004). *An Uncooperative Commodity: Privatizing Water in England and Wales*. Oxford: Oxford University Press.

Bakker, K. (2010). *Privatizing Water: Governance Failure and the World's Urban Water Crisis*. Ithaca, NY: Cornell University Press.

Barnett, C. (2010). Geography and ethics: Justice unbound. *Progress in Human Geography*, 35(2), 246–55.

Boelens, R. (2008). Water rights arenas in the Andes: Upscaling networks to strengthen local water control. *Water Alternatives*, 1(1), 48–65.

Boelens, R. (2015a). *Water, Power and Identity: The Cultural Politics of Water in the Andes*. London: Routledge.

Boelens, R. (2015b). *Water Justice in Latin America: The Politics of Difference, Equality, and Indifference*. Amsterdam: CEDLA and University of Amsterdam.

Boelens, R., Getches, D. and Guevara-Gil, A. (2010). *Out of the Mainstream: Water Rights, Politics and Identity*. New York: Routledge.

Bridge, G. and Perreault, T. (2009). Environmental governance. In N. Castree, D. Demeritt, D. Liverman and B. Rhoads (eds.), *Companion to Environmental Geography*. Oxford: Blackwell, pp. 475–97.

Budds, J. and Hinojosa, L. (2012). Restructuring and rescaling water governance in mining contexts: The co-production of waterscapes in Peru. *Water Alternatives*, 5, 119–37.

Bullard, R. D. (1983). Solid waste sites in the Black Houston community. *Sociological Inquiry*, 53, 273–88.

Bullard, R. D. (1990). *Dumping in Dixie: Race, Class, and Environmental Quality*. Boulder, CO: Westview Press.

Cole, L. W. and Foster, S. R. (2001). *From the Ground Up: Environmental Racism and the Rise of the Environmental Justice Movement*. New York: New York University Press.

Dwivedi, R. (1999). Displacement, risks and resistance: Local perceptions and actions in the Sardar Sarovar. *Development and Change*, 30(1), 43–78.

Edwards, G. A. S., Reid, L. and Hunter, C. (2016). Environmental justice, capabilities, and the theorization of wellbeing. *Progress in Human Geography*, 40(6), 754–69.

Forsyth, T. (2003). *Critical Political Ecology: The Politics of Environmental Sciences*. London and New York: Routledge.

Foucault, M. (1991). Governmentality. In G. Burchell, C. Gordon and P. Miller (eds.), *The Foucault Effect: Studies in Governmentality*. Chicago: University of Chicago Press, pp. 87–104.

Fraser, N. (1997). *Justice Interruptus: Critical Reflections on the "Postsocialist" Condition*. New York: Routledge.

Fraser, N. (2010). Injustice at intersecting scales: On "social exclusion" and the "global poor." *European Journal of Social Theory*, 13(3), 263–371.

Harvey, D. (1996). *Justice, Nature and the Geography of Difference*. Oxford: Blackwell.

Holifield, R. (2004). Neoliberalism and environmental justice in the United States Environmental Protection Agency: Translating policy into managerial practice in hazardous waste remediation. *Geoforum*, 35, 285–97.

Holifield, R. (2015). Environmental justice and political ecology. In T. Perreault, G. Bridge and J. McCarthy (eds.), *The Routledge Handbook of Political Ecology*. London: Routledge, pp. 585–97.

Israel, E. and Frenkel, A. (in press). Social justice and spatial inequality: Toward a conceptual framework. *Progress in Human Geography*.

Li, T. M. (2007). *The Will to Improve: Governmentality, Development and the Practice of Politics*. Durham, NC: Duke University Press.

Martínez-Alier, J. (2003). *The Environmentalism of the Poor: A Study of Ecological Conflicts and Valuation*. Cheltenham: Edward Elgar Publishing.

Molle, F. (2009). River-basin planning and management: The social life of a concept. *Geoforum*, 40, 484–94.

Neumann, R. (2005). *Making Political Ecology*. New York: Routledge.

Norman, E., Cook, C. and Furlong, K. (2014). *Negotiating Water Governance: Why the Politics of Scale Matter*. London: Routledge.

Nussbaum, M. C. (2006). *Frontiers of Justice: Disability, Nationality, Species Membership*. Cambridge, MA: Belknap Press.

Olson, E. and Sayer, A. (2009). Radical geography and its critical standpoints: Embracing the normative. *Antipode*, 41(1), 180–98.

Ortner, S. (1995). Resistance and the problem of ethnographic refusal. *Comparative Studies in Society and History*, 37(1), 173–93.

Perreault, T. (2005). State restructuring and the scale politics of rural water governance in Bolivia. *Environment and Planning A*, 37(2), 263–84.

Perreault, T., Bridge G. and McCarthy, J. (eds.) (2015). *The Handbook of Political Ecology*. London: Routledge.

Perreault, T., Wraight, S. and Perreault, M. (2012). Environmental justice in the Onondaga lake waterscape, New York. *Water Alternatives*, 5, 485–506.

Pulido, L. (2000). Rethinking environmental racism: White privilege and urban development in southern California. *Annals of the Association of American Geographers*, 90(1), 12–40.

Pulido, L. (2015). Geographies of race and ethnicity I: White supremacy vs. white privilege in environmental racism research. *Progress in Human Geography*, 39(6), 809–17.

Rawls, J. (1971). *A Theory of Justice*. Cambridge, MA: Belknap Press.

Rawls, J. (2001). *Justice as Fairness: A Restatement*. Cambridge, MA: Harvard University Press.

Reed, M. G. and George, C. (2011). Where in the world is environmental justice? *Progress in Human Geography*, 35(6), 835–42.

Robbins, P. (2004). *Political Ecology: A Critical Introduction*. Oxford: Blackwell.

Schlosberg, D. (2004). Reconceiving environmental justice: Global movements and political theories. *Environmental Politics*, 13, 517–40.

Schlosberg, D. (2007). *Defining Environmental Justice: Theories, Movements, and Nature*. Oxford: Oxford University Press.

Sen, A. (2009). *The Idea of Justice*. London: Allen Lane.

Sen, A. and Nussbaum, M. (1993). *The Quality of Life*. Oxford: Clarendon Press.

Sneddon, C. (2015). *Concrete Revolution: The Bureau of Reclamation, Cold War Geopolitics and Large Dams*. Chicago: University of Chicago Press.

Stewart, F. (2005). Groups and capabilities. *Journal of Human Development*, 6(2), 147–72.

Sultana, F. (2011). Suffering for water, suffering from water: Emotional geographies of resource access, control and conflict. *Geoforum*, 42, 163–72.

Swyngedouw, E. (2015). *Liquid Power: Contested Hydro-Modernities in 20th Century Spain*. Cambridge, MA: MIT Press.

Sze, J., London J., Shilling F., Gambirazzio G., Filan T. and Cadensasso M. (2009). Defining and contesting environmental justice: Socio-natures and the politics of scale. *Antipode*, 41, 807–43.

UCCCRJ (United Church of Christ Commission for Racial Justice). (1987). *Toxic Waste and Race in the United States: A National Report on the Racial and Socio-economic Characteristics on Communities with Hazardous Waste Sites*. New York: Public Access Data.

United Nations Development Program (UNDP). (2006). *Beyond Scarcity: Power, Poverty and the Global Water Crisis. Human Development Report 2006*. Houndmills: Palgrave Macmillan.

Vos, J. and Boelens, R. (2014). Sustainability standards and the water question. *Development and Change*, 45(2), 205–30.

Walker, G. (2009). Environmental justice and normative thinking. *Antipode*, 41(1), 203–05.

Walker, G. and Bulkeley, H. (2006). Geographies of environmental justice. *Geoforum*, 37, 655–59.

Wilkinson, C. (1993). *Crossing the Next Meridian: Land, Water and the Future of the American West*. Washington, DC: Island Press.

Young, I. M. (1990). *Justice and the Politics of Difference*. Princeton, NJ: Princeton University Press.

Zwarteveen, M. Z. and Boelens, R. (2014). Defining, researching and struggling for water justice: Some conceptual building blocks for research and action. *Water International*, 39(2), 143–58.

Index

15-M (anti-austerity) Movement, 239

abjection, 93
aboriginal peoples, 194, 198, 199
academic community, 16, 297
accommodation, 20, 189, 247, 250, 253, 256
accountability, 2, 15, 37, 63, 64, 76, 177, 181
accumulation, 9, 44, 46, 61, 62, 65, 92, 157, 229, 261, 278, 322, 332
accumulation by dispossession, 46
acid mine drainage, 319, 320, 323, 325
activists, 4, 38, 47, 66, 128, 162, 226–42, 246, 279, 317, 319, 325, 347, 351, 357
actor-oriented approaches, 130
afforestation, 16
Africa, 18, 34, 35, 36, 37, 60, 64, 90, 93, 99, 111, 142, 170, 188, 210–21, 255, 286, 302–12, 324, 347
African American, 34, 35, 324
Afro-Colombian, 134–47
agrarian policies, 330, 335
agribusiness, 1, 6, 7, 8, 12, 157, 283–97
agricultural
    commodities, 285, 324
    crisis, 120
    development, 142, 291, 334
    encroachment, 141
    expansion, 59, 138, 140
    exports, 286
    frontier, 308
    infrastructure, 173
    investments, 67
    policies, 330
    projects, 59, 337
agro-export companies, 7, 18, 53, 65, 284, 289, 295, 297
Ahmednagar district (India), 45
Algeria, 306
Alliance for Water Stewardship, 292
allocation principles, 51

Alta Verapaz, 173, 174, 177, 181
Amazon, 8, 111, 170, 278
Amnesty International, 12
anarchists, 118, 120, 121
Anthropocene, 170, 259–69
aquifers, 45, 59, 292
Arab Spring, 66
arap Moi, D., 308
Árbenz, J., 172, 173
Archantopoulos, Y., 232
Army Corps of Engineers, 276–7
arsenic, 156, 157, 319
Asia, 7, 64
asparagus, 7, 8
Athens, 232
atmospheric emissions, 170
Australia, 7, 9, 181
Austria, 227, 241
authority, 2, 5, 6, 13, 15, 16, 19, 20, 22, 35, 39, 43, 47, 50, 77, 78, 115, 134, 143, 164, 195, 197, 203, 206, 233, 254, 321, 349, 356
autonomy, 16, 19, 22, 52, 118, 121, 125, 126, 128, 161
Avaaz, 237
awareness-raising, 10, 241

Baja Verapaz, 177
bananas, 7, 173
Bangalore, 94
Bangladesh, 97, 98, 286
Barcelona, 128, 295
basic needs, 37, 38, 234, 262, 356
bathing, 94, 98
Belgium, 227, 237
Benjumea-Burín, R., 123
Bentham, J., 4
Berlin, 237, 238
Better Cotton Initiative, 292
biodiversity, 158, 159, 163, 260, 293, 302
biofuels, 52, 59

Bismarck, 276
Bolivia, 33, 250, 278, 316–26
Bolivian Altiplano, 316, 320
Bolivian Constitution, 38
Brazil, 11, 12, 62, 100
Bretton Woods Agreement, 171
BRICS, 62
British East Africa Protectorate, 304
British Empire, 304
Brussels, 128, 236, 237, 295
buen vivir, 154
Buenos Aires, 317, 318
Bulgaria, 227, 236, 239
Butler, J., 332

calamity polders, 143, 145
Cali, 139, 145, 146, 147, 148
Callao, 10
campesino communities, 317
Canada, 8, 9, 159, 160, 173, 189, 193–204, 205, 206, 267, 269, 276, 296
Canadian Confederation, 205
Cannonball River, 276
capability, 37, 39, 50, 52, 53, 89, 90, 91, 92, 93, 99, 100, 101, 141, 148, 169, 170, 188, 324, 348, 349, 356, 357
capability deprivation, 90, 99
capitalist transformation, 43, 44, 92, 101
carbon, 159, 170, 180, 181, 303, 309, 310
carrots, 337
cash crops, 59
caste, 44, 45, 92, 94, 97, 188, 349, 351
catchment areas, 17, 302
Cauca River, 134–47
Cauca River Corridor Project, 136, 138, 139
CENSAT Agua Viva (Colombia), 296
CEO Water Mandate (UN Global Compact), 292
Cerro de San Pedro (Mexico), 8
certification, 18, 284, 289, 292, 293, 297
Chavimochic (Peru), 290
Chennai (India), 45
Chevron-Texaco, 111
Chichupac (Guatemala), 175
children, 10, 34, 35, 37, 92, 94, 95, 96, 97, 176, 180, 210, 220, 264, 322, 326, 339, 340
Chile, 7, 266, 267, 278, 286, 290, 295, 296
China, 7, 9, 16, 62, 169, 180, 269, 285, 286, 293
Chixoy dam (Guatemala), 169
chlorine, 34
Chókwè (Mozambique), 61
Chone (Ecuador), 153, 154, 164
Chone multipurpose dam, 153, 154
churches, 220, 235
CIA, 173
CIAMSA, 144
CIDI, 215

cities, 11, 16, 34, 35, 36, 89, 94, 102, 214, 228, 320, 341
 as civilized places, 10
 electricity, 162
 expansion of, 10
 flooding, 146
 fragmentation, 94
 and gender, 90
 groundwater, 155
 labor force, 89
 megacities, 10
 metabolism, 100
 re-municipalization of services, 230
 rivers, 296
 and rural areas, 44, 53
 rust belt, 34
 sanitation, 89, 212
 and settlers, 93
 water infrastructure, 35
civil war, 116, 123, 124, 173, 182, 323
civilian population, 170, 171, 174, 181
civil-society groups, 47, 51, 226, 235, 239, 240, 242, 295, 310
class, 4, 13, 44, 45, 76, 92, 94, 95, 97, 98, 100, 101, 102, 188, 212, 218, 306, 331, 334, 335, 342, 344, 349, 351
clean development, 9, 108
cleaning, 92, 93, 95, 98, 260, 339, 340
CLES, *see* community-led urban environmental sanitation
climate, 2, 9, 11, 48, 59, 60, 62, 135, 145, 159, 173, 180, 188, 189, 204, 212, 260, 261, 264, 267, 268, 269, 287, 302, 303, 310, 318, 320, 357
climate change, 1, 2, 9, 11, 48, 59, 62, 135, 159, 188, 189, 204, 260, 261, 264, 267, 268, 269, 302, 357
cloud seeding, 260
Coca Cola, 265, 293
Cochabamba, 33, 111, 230, 318, 346
Cold War, 169, 170
collective action, 89, 100, 248, 255, 317, 325, 326
collective frameworks, 316
collective memory, 316, 317, 323
Colombia, 8, 16, 66, 134, 135, 136, 138, 141, 143, 147, 158, 160, 278, 290, 296
Colombian Government, 134, 136, 137, 160
colonial rule, 93, 94
comfort, 210, 218, 219
commensurability of values, 13
commercial agriculture, 157, 330
commoditization, 51
commodity, 39, 46, 47, 73, 74, 81, 98, 109, 121, 226, 229, 230, 266, 307, 324, 350
common good, 52, 54, 237
commons, 11, 12, 39, 46, 54, 64, 76, 77, 82, 169, 170, 266, 304
communist countries, 265
communists, 118, 121

communities, 9, 37, 73, 141, 153, 174, 176, 181, 198, 201, 202, 217, 220, 230, 263, 296, 297, 307, 311, 312, 323, 324, 325, 333
  Achi Maya, 10
  African American and Latino, 36
  Afro-Colombian, 134, 141
  agro-pastoralist, 309
  Chixoy, 179
  downstream, 49, 322
  First Nations, 196, 199, 202
  fisher, 66, 302
  forests, 303, 307, 310, 311, 312
  imagined, 110
  indigenous, 8, 74, 170, 180, 193, 194, 196, 197, 200, 201, 318, 319
  Kurdish, 162, 163
  ladino, 172
  low income, 36
  marginalized, 16
  nomadic, 161
  peasant, 65
  political, 52
  resistance, 11
  self-organization, 158
  water, 3, 18, 21, 22, 109, 189, 265, 291, 295
  water expert, 2
community leaders, 136, 174
community-led urban environmental sanitation, 216
compensation, 143, 147, 154, 155, 173, 177, 266, 293, 325
compensation schemes, 147
comunidad del milenio, 153, 154, 155
concessions, 7, 8, 11, 38, 308
concrete, 121, 122, 123, 194, 240
Conflict-Free Gold Certificate, 8
conflicts
  ecological, 128, 359
  spatial, 127
  violent, 12
  water control, 20
  water grabbing, 60
connectedness, 249, 251
conservation, 17, 72, 73, 109, 121, 158, 160, 164, 283, 289, 302, 303, 304, 305, 306, 307, 308, 309, 310, 311, 346
consumer organizations, 235
contamination, 1, 3, 18, 34, 37, 94, 156, 211, 214, 284, 286, 293, 294, 295, 316, 317, 318, 319, 320, 321, 322, 323, 324, 325, 326, 349
contestation, 43, 59, 99, 147, 252, 295
contract farmers, 47
cooking, 47, 93, 147, 339, 340
cooperation, 9, 160, 235, 247, 252, 253
Corporate Social Responsibility, 8, 18, 63, 289, 291
Corps of Engineers, 118, 120, 123
Correa, R., 154

corruption, 1, 12, 177, 308, 309
Coryndon, R., 307, 308
Costa Rica, 16
Costa, J., 119, 120
criminality, 97, 312
critical ecosystem infrastructure, 303
critical geography, 82, 246
  mining, 291
Cuba, 118
cultural heritage, 8, 135, 162, 163
cultural institutions, 117, 151
cultural justice, 5, 22, 23, 36, 348, 356
cultural landscapes, 9
cultural relativism, 2
culture, 4, 13, 18, 39, 76, 89, 161, 162, 246, 295, 296, 332
customary law systems, 10, 16, 36
Cyprus, 227
Czech Republic, 227

Dakota Access pipeline, 111, 276, 277
dalit movements, 47
dams
  affected communities, 174, 176, 177, 178, 179
  anti-dam movements, 66, 163
  building, 135, 162, 175, 180, 346
  Chixoy, 170, 174, 177
  electricity, 161
  hydro-development, 169
  Ilisu, 163
  megadams, 9, 120, 123, 151, 163, 169, 173, 176, 177, 181, 261, 290, 351
  multipurpose, 145
  negative effects, 21
  projects, 9, 135
  Santa Rita, 179
  state-based, 120
  unplanned, 302
Danone, 292
Dar-es Salaam, 11, 214
de Soto, H., 16
decentralization, 15, 126
deep tube wells, 12, 289
defecation, 95, 96, 97, 98, 215, 218, 219, 220
deforestation, 260, 302, 305, 308
Delhi, 65, 89, 97, 217
Delhi Declaration, 213, 217
demand-driven approaches, 216
democracy, 73, 125, 126, 127, 129, 188, 236, 237, 239, 246, 347
democratization, 47, 82
Denmark, 227
deprivation, 1, 90, 91, 92, 93, 96, 97, 99, 101, 102, 233, 326
desalination, 128, 260
desalination plants, 128
desert, 8, 11, 81, 284, 321

despotism, 122, 123, 126
de-territorialization, 117
Detroit, 34, 35, 36
developmental disabilities, 35
developmentalist paradigm, 135
Dhaka (Bangladesh), 97
DICSA, 144
dictatorships, 122, 173
dignity, 38, 90, 91, 93, 95, 100, 102, 154, 211, 219, 220, 221, 234
dikes, 135, 136, 137, 138, 139, 143, 145, 146, 147
disabilities, 35, 210
disasters, 2, 134, 188
disciplinary governmentality, 19, 110, 152
discourse analysis, 80
discourses, 8, 13, 20, 50, 53, 76, 78, 80, 110, 115, 122, 130, 151, 152, 153, 158, 164, 248, 279, 284, 289, 296, 324
   adaptation and mitigation, 159
   cultural, 163
   development, 154, 155
   and environmental imaginaries, 129
   foucaultian, 249
   governmental, 291
   hegemonic, 5, 250
   human rights, 51
   investment, 261
   naturalizing, 117
   neoliberal, 12, 64, 264
   policy, 17, 20, 43
   and power relations, 248
   scientific, 189
   stewardship, 293
   water footprint, 294
   and worldviews, 233
discrimination, 5, 22, 49, 76, 228, 241
disease, 34, 89, 97, 102, 260
dispossession, 5, 9, 22, 46, 61, 92, 141, 195, 291, 311, 346
dissent, 43, 52, 127
distribution, 5, 6, 11, 14, 38, 44, 47, 48, 49, 50, 52, 53, 63, 73, 110, 127, 128, 129, 153, 177, 210, 211, 212, 215, 221, 259, 261, 277, 283, 284, 348, 349, 356
   of benefits and burdens, 109
   of costs and benefits, 212, 293
   of environmental burdens, 347
   equitable, 47
   income, 6
   infrastructure, 214
   land, 172
   organization, 160
   physical, 128
   rainfall, 125
   spatial/racial, 324

distributive models, 5
diversity, 2, 5, 13, 18, 21, 50, 52, 61, 98, 111, 130, 142, 147, 152, 170, 226, 232, 234, 235, 237, 331, 347, 349
domestic duties, 340
domestic work, 91, 93, 95, 98, 102, 332, 339, 341
dominant narratives, 66
downstream users, 16
dredging, 138
drilling permits, 156
Drinking Water and Sanitation Decade, 213
Drinking Water Directive, 240
drip irrigation, 17, 289, 291, 292, 334, 337, 338
drought, 2, 45, 48, 125, 188, 202, 248, 253, 255, 260, 263, 266, 267, 292
Dublin conference, 64
Dublin Water Declaration, 65, 266
Duero River, 127
durable poverty, 92, 101

Earth Rights International, 296
Earth System Governance project, 269
Eastern Europe, 236
Ebro model, 123
Ebro River, 123, 124, 127, 128, 295
Ebro River Basin Confederation, 123
Echelons of Rights Analysis, 20, 349, 356
ECI, *see* European Citizens' Initiative
Eco Oro company, 159, 160
eco-feminism, 331
ecological defense, 22
ecological inclusiveness, 259, 268
ecological justice, 6, 22, 211, 212, 348, 356
ecological relations, 115, 117, 307
economic development, 18, 72, 73, 92, 154, 156, 169, 171, 172, 179, 306
economic growth, 9, 43, 44, 45, 46, 67, 92, 101, 156, 157, 261, 268, 269, 278, 303, 310
economic theory, 4
economy of affection, 252
ecosystem services, 17, 74, 260, 262, 263, 267, 268, 269, 293, 302, 303, 310, 311
ecosystems, 6, 22, 47, 60, 62, 74, 84, 135, 140, 159, 170, 195, 205, 260, 302, 303
Ecuador, 7, 8, 16, 111, 153, 154, 162, 278, 290, 295, 296, 346
Ecuador Estratégico, 154, 155
education, 4, 44, 89, 91, 92, 175, 178, 205, 215
efficiency, 2, 7, 11, 17, 18, 35, 46, 47, 53, 72, 73, 74, 75, 76, 83, 129, 155, 213, 216, 228, 229, 230, 266, 284, 286, 289, 290, 291, 292, 293, 295, 296
Egypt, 286
elderly people, 219, 264, 337
electricity, 138, 161, 162, 173, 178, 179, 228, 290, 303
embeddedness, 5, 21, 50, 52, 251
embodied subjectivities, 97, 101

Emergency Manager Law (Michigan), 35
empathy, 53
empowerment, 21, 81, 189, 217, 357
End of Open Defecation Campaign, 210
energy, 12, 59, 60, 62, 65, 96, 100, 138, 159, 161, 162, 170, 171, 172, 173, 177, 180, 228, 241, 260, 290, 292, 302
Energy Transfer Partners, 276, 277
engineering technologies, 122
England, 228, 252
entitlements, 45, 46, 51, 52, 54, 92, 93, 201
entrepreneurial stereotypes, 343
environmental burdens and benefits, 347
environmental citizenship, 135, 142, 144, 147
Environmental Flow Needs, 205
environmental governance, 3, 65, 142, 346
environmental health, 215
environmental impacts, 9, 211, 267, 292, 294
environmental justice, 9, 20, 36, 37, 44, 49, 50, 53, 54, 211, 212, 304, 316, 323, 346, 347, 348, 357
environmental laws, 37
environmental restoration, 178
environmental struggles, 346
environmental warfare, 169, 180
epistemic forms, 82
   discursive, 17, 18, 20, 60, 80, 81, 82, 83, 84, 116, 122, 288, 305
   practical, 82
   prescriptive, 82
epistemology, 3, 81, 82, 284
equality, 5, 13, 14, 22, 77, 125, 128, 268
equity, 5, 38, 39, 45, 47, 48, 50, 53, 65, 71, 73, 74, 75, 76, 77, 82, 83, 84, 157, 179, 221, 246, 254, 324
erosion control, 16
Estonia, 227
ethanol, 62, 137
ethics, 4, 14, 77, 82, 204
Ethiopia, 295
Ethiopian Government, 7
ethnic inequality structures, 4
ethnicity, 44, 92, 93, 94, 102, 188, 351
ethnographic methods, 333
Eurasia, 64
Europe, 8, 9, 18, 80, 119, 126, 135, 190, 226, 228, 230, 231, 232, 234, 236, 238, 241, 265, 284, 285, 341
European Anti-Poverty Network, 234
European Central Bank, 231
European Citizens' Initiative, 226, 236, 237, 238
European Commission, 226, 228, 231, 232, 237, 240
European Environmental Bureau, 234
European Federation of Public Service Unions, 226
European Parliament, 240, 241
European Public Health Alliance, 234

European Union, 78, 180, 221, 226, 228, 232
Eurozone, 231
evictions, 175, 180, 303, 309, 310, 312
exclusion, 3, 5, 14, 22, 38, 43, 53, 64, 89, 97, 115, 135, 188, 193, 201, 203, 212, 248, 307, 325, 348
expert knowledge systems, 3
expertocracy, 3, 4, 14
export agriculture, 7, 283, 286, 289, 290, 291, 297
extractive industries, 8, 21, 61, 283, 295

fairness, 2, 4, 22, 52, 53, 76, 77, 83, 211, 212, 251, 253, 254, 255, 256, 290, 324, 348
Fascism, 125, 128
fava beans, 322, 324
fecal coliform bacteria, 34
Federation of Young European Greens, 235
female body, 90, 95
female identities, 339, 341
femininities, 339, 340, 341, 343
feminism, 96, 330, 331, 332, 333, 334, 343, 348
fertilizers, 286, 338
fetishization of law, 51
financial institutions, 9, 39, 46, 177
Finland, 227
First in Time, First in Right, 38, 194, 197, 202
First National Plan of Hydraulic Works (Spain), 123
First National Water Plan (Spain), 124
First Nations, 189, 193, 194, 196, 197, 198, 199, 200, 201, 202, 204, 205
fisheries, 146, 170, 205
fishers, 66, 142
FITFIR, *see* First in Time, First in Right
flex crops, 59, 62
Flint, 34–6, 37, 39
Flint River (USA), 34
flood-control projects, 135
flooding, 1, 3, 64, 134, 135, 136, 137, 138, 140, 141, 142, 143, 145, 146, 154, 162, 188, 263, 267, 277, 349, 350, 357
floodplains, 64, 135, 137, 138, 139, 140, 141, 142, 145, 146, 147
flowers, 6
fluoride, 156, 157
fluviofelicidad, 296
food, 1, 6, 7, 16, 17, 18, 37, 44, 45, 47, 52, 53, 59, 60, 62, 65, 74, 124, 158, 169, 180, 189, 212, 252, 261, 291, 292, 293, 294, 302, 321, 322, 324
Food and Water Europe, 234
food production, 18, 124, 158, 261
food security, 1, 6, 7, 62
Forest Act (Kenya), 309
Forest Department (Kenya), 305, 306, 307, 309
forest governance, 303
Forum for Policy Dialogue on Water Conflicts (India), 296
Foucault, M., 14, 19, 108, 110, 152, 287, 294, 333

FPIC, *see* Free, Prior and Informed Consent
frame analysis, 80
frames of reference, 116
France, 90, 172, 227, 236, 237, 239, 240
Franco, F., 59, 60, 64, 123, 124, 125, 126, 127, 158
Fraser, N., 6, 54, 348
free market, 13, 17, 172, 228, 288
Free, Prior and Informed Consent, 200
freedoms, 37, 53, 91, 228, 278, 324, 347, 356
free-market rules, 3
French Communist Party, 172
Friends of the Earth International, 296
fruits, 6, 45, 137, 285
fuel, 60, 62, 170, 215, 261, 267, 277, 306, 334, 344
fuelwood, 307
Fundación de Antropología Forense de Guatemala, 182
Fundación Nueva Cultura del Agua, 295
Future Earth, 262, 269

Galeano, E., 13
GAP, *see* Southeastern Anatolia Project
gas, 154, 171, 182, 189, 268, 277, 278, 320
gay, lesbian and transgender rights, 233
gender, 4, 47, 77, 91, 92, 93, 95, 96, 97, 99, 101, 188, 210, 211, 212, 331, 332, 335, 344, 349, 351
 discrimination, 5
 division of labor, 95
 divisions of labor, 101
 and domestic labor, 95
 and farming, 336
 identities, 331, 338, 342
 inequalities, 98
 relations, 97, 101, 330, 331
 and sanitation practices, 90
 social relations of work, 97
 standards, 13
 subjectivities, 331, 333
 subordination, 90, 95, 96, 97, 98, 99, 100, 101
 violence, 219
gendered institutions, 45
General Motors, 34
general public, 22, 226, 239
genocide, 5, 177, 178, 182
Germany, 163, 169, 227, 238, 241
GIZ, 221, 294, 297
glaciers, 260
global commodity transfers, 109, 121
global financial crisis, 36
global markets, 230
Global North, 18, 90, 95, 189
global policy, 2, 12, 102, 221
Global South, 1, 62, 90, 93, 94, 96, 109, 213
GlobalGAP, 292
globalization, 1, 43, 44, 111, 160, 265
glocal, 261, 262, 266, 268, 269
God, 119, 266

gold, 8, 158, 159, 171, 179, 277, 291, 319
golf estates, 266
good governance, 14, 15, 115, 158, 188
governance models, 15
government rationalities, 151
governmental agents, 13
governmentality, 19, 21, 108, 109, 110, 111, 152, 153, 154, 155, 157, 161, 163, 284, 285, 287, 288, 296
 disciplinary, 19
 dominant schemes, 110
 hydro-political, 154
 hydrosocial, 153
 neoliberal, 19, 152, 160, 284, 285, 286, 288, 289, 294, 297
 state and marked based, 21
 transnational, 111
 water stewardship, 292, 293
Gramsci, A., 249
grassroots, 20, 21, 22, 65, 102, 136, 144, 147, 190, 216, 226, 241, 295, 304, 311, 347
grazing corridors, 64
Great Lakes, 34
Great Plains and Midwest, 277
Greece, 227, 231, 232, 239, 240, 241
green accounting, 284
green beans, 155
green capitalism, 160
green economy, 9, 302, 310
green grabbing, 59, 61
Green Morocco Plan, 291, 336
green spaces, 36, 347
green water, 260
greenhouse gases, 188, 277
greenwashing, 292
Grogan, E., 306
Gross Domestic Product, 91
groundwater, 1, 7, 45, 60, 65, 94, 155, 157, 201, 218, 260, 266, 286, 290, 291, 330, 331, 334, 337, 338, 339, 343
groundwater drilling ban, 156
Grupo Gloria (company), 12
Guanajuato, 155, 156, 157
Guatemala, 170, 172, 173, 174, 175, 176, 178, 179, 180, 181, 182
guerrilla, 162
Guinea, 142

hand washing, 219, 220
happiness, 4, 98, 99, 342
harvesting, 157, 182, 249, 339, 340, 343
Harvey, D., 46, 61
headwater catchments, 8
heavy metals, 34, 319, 321
hegemony, 110, 248, 249, 253, 279
Heineken, 293
herbicides, 286
highest economic water-use value, 289, 290

highland forests (Kenya), 306, 307, 308
Hindu law, 265, 266
Hindu peasants, 253
Hirakud dam (India), 45
Historical Verification Commission (Guatemala), 177, 182
Houston, 347
Huanuni Mining Company, 325
Huanuni river (Bolivia), 316, 317, 319, 321, 324, 326
Huanuni Valley, 317, 326
human right to water, 10, 38, 39, 50, 51, 159, 226, 240, 241, 255, 268
human rights, 10, 12, 38, 39, 47, 50, 51, 53, 66, 159, 176, 177, 178, 181, 182, 220, 226, 228, 234, 240, 241, 242, 255, 262, 263, 264, 268, 350
human welfare, 82
human-material-natural network, 110, 122
Hungary, 227
hunger, 53
hydraulic development,, 9, 122
hydraulic engineering, 1, 127
hydraulic fracturing, 277
hydraulic infrastructure, 20, 117, 151, 152, 153, 154, 349, 351
hydraulic technologies, 116
hydraulic works, 9, 120, 125, 126
hydraulics, 109
hydro-colonialism, 7
hydrocracies, 297
hydroelectric power, 161, 172
hydroelectricity, 169, 181
hydrology, 72, 125
hydro-political dream schemes, 152
hydropower, 9, 47, 60, 61, 66, 78, 164, 172, 181, 289
hydrosocial cycle, 48, 124, 128, 151
hydrosocial imaginaries, 109
hydrosocial networks, 2, 4
hydrosocial territory, 11, 80, 108, 109, 110, 116, 117, 121, 122, 128, 129, 151, 152, 153, 156, 157, 160, 162, 163, 164, 283, 284, 287, 289, 290, 296, 346
hydrosphere, 260
hydro-territorial politics, 115
hygiene, 47, 216, 219, 220

Ica (Peru), 7, 107, 290
identity, 13, 20, 22, 78, 81, 89, 121, 128, 138, 158, 161, 163, 233, 278, 294, 338, 339, 343
ideologies, 14, 115, 116, 118, 120, 153, 249, 350
Ilisu Dam, 162, 163, 164
illness, 92, 97, 102
Imbabura Province (Ecuador), 7
inclusion, 13, 38, 115, 135, 144, 153, 154, 161, 201, 210, 212, 248, 268, 296
India, 7, 11, 43, 44, 45, 46, 48, 51, 52, 53, 54, 62, 66, 94, 97, 99, 194, 195, 196, 198, 199, 201, 205, 249, 251, 253, 265, 266, 267, 284, 285, 286, 296, 334, 346
indigenous communities, 8, 9, 158, 177, 179
Indigenous peoples, 193, 194, 195, 196, 197, 198, 200, 201, 203, 204, 205
indigenous rights, 5, 7, 22, 50, 179, 233
indignities, 90, 95, 96, 101
individual freedom, 5, 52
individualism, 251
industrial development, 63, 144
industrial waste, 34
industrialization, 34, 36, 44, 157
industry, 1, 18, 44, 46, 78, 123, 137, 138, 140, 142, 144, 170, 172, 177, 179, 180, 199, 228, 266, 283, 284, 288, 302, 317
informal settlements, 90, 92, 93, 94, 95, 96, 97, 215, 219
  bustees, 90
  favelas, 90
  homeless encampments, 90
  shantytowns, 90
insecticides, 286
institutional reforms, 109, 311, 312
Integrated System of National Hydraulic Equilibrium (Spain), 127
Integrated Water Resources Management, 48, 65, 78, 83
Inter-American Development Bank, 9, 16, 145, 173, 175, 179
Inter-American Human Rights Commission, 9, 175, 178
Inter-American Human Rights Court, 178, 180
inter-basin transfers, 126, 261, 289
inter-legality, 15
internal colonization, 119
International Bank for Reconstruction and Development, 171
International Rivers, Food & Water Watch, 296
International Union for Conservation of Nature, 292
interpretative policy analysis, 80
interventions, 1, 2, 3, 4, 8, 14, 21, 43, 47, 53, 108, 109, 118, 120, 124, 126, 128, 134–47, 173, 210, 211, 213, 216, 218, 219, 220, 288, 307, 343
Inuit, 204
invisibility, 2, 60, 108, 157, 230, 249, 254, 286, 290, 336, 340, 342, 343
Iran, 147
Ireland, 227, 231
irrigated agriculture, 157, 173
irrigation, 7, 18, 39, 44, 46, 72, 78, 89, 119, 122, 124, 138, 145, 146, 151, 154, 249, 260, 277, 286, 287, 289, 291, 292, 295, 302, 322, 330
  efficiency, 17
  management, 248
  modernization, 5, 18
  networks, 261
  privatization, 188
  projects, 126, 140

irrigation (*cont.*)
 schemes, 9, 45, 61, 162
 systems, 7, 18, 45, 46, 109, 291
 Turkey, 161
Islamic law, 264, 265, 266
ISO, 294, 297
Italian Water Movement, 236, 237
Italy, 227, 231, 236, 237
IWRM, *see* Integrated Water Resources Management

Jamundí (Colombia), 144, 145, 147
Japan, 169
justice, 2–6, 13–22, 36–40, 43–5, 47–54, 65, 82–3, 90–2, 99–102, 108, 111–12, 123–9, 158–64, 176, 177–81, 187–91, 200–4, 210–21, 226–42, 267–9, 310–12, 323–6, 343–4
 activism, 66, 242
 adaptive, 268
 climate, 189
 conceptions of, 53
 dimensions of, 22, 211, 348, 356
 distributional, 4, 5, 6, 20, 211, 218
 distributive, 22
 domains of, 6
 ecological, 6, 22, 128
 economic, 252
 environmental, 9, 20, 49, 111, 181, 212, 234, 253, 279, 343, 347, 357
 formal, 5
 as inclusion, 200
 paradigms, 129
 procedural, 218, 350
 as recognition, 5, 6, 13, 16, 23, 201
 recognitional, 220
 restorative, 169
 sanitation, 90, 100, 210, 212
 situated, 161
 social, 53, 91, 92, 160
 theories, 50, 52
 trivalent conception of, 22, 49, 348
 understandings of, 5
 urban water, 101
 water allocation, 53
Justicia Hídrica/Water Justice alliance, 296

Kampala, 188, 211, 214, 215, 217, 219, 221, 222
Kawempe (Uganda), 211, 215, 218, 220, 222
Kazakhstan, 7
Kenya, 8, 93, 96, 97, 99, 302, 303, 304, 305, 306, 307, 308, 309, 310, 311, 312
Kenya Colony and Protectorate, 304
Kenya Forest Service, 303, 309, 311
Kenya Water Towers Agency, 302, 310
Keynes, J.M., 171
Khartoum, 251
Kibaki, M., 309

kinship, 121, 252, 254
knowledge, 2, 14, 15, 18, 19, 53, 73, 81, 94, 108, 110, 115, 116, 121, 145, 152, 153, 160, 163, 164, 175, 189, 200, 249, 250, 287, 288, 292, 357
 collective, 53
 contested, 349
 expert, 3, 72, 75, 190
 and discourses, 43, 152
 indigenous, 204, 253
 local, 199
 local property, 14
 and power, 20, 47
 privileged sources, 83
 professional, 14
 scientific, 14, 109, 121
 situated, 14
 subjective, 324
 systems, 153
 technocratic, 79
 traditional, 200, 203
 transdisciplinary, 23
 and truths, 13
 and values, 82
 water communities, 18
 and worldviews, 158
Kukdi (India), 45, 46
Kurdish diaspora, 163
Kurdish question, 161
Kurds, 161, 163
Kutch region (India), 48

La Marea Azul, 239
La Paz, 318
labor, 19, 66, 96, 98, 120, 231, 233, 249, 251, 265, 318, 331, 344
 collective, 351
 division of, 90, 95, 97, 101, 172, 339
 domestic, 92, 93, 95, 100
 and farming, 154
 female, 335, 343
 forced, 175, 323
 gender division of, 95
 and land tenure, 45
 low-cost, 46
 organization of, 330
 reproductive, 100
 unions, 235
 unskilled, 96
 wage, 324, 343
labor-intensive activities, 343
Lac-Mégantic (Quebec), 277
ladinos, 181
Lagunas de Laminación, 143, 144
Lake Huron, 34
Lake Oahe, 277
Lake Uru Uru, 319, 321, 327

lakes, 64, 74, 120, 141, 319
land grabbing, 59, 60, 61, 62, 63, 64, 247, 304
land grabs, 6
land rights, 45, 65, 66, 193, 263
land tenure, 62, 357
landless people, 46
landscapes, 67, 135, 291, 296, 316, 317, 321, 323, 326
language, 43, 51, 79, 80, 81, 82, 83, 159, 160, 161, 178, 237, 343
Latin America, 64, 100, 111, 164, 278, 316, 322, 323
latrines, 212, 214, 215, 218, 220
Latvia, 227
laundry, 92, 93, 94, 98
laws, 2, 3, 8, 19, 20, 22, 43, 53, 66, 83, 108, 179, 200, 203, 205, 229, 240, 261, 269, 278, 283, 303, 324, 349, 351
Laws of Manu, 265, 266, 267
lead poisoning, 35, 37
legal frameworks, 19, 46
legal plurality, 13, 15, 21, 22, 111, 158, 161, 162, 163, 349
leisure, 98, 99, 337
Lent (The Netherlands), 143
levees, 349
liberal bourgeoisie, 120
Life-Cycle Analysis, 294
Lilienthal Plan, 138, 141, 145
Lilienthal, D., 147
Lima, 7, 10, 11
Lisbon Treaty, 226, 232
Lithuania, 227, 240
livelihoods, 1, 8, 9, 17, 21, 45, 46, 47, 62, 66, 74, 93, 97, 99, 102, 122, 134, 141, 142, 154, 155, 158, 159, 173, 177, 180, 188, 195, 199, 252, 283, 291, 302, 311, 324, 325, 337
livestock, 47, 99, 158, 249, 253, 321, 322, 324
Living River approach, 135
local governments, 16, 35, 230, 320
local water societies, 21, 109
logic of capital, 38, 61
Lorenzo Pardo, M., 123, 124
Luxembourg, 227
Luxemburg, R., 344

Maasai, 307
Maastricht Treaty, 228
Maathai, W., 309
Machacamarca, 321
Macías Picavea, R., 118, 119
Madrid, 128, 237, 239, 318
Maharashtra (India), 44, 45, 46
Maine, 265
Majes (Peru), 18
Making space for water approach, 135
Malawi, 250
male dominance, 95, 97

Mali, 5
malnutrition, 96, 97, 102
Malta, 227
Manabí (Ecuador), 153
Marathwada (India), 45
marginalization, 5, 115, 219, 283, 324, 347
marine protected areas, 267
market, 19, 38, 39, 98, 110, 117, 218, 226, 230, 238, 241, 264, 291
  actors, 5
  and human rights, 242
  deregulation, 231
  export, 18, 171
  ideology, 229
  incentive structures, 229
  operations, 288
  policies, 142
  principles of supply and demand, 324
market economy, 13, 14, 171, 236
market forces, 1, 3, 47, 229, 230, 284
market's invisible hand, 290
marshes, 120
Marxism, 61, 120, 233, 248, 333
masculinities, 336, 338, 340, 343
mass mobilizations, 47
Mauritania, 6
Maya Achi, 174, 176
MDGs, *see* Millennium Development Goals
meanings and values, 14
Mediterranean, 124, 128
mega hydraulic infrastructure, 9
megacities, 1, 11
memorial sites, 316
memory, 35, 119, 316, 317, 319, 320, 323, 326, 350
menstrual hygiene, 95, 219
methane, 170
Métis, 204
Mexico, 7, 8, 16, 155, 180, 286, 291, 295
Michigan, 34, 35, 37
micro-credit institutions, 45
micro-politics, 233
middle class, 44
Middle East, 64, 161
Mies, M., 332
migrant populations, 90
military bases, 126
Millennium Development Goals, 10, 210, 264
Miller, D., 54, 195
mineral resources, 159, 172
minimum wage, 96
mining, 1, 8, 9, 60, 159, 160, 164, 179, 180, 278, 283, 289, 290, 291, 293, 295, 317, 318, 319, 320, 321, 322, 323, 325, 326, 346
  Bolivia, 316
  Canada, 9
  Colombia, 158
  Ecuador, 154

mining (*cont.*)
  Guatemala, 176
  Latin America, 111, 278
  Mexico, 8
  open pit, 319
misrecognition, 3, 5, 15, 212, 249
Mississippi River, 277
Missouri River, 276, 277, 296
mixed-cropping patterns, 337
MMISF Act, 47
mobilizations, 84, 246, 295, 304, 311
modernization, 116, 119, 120, 121, 122, 124, 126, 128, 147, 154, 158, 176, 189, 194, 201, 284, 286, 307, 308, 330, 331, 333, 335, 343
monetary value, 17, 218
monoculture, 155
moral ecological rationalities, 248, 249, 253
Morales, E., 318, 319
moralization of technology, 130
Morocco, 66, 98, 99, 291, 330, 336, 341, 343
mortality, 99
Mozambique, 54, 61
multiculturalism, 13, 153, 161
multifunctionality, 147
multilateral financial institutions, 2
multinational companies, 2, 159, 284, 287, 293
multinational retailers, 284
multi-stakeholder approaches, 48, 147
Mumbai, 93, 94, 251
municipal infrastructure, 89
Murcia, 128
Muslim settlers, 93, 251
MWRRA Act, 54

NAFTA, 8
Nairobi, 306, 311
Namibia, 253, 254
Nandi population (Kenya), 305
Narmada dam (India), 346
narratives, 60, 66, 78, 83, 122, 134, 135, 136, 139, 141, 142, 145, 147, 316, 319, 320, 321, 326
Nasik district (India), 45
National Development Plan (Guatemala), 173
National Hydraulic Plan (Spain), 127
National Land Commission (Kenya), 307
National Power Authority (Guatemala), 172
National Water and Sewerage Corporation (Uganda), 214, 222
National Water Forum (Ecuador), 296
nation-state, 2, 19, 129
Native American tribes, 276
Native Americans, 81, 276, 277
natural capital accounting, 302
natural resources, 19, 61, 62, 63, 75, 115, 164, 180, 268, 269, 286, 288
natural sciences, 123, 351
naturalization, 2, 10, 12, 15, 47, 48, 53

nature, 2, 21, 23, 43, 46, 53, 74, 80, 83, 108, 116, 119, 120, 121, 127, 129, 134, 260, 261, 310, 332, 351
nature reserves, 16
Nduruma River (Tanzania), 65
negative externalities, 230
negotiation, 60, 61, 66, 76, 100, 117, 135, 152, 155, 160, 171, 178, 188, 198, 255, 293
neoliberal incentives, 19
neoliberal paradigm, 5, 7, 11, 12, 14, 15, 19
neoliberal water governance, 19, 283–301
neoliberalism, 2, 16, 17, 19, 36, 43, 230
Nestlé, 293
Netherlands, 147
Netherlands, the, 8, 134, 135, 136, 139, 141, 142, 143, 144, 145, 147, 148, 169, 227, 278
networks, 11, 21, 80, 83, 108, 111, 117, 121, 125, 126, 129, 151, 152, 171, 221, 233, 241, 261, 269, 279, 289, 290, 308
New Hampshire, 171
New Zealand, 181
new-institutionalism, 20, 229
NGOs, 79, 111, 144, 157, 159, 162, 163, 175, 211, 213, 214, 215, 218, 220, 226, 234, 248, 287, 292, 296
nickel, 179
Niger Delta, 111, 278
Nigeria, 278
Nile, 7
non-human actors, 82, 116, 117, 152, 249
non-movements, 100
non-recognition, 189, 355
non-state actors, 61, 216, 287
norms, 2, 3, 14, 18, 19, 37, 111, 115, 129, 130, 152, 155, 164, 204, 233, 248, 252, 254, 255, 283, 288, 294, 296, 297, 324, 331, 339, 349, 350, 351
  and fairness, 2
  behavioral, 6
  distributional, 129, 253
  gendered, 335
  global, 53
  governance, 289
  local, 110
  moral, 252
  neoliberal, 297
  operational, 356
  social, 12, 90, 109, 219, 279, 334
  societal, 37
  universal, 16
North America, 34, 36, 38
North Dakota, 276, 277
North Korea, 169
Northwest Territories, 204

Obama, B., 276, 277
ocean, 61, 66, 118, 286
ocean grabbing, 61
Odebrecht (company), 12

Odisha, 44, 45
Office du Niger, 6, 63
Office of Human Resettlements (Guatemala), 174
oil, 8, 62, 111, 154, 171, 172, 176, 276, 277, 278, 296, 322
oil companies, 8, 111
Olmos (Peru), 11, 12, 290
operation and maintenance, 11, 35, 214, 229
operational norms and rules, 350
Oruro, 317, 318, 319, 320, 322, 325, 326
Oslo Peace Accords, 176, 182
Ostrom, E., 75, 76, 77, 82
outsourcing, 268, 269
Overdiepse Polder, 143

Paco El Rana, 125
Pact of Madrid, 126
Pakistan, 286, 291
palm oil, 59
Pangani Basin (Tanzania), 65
Papua New Guinea, 346
paramilitary forces, 303
páramos, 158, 159, 160, 164
Pardo, L., 124
Paris, 90, 267
Paris Agreement, 267
participation, 6, 13, 36, 38, 49, 65, 71, 75, 76, 77, 79, 83, 84, 142, 153, 154, 163, 198, 200, 201, 212, 214, 216, 217, 240, 253, 264, 296, 347, 348, 355, 356
    citizen, 136
    collective, 52
    democratic, 75, 350
    indigenous, 198
    institutional, 144
    local, 188, 218
    market, 98
    political, 54, 81
    political, 53
    public, 71, 147, 190
    and socio-environmental justice, 349
    stakeholders, 80, 292
    values of, 77
participatory approaches, 216, 218
pastoralists, 248, 253, 254, 302, 307
Patagonia Without Dams (Chile), 296
Payment for Environmental Services, 16, 160
Paz-Estenssoro, V., 318
peasants, 5, 142, 155, 175, 252
Pelzig, E., 238
peppers, 337
personal hygiene, 95
Peru, 7, 8, 10, 11, 16, 18, 107, 140, 278, 284, 286, 290, 291, 292, 295, 346
PES, *see* Payment for Environmental Services
pesticides, 16, 342
Petén, 181

Philippines, the, 16, 118, 333
pigs, 155
PIM (Participatory Irrigation Management), 47
Plan International, 215
planning, 17, 72, 75, 78, 79, 82, 84, 93, 125, 129, 134, 142, 147, 161, 177, 202, 213, 267, 291
Platform against privatizing Canal Isabel II, 239
Platform to Defend the Ebro River, 127
Plato's Republic, 115
plurality, 2, 23, 51, 52, 53, 234, 255, 268
Poland, 227
police, 127, 179, 182, 276, 277, 278, 288
policy analysis, 83
policy approaches, 71, 83, 84
    Discursive Policy Analysis, 71, 83
    Efficiency-based Analysis, 71, 83
    Institutional Analysis and Development, 71, 75, 83
    Physically-based Watershed and River Basin Approaches, 71, 83
policy frameworks, 71, 83, 335
policy-makers, 3, 11, 23, 75, 109
policy-making, 13, 21, 71, 72, 80, 82, 83, 188
political ecology, 2, 4, 23, 63, 82, 94, 116, 211, 246, 310, 346
    feminist, 96
political economy, 48, 61, 63, 101, 157, 311
political neutrality, 123
political representation, 5, 6, 21, 22, 92, 101
politics, 4, 49, 76, 79, 82, 101, 109, 153, 249, 276, 308
    bureaucratic, 128
    conservation, 307
    cultural, 97
    hydraulic, 119, 122
    of CSR, 18
    of identity, 233
    of inclusion, 161
    and IWRM, 82
    positivist, 80
    of recognition, 13, 130, 204
    of scale, 261, 351
    special-interest, 135
    of truth, 21, 121
    territorial, 61, 129
pollution, *see* contamination
polycentrism, 22, 77
Portugal, 227, 231
positivism, 47
post-colonial South, 93
post-industrial societies, 93
potatoes, 147, 158, 318, 322, 324, 337
potential agricultural lands, 304
Potosí (Bolivia), 8, 318, 319, 320, 326
poverty (the poor), 13, 44, 51, 61, 80, 90, 94, 96, 101, 102, 188, 216, 234, 249, 250, 252, 266, 269, 297, 323
    access to drinking water, 214

poverty (the poor) (*cont.*)
　access to water, 39
　as capability deprivation, 92
　and climate change, 261
　and contamination, 284
　Bolivia, 320
　exclusion, 65
　GAP, 162
　gendered, 98
　landlessness, 5
　pragmatism, 250
　relational, 93, 99, 101
　rural communities, 157
　sanitation, 39
　Sen´s concept, 91
　subsidies for, 230
　urban, 44, 89, 94, 99, 100, 210, 213
　violence against, 97
　voices of, 91
　water security, 1
　and wetlands, 267
　women, 97, 219, 335
poverty line, 34
power, 13, 14, 15, 39, 53, 61, 62, 92, 101, 111, 115, 117, 120, 129, 134, 142, 143, 144, 148, 152, 164, 182, 188, 201, 229, 236, 237, 251, 287, 333, 351, 356
　and authority, 43
　bargaining, 65, 254
　capillary, 108, 155, 162
　Cold War, 170
　economic, 118, 176, 180
　and ethics, 82
　geometries, 3
　hegemonic, 248, 255
　invisibility, 2, 249, 254
　and knowledge, 20, 47
　military, 110
　normalizing, 2, 19, 288
　and politics, 76, 101
　purchasing, 285, 334
　relations, 4, 6, 10, 11, 20, 38, 44, 48, 60, 108, 116, 129, 248, 251, 347, 350, 351
　and science, 3
　social, 38, 279
　social relations of, 50
　sovereign, 19
　state, 143, 278, 287
　structures, 2, 130, 255, 334, 344, 350
　territorial, 123
　workings of, 19, 52
powerlessness, 91
PPP, *see* public-private partnerships
pragmatism, 248, 250, 251
Primo de Rivera, M., 122
privacy, 95, 98, 99
private actors, 14
private security, 277

private-home-women's domain, 339, 341
privatization, 1, 11, 38, 44, 47, 51, 65, 177, 182, 228, 229, 230, 231, 232, 234, 236, 239, 241, 242, 246, 263, 264, 268, 330, 337, 340, 350
Procaña, 144
procedural justice, 37, 212, 217, 326, 347
Procter and Gamble, 294
production standards, 18
productive efficiency, 18
productive nostalgia, 317
profit maximization, 229, 331
property, 5, 13, 14, 15, 16, 38, 39, 47, 52, 54, 63, 64, 128, 156, 180, 218, 229, 262, 263, 268, 303, 331, 344
pro-poor approaches, 264
public funds, 44, 229
public good, 46, 47, 77, 127, 226, 229, 239, 240, 262, 268
public health, 94, 230, 235
public policy, 11, 71, 75
public shame, 96
public water utilities, 1, 190
public-private partnerships, 11, 16, 229, 265, 266
public-work-men's domain, 339, 341
Puerto Rico, 147
pumps, 7, 45, 260

Quebec, 277
Quechua communities, 318, 320
quinoa, 318, 322, 324
Quispe Condori, M., 321

race, 49, 77, 92, 93, 94, 119, 188, 212, 269, 349, 351
racial segregation, 36, 93
racism, 36, 193, 323, 347
rainfall, 7, 119, 125, 128, 267
rainwater, 94, 249, 253
rainwater harvesting, 249, 253
Ramamurthy, R., 333
rape, 97
rational and efficient water use, 15, 115
rational market actors, 5
rationality, 3, 17, 19, 52, 76, 81, 123, 130, 151, 154, 159, 213, 253, 284, 287, 288, 289, 291, 294, 297, 332
Rawls, J., 4, 52, 54, 324, 347, 348
reciprocity, 17, 251, 252, 254
recognition, 6, 13, 16, 23, 36, 37, 38, 49, 53, 66, 80, 153, 161, 163, 170, 189, 190, 193, 196, 200, 201, 202, 204, 211, 219, 221, 233, 348, 349, 356
　cultural, 5, 6, 22, 348
　of social difference, 348
recognitional justice, 6, 22, 211, 221, 348
recycling, 292
REDD+, 310, 311
Reducing Emissions from Deforestation and Forest Degradation (REDD+) program, 310

reforestation, 140
refugee camps, 90
regeneracionismo, 118
relational inclusiveness, 259
relational poverty, 92, 101
religion, 102, 119, 188, 235, 252
re-municipalization, 65
representation regimes, 350
representational justice, 6, 19, 22, 36, 347, 348
Republic (Spain), 124, 311
Republican Party, 277
research and action, 23, 49, 163
resistance, 10, 11, 12, 39, 60, 63, 64, 66, 100, 128, 130, 136, 143, 144, 154, 155, 195, 232, 246, 247, 248, 250, 251, 252, 254, 295, 305, 332
resource extraction, 3, 67, 109, 121, 278, 279
resource grabbing, 60, 63, 66, 67
resource regimes, 279
retail companies, 18
Revolutionary Armed Forces (Colombia), 141
rice, 59, 147, 169
Right2Water, 226, 227, 228, 229, 231, 232, 234, 235, 236, 237, 238, 239, 240, 241
rights, 4, 9, 13, 15, 16, 23, 37, 43, 50, 51, 52, 97, 121, 129, 153, 159, 170, 180, 194, 199, 201, 252, 255, 256, 268, 348, 356
　accumulation of, 9
　and equity, 38
　civil, 179, 233
　concessionary, 306
　contents of, 20
　dalit movements, 47
　distribution of, 264
　environmental, 135, 144
　First Nations, 194, 199
　forest access and use, 307
　human, 176, 229, 262
　indigenous, 193, 195, 197, 199, 200, 201, 203, 204, 277
　individual, 212, 347
　informal, 60
　land tenure, 62, 63, 334, 337
　liberal conception of, 38, 349
　marginalized populations´, 355
　pluralism, 22, 53
　property, 47, 64, 128, 229, 268, 303
　Rawls notion of, 52
　recognition of, 49
　right to have rights, 37
　sovereign, 265
　standardization, 12
　systems, 2, 3, 12, 13, 14, 158
　universal, 51, 53
　women, 97, 330
Río Grande (Ecuador), 154, 155
Río Negro (Guatemala), 173, 174, 175, 176, 177, 178, 182

Ríos Montt, E., 174, 176, 178
riparian vegetation, 141, 142
risk control, 135
River Basin Confederations (Spain), 123
river systems, 218
riverbanks, 138
riverine communities, 141
rivers, 9, 64, 72, 74, 124, 125, 135, 138, 157, 162, 172, 181, 260, 265, 266, 286, 295, 319, 321, 326
road construction, 172, 173
Roman Law, 265
Romania, 227, 236, 239, 241
Room for the River approach, 134, 135, 142, 147, 148
Roundtable on Responsible Soy, 292
Roundtable on Sustainable Biomaterials, 292
rules, 21, 76, 77, 83, 129, 152, 194, 233, 269, 288, 291, 350, 351
　community, 13
　consultation, 201
　elegibility, 72
　formal, 14
　institutionalized, 264
　market, 229
　on water protection, 266
　operational, 349
　social, 97, 252
　state, 111
　territorial, 111
　water governance, 262
rural areas, 10, 44, 51, 53, 98, 172, 173
Russia, 62, 181

Sabana de Bogotá, 136
SABMiller, 293
Saïss, 330, 331, 334, 341, 344
salt water, 260, 267
Salvajina dam (Colombia), 138, 141, 145
San Luis Potosí, 8
sanitation, 11, 36, 39, 47, 89, 91, 92, 93, 94, 96, 97, 98, 99, 101, 189, 211, 212, 213, 214, 215, 216, 218, 219, 220, 221, 236, 346, 350
　access to, 93, 94, 95, 96, 97, 102, 218, 264
　and female body, 90
　gendered, 95, 97
　household, 89, 91, 96
　human right to, 66, 226, 240, 241, 268
　interventions, 219, 220
　justice, 91, 100, 102, 190
　marketing, 218
　onsite approaches, 215
　operational and management costs, 73
　political economy of, 101
　and poverty, 96
　and social reproduction, 93
　urban, 89, 90, 101, 210, 213
　women, 96
　World Bank strategy, 213, 217

Santa Rita Dam (Guatemala), 179, 180, 181
Santurbán (Colombia), 158, 159, 160
São Paulo, 318
Sardar Sarovar Narmada project (India), 48
scale, 109
   collective action, 325
   governance, 16
   hydraulic ordering, 121
   land grabbing, 61
   local protests, 295
   manipulation of, 53
   networks, 129
   politics of, 261, 297, 351
   and power hierarchies, 122
   river basin, 78, 125
   and spatiality, 49
   and technopolitical water society, 122
   and territories, 117
   universalization, 48
   and water flows, 63
   wartershed, 350
   water governance, 2, 16
scales, 4, 50, 66, 109, 117, 127, 128, 160, 234, 241, 348
Scandinavia, 235, 237
Schlosberg, D., 5, 22, 23, 36, 49, 50, 51, 53, 54, 153, 202, 212, 233, 234, 324, 346, 348
scholars, 4, 16, 23, 66, 75, 77, 79, 80, 82, 83, 135, 193, 205, 212, 213, 332, 333, 348, 351, 357
science, 2, 3, 22, 47, 75, 121, 261, 263, 269, 332
Scott, J., 252, 324
SDGs, *see* Sustainable Development Goals
sea-level rise, 260
seawater, 128, 292, 295
securitization, 265, 268, 284
security, 6, 16, 62, 73, 83, 89, 91, 95, 111, 135, 162, 174, 176, 189, 219, 287, 288, 304, 309, 348
Segura basin, 125
self-sufficiency, 124, 306
semi-subsistence agriculture, 318
Sen, A., 37, 39, 50, 52, 91, 92, 324, 348, 349, 356, 357
Senegal, 6, 99, 295
settlements, 90, 92, 93, 161, 162, 179, 210, 211, 215
sewerage, 212, 213, 218, 221
sexism, 97
sexuality, 211, 212, 351
SFR, *see* Strategic Framework for Reform (Uganda)
sheep, 322
Shell, 111, 278
shelter, 37, 89, 92, 100
Sheonath river (India), 266
silver, 172, 179, 318
Similkameen Indian Band, 196, 204
Sioux, 276, 277, 296
skill acquisition, 92
slaves, 142

sleep, 99
Slovakia, 227, 240, 241
Slovenia, 227, 241
sludge, 210, 212, 214
slum demolitions, 89
smallholder families, 10
Smith, A., 286
social contract, 4, 230, 348
social differentiation, 53, 109, 129
social inequalities, 136, 210
social reproduction, 93, 95
social stigmas, 220
socialist economy, 172
socialists, 120, 121
socio-ecological integrity, 22, 153, 268, 283, 348, 356
socio-ecological justice, 6, 22, 348
socio-economic justice, 36, 233
socio-legal systems, 15
socio-nature, 4, 14, 19, 22, 116, 117, 151, 278, 279, 330
socio-technical configurations, 129
Socrates, 115
soils, 7, 141, 260, 318, 319, 320
solid waste management, 212
Sonso Lagoon (Colombia), 145
SOSteTONERO, 232, 239
South Africa, 11, 16, 62, 304, 306
South America, 188, 285
South Dakota, 277
South East Asia, 170
South Vietnam, 147
Southeastern Anatolia Project, 161
Southern Rhodesia, 304, 306
sovereign power, 19, 110
sovereignty principles, 259
soya, 59, 62
Spain, 11, 118, 119, 120, 121, 123, 124, 125, 126, 127, 128, 237, 240, 241, 265
   20th-century, 116
   and Arabs, 119
   civil war, 123
   liberal state, 120
   national water system, 123
   new water culture movement, 295
   Platform against privatizing Canal Isabel II, 239
   territorial governmentalization projects, 121
sprinkler technology, 289
SSWARS, 215
St. Louis, 277
Standing Rock reservation, 111, 276
state, 20, 35, 39, 45, 46, 52, 109, 116, 120, 157, 189, 230
   accountability, 269
   actors, 61, 216
   authoritarian, 125
   and civil society, 79
   cooperatives, 337
   governmentality, 21, 161

Guatemala, 178
hydraulic interventions, 124
Kenya, 308
land-use planning, 17
law, 62, 63
and liberal bourgeoisie, 120
military violence, 10
monopoly on water, 13
neoliberalism, 19
power, 287
privatization of water rights, 44
reterritorialization, 309
rights protection, 266
territorial control, 157
top-down policy interventions, 14
state-administrative networks, 109
state-capital alliances, 63
steel, 121, 122, 126
stock market, 239
street hawkers, 89
street-level hawking, 100
subaltern groups, 283
subjugation, 3
sub-Saharan Africa, 60, 90, 93, 99, 210, 211, 213, 214, 219, 221
subsidies, 12, 17, 157, 218, 233, 284, 289, 290, 291, 331, 343
subsistence, 45, 172, 195, 252, 303, 307, 318, 324
subsistence agriculture, 172, 324
Suez, 12
Suez (multinational), 230
sugarcane, 45, 59, 62, 136, 137, 138, 139, 140, 141, 142, 143, 144, 145, 146, 147
suicide, 7, 45
Sui-ho Dam (N. Korea), 169
sustainability, 1, 18, 47, 177, 204, 217, 218, 221, 305, 306
sustainable agriculture, 160
Sustainable Agriculture Initiative, 292
Sustainable Development Goals, 210, 220, 264, 269
Sustainable Development Policy Institute (Pakistan), 291
Sweden, 227
Switzerland, 8

Tajo River, 124, 125, 126, 127
Tana Delta (Kenya), 64
Tangiers, 98, 99
Tanzania, 7, 11, 65, 250
technification, 47, 48, 53
technocratization, 268
technology, 2, 3, 17, 19, 53, 116, 117, 129, 130, 139, 143, 265, 290, 293, 294, 295, 296, 323, 338, 340, 350, 351
Tecnicaña, 144
Tenneessee Valley Authority, 147
territorial ordering, 108, 130

territorial rights, 142, 159
territoriality, 78, 108, 110, 111, 122, 125, 287
territorialization, 110, 115, 116, 128, 156, 164
territory, 11, 13, 22, 116, 117, 129, 140, 143, 147, 152, 156, 157, 163, 200, 204, 205, 265, 287, 291, 295, 296, 338
  of abundance, 157
  Colombia, 141
  communities, 64
  defense of, 22, 159
  depopulation, 305
  forest, 309, 311
  GAP, 162
  geopolitical, 115
  governmentalizing, 121
  hydraulic grid, 115, 164
  hydrosocial, 108, 111, 117, 126, 129, 151, 152, 153, 158
  imaginaries, 135
  indigenous peoples, 195
  Kurds, 163
  modernist government-rationalities, 110
  páramo, 158, 159
  re-patterning, 117, 155, 163
  Sioux, 276, 277, 296
  Standing Rock reservation, 111
  vernacular, 289
  wetlands, 134
Thailand, 54, 286
Thatcher, M., 228
The Nature Conservancy, 292
The Rainforest Alliance, 292
Thessaloniki, 232, 239
Thompson, E.P., 252
tidal surges, 188
time savings, 96, 98
toilets, 97, 212, 215, 219, 220
totora, 321, 322, 323
tourism, 143, 289, 295
toxic waste disposal, 49
trade liberalization, 7
trade unions, 226, 234, 235, 239, 242
tradeoffs, 180
traditionalism, 120, 339
transdisciplinary social action, 4
transparency, 15, 62, 67
transport, 122, 214, 218, 228, 230, 251, 277, 284, 289
tribunals, 13
True Pricing, 294
Trump, D., 277
truth, 4, 14, 18, 19, 21, 47, 51, 108, 110, 121, 129, 177, 288, 323
Tungen families (Kenya), 307
Tunisia, 295
Turkey, 161, 162, 163, 164
TVA, *see* Tenneessee Valley Authority
TVA model, 138, 141, 147

Uganda, 211, 214, 217
UK (United Kingdom), 8, 163, 171, 226, 227, 228, 230, 236, 237, 239, 241, 267
UN Climate Change Treaty, 181
UN Economic and Social Council, 213
UN General Assembly, 66, 264
UN Human Rights Commission, 264
UN Human Rights Council, 66
UN International Watercourses Convention, 264
UN Ramsar Convention, 267
UNECE Convention, 267
unemployment, 97, 102, 232
Unilever, 292, 293
United Fruit Company, 173
United Nations Monetary and Financial Conference, 171
United States, *see* USA
United States of America, 80, 126, 171, 176, 205
universalization, 47, 48, 53
UPE, *see* urban political ecology
Upper Mesopotamia, 161
upstream farmers, 17
urban abandonment, 36
urban areas, 10, 89, 94, 96, 98, 159, 216
urban citizenship, 1, 100
urban development, 135
urban political ecology, 211
US civil-rights movement, 347
US Environmental Protection Agency, 276
US Foreign Financial Services Act of 1977, 174
US State Department, 276
USA (United States of America), 8, 49, 265, 285
user collectives, 15, 19, 296
user groups, 2, 3, 15, 16, 20, 21, 22, 152, 190, 294
USSR, 181
usufructuary rights, 259
utilitarian philosophy, 4
utopian hydraulism, 3, 119, 120, 123, 126

Valencia, 128, 151
vegetables, 6, 137, 155, 285, 318, 330
Venezuela, 173, 278
venture capital firms, 181
Veolia, 230, 238
Ver.di trade union, 238
Vietnam, 99, 250
violence, 10, 14, 20, 62, 94, 96, 97, 102, 122, 129, 142, 174, 178, 179, 180, 181, 219, 287, 304, 305, 306, 310, 322, 323
virtual water, 6, 7, 74, 75, 283, 284, 285, 286, 288, 289, 290, 295, 297
Volta Valley Authority, 6

WaFNE (UNEP), 294, 297
Waghad (India), 46
Wales, 228

Wami-Ruvu River Basin (Tanzania), 65
Washington Consensus, 65
waste disposal, 95
wastewaters, 210, 211, 212, 214, 218
water
  availability, 7, 44, 65, 94, 125, 260
  bills, 36
  bureaucracies, 14
  collective good, 39, 82
  commodification, 230
  companies, 11, 226, 228, 232, 239, 240
  conservation, 16
  crisis, 3, 246
  culture, 16, 19, 21, 110
  cycle, 64
  degradation, 3
  dispossession and accumulation, 4
  distribution, 7, 20, 43, 50, 51, 60, 63, 128, 212, 249
  drinking, 10, 11, 16, 34, 35, 36, 39, 45, 98, 101, 154, 157, 159, 188, 193, 214, 226, 264, 276, 302, 346
  economic good, 82
  ecospace, 259, 260, 261, 262, 263, 264, 265, 268, 269
  extraction, 8, 10, 156, 283, 293, 295
  footprint, 289, 294
  gifting, 251
  grabbing, 2, 6–8, 14, 59, 60, 61, 62, 63, 64, 65, 66, 111, 265, 346, 355
  knowledge, 2, 4, 6, 14, 15, 19, 22, 23, 109, 121, 275, 279, 350, 356
  laws, 3, 13
  liberalization, 262, 268
  marketization, 230
  markets, 38
  meanings, 82
  monopolization, 262
  negotiations, 49
  ontology, 44
  pricing, 11, 17, 19, 73, 75, 78, 127, 189, 229, 253, 264, 266
  privatization, 230, 262, 269
  productivity, 17, 290
  quality, 16, 37, 64, 78, 80, 159, 193, 199, 212, 261, 296, 319, 323, 349
  reality, 4, 14, 15, 18, 95, 110
  reallocation, 60, 66, 290
  reforms, 3, 108
  saving, 1, 17
  scarcity, 81
  securitization, 262
  security, 1, 23, 46, 81
  tariffs, 11, 73, 231, 233, 253
  technologies, 262
  transfers, 10, 11, 124, 127, 128, 129, 290, 295
  uses, 1, 46, 52, 74, 78, 109, 290

valuation, 53
value of, 74, 81, 82
water allocation, 6, 14, 20, 43, 47, 48, 51, 52, 53, 60, 74, 190, 194, 196, 197, 230, 261, 268, 269, 290, 349
  efficiency, 17
  First Nations, 202
  FITFIR, 197, 201, 202
  and neoliberal water policies, 289
  prioritazation, 269
  and private se ctor, 14
  reserve, 196
Water Footprint Network, 294
Water Framework Directive, 78, 228
Water Futures Partnership, 294
water justice, 3, 22, 23, 36, 49, 50, 53, 82, 157, 181, 204, 246, 248, 255, 260, 262, 265, 268, 283, 284, 326, 334, 346, 351, 357
  activists and academics, 246
  conceptualization, 4, 6, 20, 21, 51, 346–9, 355
  definition, 4, 22, 49
  and discourses, 249
  and ecological inclusiveness, 259
  and feminism, 331, 334
  glocal approach, 269
  and governmentality, 153
  and memory narratives, 316
  and national imaginaries, 278
  place sensitive assessments, 53
  Plato´s Republic, 115
  and social movements, 129, 163, 226, 232, 234
  and virtual trade, 283
water resources
  contamination, 283
water rights, 8, 15, 16, 20, 21, 38, 39, 43, 44, 45, 46, 50, 51, 63, 75, 145, 156, 194, 196, 229, 233, 268, 286, 292, 350
  administration of, 156
  allocation of, 5, 262
  cultural embeddedness, 21
  customary, 10, 16
  definitions of, 15, 16, 38, 51
  and echelons of right analysis, 20, 349
  and ecological arguments, 128
  formalization of, 15, 16, 65
  indigenous, 193, 195, 196, 201
  individualized, 39
  and mining companies, 8
  prioritization of, 290
  private, 19
  privatization of, 44
  recognition of, 193
  and reserve lands, 195, 196
  security, 16
  seen as standard black boxes, 15
  state-centric, 15
  tradable, 290
  transfer of, 156
  transferability of, 12
  and transnational actors, 1
  and technology, 265
  uniform, 46
  and World Bank, 189
water stewardship, 204, 289, 291, 292, 293, 294
Water Sustainability Act, 189, 194, 197, 201, 203
Water War (Bolivia), 111, 230
WaterAid, 215
waterscapes, 8, 48, 53, 64, 67, 80, 129, 151
watershed conservation, 312
weeding, 339, 340, 343
well-being, 4, 46, 72, 91, 97, 98, 261, 268, 303, 349, 356, 357
wells, 7, 45, 156, 157, 249, 291, 293, 321, 326, 330, 334, 335, 337
West Germany, 173
wetlands, 64, 74, 134, 136, 138, 140, 141, 142, 144, 147, 158, 267, 323
white supremacy, 347
WHO/UNICEF Joint Monitoring Committee, 101
wind storms, 188
Winters Doctrine, 201
womanhood, 332, 342, 343
women, 47, 95, 96, 97, 98, 100, 291, 332, 336, 339, 341, 343
  agricultural duties, 339
  collective action, 100
  Dalit, 97
  domestic labor, 95
  empowerment, 161
  farm work, 340
  feminism, 333
  First Nations, 199
  household water collection, 90
  identity, 339
  and irrigation modernization, 5
  Morocco, 98
  practical needs, 101
  public health, 235
  rights, 233
  and social rules, 97
  subjectivities, 96
  taboos and emotional stress, 219
  urban poor, 99
  wage workers, 342
  Women in Europe for a Common Future, 234
  young, 341
World Bank, 10, 14, 15, 16, 17, 46, 59, 60, 67, 78, 137, 171, 172, 173, 175, 176, 177, 178, 179, 181, 182, 188, 189, 211, 213, 214, 217, 228, 230, 290, 291, 293, 302
World Business Council for Sustainable Development, 292
World Economic Forum, 294

World Toilet Day, 210
World War II, 94, 169, 170
World Wildlife Fund, 292

Xesiguan (Guatemala), 175
Xinjiang Turpan Water Conservation Project, 293

Yaku Pampa, 319
Young, I., 5, 22, 54

Zambia, 250
Zaragoza, 128, 295
Zimbabwe, 250, 304, 306